Texts and Monographs in Physics

Springer
Berlin
Heidelberg
New York
Barcelona
Hong Kong
London
Milan
Paris
Singapore
Tokyo

Texts and Monographs in Physics

From Microphysics to Macrophysics I + II Methods and Applications of Statistical Physics By R. Balian

Multi-Hamiltonian Theory of Dynamical Systems
By M. Błaszak

Quantum Mechanics: Foundations and Applications
3rd enlarged edition By A. Böhm

Operator Algebras and Quantum Statistical Mechanics I + II 2nd edition
By O. Bratteli and D. W. Robinson

Statistical Methods in Quantum Optics 1. Master Equations and Fokker–Planck Equations
By H. J. Carmichael

Geometry of the Standard Model of Elementary Particles
By. A. Derdzinski

Scattering Theory of Classical and Quantum *N*-Particle Systems
By. J. Derezinski and C. Gérard

Effective Lagrangians for the Standard Model
By A. Dobado, A. Gómez-Nicola, A. L. Maroto and J. R. Peláez

Quantum
The Quantum Theory of Particles, Fields, and Cosmology By E. Elbaz

Quantum Relativity
A Synthesis of the Ideas of Einstein and Heisenberg
By D. R. Finkelstein

Quantum Mechanics I + II
By A. Galindo and P. Pascual

The Elements of Mechanics
By G. Gallavotti

Local Quantum Physics
Fields, Particles, Algebras
2nd revised and enlarged edition
By R. Haag

Supersymmetric Methods in Quantum and Statistical Physics By G. Junker

***CP* Violation Without Strangeness**
Electric Dipole Moments of Particles, Atoms, and Molecules
By I. B. Khriplovich and S. K. Lamoreaux

Quantum Groups and Their Representations
By A. Klimyk and K. Schmüdgen

Statistical Mechanics of Lattice Systems
Volume 1: Closed-Form and Exact Solutions 2nd, revised and enlarged edition
By D. A. Lavis and G. M. Bell

Statistical Mechanics of Lattice Systems
Volume 2: Exact, Series and Renormalization Group Methods
By D. A. Lavis and G. M. Bell

Fields, Symmetries, and Quarks 2nd, revised and enlarged edition By U. Mosel

Quantum Entropy and Its Use
By M. Ohya and D. Petz

Generalized Coherent States and Their Applications
By A. Perelomov

Path Integral Approach to Quantum Physics An Introduction
2nd printing By G. Roepstorff

Renormalization An Introduction
By M. Salmhofer

Finite Quantum Electrodynamics
The Causal Approach 2nd edition
By G. Scharf

From Electrostatics to Optics
A Concise Electrodynamics Course
By G. Scharf

The Mechanics and Thermodynamics of Continuous Media By M. Šilhavý

Large Scale Dynamics of Interacting Particles By H. Spohn

The Theory of Quark and Gluon Interactions 2nd completely revised and enlarged edition By F. J. Ynduráin

Relativistic Quantum Mechanics and Introduction to Field Theory
By F. J. Ynduráin

David A. Lavis George M. Bell

Statistical Mechanics of Lattice Systems

Volume 1: Closed-Form and Exact Solutions

Second, Revised and Enlarged Edition
With 102 Figures and 5 Tables

Springer

Dr. David A. Lavis
Department of Mathematics
King's College, University of London
Strand
London WC2R 2LS, United Kingdom

Professor George M. Bell †
Department of Mathematics
King's College, University of London
Strand
London WC2R 2LS, United Kingdom

The 1st edition was published under the title: Statistical Mechanics of Lattice Models. Volume 1: Closed Form and Exact Theories of Cooperative Phenomena
© 1989 Ellis Horwood Ltd., Chichester, UK

ISBN 978-3-642-08411-9

Library of Congress Cataloging-in-Publication Data applied for.

Die Deutsche Bibliothek - CIP-Einheitsaufnahme
Lavis, David A.: Statistical mechanics of lattice systems D.A. Lavis ; G.M. Bell. - Berlin ; Heidelberg ; New York ; Barcelona ; Hong Kong ; London ; Milan ; Paris ; Singapore ; Tokyo : Springer
(Texts and monographs in physics) 1. Closed form and exact solutions : with 5 tables.
- 2., rev. and enl. ed. - 1999

Preface

Most of the interesting and difficult problems in statistical mechanics arise when the constituent particles of the system interact with each other with pair or multiparticle energies. The types of behaviour which occur in systems because of these interactions are referred to as *cooperative phenomena* giving rise in many cases to phase transitions. This book and its companion volume (Lavis and Bell 1999, referred to in the text simply as Volume 2) are principally concerned with phase transitions in lattice systems. Due mainly to the insights gained from scaling theory and renormalization group methods, this subject has developed very rapidly over the last thirty years.[1] In our choice of topics we have tried to present a good range of fundamental theory and of applications, some of which reflect our own interests.

A broad division of material can be made between exact results and approximation methods. We have found it appropriate to include some of our discussion of exact results in this volume and some in Volume 2. The other main area of discussion in this volume is mean-field theory leading to closed-form approximations. Although this is known not to give reliable results close to a critical region, it often provides a good qualitative picture for phase diagrams as a whole. For complicated systems some kind of mean-field method is often the only tractable method available. In Volume 2 we complement the work in this text by discussions of scaling theory, algebraic methods and the renormalization group. Although these two volumes individually contain a less than comprehensive range of topics, we have attempted to ensure that each stands alone as treatment of its material. References to Volume 2 in this volume are normally given alongside other alternative sources.

The work in these two volumes was developed in part from courses given by George Bell and myself in the Mathematics Departments of the University of Manchester Institute of Science and Technology and Chelsea and King's Colleges, London. Questions and comments by students in these courses have helped to reduce errors and obscurities. We have also benefited from the exchange of ideas with friends and co-workers. We are grateful to Professor G. S. Rushbrooke for his careful reading of the original typescript and for his many helpful and instructive comments.

[1] An account of the historical development of the subject from its beginnings in the nineteenth century is given by Domb (1985).

This volume is a revised version of the text first published by Ellis Horwood (Bell and Lavis 1989). In April 1993, before the second volume was complete, my friend and co-author George Bell suddenly died. The consequent delay in the completion of Volume 2 made the production of a revised version of Volume 1 desirable. I am very grateful to Professor Elliott Lieb, who encouraged and facilitated this new edition by Springer. I should also like to thank Rachel George, who produced a working LaTeX version from the published text of the first edition and created postscript diagrams based on the original drawings by Debbie Bell. Geoffrey Bellringer of King's College Library provided me with valuable help in compiling the bibliography. Finally I should like to express my gratitude to Professor Wolf Beiglböck and the staff at Springer for their patience and advice.

London, December 1998. *David Lavis*

Table of Contents

1. Introduction to Thermodynamics and Phase Transitions

1.1 Thermodynamic Variables: Simple Fluids

We assume that the reader has some previous knowledge of thermodynamics.[1] Our intention here is to revise briefly the essential part of the theory, to introduce critical phenomena and, at the end of this chapter, to put the theory into a suitably generalized form for later use.

In the first instance, thermodynamics deals with systems in equilibrium. This means systems in which, under certain given conditions, all natural processes of change have come to an end. The observable state of such a system is therefore fixed and constant in time. In a continuous mass of fluid, for example, any initial differences in temperature or density between different regions have been smeared out by convection or diffusion when the equilibrium state is reached. Because of this uniformity the state of a system in equilibrium can be specified by a small number of variables. For a fluid or isotropic solid composed of one type of molecule the state can be specified by three independent variables, such as the number of molecules M, the volume $\widetilde{V}$ and the temperature $\widetilde{T}$. Other possible sets of independent variables are $M, \widetilde{T}$ and the pressure $\widetilde{P}$; $M, \widetilde{T}$ and the volume per molecule $\tilde{v}$; $\widetilde{V}, \tilde{v}$ and $\widetilde{P}$.[2] Once the independent variables are chosen, the other variables are regarded as functions of the independent set. For example, when the state of the fluid is specified by $M, \widetilde{V}$ and $\widetilde{T}$ the volume per molecule is given by $\tilde{v} = \widetilde{V}/M$.

Thermodynamic variables fall into two classes, intensive and extensive. Intensive variables, for instance $\widetilde{T}$, $\widetilde{P}$ and $\tilde{v}$, express the internal constitution of the system. For fixed values of the intensive variables, the extensive variables such as M and $\widetilde{V}$ are proportional to the size of the system. To completely specify a system, at least one extensive variable must be used, but the others can be intensive, as when the set $(M, \widetilde{P}, \widetilde{T})$ of independent variables is chosen for the fluid. Any intensive variable depends only on other intensive variables. A fluid or isotropic solid has an *equation of state* $f(\widetilde{P}, \tilde{v}, \widetilde{T}) = 0$ which is a relation between the intensive variables $\widetilde{P}$, $\tilde{v}$ and $\widetilde{T}$.

[1] Such as, for instance, is contained in Chaps. 1–5 of Pippard (1967). See also Adkins (1983), Landsberg (1978) and Buchdahl (1966).

[2] At this stage it is convenient to attach the tilde symbol to certain thermodynamic variables. This allows a more compact notation to be introduced in Sect. 1.7 and used extensively in this volume and Volume 2.

The internal energy U of a system arises from the kinetic energies of the molecules and the energy of interaction between them. It is an extensive variable, and from the law of energy conservation its differential may be written as

$$dU = d'Q - \widetilde{P}d\widetilde{V} + \mu dM\,, \tag{1.1}$$

where μ is an intensive variable, called the *chemical potential*, and $d'Q$ is the heat absorbed. Now suppose that the system is brought from equilibrium state $\mathfrak{A}$ to equilibrium state $\mathfrak{B}$ by a 'reversible' process; that is one which is slow enough for the intermediate states through which the system passes to be themselves effectively equilibrium states. Since U is a function of the state, the change in U is independent of the path taken from $\mathfrak{A}$ to $\mathfrak{B}$ and can be written $U_{\mathfrak{B}} - U_{\mathfrak{A}}$. Now Q is not a function of the state, and the heat absorbed during the process depends on the path taken from $\mathfrak{A}$ to $\mathfrak{B}$. This is the reason that the symbol 'd′' rather than 'd' has been used for its differential. However, by the *second law of thermodynamics* $d'Q = \widetilde{T}d\widetilde{S}$ for an infinitesimal change in equilibrium state, where $\widetilde{S}$ is an extensive state function, called the *entropy*. Hence, for such a change

$$dU = \widetilde{T}d\widetilde{S} - \widetilde{P}d\widetilde{V} + \mu dM\,. \tag{1.2}$$

Since the state of the fluid or isotropic solid can be completely specified by the values of $\widetilde{S}$, $\widetilde{V}$ and M, (1.2) gives dU as a total differential. The pairs of variables $(\widetilde{S},\widetilde{T})$, $(\widetilde{V},-\widetilde{P})$ and (M,μ), one extensive and the other intensive in each case, occurring in the differential are called *conjugate*. From (1.2) it follows that[3]

$$\widetilde{T} = \left(\frac{\partial U}{\partial \widetilde{S}}\right)_{M,V}, \qquad \widetilde{P} = -\left(\frac{\partial U}{\partial \widetilde{V}}\right)_{S,M}, \qquad \mu = \left(\frac{\partial U}{\partial M}\right)_{S,V}. \tag{1.3}$$

Now suppose that, with its internal constitution fixed, the size of the system is increased by a factor λ. Since all extensive variables will increase by the factor λ,

$$U(\lambda\widetilde{S},\lambda\widetilde{V},\lambda M) = \lambda U(\widetilde{S},\widetilde{V},M)\,. \tag{1.4}$$

Differentiation with respect to λ, and substitution from (1.3) gives

$$U = \widetilde{T}\widetilde{S} - \widetilde{P}\widetilde{V} + \mu M\,, \tag{1.5}$$

when λ is put equal to 1.

[3] Because of the various choices possible for the independent variables in any thermodynamic system, it is often necessary to specify the variable or variables kept constant during a partial differentiation by using subscripts.

1.2 Change of Variable and Thermodynamic Potentials

An extensive variable, termed the *enthalpy*, is defined by

$$H = U + \widetilde{P}\widetilde{V}\,. \tag{1.6}$$

Substitution from (1.2) yields

$$\mathrm{d}H = \mathrm{d}U + \widetilde{P}\mathrm{d}\widetilde{V} + \widetilde{V}\mathrm{d}\widetilde{P} = \widetilde{T}\mathrm{d}\widetilde{S} + \widetilde{V}\mathrm{d}\widetilde{P} + \mu\mathrm{d}M\,. \tag{1.7}$$

It can be seen that (1.6) represents a *Legendre transformation* associated with replacement of $\widetilde{V}$ as an independent variable by its conjugate $-\widetilde{P}$. From an experimental point of view it is feasible to treat either $\widetilde{P}$ or $\widetilde{V}$ as an independent variable. However, although an experimenter can control $\widetilde{T}$ directly, this is hardly possible for $\widetilde{S}$. Hence to replace $\widetilde{S}$ as an independent variable by its conjugate $\widetilde{T}$ we employ the Legendre transformation

$$A = U - \widetilde{T}\widetilde{S}\,, \tag{1.8}$$

which defines the *Helmholtz free energy* A. The latter is an example of a *thermodynamic potential*; the reason for the use of this term will appear in Sect. 1.5. In a similar way to (1.7),

$$\mathrm{d}A = -\widetilde{S}\mathrm{d}\widetilde{T} - \widetilde{P}\mathrm{d}\widetilde{V} + \mu\mathrm{d}M\,. \tag{1.9}$$

The *Gibbs free energy* G is obtained by the Legendre transformation

$$G = A + \widetilde{P}\widetilde{V} = U + \widetilde{P}\widetilde{V} - \widetilde{T}\widetilde{S} \tag{1.10}$$

and, either from (1.2) or from (1.9),

$$\mathrm{d}G = -\widetilde{S}\mathrm{d}\widetilde{T} + \widetilde{V}\mathrm{d}\widetilde{P} + \mu\mathrm{d}M\,. \tag{1.11}$$

It can be seen that, just as A is appropriate to the set of independent variables $(\widetilde{T}, \widetilde{V}, M)$, so G is the thermodynamic potential for the set $(\widetilde{T}, \widetilde{P}, M)$.

As indicated above, any intensive variable is dependent only on other intensive variables. For a system specified by $n + 1$ independent variables the maximum number of independent intensive variables is n. We now derive an important relation among such variables. By comparing (1.5) and (1.10) it can be seen that

$$G = \mu M\,, \tag{1.12}$$

and hence

$$\mathrm{d}G = \mu\mathrm{d}M + M\mathrm{d}\mu\,. \tag{1.13}$$

Then substitution in (1.11) yields

$$-\widetilde{S}\mathrm{d}\widetilde{T} + \widetilde{V}\mathrm{d}\widetilde{P} - M\mathrm{d}\mu = 0\,. \tag{1.14}$$

Defining the intensive variables

$$\tilde{s} = \widetilde{S}/M\,, \qquad \tilde{v} = \widetilde{V}/M\,, \tag{1.15}$$

which are respectively the entropy and volume per molecule, we have the *Gibbs–Duhem equation*

$$-\tilde{s}\mathrm{d}\widetilde{T} + \tilde{v}\mathrm{d}\widetilde{P} - \mathrm{d}\mu = 0\,. \tag{1.16}$$

The *grand potential* D is appropriate to the set of independent variables $(\widetilde{V}, \widetilde{T}, \mu)$ and is important in connection with statistical mechanics, although not often mentioned in thermodynamics texts. Using the Legendre transformation

$$D = A - \mu M = U - \widetilde{T}\widetilde{S} - \mu M\,, \tag{1.17}$$

it can be shown, either from (1.2) or from (1.9), that

$$\mathrm{d}D = -\widetilde{S}\mathrm{d}\widetilde{T} - \widetilde{P}\mathrm{d}\widetilde{V} - M\mathrm{d}\mu \tag{1.18}$$

and (1.17) with (1.5) gives

$$D = -\widetilde{P}\widetilde{V}\,. \tag{1.19}$$

1.3 Response Functions and Thermodynamic Relations

Response functions specify the reaction of the system to externally imposed changes in the state. The common ones defined here implicitly assume a fixed value for M. They can be expressed in terms of second derivatives of thermodynamic potentials. For instance the isobaric (constant-pressure) *coefficient of thermal expansion* of a fluid or isotropic solid satisfies the relations

$$\alpha_P = \frac{1}{\widetilde{V}}\left(\frac{\partial \widetilde{V}}{\partial \widetilde{T}}\right)_P = \frac{1}{\widetilde{V}}\frac{\partial^2 G}{\partial \widetilde{T}\partial \widetilde{P}}\,, \tag{1.20}$$

where the first member defines α_P, and the second follows from (1.11). In a similar way the isothermal (constant-temperature) *compressibility* is given by

$$\mathcal{K}_T = -\frac{1}{\widetilde{V}}\left(\frac{\partial \widetilde{V}}{\partial \widetilde{P}}\right)_T = -\frac{1}{\widetilde{V}}\frac{\partial^2 G}{\partial \widetilde{P}^2}\,. \tag{1.21}$$

Relations between derivatives of thermodynamic functions arise from the fact that a state can be specified by a small number of variables and from the first and second laws of thermodynamics, expressed by such relations as (1.2). For instance from (1.20) and (1.21), we may write for a fluid of constant composition

$$\mathrm{d}\widetilde{V} = \widetilde{V}\alpha_P\mathrm{d}\widetilde{T} - \widetilde{V}\mathcal{K}_T\mathrm{d}\widetilde{P} \tag{1.22}$$

and hence

$$\left(\frac{\partial \widetilde{P}}{\partial \widetilde{T}}\right)_V = \frac{\alpha_P}{\mathcal{K}_T}\,. \tag{1.23}$$

There is thus no need to define a separate response function to express the variation of pressure with temperature at constant volume. From their definitions, α_P and $\mathcal{K}_T$ are intensive variables and, using (1.12),

$$\alpha_P = \frac{1}{\tilde{v}}\frac{\partial^2\mu}{\partial\tilde{T}\partial\tilde{P}}\,, \qquad \mathcal{K}_T = -\frac{1}{\tilde{v}}\frac{\partial^2\mu}{\partial\tilde{P}^2}\,. \tag{1.24}$$

A *heat capacity* gives the amount of heat needed to raise the temperature of the system by 1 K, under stated conditions. For *molar* heat capacities the system is taken to contain one mole of matter. ($M = M_\mathrm{A} = 6.022 \times 10^{23}$ molecules, M_A being Avogadro's number.) Since $\mathrm{d}'Q = \tilde{T}\mathrm{d}\tilde{S}$, the molar heat capacity (molar specific heat) at constant pressure is given by

$$C_P = \tilde{T}\left(\frac{\partial\tilde{S}}{\partial\tilde{T}}\right)_P = \left(\frac{\partial H}{\partial\tilde{T}}\right)_P = -\tilde{T}\frac{\partial^2 G}{\partial\tilde{T}^2} = -M_\mathrm{A}\tilde{T}\frac{\partial^2\mu}{\partial\tilde{T}^2}\,. \tag{1.25}$$

From the last expression it can be seen that the molar heat capacity is an intensive variable. Similarly the molar heat capacity at constant volume is

$$C_V = \tilde{T}\left(\frac{\partial\tilde{S}}{\partial\tilde{T}}\right)_V = \left(\frac{\partial U}{\partial\tilde{T}}\right)_V = -\tilde{T}\frac{\partial^2 A}{\partial\tilde{T}^2} = -M_\mathrm{A}\tilde{T}\frac{\partial^2 a}{\partial\tilde{T}^2}\,, \tag{1.26}$$

where a is the Helmholtz free energy per molecule. These four response functions are not independent. To show this, note first that, at constant composition,

$$\mathrm{d}\tilde{S} = \left(\frac{\partial\tilde{S}}{\partial\tilde{T}}\right)_V \mathrm{d}\tilde{T} + \left(\frac{\partial\tilde{S}}{\partial\tilde{V}}\right)_T \mathrm{d}\tilde{V} \tag{1.27}$$

and hence

$$\left(\frac{\partial\tilde{S}}{\partial\tilde{T}}\right)_P - \left(\frac{\partial\tilde{S}}{\partial\tilde{T}}\right)_V = \left(\frac{\partial\tilde{S}}{\partial\tilde{V}}\right)_T \left(\frac{\partial\tilde{V}}{\partial\tilde{T}}\right)_P\,. \tag{1.28}$$

The importance, pointed out above, of specifying which other variables are held constant when differentiating with respect to a given variable is clearly displayed by this relation. Now, from (1.9),

$$\left(\frac{\partial\tilde{S}}{\partial\tilde{V}}\right)_T = -\frac{\partial^2 A}{\partial\tilde{V}\partial\tilde{T}} = \left(\frac{\partial\tilde{P}}{\partial\tilde{T}}\right)_V\,. \tag{1.29}$$

From (1.25), (1.26), (1.20) and (1.23)

$$C_P - C_V = \tilde{V}\tilde{T}\frac{\alpha_P^2}{\mathcal{K}_T}\,, \tag{1.30}$$

where $\tilde{V}$ is the molar volume $M_\mathrm{A}\tilde{v}$.

1.4 Magnetic Systems

Until now we have considered energy transfer by heat or matter flow and mechanical work, but energy can also be transferred by the action of a magnetic field. It will be supposed that the latter has a fixed direction and can be represented by a scalar $\widetilde{\mathcal{H}}$ of dimension ampère/metre (A/m). The magnetization or magnetic moment is taken as parallel to the field and can be represented by a scalar $\widetilde{\mathcal{M}}$ of dimension weber × metre (Wb m). Obviously $\widetilde{\mathcal{H}}$ is an intensive variable and $\widetilde{\mathcal{M}}$ an extensive variable. If the configuration (i.e. size and shape) and composition are fixed then

$$\mathrm{d}U = \mathrm{d}'Q + \widetilde{\mathcal{H}}\mathrm{d}\widetilde{\mathcal{M}} = \widetilde{T}\mathrm{d}\widetilde{S} + \widetilde{\mathcal{H}}\mathrm{d}\widetilde{\mathcal{M}}\,. \tag{1.31}$$

Experimentally it is hard to specify $\widetilde{\mathcal{M}}$ initially, and it is usually better to use $\widetilde{\mathcal{H}}$ as an independent variable. A magnetic enthalpy H_1 may be defined by the Legendre transformation

$$H_1 = U - \widetilde{\mathcal{H}}\widetilde{\mathcal{M}}\,. \tag{1.32}$$

The thermodynamic potential A may again be defined by (1.8) and a potential F_1, corresponding to independent variables $\widetilde{T}$ and $\widetilde{\mathcal{H}}$, by

$$F_1 = U - \widetilde{T}\widetilde{S} - \widetilde{\mathcal{H}}\widetilde{\mathcal{M}}\,. \tag{1.33}$$

From (1.31)–(1.33) and (1.8) we have

$$\begin{aligned} \mathrm{d}H_1 &= \widetilde{T}\mathrm{d}\widetilde{S} - \widetilde{\mathcal{M}}\mathrm{d}\widetilde{\mathcal{H}}\,, \\ \mathrm{d}A &= -\widetilde{S}\mathrm{d}\widetilde{T} + \widetilde{\mathcal{H}}\mathrm{d}\widetilde{\mathcal{M}}\,, \\ \mathrm{d}F_1 &= -\widetilde{S}\mathrm{d}\widetilde{T} - \widetilde{\mathcal{M}}\mathrm{d}\widetilde{\mathcal{H}}\,. \end{aligned} \tag{1.34}$$

Again standardizing to one mole ($M = M_{\mathrm{A}}$), we can define heat capacities $C_{\mathcal{H}}$ and $C_{\mathcal{M}}$ at constant field and magnetization respectively, the latter being difficult to realize experimentally. We have

$$C_{\mathcal{H}} = \widetilde{T}\left(\frac{\partial \widetilde{S}}{\partial \widetilde{T}}\right)_{\mathcal{H}} = -\widetilde{T}\frac{\partial^2 F_1}{\partial \widetilde{T}^2}\,. \tag{1.35}$$

Another important response function for magnetic systems is the *isothermal magnetic susceptibility* χ_T which is defined for unit volume of matter. Noting that the magnetization for unit volume is $\widetilde{\mathcal{M}}/(M\tilde{v})$ where M is the number of atoms and $\tilde{v}$ the volume per atom,

$$\chi_T = \frac{1}{M\tilde{v}\mu_0}\left(\frac{\partial \widetilde{\mathcal{M}}}{\partial \widetilde{\mathcal{H}}}\right)_T = -\frac{1}{M\tilde{v}\mu_0}\frac{\partial^2 F_1}{\partial \widetilde{\mathcal{H}}^2}\,. \tag{1.36}$$

Here μ_0 is the permeability of free space, $4\pi 10^{-7}$ Wb/(A m), and its inclusion makes χ_T dimensionless.

1.5 Stationary Properties of Thermodynamic Functions

The thermodynamics of irreversible processes is based on the principle that, when a completely isolated system is not in thermodynamic equilibrium, all processes of change increase its entropy. It follows that when an equilibrium state is reached, the entropy is at a maximum and, for small deviations from equilibrium, $\mathrm{d}\widetilde{S} = 0$. Note that, while $\mathrm{d}\widetilde{S} = 0$, if we put $\mathrm{d}U = \mathrm{d}\widetilde{V} = \mathrm{d}M = 0$ in (1.2), this simply results from the functional relationship which, for equilibrium states of the fluid, exists between $\widetilde{S}$ and the independent thermodynamic variables U, $\widetilde{V}$ and M. The stationary property $\mathrm{d}\widetilde{S} = 0$ is more general since it is true for small deviations from equilibrium.

We now derive an application of the stationary property. Consider a composite system consisting of two subsystems, each in a rigid container, which are in thermal contact with each other but thermally isolated from the outside world. The subsystems are in internal equilibrium and their temperatures are $\widetilde{T}_1$ and $\widetilde{T}_2$ respectively. Assuming that boundary effects are negligible the entropy of the composite system $\widetilde{S} = \widetilde{S}_1 + \widetilde{S}_2$ where $\widetilde{S}_1$ and $\widetilde{S}_2$ are the entropies of the separate subsystems. Now suppose that heat $\mathrm{d}'Q$ is transferred from subsystem 1 to subsystem 2; then

$$
\begin{aligned}
&\widetilde{T}_1 \mathrm{d}\widetilde{S}_1 = \mathrm{d}'Q_1 = -\mathrm{d}'Q\,, \\
&\widetilde{T}_2 \mathrm{d}\widetilde{S}_2 = \mathrm{d}'Q_2 = \mathrm{d}'Q\,, \\
&\mathrm{d}\widetilde{S} = \mathrm{d}\widetilde{S}_1 + \mathrm{d}\widetilde{S}_2 = \left(\frac{1}{\widetilde{T}_2} - \frac{1}{\widetilde{T}_1}\right)\mathrm{d}'Q\,.
\end{aligned}
\tag{1.37}
$$

If there is thermodynamic equilibrium in the whole system before the heat transfer takes place, then, by the stationary property, $\mathrm{d}\widetilde{S} = 0$. Hence a condition for such equilibrium is $\widetilde{T}_1 = \widetilde{T}_2$. If $\widetilde{T}_1 > \widetilde{T}_2$, so that the subsystems are not in thermal equilibrium, then $\mathrm{d}\widetilde{S} > 0$ if $\mathrm{d}'Q > 0$. Hence the increasing entropy principle implies that heat is transferred from the higher temperature to the lower temperature.

Now suppose that the system is no longer thermally isolated but remains at fixed configuration and composition. An example is a fluid in a rigid container with heat-conducting walls immersed in a heat bath at given temperature $\widetilde{T}$, the independent thermodynamic variables now being $\widetilde{T}$, the container volume $\widetilde{V}$ and the number M of molecules enclosed. For a system of this type it can be shown that any irreversible process decreases the Helmholtz free energy A. Hence A is at a minimum when the equilibrium state for temperature $\widetilde{T}$ is reached and $\mathrm{d}A = 0$ for small deviations from equilibrium. Thus A is the thermodynamic potential for a system which is mechanically isolated but can exchange heat with the outside world.

For a fluid the condition of mechanical isolation can be removed by supposing that the rigid container is penetrated by a tube with a close-fitting movable piston of fixed area on which a constant force is exerted from the outside. The independent variables are now $\widetilde{T}$, the pressure $\widetilde{P}$ (i.e. the piston force divided by area) and M. The Gibbs free energy G can be shown to be the appropriate thermodynamic potential, G being at a minimum in the equilibrium state for the given values of $\widetilde{T}$ and $\widetilde{P}$, and the condition $\mathrm{d}G = 0$ applying for small deviations from equilibrium. Alternatively the condition of molecular isolation can be removed by supposing that the rigid container walls are permeable to the type of molecule composing the fluid. The environment must now be regarded as both a heat bath at given temperature $\widetilde{T}$ and a 'molecule bath' at given chemical potential μ, the independent thermodynamic variables being $\widetilde{T}$, $\widetilde{V}$ and μ. The appropriate thermodynamic potential is now D, as defined by (1.17), and $\mathrm{d}D = 0$ for small deviations from the equilibrium state for the given values of $\widetilde{T}$ and μ.

If a volume of fluid at an appropriate density is placed in a rigid transparent container and the temperature is lowered, then at a certain point the appearance of *critical opalescence* indicates that light is being scattered by large-density inhomogeneities. If the temperature is lowered still further, there is a separation into distinct regions or *phases* of liquid and vapour respectively with a well defined boundary between them. We shall now derive conditions for such phases to exist in thermodynamic equilibrium at a given temperature $\widetilde{T}$. Let one phase contain M_1 molecules at volume $\widetilde{V}_1$ and the other M_2 molecules at volume $\widetilde{V}_2$, where $\widetilde{V}_1 + \widetilde{V}_2 = \widetilde{V}$, the volume of the rigid container, and $M_1 + M_2 = M$, the total number of molecules in the container. Regarding boundary contributions as negligible, the thermodynamic potential A of the system is

$$A = A_1(\widetilde{T}, \widetilde{V}_1, M_1) + A_2(\widetilde{T}, \widetilde{V}_2, M_2) . \tag{1.38}$$

Now suppose that there is a small deviation from equilibrium in which $\widetilde{V}_1$ increases by a positive amount $\mathrm{d}\widetilde{V}$, and $\widetilde{V}_2$ decreases by $\mathrm{d}\widetilde{V}$, while M_1 and M_2 remain fixed. Then, from (1.38) and (1.9),

$$\mathrm{d}A = \mathrm{d}A_1 + \mathrm{d}A_2 = -\widetilde{P}_1 \mathrm{d}\widetilde{V}_1 - \widetilde{P}_2 \mathrm{d}\widetilde{V}_2 = (\widetilde{P}_2 - \widetilde{P}_1)\mathrm{d}\widetilde{V} . \tag{1.39}$$

Since the phases are assumed to be in thermodynamic equilibrium before the small volume changes occur, $\mathrm{d}A = 0$ and hence $\widetilde{P}_1 = \widetilde{P}_2$. Similarly, if M_1 increases by $\mathrm{d}M$ and M_2 decreases by $\mathrm{d}M$ with $\widetilde{V}_1$ and $\widetilde{V}_2$ fixed then

$$\mathrm{d}A = \mathrm{d}A_1 + \mathrm{d}A_2 = \mu_1 \mathrm{d}M_1 + \mu_2 \mathrm{d}M_2 = (\mu_1 - \mu_2)\mathrm{d}M \tag{1.40}$$

but, again, $\mathrm{d}A = 0$ implying $\mu_1 = \mu_2$. Hence, for phase equilibrium at temperature $\widetilde{T}$, we have the conditions that the pressures and the chemical potentials respectively in the two phases must be equal:

$$\widetilde{P}_1 = \widetilde{P}_2 , \qquad \mu_1 = \mu_2 . \tag{1.41}$$

1.6 Phase Equilibrium in the Van der Waals Gas

A perfect gas in which the molecules are mass points with no potential energy of interaction has the equation of state

$$\widetilde{P}\tilde{v} = k_{\mathrm{B}}\widetilde{T}\,, \qquad \tilde{v} = \widetilde{V}/M\,, \tag{1.42}$$

where k_{B} is Boltzmann's constant, equal to 1.381×10^{-23} J/K. This is obeyed to a good approximation by dilute real gases, but at higher densities there are large deviations, and the phenomenon of phase separation, discussed in Sect. 1.5, occurs. In a real fluid there are strong repulsive forces between molecules when very close together, giving a *finite size* or *hard core* effect, and attractive forces at larger distances apart. Displaying profound physical intuition, van der Waals (1873)[4] modified (1.42) in a heuristic fashion by replacing $\tilde{v}$ by $\tilde{v} - b$ to allow for finite molecular size, and $\widetilde{P}$ by $\widetilde{P} + c\tilde{v}^{-2}$. The latter change allows for the attractive forces since it implies a smaller external force to hold the M molecules in volume $M\tilde{v}$ than would otherwise be needed. We thus have

$$\left(\widetilde{P} + \frac{c}{\tilde{v}^2}\right)(\tilde{v} - b) = k_{\mathrm{B}}\widetilde{T}\,. \tag{1.43}$$

If curves of constant temperature (isotherms) are plotted in the P–v plane for (1.43) it is found that below a temperature $\widetilde{T}_{\mathrm{c}}$, to be deduced below, they have the 's-shape' shown in Fig. 1.1. Now let points A and B on a horizontal line intersecting the isotherm correspond to states labelled 1 and 2 respectively. It is obvious that $\widetilde{P}_1 = \widetilde{P}_2$ and it will now be shown that the second phase equilibrium condition of (1.41) is satisfied if the areas of the two loops between the line segment AB and the isotherm are equal (*Maxwell's equal-areas rule*). To prove this, put $\mathrm{d}\widetilde{T} = 0$ in (1.16), giving

$$\mathrm{d}\mu = \tilde{v}\mathrm{d}\widetilde{P} = \mathrm{d}(\widetilde{P}\tilde{v}) - \widetilde{P}\mathrm{d}\tilde{v} \tag{1.44}$$

and, integrating over the part of the isotherm between A and B,

$$\mu_2 - \mu_1 = \widetilde{P}_1(\tilde{v}_2 - \tilde{v}_1) - \int_{\tilde{v}_2}^{\tilde{v}_1} \widetilde{P}\mathrm{d}\tilde{v}\,. \tag{1.45}$$

The first term on the right-hand side of (1.45) is the area between the line segment AB and the $\tilde{v}$-axis, whereas the integral is the area between the corresponding part of the isotherm and the $\tilde{v}$-axis. These two quantities are equal when the equal-areas rule applies and hence the latter implies $\mu_1 = \mu_2$.

Now consider a fixed number M of molecules in a rigid container of volume $\widetilde{V} = M\tilde{v}_{\mathrm{m}}$ at the temperature $\widetilde{T}$ of the isotherm shown in Fig. 1.1. If $\tilde{v}_1 < \tilde{v}_{\mathrm{m}} < \tilde{v}_2$ the fluid can separate into the equilibrium phases 1 and 2, the number of molecules in the two phases being respectively M_1 and M_2, where

[4] For accounts of the impact of van der Waals' work on the development of the theory of phase transitions see Rigby (1970), Rowlinson (1973), de Boer (1974) and Klein (1974).

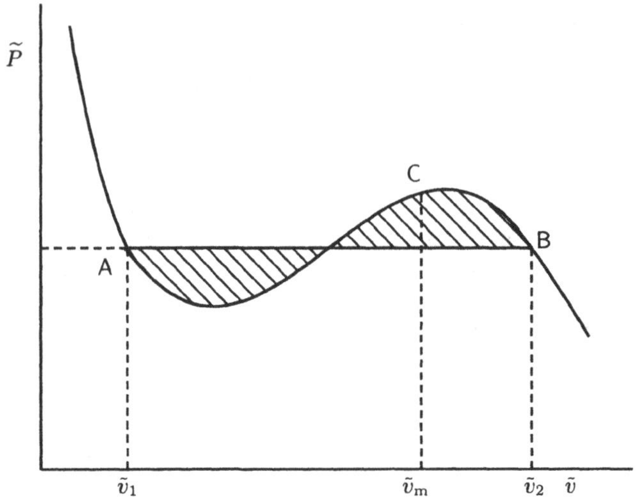

Fig. 1.1. Van der Waals gas isotherm for $\widetilde{T} < \widetilde{T}_{\mathrm{c}}$; equal-areas rule for conjugate phases A and B.

$$M\tilde{v}_{\mathrm{m}} = M_1\tilde{v}_1 + M_2\tilde{v}_2\,, \qquad M = M_1 + M_2\,. \tag{1.46}$$

Condition (1.41) ensures that the two-phase system is locally stable, but for complete thermodynamic stability it must shown that the thermodynamic potential A of the two-phase system is lower than that of the homogeneous system of molecular volume $\tilde{v}_{\mathrm{m}}$, whose state is represented by point C on the isotherm in Fig. 1.1. Since M is fixed we can use instead of A the thermodynamic potential a per molecule, which satisfies the relations

$$\widetilde{P} = -\frac{\partial a}{\partial \tilde{v}}\,, \qquad \mu = a + \widetilde{P}\tilde{v} = a - \tilde{v}\frac{\partial a}{\partial \tilde{v}}\,, \tag{1.47}$$

derived from (1.9), (1.10) and (1.12). Now consider the $(a, \tilde{v})$ isotherm corresponding to the $(\widetilde{P}, \tilde{v})$ isotherm of Fig. 1.1. From (1.47),

$$\frac{\partial \widetilde{P}}{\partial \tilde{v}} = -\frac{\partial^2 a}{\partial \tilde{v}^2}\,. \tag{1.48}$$

Hence where $\partial \widetilde{P}/\partial \tilde{v} > 0$, that is between the minimum and the maximum on the $(\widetilde{P}, \tilde{v})$ isotherm, $\partial^2 a/\partial \tilde{v}^2 < 0$ and the $(a, \tilde{v})$ isotherm is concave to the axes. At all other points, $\partial \widetilde{P}/\partial \tilde{v} < 0$ and the $(a, \tilde{v})$ isotherm is convex to the axes, giving it the form shown in Fig. 1.2. From (1.47) the slope of the tangent at any value of v gives the corresponding value of $-\widetilde{P}$ whereas the intercept of the tangent with the a-axis gives μ. The equilibrium conditions (1.41) are thus geometrically equivalent to a *double tangent rule*. The points on the $(a, \tilde{v})$ curve corresponding to the two equilibrium phases lie on a double tangent, as shown in Fig. 1.2. Assuming that the contribution of the phase

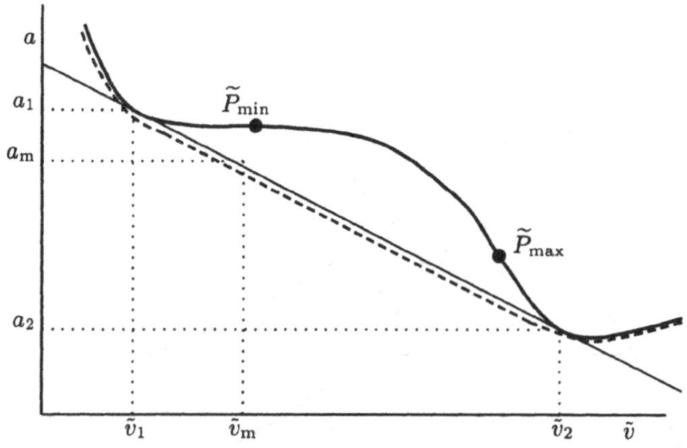

Fig. 1.2. Helmholtz free energy–volume isotherm for $\widetilde{T} < \widetilde{T}_c$ showing double tangent.

boundaries is negligible, the free energy density value a_s for the two-phase system is given by

$$Ma_s = M_1a_1 + M_2a_2 \,. \tag{1.49}$$

Comparison with the first relation of (1.46) shows that a_s is the ordinate of the point on the double tangent lying vertically above the point $\tilde{v}_m$ on the $\tilde{v}$-axis (see Fig. 1.2). Since the $(a, \tilde{v})$ isotherm lies above the double tangent for the whole range from $\tilde{v}_1$ to $\tilde{v}_2$, it follows that a_s is less than the value of a for a homogeneous fluid with the same molecular volume. Thus replacing the loop in the homogeneous fluid $(a, \tilde{v})$ curve between $\tilde{v}_1$ and $\tilde{v}_2$ by the double tangent segment gives the equilibrium $(a, \tilde{v})$ curve, indicated by the chain line in Fig. 1.2. This is mathematically the convex envelope of the homogeneous fluid $(a, \tilde{v})$ curve.

Phase separation thus occurs whenever there are maxima and minima on the isotherms given by (1.43) and these appear at the *critical point* where

$$\left(\frac{\partial \widetilde{P}}{\partial \tilde{v}}\right)_T = 0\,, \qquad \left(\frac{\partial^2 \widetilde{P}}{\partial \tilde{v}^2}\right)_T = 0\,. \tag{1.50}$$

The critical values of $\widetilde{T}$, $\tilde{v}$ and $\widetilde{P}$ are denoted by $\widetilde{T}_c$, $\tilde{v}_c$ and $\widetilde{P}_c$ respectively, and the isotherm for which $\widetilde{T} = \widetilde{T}_c$ is called the *critical isotherm*. From (1.43),

$$\left(\frac{\partial \widetilde{P}}{\partial \tilde{v}}\right)_T = -\frac{k_B\widetilde{T}}{(\tilde{v}-b)^2} + \frac{2c}{\tilde{v}^3}\,, \qquad \left(\frac{\partial^2 \widetilde{P}}{\partial \tilde{v}^2}\right)_T = \frac{2k_B\widetilde{T}}{(\tilde{v}-b)^3} - \frac{6c}{\tilde{v}^4}\,, \tag{1.51}$$

and, by substituting (1.51) in (1.50), it is easy to show that

$$k_B\widetilde{T}_c = \frac{8c}{27b}\,, \qquad \tilde{v}_c = 3b\,, \qquad \widetilde{P}_c = \frac{c}{27b^2}\,. \tag{1.52}$$

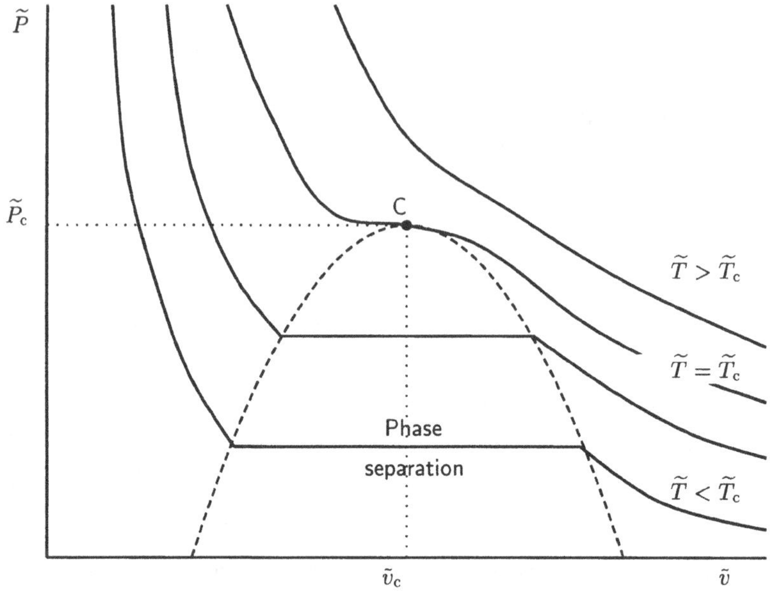

Fig. 1.3. Equilibrium pressure–volume isotherms; C is the critical point.

Typical equilibrium isotherms are shown in Fig. 1.3. For $\widetilde{T} < \widetilde{T}_c$, an equilibrium isotherm consists of a horizontal segment or 'tie line', like AB in Fig. 1.1, and monotonically decreasing portions linked with each end. The equilibrium phases at the end-points of a tie line are known as *conjugate phases.* The broken curve in Fig. 1.3 is called the *coexistence curve*, and the liquid and vapour regions lie to its left and right respectively. For $\widetilde{T} > \widetilde{T}_c$ the isotherms are monotonic decreasing and for large $\widetilde{T}/\widetilde{T}_c$ or $\tilde{v}/\tilde{v}_c$ their shape is close to the perfect gas form given by (1.42).

Though the critical values given by (1.52) are dependent on the particular equation of state (1.43) it should be noted that the thermodynamic arguments above require only that s-shaped isotherms should appear below a certain temperature. A number of other 'classical' equations of state, satisfying this condition, have been postulated (see Example 1.6 and Chap. 5).

1.7 The Field-Extensive Variable Representation of Thermodynamics

The basic thermodynamic theory will now be expressed in a form applicable to any system and useful for the development of statistical mechanics. The expression $-\widetilde{P}\mathrm{d}\widetilde{V}$ for mechanical work used until now is adequate for fluids or isotropic solids, but for non-isotropic solids (crystals) it must be replaced by a sum over stress and strain components. When a magnetic field acts on the system a magnetic work term $\widetilde{\mathcal{H}}\mathrm{d}\widetilde{\mathcal{M}}$ (see Sect. 1.4) is added to the mechanical work, and when an electrical field acts, there is a similar electrical work term. In general the composition of the system must be specified by the numbers M_j of molecules or atoms of several *components*[5] labelled with the index j, whereas up to now we have assumed a one-component system with composition specified by a single number M. Thus (1.2) or (1.31) must be replaced by

$$\mathrm{d}U = \widetilde{T}\mathrm{d}\widetilde{S} + \sum_{\{i\}} Y_i \mathrm{d}X_i + \sum_{\{j\}} \mu_j \mathrm{d}M_j \,, \tag{1.53}$$

where the intensive variables Y_i are generalized forces, the X_i are the conjugate extensive mechanical variables, and each μ_j is the chemical potential of a component j. The difference in sign between $-\widetilde{P}\mathrm{d}\widetilde{V}$ and $Y_i\mathrm{d}X_i$ arises from the fact that the work done by the pressure reduces the volume of the system. Here the internal energy U and the chemical potentials μ_j have the dimensions of energy, joule (J) in SI units, whereas the numbers M_j conjugate to the μ_j are dimensionless. To treat all terms on the right-hand side of (1.53) on the same basis it is convenient to make all the extensive variables whose differentials appear there dimensionless. Thus we define

$$S = \widetilde{S}/k_{\mathrm{B}} \,, \qquad V = \widetilde{V}/v_0, \qquad \mathcal{M} = \widetilde{\mathcal{M}}/m_0 \,, \tag{1.54}$$

where k_{B} is Boltzmann's constant, v_0 is a standard volume and m_0 is the dipole moment of a single molecule. The transformation from $\widetilde{V}$ to V may appear artificial, but in lattice models, where the position of the molecules is restricted to the lattice sites, v_0 can be taken to be the volume per site, and V then becomes the number of sites in the lattice. The transformation of the extensive variables to dimensionless form implies a corresponding transformation in their conjugates. These now all take on the dimensions of energy and will be known as *fields*. From (1.54), the conjugates of S, V and $\mathcal{M}$ respectively are

$$T = k_{\mathrm{B}}\widetilde{T} \,, \qquad P = \widetilde{P}v_0 \,, \qquad \mathcal{H} = \widetilde{\mathcal{H}}m_0 \,. \tag{1.55}$$

In place of (1.53) we can now write, for a system with $n+1$ independent variables,

$$\mathrm{d}U = T\mathrm{d}S + \sum_{i=1}^{n} \xi_i \mathrm{d}Q_i \,, \tag{1.56}$$

where T and the ξ_i are all fields with the dimensions of energy, while their conjugates S and Q_i are dimensionless. (The Q_i here are not heat terms.) The internal energy U is regarded as a function of the $n+1$ extensive variables S, Q_i, $i = 1, \ldots, n$ and, in a similar way to (1.5),

[5] A component is a set of atoms or molecules which are not convertible to atoms or molecules of another component by any process occurring in the system, e.g. oxygen, nitrogen and carbon dioxide in the atmosphere.

$$U = TS + \sum_{i=1}^{n} \xi_i Q_i \,. \tag{1.57}$$

By taking the differential of (1.57) and comparing it with (1.56)

$$S\mathrm{d}T + \sum_{i=1}^{n} Q_i \mathrm{d}\xi_i = 0 \,. \tag{1.58}$$

A subset $Q_1, \ldots, Q_\eta$ of the extensive variables can now be replaced as independent variables by their intensive conjugates $\xi_1, \ldots, \xi_\eta$ using the Legendre transformation

$$H_\eta = U - \sum_{i=1}^{\eta} \xi_i Q_i = TS + \sum_{i=\eta+1}^{n} \xi_i Q_i \,, \tag{1.59}$$

where H_η is an enthalpy and the last expression of (1.59) follows from (1.57). Taking the differential of (1.59) and comparing with (1.56) gives

$$\mathrm{d}H_\eta = T\mathrm{d}S - \sum_{i=1}^{\eta} Q_i \mathrm{d}\xi_i + \sum_{i=\eta+1}^{n} \xi_i \mathrm{d}Q_i \,. \tag{1.60}$$

The enthalpy H_η is not uniquely defined since it depends on the order in which we attach the indices $1, \ldots, n$ to the n variables Q_i. For instance, H as defined for a fluid in (1.6) and H_1 as defined for a magnetic system in (1.32) are both examples of the general H_1, with Q_1 taken as V and $\mathcal{M}$ respectively. The internal energy U could be termed H_0 and in a sense the concept of enthalpy is more fundamental than that of internal energy since it is to some extent arbitrary whether energy terms involving independent fields are included in U or treated separately, as in (1.59).

The Helmholtz free energy A is defined as in (1.8) so that

$$A = U - TS = \sum_{i=1}^{n} \xi_i Q_i \,, \tag{1.61}$$

$$\mathrm{d}A = -S\mathrm{d}T + \sum_{i=1}^{n} \xi_i \mathrm{d}Q_i \,. \tag{1.62}$$

A is the thermodynamic potential which attains a minimum value when the state of thermodynamic equilibrium is reached in a system specified by T and the extensive variables $Q_1, \ldots, Q_n$; that is, a system where the only interchange of energy with the outside world is by heat transfer.

The replacement of $Q_1, \ldots, Q_\eta$ as independent variables by their conjugate fields $\xi_1, \ldots, \xi_\eta$ means that the system is opened to energy interchange by way of these fields. The appropriate thermodynamic potential is now F_η where

$$F_\eta = H_\eta - TS = U - TS - \sum_{i=1}^{\eta} \xi_i Q_i = \sum_{i=\eta+1}^{n} \xi_i Q_i, \tag{1.63}$$

$$\mathrm{d}F_\eta = -S\mathrm{d}T - \sum_{i=1}^{\eta} Q_i \mathrm{d}\xi_i + \sum_{i=\eta+1}^{n} \xi_i \mathrm{d}Q_i , \tag{1.64}$$

giving

$$\begin{aligned} S &= -\frac{\partial F_\eta}{\partial T}\,, \qquad Q_i = -\frac{\partial F_\eta}{\partial \xi_i}\,, \qquad i = 1,\ldots,\eta\,, \\ \xi_i &= \frac{\partial F_\eta}{\partial Q_i}\,, \qquad i = \eta+1,\ldots,n\,. \end{aligned} \tag{1.65}$$

From the first relation of (1.63), $S = (H_\eta - F_\eta)/T$ and hence the first relation of (1.65) can be expressed as

$$H_\eta = -T^2 \frac{\partial (F_\eta/T)}{\partial T} = \frac{\partial (F_\eta/T)}{\partial (1/T)}\,, \tag{1.66}$$

which is the *generalized Gibbs–Helmholtz relation.* The remarks made above about the specification of H_η also apply to F_η. In Sect. 1.2, G and D are both examples of F_1 with Q_1 taken as V and M, respectively, whereas in Sect. 1.4 the quantity denoted by F_1 provides another example of the general F_1 with $Q_1 = \mathcal{M}$. The Helmholtz free energy A could be denoted as F_0 and the relations given above for A and $\mathrm{d}A$ can be obtained from (1.63) and (1.64) by putting $\eta = 0$.

For a system with components $1,\ldots,\kappa$, a generalized grand potential may be obtained by assigning $\eta = \kappa$,

$$Q_j = M_j\,, \qquad \xi_j = \mu_j\,, \qquad j = 1,\ldots,\kappa \tag{1.67}$$

and then identifying D with F_κ. From (1.63) and (1.64),

$$D = U - TS - \sum_{j=1}^{\kappa} \mu_j M_j = \sum_{i=\kappa+1}^{n} \xi_i Q_i\,, \tag{1.68}$$

$$\mathrm{d}D = -S\mathrm{d}T - \sum_{j=1}^{\kappa} M_j \mathrm{d}\mu_j + \sum_{i=\kappa+1}^{n} \xi_i \mathrm{d}Q_i\,. \tag{1.69}$$

For the one-component simple fluid where the only mechanical field ξ_1 is $-P$, (1.68) and (1.69) reduce to (1.17)–(1.19).

From (1.56) a degree of freedom cannot be a means of transmitting energy to the system if Q_i is constant or, on the other hand, if the corresponding field ξ_i is zero. In the first case the system is closed with respect to energy transmission by way of field ξ_i, and in the second this particular field is not acting in the system's environment. Clearly if Q_i is constant then it must be regarded as one of the independent variables and, in the formalism above, $\eta + 1 \le i \le n$ whereas if ξ_i is zero then it is an independent variable and $1 \le i \le \eta$. In neither case does the corresponding term appear on the right-hand side of (1.60) or (1.64). Sometimes it is feasible experimentally to put Q_i = constant; for instance by enclosing a fluid in a rigid vessel we can have, effectively, V = constant, but it is impossible to put the conjugate field

$-P = 0$. On the other hand, a system can be isolated magnetically by putting the field $\mathcal{H} = 0$.

If we try to use only the fields as independent variables by putting $\eta = n$, then from the last expression of (1.63) the corresponding potential $F_n = 0$ and (1.64) reduces to (1.58). This 'catastrophe' clearly results from an attempt to define an extensive thermodynamic potential in terms of intensive quantities only. However, by adopting a different approach it is possible to have a representation of thermodynamics in which only intensive variables occur, and this will be done in the next section.

1.8 The Field-Density Representation of Thermodynamics

In this representation, the systematic theory of which is due to Griffiths and Wheeler (1970), extensive variables are replaced by a new set of intensive variables called *densities* which are ratios of extensive variables. This formulation is useful in relating thermodynamics to statistical mechanics where it is necessary to consider systems in which every extensive variable tends to infinity. The advantage of using properly defined densities is that they remain finite in the *thermodynamic limit.*

Choosing a particular extensive variable as Q_n we define the $n-1$ densities ρ_i by

$$\rho_i = Q_i/Q_n\,, \qquad i = 1, \ldots, n-1\,. \tag{1.70}$$

The entropy, internal energy and Helmholtz free energy densities are defined in a corresponding way:

$$s = S/Q_n\,, \qquad u = U/Q_n\,, \qquad a = A/Q_n\,. \tag{1.71}$$

Most frequently either $Q_n = V$, so that the densities are 'per unit of reduced volume', or $Q_n = M$, where M is the total number of molecules in the system, so that the densities are 'per molecule'. We have in fact already used 'per molecule' densities for a simple fluid in deriving (1.16) and in the treatment of the van der Waals gas in Sect. 1.6. Equations (1.57) and (1.58) can be converted to field density form by dividing through by Q_n to yield

$$u = sT + \sum_{i=1}^{n-1} \rho_i \xi_i + \xi_n\,, \tag{1.72}$$

$$0 = s\mathrm{d}T + \sum_{i=1}^{n-1} \rho_i \mathrm{d}\xi_i + \mathrm{d}\xi_n\,. \tag{1.73}$$

Equation (1.73) is a generalization of the simple fluid relation (1.16) and shows that there cannot be more than n independent intensive variables for a system where there can be as many as $n+1$ independent extensive variables. Taking the differential of (1.72) yields

$$du = sdT + Tds + \sum_{i=1}^{n-1} \rho_i d\xi_i + \sum_{i=1}^{n-1} xi_i d\rho_i + d\xi_n \,, \tag{1.74}$$

which can be combined with (1.73) to give

$$du = Tds + \sum_{i=1}^{n-1} \xi_i d\rho_i \,, \tag{1.75}$$

which is the field-density analogue of (1.56).

Free energy (or thermodynamic potential) and enthalpy densities may be defined by

$$f_\eta = F_\eta / Q_n \,, \qquad h_\eta = H_\eta / Q_n \,, \qquad \eta = 0, 1, 2, \ldots, n-1 \,, \tag{1.76}$$

where $f_0 = a$, the Helmholtz free energy density. Dividing (1.63) by Q_n yields immediately

$$f_\eta = h_\eta - Ts = u - Ts - \sum_{i=1}^{\eta} \xi_i \rho_i = \sum_{i=\eta+1}^{n-1} \xi_i \rho_i + \xi_n \,. \tag{1.77}$$

From (1.77) and (1.75) it is easy to show that

$$df_\eta = -sdT - \sum_{i=1}^{\eta} \rho_i d\xi_i + \sum_{i=\eta+1}^{n-1} \xi_i d\rho_i \,. \tag{1.78}$$

Since F_η has a minimum value for thermodynamic equilibrium with fixed values of T, $\xi_1, \ldots, \xi_\eta$, $Q_{\eta+1}, \ldots, Q_n$ it follows that f_η has a minimum value for thermodynamic equilibrium with fixed values of T, $\xi_1, \ldots, \xi_\eta$, $\rho_{\eta+1}, \ldots, \rho_{n-1}$. We have already used the minimum property of a for a one-component fluid with the ratio V/M fixed in Sect. 1.6. A free energy density, which is a function of T and the $n-1$ independent fields $\xi_1, \ldots, \xi_{n-1}$ can be obtained by putting $\eta = n-1$. Equations (1.77) and (1.78) then become

$$\begin{aligned} f_{n-1} &= u - Ts - \sum_{i=1}^{n-1} \rho_i \xi_i = \xi_n \,, \\ df_{n-1} &= -sdT - \sum_{i=1}^{n-1} \rho_i d\xi_i \,. \end{aligned} \tag{1.79}$$

The second formula is equivalent to (1.73).

For many-component systems it is often convenient to express the composition in terms of the mole fractions x_j, given by

$$x_j = M_j / M \,, \qquad j = 1, \ldots, \kappa \,, \qquad \sum_{j=1}^{\kappa} x_j = 1 \,, \tag{1.80}$$

where

$$M = \sum_{j=1}^{\kappa} M_j \,. \tag{1.81}$$

To use the total number of molecules as an extensive variable we must first note that the energy of the molecular transfer term can be written

$$\sum_{j=1}^{\kappa} \mu_j \mathrm{d}M_j = \sum_{j=1}^{\kappa-1} (\mu_j - \mu_\kappa)\mathrm{d}M_j + \mu_\kappa \mathrm{d}M \,. \tag{1.82}$$

Let us now assign

$$\begin{aligned} &Q_{n-\kappa+j} = M_j \,, \qquad &&\xi_{n-\kappa+j} = \mu_j - \mu_\kappa \,, \qquad &&j = 1, \ldots, \kappa - 1 \,, \\ &Q_n = M \,, &&\xi_n = \mu_\kappa \,. \end{aligned} \tag{1.83}$$

With these variables a 'per molecule' free energy density can be defined for any value $\eta \le n-1$, but we shall consider only f_{n-1} explicitly. From (1.79) and (1.83)

$$f_{n-1} = \mu_\kappa = u - Ts - \sum_{i=1}^{n-\kappa} \rho_i \xi_i - \sum_{j=1}^{\kappa-1} x_j (\mu_j - \mu_\kappa) \,, \tag{1.84}$$

$$\mathrm{d}f_{n-1} = \mathrm{d}\mu_\kappa = -s\mathrm{d}T - \sum_{i=1}^{n-\kappa} \rho_i \mathrm{d}\xi_i - \sum_{j=1}^{\kappa-1} x_j \mathrm{d}(\mu_j - \mu_\kappa) \,. \tag{1.85}$$

The last relation can be re-expressed in the form

$$s\mathrm{d}T + \sum_{i=1}^{n-\kappa} \rho_i \mathrm{d}\xi_i + \sum_{j=1}^{\kappa} x_j \mathrm{d}\mu_j = 0 \,, \tag{1.86}$$

which is a generalized form of the Gibbs–Duhem equation (1.16).

1.9 General Theory of Phase Equilibrium

Liquid–vapour phase separation has been discussed in the context of the van der Waals equation of state although any 'classical' theory giving pressure–volume isotherms of similar form will yield a critical point and coexistence curve in the same way. We now consider phase equilibrium in the system defined in Sect. 1.7 with $n+1$ extensive variables $S, Q_1, \ldots, Q_n$. It will be supposed that the state of the system is specified by the fields $T, \xi_1, \ldots, \xi_\eta$ and the extensive variables $Q_{\eta+1}, \ldots, Q_n$. The thermodynamic potential with a stationary value at equilibrium, when the above variables have fixed values, is F_η, defined by (1.63). We now regard the system as split into two homogeneous phases in equilibrium, distinguished by the indices 1 and 2. Assuming as usual that boundary contributions to the extensive variables are negligible,

$$\begin{aligned} F_\eta &= F_{\eta 1} + F_{\eta 2}\,, \\ Q_i &= Q_{i1} + Q_{i2}\,, \qquad i = \eta + 1, \ldots, n\,. \end{aligned} \tag{1.87}$$

The values of the dependent fields in the conjugate phases will be denoted by ξ_{i1} and ξ_{i2} respectively ($i = \eta + 1, \ldots, n$). Now suppose that there is a small deviation from equilibrium in which

$$\begin{aligned} Q_{i1} &\to Q_{i1} + \mathrm{d}Q_i, \\ Q_{i2} &\to Q_{i2} - \mathrm{d}Q_i\,, \end{aligned} \qquad i = \eta + 1, \ldots, n\,. \tag{1.88}$$

Then, by (1.64) and (1.87),

$$\begin{aligned} \mathrm{d}F_\eta = \mathrm{d}F_{\eta 1} + \mathrm{d}F_{\eta 2} &= \sum_{i=\eta+1}^{n} \xi_{i1}\mathrm{d}Q_i + \sum_{i=\eta+1}^{n} \xi_{i2}(-\mathrm{d}Q_i) \\ &= \sum_{i=\eta+1}^{n} (\xi_{i1} - \xi_{i2})\mathrm{d}Q_i\,. \end{aligned} \tag{1.89}$$

Since F_η has a stationary value at equilibrium, $\mathrm{d}F_\eta = 0$ for small deviations from equilibrium and hence the coefficients of the $\mathrm{d}Q_i$ in the last expression of (1.89) are zero. The independent fields $\xi_i, \ldots, \xi_\eta$ and T are the same throughout the system by hypothesis. Hence each of the n fields must have the same value in the two phases in equilibrium. This result is seen to be independent of the original choice of independent variables. It can easily be extended to a situation where there are p phases in equilibrium, and we can then write

$$\xi_{i1} = \xi_{i2} = \cdots = \xi_{ip-1} = \xi_{ip}\,, \qquad i = 1, \ldots, n\,, \tag{1.90}$$

$$T_1 = T_2 = \cdots = T_{p-1} = T_p\,, \tag{1.91}$$

where T_k is the temperature of phase k. It should be noted that phase equilibrium conditions involve intensive variables only; the sizes of phases in equilibrium are arbitrary.

1.9.1 A One-Component Fluid

For a simple one-component fluid system, $n = 2$ since, apart from T, the only fields are $-P$ and μ. Hence, for liquid–vapour equilibrium, conditions (1.90) reduce to the relations (1.41), derived independently in Sect. 1.5. Fig. 1.3, obtained for a van der Waals gas, qualitatively resembles the liquid–vapour part of the phase diagram for a real substance. It can be seen that on any isotherm the pressure at which liquid and vapour phases coexist is determined. The same must therefore be true of the chemical potential, since μ is a function of P and T. Hence if the temperature of liquid–vapour equilibrium is altered, the corresponding changes in pressure and chemical potential are determined. The one-component, two-phase system is said to have one degree of freedom. We now ask: how many degrees of freedom $\mathcal{F}(n,p)$ are possessed by the p-phase system whose equilibrium conditions are (1.90) and (1.91)? If the phases were isolated from each other there would be n independent intensive variables in each phase, making pn in all. However (1.90) imposes $n(p-1)$ conditions on the intensive variables and (1.91) a further $p-1$ conditions. Thus

$$\mathcal{F}(n,p) = pn - (p-1)(n+1) = n - p + 1\,. \tag{1.92}$$

As an alternative to the term 'degrees of freedom' we can say that $\mathcal{F}(n,p)$ is the dimension of the region of coexistence of the p phases. The relation (1.92) is better known in the form of the *Gibbs phase rule* which applies to an isotropic system with κ components. Apart from T, the fields in this case are $\mu_1, \mu_2, \ldots, \mu_\kappa$ and $-P$. Thus from (1.92) the dimension of the region of coexistence is

$$\mathcal{D}(\kappa,p) = \mathcal{F}(\kappa+1,p) = \kappa - p + 2\,. \tag{1.93}$$

For two coexisting one-component phases $\mathcal{D}(1,2) = 1$ in agreement with the result obtained above for liquid–vapour equilibrium. Since $\mathcal{D}(\kappa,p) \geq 0$ we conclude that the largest number of phases of a κ component system which can coexist is $\kappa+2$. There are exceptions to this, which arise when the phase diagram has some special symmetry. One of these is the triangular ferrimagnetic Ising model discussed in Sect. 4.5 and another is the metamagnet in a staggered field (see Sects. 4.3 and 4.4 and Volume 2, Sects. 2.2 and 2.12). The most well-known example of a single point of three phase coexistence of a one component system is the solid–liquid–vapour *triple point* in the (T,P) plane. Liquid–vapour, solid–vapour and solid–liquid equilibrium respectively exist along curves in the (T,P) plane, and these curves meet at the triple point (see Fig. 1.4). If the pressure on a fluid is steadily reduced at a fixed temperature $T < T_c$ there is a discontinuity in the molar volume when the liquid–vapour transition point is reached (see Fig. 1.3). This is an example of a *first-order transition*, the term first-order implying a discontinuity in some of the densities ρ_i defined in Sect. 1.8. The terms 'first-order transition' and 'conjugate phases' refer to the same phenomenon. Whether a transition from one phase to another or separation of a homogeneous medium into conjugate phases is observed depends on which thermodynamic quantities are fixed and which are allowed to vary. We now consider the behaviour of a free energy (thermodynamic potential) density function at a first-order transition. The free energy density $f_{n-1}(T,\xi_1,\ldots,\xi_{n-1})$, defined by (1.79), is equal to the field ξ_n. Hence f_{n-1} is continuous by (1.90). However since, by (1.79),

$$\rho_i = -\frac{\partial f_{n-1}}{\partial \xi_i}\,, \qquad i = 1,\ldots,n-1\,, \tag{1.94}$$

and since some or all of the densities ρ_i are unequal in the conjugate phases, there will be discontinuities in the corresponding first derivatives of f_{n-1}. Geometrically, the function f_{n-1}, (i.e. ξ_n) can be regarded as a surface in

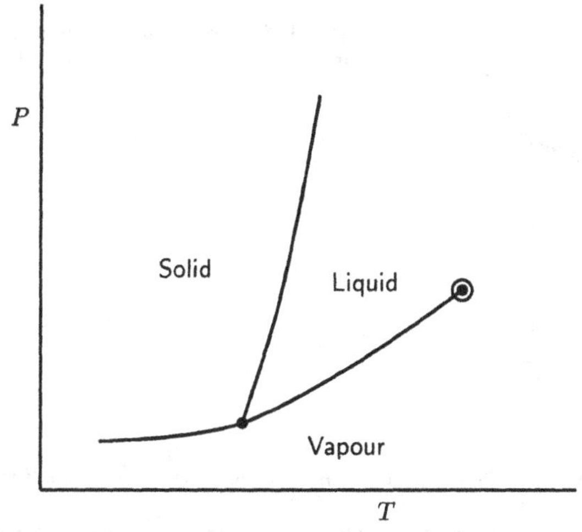

Fig. 1.4. Phase equilibrium curves in pressure–temperature plane; the critical point is circled.

an $(n+1)$-dimensional space with coordinates $(T, \xi_1, \ldots, \xi_n)$. The first-order transition will appear as a 'crease' in this surface. Taking a plane through the surface parallel to the f_{n-1}-axis and the axis corresponding to a field ξ_i we obtain a diagram like that shown in Fig. 1.5. For the simple one-component system, with $n = 2$, $f_{n-1} = \mu(T, P)$ if 'per molecule' densities are used. The phase equilibrium lines in Fig. 1.4 are the projections into the (T, P) plane of the first-order transition creases in the $\mu(T, P)$ surface. We now consider the behaviour of f_{n-2}, which is a function of the fields $T, \xi_1, \ldots, \xi_{n-2}$ and the density ρ_{n-1}, at a first-order transition where ρ_{n-1} has a discontinuity. Since f_{n-2} depends on ρ_{n-1}, f_{n-2} will normally be discontinuous at the transition. A plot of f_{n-2} against ρ_{n-1} for homogeneous phases with the fields $T, \xi_1, \ldots, \xi_{n-2}$ held constant thus consists of two disconnected parts, corresponding to the phases on the two sides of the transition (Fig. 1.6). However, putting $\eta = n - 2$ in (1.77) and (1.78) gives

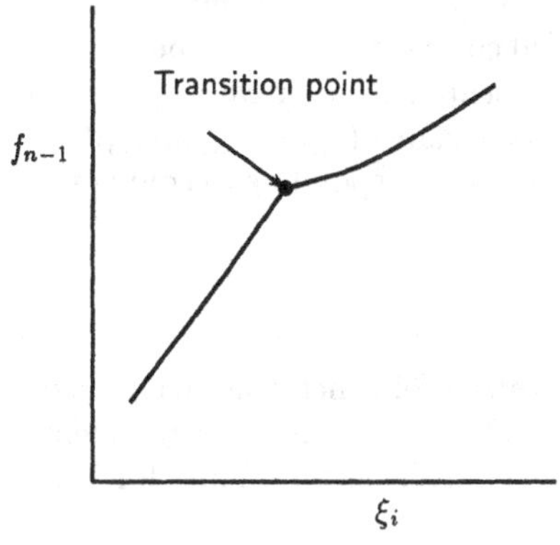

Fig. 1.5. Free energy–field isotherm showing a first-order transition point.

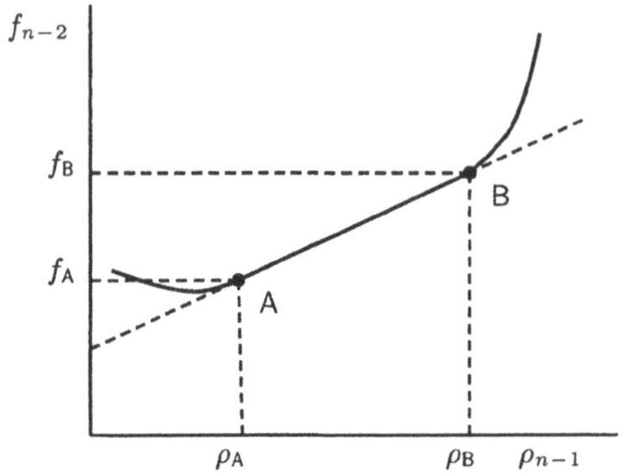

Fig. 1.6. Free energy-density isotherm with double tangent AB.

$$\xi_{n-1} = \frac{\partial f_{n-2}}{\partial \rho_{n-1}}, \quad \xi_n = f_{n-2} - \xi_{n-1}\rho_{n-1} = f_{n-2} - \rho_{n-1}\frac{\partial f_{n-2}}{\partial \rho_{n-1}}. \tag{1.95}$$

The end-points A and B of the two parts of the f_{n-2} curve correspond to conjugate phases. Since fields are continuous at a first-order transition the right-hand sides of the two relations of (1.95) are equal at A and B. Geometrically (1.95) thus implies a double tangent at the points A and B in Fig. 1.6. Let us denote the values of f_{n-2} at the points A and B by f_A and f_B respectively and the corresponding values of ρ_{n-1} by ρ_A and ρ_B. If the prescribed value of ρ_{n-1} lies between ρ_A and ρ_B then the system will split into conjugate phases A and B. If boundary effects are disregarded, the overall value of the free energy density f_{n-2} for the system will be a linear combination of f_A and f_B. Hence the plot of f_{n-2} for the whole range of ρ_{n-1} consists of the two curved portions terminating at A and B respectively and the segment AB of the common tangent. For the simple one-component system, with $n = 2$, $f_{n-2} = a(T, v)$ if 'per molecule' densities are used. Hence the relations (1.47), derived above, are a special case of (1.95). In Fig. 1.2, derived from the approximate van der Waals theory, the convex envelope indicated by the chain line corresponds to Fig. 1.6.

Returning to the general system, the free energy density f_η, defined by (1.77), is a function of the independent densities $\rho_{\eta+1}, \ldots, \rho_{n-1}$ and the independent fields $T, \xi_1, \ldots, \xi_\eta$. If the latter are held constant then the dependence of f_η on the independent densities can be represented geometrically by a surface in the $(n-\eta)$-dimensional space with coordinates $(\rho_{\eta+1}, \ldots, \rho_{n-1}, f_\eta)$. By generalizing (1.95) it can be shown that points corresponding to conjugate phases lie on the same tangent hyperplane.

1.9.2 Azeotropy

We now discuss briefly the phenomenon of *azeotropy* of which a more detailed account is given by Rowlinson and Swinton (1982). Consider a two-component fluid system with mole fractions $x_1 = x$, $x_2 = 1 - x$. Then $n = 3$ and for 'per molecule' densities we use (1.83), putting

$$\begin{aligned} \xi_1 &= -P\,, & \rho_1 &= v\,, \\ \xi_2 &= \mu_1 - \mu_2\,, & \rho_2 &= x\,, \\ \xi_3 &= \mu_2\,. \end{aligned} \tag{1.96}$$

Then, from (1.85),

$$\mathrm{d}f_2 = \mathrm{d}\xi_3 = -s\mathrm{d}T - v\mathrm{d}\xi_1 - x\mathrm{d}\xi_2\,. \tag{1.97}$$

With $\kappa = 2$ and $p = 2$ (two phase equilibrium of a two-component mixture) (1.93) yields $\mathcal{D}(2,2) = 2$, giving a surface in (ξ_1, ξ_2, T) space. Consider the curve of intersection of this surface with a plane $\xi_1 = -P =$ constant. For a small displacement along such a curve, a relation like (1.97), with $\mathrm{d}\xi_1 = 0$, is satisfied by each of the two equilibrium phases. Subtracting one of these relations from the other and using the fact that, by (1.90), $\mathrm{d}\xi_3$ is the same for both phases we have

$$\frac{\mathrm{d}T}{\mathrm{d}\xi_2} = -\frac{\Delta x}{\Delta s}\,, \tag{1.98}$$

where Δs and Δx respectively are the differences between the values of entropy per molecule and mole fraction between the conjugate phases. An *azeotropic point* occurs when the conjugate phases have the same composition ($\Delta x = 0$) but different molecular entropies ($\Delta s \neq 0$). Then, by (1.98), $\mathrm{d}T/\mathrm{d}\xi_2 = 0$, so that the phase equilibrium curve in the (ξ_2, T) plane has a horizontal tangent at the azeotropic point. If the coordinates are changed from ξ_2 and T to x and T then the phase equilibrium curve is replaced by two curves each representing one of the conjugate phases. For each of these phases we can write

$$\frac{\mathrm{d}x}{\mathrm{d}T} = \left(\frac{\partial x}{\partial \xi_2}\right)_{T,\xi_1} \frac{\mathrm{d}\xi_2}{\mathrm{d}T} + \left(\frac{\partial x}{\partial T}\right)_{\xi_1,\xi_2}\,, \tag{1.99}$$

where $(\partial x/\partial \xi_2)_{T,\xi_1} > 0$ by the stability condition (1.107) and $\mathrm{d}\xi_2/\mathrm{d}T$ is given by (1.98). The conjugate phase curves meet at the azeotropic point where $\Delta x = 0$ and, since $\mathrm{d}\xi_2/\mathrm{d}T = \infty$, it follows from (1.99) that $\mathrm{d}T/\mathrm{d}x = 0$. Hence, in the (x, T) plane, the conjugate phase curves touch, with a common horizontal tangent at the azeotropic point. Fig. 1.7 shows positive azeotropy; for negative azeotropy the touching curves have minima at the azeotropic point. Azeotropic points in the plane $\xi_1 = -P =$ constant form a curve in the (ξ_1, ξ_2, T) or (ξ_1, x, T) space, and hence, azeotropic behaviour, similar to that discussed above, occurs in $T =$ constant planes. The difference between an azeotropic point and a critical point for phase separation should be noted. At a critical point, Δs and Δv, as well as Δx, are zero.

1.10 Classical Theory and Metastability

Exact solutions for the thermodynamic functions of lattice models exist only for equilibrium states (see Chap. 8) and assume the forms shown in Figs. 1.5

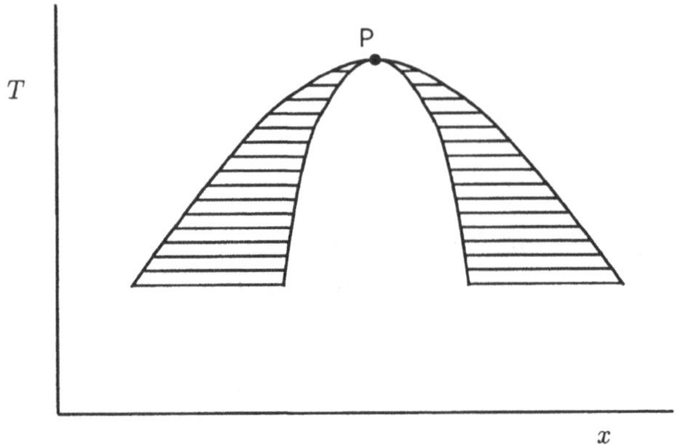

Fig. 1.7. Positive azeotropy; P is azeotropic point and horizontals are tie lines.

and 1.6, when a first-order transition occurs. We now consider how such a transition appears in classical approximations, taking as our guide the van der Waals gas. In the latter we have seen that intensive thermodynamic variables are continuous over the entire range of the volume per molecule v, including the interval where the homogeneous fluid is unstable and separates into conjugate phases of different density. From (1.79)

$$\mathrm{d}f_{n-1} = \mathrm{d}\xi_n = -\rho_i \mathrm{d}\xi_i \,, \tag{1.100}$$

given that T and ξ_j for $j \neq i$, $j \neq n$ are held constant, which will be assumed for the rest of this paragraph. For a one-component fluid with 'per molecule' densities, $\xi_n = \mu$ and since $n = 2$ we have to put $\xi_i = \xi_{n-1} = -P$, $\rho_i = v$. In a classical theory, where ξ_n and ξ_i are continuous over the whole range of ρ_i, the conditions for phase separation can be satisfied only if the curve of ξ_i against ρ_i is s-shaped, as in Fig. 1.1. Thus the derivative $\partial \xi_i/\partial \rho_i > 0$ for large or small values of ρ_i, but there is an interval where $\partial \xi_i/\partial \rho_i < 0$. From (1.100), $\partial f_{n-1}/\partial \rho_i$ changes sign when $\partial \xi_i/\partial \rho_i$ does, assuming $\rho_i \neq 0$ at this point. Hence a plot of f_{n-1} against ξ_i yields a curve like ABCDE in Fig. 1.8. The part ABE represents stable states, since it corresponds to the least value of f_{n-1} for each ξ_i and is equivalent to the equilibrium curve of Fig. 1.5. At the point B, where the lines AC and DE intersect, the slope of the equilibrium curve ABE is discontinuous and hence B corresponds to the first-order transition. The loop BCD represents unstable states, and the cusps at C and D correspond to the (unstable) values of ρ_i, where $\partial f_{n-1}/\partial \rho_i$ and $\partial \xi_i/\partial \rho_i$ change sign.

There are thus three equivalent ways of locating a first-order transition and determining the corresponding phases in a classical theory:

(i) Plot f_{n-1} against ξ_i (with T and the fields ξ_j for $j \neq i$, $j \neq n$ held constant) obtaining enough points on the curves AC and ED (see Fig. 1.8) to determine the point of intersection B.

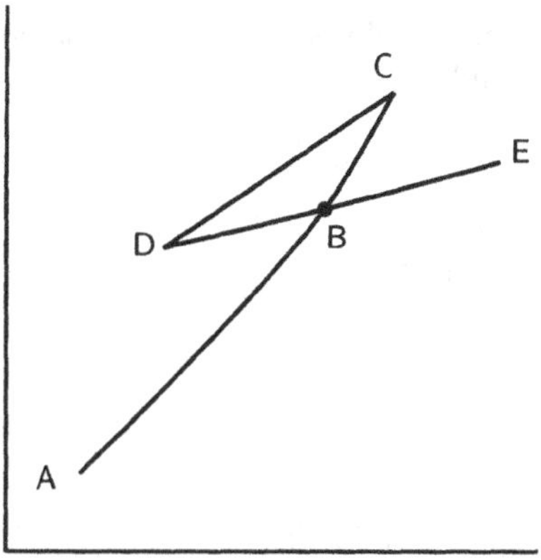

Fig. 1.8. Classical free energy–field isotherm; AB and BE represent stable states.

(ii) Plot ξ_i against ρ_i (with T and the fields ξ_j for $j \neq i$, $j \neq n$ held constant) and draw the equal-areas tie line, as in Fig. 1.1. Numerically this is not as convenient as (i). However, the existence of a range of ρ_i where

$$\frac{\partial \xi_i}{\partial \rho_i} < 0 \tag{1.101}$$

is a sufficient condition for an instability loop leading to phase separation. Also, the relations

$$\frac{\partial \xi_i}{\partial \rho_i} = 0, \qquad \frac{\partial^2 \xi_i}{\partial \rho_i^2} = 0 \tag{1.102}$$

for the disappearance of such a range of ρ_i determine the position of a critical point (end of a first-order transition line).

(iii) Plot $f_{n-2}(= f_{n-1} + \rho_{n-1}\xi_{n-1})$ against ρ_{n-1} (with T and the fields ξ_i, for $j = 1, \ldots, n-2$, held constant). If phase separation occurs a part of the curve will be convex upwards and, from (1.95), the conjugate phases can be determined by drawing a double tangent, as in Fig. 1.2. The convex envelope is the equilibrium curve corresponding to Fig. 1.6. This method is as general as (i) and (ii) since the fields can be relabelled to make any ξ_i for $i < n$ into ξ_{n-1}.

1.10.1 Metastability in a One-Component Fluid

So far we have not differentiated between unstable states, which we have defined as states where the appropriate thermodynamic potential is greater than its least attainable value. Taking the one-component fluid as an example, suppose that M molecules occupy a fixed volume V at a temperature T. Now assume that half the mass of fluid undergoes a small contraction, taking the molecular volume to $v-\delta v$. The other half must then undergo a small dilation, taking its molecular volume to $v+\delta v$. Using Taylor's theorem the change in the Helmholtz free energy of the whole fluid is

$$\begin{aligned}\delta A &= \frac{1}{2}M\left[\frac{\partial a}{\partial v}\delta v + \tfrac{1}{2}\frac{\partial^2 a}{\partial v^2}(\delta v)^2\right]\\ &\quad+\frac{1}{2}M\left[\frac{\partial a}{\partial v}(-\delta v) + \frac{1}{2}\frac{\partial^2 a}{\partial v^2}(-\delta v)^2\right] + \mathrm{O}(\delta v^3)\\ &= \frac{1}{2}M\frac{\partial^2 a}{\partial v^2}(\delta v)^2 + \mathrm{O}(\delta v^3)\\ &= -\frac{1}{2}M\frac{\partial P}{\partial v}(\delta v)^2 + \mathrm{O}(\delta v^3),\end{aligned} \tag{1.103}$$

using (1.48) for the last expression. Hence, if $(\partial P/\partial v)_T > 0$ then $\delta A < 0$. Thus, the free energy of the fluid is not a minimum with respect to small amplitude, long-wavelength fluctuations in the density. Such a situation will be termed *internal* or *intrinsic instability.* It follows that for internal stability the condition

$$\left(\frac{\partial P}{\partial v}\right)_T < 0 \tag{1.104}$$

must apply. Many other *thermodynamic inequalities* can be derived. For instance by considering a temperature fluctuation at constant volume it can be deduced that

$$\frac{C_V}{M_A} = Tk_B\left(\frac{\partial s}{\partial T}\right)_V > 0\,, \tag{1.105}$$

where C_V is the molar heat capacity at constant volume, M_A is Avogadro's number and s is (according to the conventions we adopted in Sects. 1.7 and 1.8) the dimensionless entropy per molecule. Since (1.104) implies that the isothermal compressibility $\mathcal{K}_T > 0$ it follows from (1.30) and (1.105) that

$$C_P > C_V > 0\,, \tag{1.106}$$

where C_P is the molar heat capacity at constant pressure. By a similar proof to that of (1.104) it can be shown for the general system that, for internal stability,

$$\frac{\partial \xi_i}{\partial \rho_i} > 0\,, \tag{1.107}$$

other independent fields or densities and the temperature T being fixed. By considering simultaneous fluctuations of several variables, many more complicated thermodynamic inequalities can be derived (Pippard 1967).

We now return to the van der Waals gas and consider the parts of the range (v_1, v_2) in Fig. 1.1 where the condition (1.104) is obeyed. For such a value of v the free energy of a homogeneous fluid cannot be diminished by small density fluctuations, but it can be reduced by separation into conjugate phases with

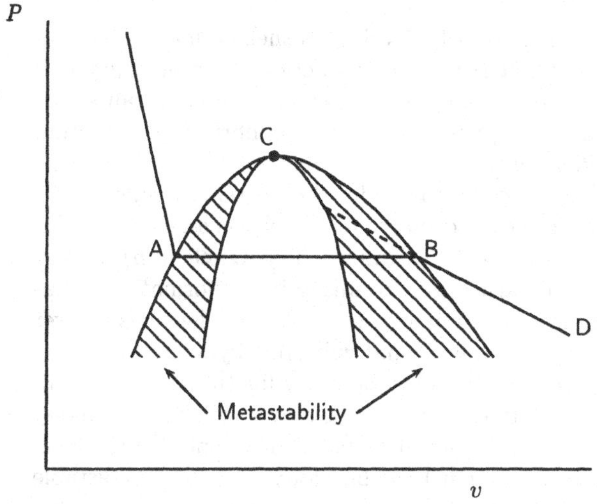

Fig. 1.9. Liquid-vapour system: metastable region is hatched and C is the critical point.

molecular volumes v_1 and v_2 respectively. However, for such a separation, large changes in density must occur. A homogeneous fluid in such a state is said to be *metastable*. We have thus distinguished two kinds of unstable state which are termed intrinsically unstable and metastable respectively. Intrinsically unstable states are infinitesimally close to states of lower free energy whereas metastable states are separated from them by finite intervals of certain variables. On Fig. 1.3 the coexistence (broken) curve divides stable from unstable regions. The curve defined by $(\partial P/\partial v)_T = 0$, which touches the coexistence curve at the critical point, subdivides the unstable region into domains of metastability and intrinsic instability respectively (see Fig. 1.9). Such a curve is known as a *spinodal*. On the diagram, for a general classical system shown in Fig. 1.8, the segments BC and DB of the instability loop BCD correspond to metastable states whereas the segment CD corresponds to intrinsically unstable states. This latter are said to be subject to *spinodal decomposition* by long-wavelength fluctuations.

1.10.2 The Experimental Situation

Metastable states are not simply artifacts of classical theory, but are frequently encountered experimentally. Suppose that on Fig. 1.9 a vapour is compressed from point D to a point B on the coexistence curve. If the system remains in thermodynamic equilibrium, any further increase in pressure will cause a transition to the liquid part of the isotherm starting at point A, AB being a tie line. However, when Fig. 1.9 represents the behaviour of a real fluid, a dust-free vapour may progress along the broken curve continuous with DB into the metastable region in Fig. 1.9. It is then termed *supersaturated.* Metastable states can also be reached by reduction of temperature, and the substance is then said to be *supercooled.* Supercooled liquids can sometimes exist at temperatures considerably below the equilibrium temperature of transition to the crystalline form.

We now discuss briefly the physical reasons for the existence of metastable states in real substances, using the supersaturated vapour as an example. For the latter to go over to the liquid state, localized (low-wavelength) density fluctuations of large amplitude must occur, resulting in the formation of nuclei, each containing a large number of molecules. Such a nucleus has a lower free energy than an equal mass of vapour and will grow by accretion. However, a nucleus created by a spontaneous local density fluctuation is likely to be small and because of its relatively large surface area will have a higher free energy than an equal mass of vapour. This is another aspect of the local stability of metastable states. Nevertheless thermodynamically metastable states are different in character from static mechanical metastable states, like that of a ball on a surface at the lowest point of a depression separated by a ridge from a deeper depression.

The degree of permanence of a thermodynamically metastable state depends on the *relaxation time*, the mean waiting time for a sufficiently large fluctuation to initiate the transformation to the stable state. For the mechanical analogy to be appropriate, in fact, the ball must be regarded as subject to impulses of random strength at random intervals. However, relaxation times can be very large. For instance, glasses are supercooled liquids with higher free energies than the crystalline forms whereas diamond is a metastable crystalline form of carbon. Thus, contrary to the common belief, diamonds are not 'for ever', just for a very long time. In a supersaturated vapour dust particles drastically reduce the relaxation time by facilitating the formulation of large nuclei. Charged particles (ions) have the same effect, which is used in the cloud chamber apparatus. Here the tracks of charged particles through supersaturated water vapour become visible because of the formation of liquid water droplets. Work on first-order transition mechanisms is reviewed by Gunton et al. (1983) and Binder (1987).

Examples

1.1 From relations given in Sects. 1.1 and 1.2, show that for a fluid of fixed composition:

$$\text{(a)}\quad \left(\frac{\partial \widetilde{T}}{\partial \widetilde{V}}\right)_S = -\left(\frac{\partial \widetilde{P}}{\partial \widetilde{S}}\right)_V, \qquad \text{(b)}\quad \left(\frac{\partial \widetilde{T}}{\partial \widetilde{P}}\right)_S = \left(\frac{\partial \widetilde{V}}{\partial \widetilde{S}}\right)_P,$$

$$\text{(c)}\quad \left(\frac{\partial \widetilde{P}}{\partial \widetilde{T}}\right)_V = \left(\frac{\partial \widetilde{S}}{\partial \widetilde{V}}\right)_T, \qquad \text{(d)}\quad \left(\frac{\partial \widetilde{V}}{\partial \widetilde{T}}\right)_P = -\left(\frac{\partial \widetilde{S}}{\partial \widetilde{P}}\right)_T.$$

((c) is proved in Sect. 1.3 and the other relations can be obtained in a similar way.) These are *Maxwell's thermodynamic relations.* It can be seen that they have the same form if the variables T, S, P and V are used in place of $\widetilde{T}$, $\widetilde{S}$, $\widetilde{P}$ and $\widetilde{V}$.

1.2 From relations given in Sect. 1.2, show that, for a fluid of fixed composition,

$$U = -\widetilde{T}^2\left(\frac{\partial(A/\widetilde{T})}{\partial\widetilde{T}}\right)_V, \qquad H = -\widetilde{T}^2\left(\frac{\partial(G/\widetilde{T})}{\partial\widetilde{T}}\right)_P,$$

$$G = -\widetilde{V}^2\left(\frac{\partial(A/\widetilde{V})}{\partial\widetilde{V}}\right)_T.$$

The first two relations here are known as *Gibbs–Helmholtz relations* and are special cases of (1.66). They can be seen to have the same form if T and V replace $\widetilde{T}$ and $\widetilde{V}$.

1.3 In terms of the variables T, S, P and V the formulae (1.25) and (1.26) for the heat capacities C_P and C_V are modified to

$$\begin{aligned} C_P &= k_B T\left(\frac{\partial S}{\partial T}\right)_P = k_B\left(\frac{\partial H}{\partial T}\right)_P \\ &= -k_B T\frac{\partial^2 G}{\partial T^2} = -k_B M_A T\frac{\partial^2 \mu}{\partial T^2}\,. \end{aligned}$$

and

$$\begin{aligned} C_V &= k_B T\left(\frac{\partial S}{\partial T}\right)_V = k_B\left(\frac{\partial U}{\partial T}\right)_V \\ &= -k_B T\frac{\partial^2 A}{\partial T^2} = -k_B M_A T\frac{\partial^2 a}{\partial T^2}\,. \end{aligned}$$

Determine the changes which are needed in the formulae (1.20) and (1.21) for α_P and $\mathcal{K}_T$.

1.4 Defining $C_{\mathcal{H}}$ as in (1.35) and $C_{\mathcal{M}} = \widetilde{T}(\partial\widetilde{S}/\partial\widetilde{T})_{\mathcal{M}}$ show that

$$C_{\mathcal{H}} - C_{\mathcal{M}} = T\left(\frac{\partial\mathcal{M}}{\partial\widetilde{T}}\right)_{\mathcal{H}}^2\left[\left(\frac{\partial\widetilde{\mathcal{M}}}{\partial\widetilde{\mathcal{H}}}\right)_T\right]^{-1}.$$

(Reference to the proof of (1.30) may be helpful.) Determine what changes are needed in this formula if $\mathcal{M}$ and $\mathcal{H}$ are used in place of $\widetilde{\mathcal{M}}$ and $\widetilde{\mathcal{H}}$.

1.5 For a one component fluid, where $n = 2$, show that, putting $Q_n = M$, equations (1.77) and (1.78) can be written

$$f_1 = u - Ts + Pv = a + Pv = \mu\,,$$
$$\mathrm{d}f_1 = -s\mathrm{d}T + v\mathrm{d}P\,,$$
$$\mathrm{d}a = -s\mathrm{d}T - P\mathrm{d}v\,.$$

Show that van der Waals equation of state (1.43) can be expressed in the form

$$\left(P + \frac{\varepsilon}{v^2}\right)(v-1) = T$$

if b is taken as the standard volume v_0 of equations (1.54) and (1.55) and $\varepsilon = c/b$. Using this equation, show that the constant temperature relation $\mathrm{d}a = -P\mathrm{d}v$ can be integrated to give

$$a = -\frac{\varepsilon}{v} - T\ln(v-1) + \psi(T)\,,$$

where ψ is an arbitrary function of T. Obtain μ by integrating the constant temperature relation $\mathrm{d}\mu = v\mathrm{d}P$ and verify that the expressions for μ and a are compatible with the relation $\mu = a + Pv$.

1.6 An alternative classical state equation to van der Waals' is that of *Dieterici:*

$$\widetilde{P} = \frac{k_{\mathrm{B}}\widetilde{T}\exp(-c/k_{\mathrm{B}}\widetilde{T}\tilde{v})}{\tilde{v} - b}\,.$$

Show that the critical values of $\tilde{v}$, $\widetilde{T}$ and $\widetilde{P}$ are given by

$$k_{\mathrm{B}}\widetilde{T}_{\mathrm{c}} = \frac{c}{4b}\,, \qquad \tilde{v}_{\mathrm{c}} = 2b\,, \qquad \widetilde{P}_{\mathrm{c}} = \frac{c}{4b^2}\exp(-2)$$

and that the equation of state can be expressed in the form

$$\frac{\widetilde{P}}{\widetilde{P}_{\mathrm{c}}} = \frac{\widetilde{T}}{\widetilde{T}_{\mathrm{c}}}\frac{1}{2(\tilde{v}/\tilde{v}_{\mathrm{c}}) - 1}\exp\left[2\left(1 - \frac{\tilde{v}_{\mathrm{c}}\widetilde{T}_{\mathrm{c}}}{\tilde{v}\widetilde{T}}\right)\right].$$

1.7 Show that the spinodal curves for van der Waals' equation and Dieterici's equation are given respectively by:

(i) $k_{\mathrm{B}}\widetilde{T}\tilde{v}^3 - 2c(\tilde{v} - b)^2 = 0$,

(ii) $k_{\mathrm{B}}\widetilde{T}\tilde{v}^2 - c(\tilde{v} - b) = 0$.

Show that for Dieterici's equation the critical temperature occurs when (ii) starts to give real values for $\tilde{v}$.

2. Statistical Mechanics and the One-Dimensional Ising Model

2.1 The Canonical Distribution

In this chapter we present and discuss necessary results in equilibrium statistical mechanics. After introducing Ising lattice models we illustrate the theory by treating a one-dimensional Ising ferromagnet and finish with an application of the latter to a biophysical system. There are many texts on statistical mechanics which develop the basic theory; a brief selection is Huang (1963), Ma (1985), Chandler (1987), and Thompson (1988).

We consider a collection or *assembly* of a number of atoms or molecules, which are not regarded as directly observable. The atoms or molecules are termed *microsystems* whereas the whole collection is a *macrosystem*. The macrosystem corresponds to an observable thermodynamic system, as studied in the previous chapter. It is supposed that the temperature T and the extensive variables $Q_1, \ldots, Q_n$ are specified, the numbers of molecules of the various components being included in the Q_i. Such a system is mechanically and chemically isolated, but is in thermal equilibrium with its environment, which can be thought of as a *heat bath* at temperature T. The given values of the extensive variables are compatible with a number of different states of the microsystems or *microstates*. For instance, in an assembly of molecules in the gaseous state, many different sets of values of the position coordinates of the molecules are possible with a given volume V. The internal variables specifying the microstates will be denoted by $\boldsymbol{\sigma}$ and the space of microstates for prescribed values of $Q_1, \ldots, Q_n$ will be denoted by Λ. Each set of values of $\boldsymbol{\sigma}$ determines a value $E(\boldsymbol{\sigma})$ of the energy of the assembly. The microstate is free to vary with time over all the states $\boldsymbol{\sigma}$ in the space Λ and, as it does so, the energy $E(\boldsymbol{\sigma})$ fluctuates in value. However, the microstates successively occupied by the assembly are not uniformly distributed over Λ because a fundamental probability distribution rule is obeyed.

For discrete-valued $\boldsymbol{\sigma}$, the probability of a given set of values of $\boldsymbol{\sigma}$ is given by the *Boltzmann distribution*

$$p(\boldsymbol{\sigma}) = \frac{\exp[-E(\boldsymbol{\sigma})/T]}{Z} , \tag{2.1}$$

where

$$Z = \sum_{\Lambda} \exp[-E(\boldsymbol{\sigma})/T] . \tag{2.2}$$

This defines the *canonical distribution* with Z, defined by (2.2), being the *canonical partition function*. Its form ensures that the necessary consistency condition

$$\sum_{\Lambda} p(\boldsymbol{\sigma}) = 1 \tag{2.3}$$

is satisfied. For a quantum system, each set of $\boldsymbol{\sigma}$ specifies an eigenstate; the $E(\boldsymbol{\sigma})$ are energy eigenvalues and Z can be regarded as the trace of $\exp(-\widehat{H}/T)$, where $\widehat{H}$ is the Hamiltonian operator. For continuous $\boldsymbol{\sigma}$, $p(\boldsymbol{\sigma})\mathrm{d}\boldsymbol{\sigma}$, with $p(\boldsymbol{\sigma})$ given by (2.1), is the probability of the microstate lying in the element $\mathrm{d}\boldsymbol{\sigma}$ of Λ. Correspondingly, the summations in (2.2) and (2.3) are replaced by integrations. We shall use the discrete rather than the continuous formulation since our interest is in discrete lattice systems. In justifying equation (2.1) it is customary to think of a collection or *ensemble* of a large number $\mathcal{N}$ of replicas of the given assembly, and $\mathcal{N}p(\boldsymbol{\sigma})$ then gives the number of members of the ensemble in microstate $\boldsymbol{\sigma}$. The *ergodic hypothesis* identifies averages taken over the ensemble using (2.1) with time averages over the fluctuating microstate of a single assembly. We shall simply assume the truth of (2.1) and of the corresponding expressions for different sets of independent thermodynamic variables given in Sect. 2.2. (For further discussion, see Jancel (1969).) The mean (or expectation) value of the energy $E(\sigma)$ is, by (2.1),

$$\langle E(\boldsymbol{\sigma})\rangle = \sum_{\Lambda} p(\boldsymbol{\sigma}) E(\boldsymbol{\sigma}) = \frac{\sum_{\Lambda} E(\boldsymbol{\sigma}) \exp[-E(\boldsymbol{\sigma})/T]}{Z} , \tag{2.4}$$

which is termed the *ensemble average*. If $\langle E(\boldsymbol{\sigma})\rangle$ is identified with the equilibrium internal energy U then, from (2.2) and (2.4),

$$U = \langle E(\boldsymbol{\sigma})\rangle = -\frac{1}{Z}\frac{\partial Z}{\partial(T^{-1})} = -\frac{\partial \ln Z}{\partial(T^{-1})} . \tag{2.5}$$

The energy $E(\boldsymbol{\sigma})$ depends on the prescribed values of $Q_1, \ldots, Q_n$ and the response of the system in state $\boldsymbol{\sigma}$ to variation of Q_i is given by $\partial E(\boldsymbol{\sigma})/\partial Q_i$. We identify the field ξ_i with the mean value of this derivative and hence write

$$\xi_j = \left\langle \frac{\partial E(\boldsymbol{\sigma})}{\partial Q_i} \right\rangle = \sum_{\Lambda} p(\boldsymbol{\sigma}) \frac{\partial E(\boldsymbol{\sigma})}{\partial Q_i} = -T \frac{\partial \ln Z}{\partial Q_i} , \qquad i = 1, \ldots, n . \tag{2.6}$$

It can be seen that (2.5) becomes identical with the Gibbs–Helmholtz relation for the Helmholtz free energy A ((1.66) with $\eta = 0$) if

$$A = -T \ln Z . \tag{2.7}$$

Equation (2.6) then becomes

$$\xi_i = \frac{\partial A}{\partial Q_i} , \qquad i = 1, \ldots, n , \tag{2.8}$$

which gives the same relation between fields, thermodynamic potential and extensive variables as (1.65). Hence the basic relations of thermodynamics are consistent with statistical mechanics if the identifications (2.5), (2.6) and (2.7) are made.

2.1.1 The Thermodynamic Limit

The formulae (2.5) and (2.6) are examples of the general principle that dependent thermodynamic quantities are identified in statistical mechanics with the mean (or expectation) values of fluctuating variables. As with any application of probability theory expectation values increase in significance as the spread, or variance, of the fluctuating variable decreases. As we shall see in Sect. 2.5 this convergence towards 'thermodynamic behaviour' occurs, in general, as the size of the system increases. As in Sect. 1.8, we now divide the extensive variables Q_i, $(i = 1, \ldots, n-1)$ by Q_n to obtain densities ρ_i and for definiteness identify Q_n with M, the total number of molecules. Then A is a function of $T, \rho_1, \ldots, \rho_{n-1}$ and M. Since A is an extensive variable we can put $A = Ma(T, \rho_1, \ldots, \rho_{n-1})$ where a is a thermodynamic potential density as defined in Sect. 1.8. Hence for statistical mechanics to yield thermodynamic behaviour, $\ln Z$ should, like A, be proportional to M but, generally speaking, this will not be true for small M. However, thermodynamic systems consist of a very large number of molecules, roughly of the order of Avogadro's number M_{A}. In principle, we can go to what is known as the *thermodynamic limit* and replace (2.7) by

$$a = -T \lim_{M \to \infty} \frac{\ln Z}{M} . \tag{2.9}$$

The model used is a valid one only if the mathematical limit on the right-hand side exists. In practice, we frequently find that

$$\ln Z = M f(T, \rho_1, \ldots, \rho_{n-1}) + f' , \tag{2.10}$$

where $f'/M \to 0$ as $M \to \infty$. Discarding f' before substitution in (2.7) is then equivalent to using the thermodynamic limit. We can in fact discard factors in Z which are as large as M^k, where k is a constant, because $(k \ln M)/M \to 0$ as $M \to \infty$.[1] Size effects will be discussed again in Sects. 2.5 and 2.6.

2.1.2 Kinetic and Configuration Variables

It may happen that the internal variables $\boldsymbol{\sigma}$ can be divided into two sets $\boldsymbol{\sigma}_1$ and $\boldsymbol{\sigma}_2$, corresponding to a separation of Λ into subspaces Λ_1 and Λ_2, such that

[1] For further discussion of the thermodynamic limit see Volume 2, Chap. 4 and for rigorous results see Ruelle (1969) and Griffiths (1972).

$$E(\boldsymbol{\sigma}) = E_1(\boldsymbol{\sigma}_1) + E_2(\boldsymbol{\sigma}_2) . \tag{2.11}$$

Then, from (2.2),

$$Z = Z_1 Z_2 , \tag{2.12}$$

where

$$Z_i = \sum_{\Lambda_i} \exp[-E_i(\boldsymbol{\sigma}_i)/T] , \qquad i = 1, 2 . \tag{2.13}$$

Thus the probability of a given set of values of $\boldsymbol{\sigma}_2$, irrespective of $\boldsymbol{\sigma}_1$, is given by

$$\begin{aligned} p(\boldsymbol{\sigma}_2) &= \sum_{\Lambda_1} p(\boldsymbol{\sigma}) \\ &= \frac{\exp[-E_2(\boldsymbol{\sigma}_2)/T] \sum_{\Lambda_1} \exp[-E_1(\boldsymbol{\sigma}_1)/T]}{Z} \\ &= \frac{\exp[-E_2(\boldsymbol{\sigma}_2)/T]}{Z_2} \end{aligned} \tag{2.14}$$

and there is a similar expression for $p(\boldsymbol{\sigma}_1)$. It can be seen that, provided (2.11) applies, Λ_1 and Λ_2 can be treated as if they corresponded to different systems.

We shall assume a separation between momentum and configurational variables such that (2.11) applies with $E_1 = E_{\text{kin}}$, the kinetic energy of the assembly, and $E_2 = E_{\text{c}}$, the configurational energy, which usually derives from intermolecular interactions. Then, from (2.12) and (2.13),

$$Z = Z_{\text{kin}} Z_{\text{c}} , \tag{2.15}$$

where

$$Z_{\text{c}} = \sum_{\Lambda_{\text{c}}} \exp[-E_{\text{c}}(\boldsymbol{\sigma}_{\text{c}})/T] , \tag{2.16}$$

is the *canonical configurational partition function.* Now suppose, as in Sect. 1.7, that the system is composed of κ different types of microsystems with M_j microsystems of type j for $j = 1, \ldots, \kappa$. Then, if E_{kin} is the sum of the kinetic energies of the individual microsystems,

$$Z_{\text{kin}} = \prod_{j=1}^{\kappa} [\psi_j(T)]^{M_j} . \tag{2.17}$$

Thus if we omit Z_{kin} from consideration, as we usually shall, we are omitting terms dependent on temperature alone from U and A.

We next define the configurational fields. For the chemical potentials corresponding to components $j = 1, \ldots, \kappa$ let

$$\xi_{j\text{c}} = \mu_{j\text{c}} = \mu_j + T \ln \psi_j(T) \tag{2.18}$$

and, for fields which are not chemical potentials, we have simply

$$\xi_{ic} = \xi_i \, . \tag{2.19}$$

Then, from (2.6), (2.15) and (2.17), for all configurational fields

$$\xi_{ic} = -T\frac{\partial \ln Z_c}{\partial Q_i} = \frac{\partial A_c}{\partial Q_i} \, , \qquad i = 1, \ldots, n \, , \tag{2.20}$$

where the configurational Helmholtz free energy is

$$A_c = -T \ln Z_c \, . \tag{2.21}$$

By (2.18) and (2.19) the configurational fields ξ_{ic} are either the same as the original fields ξ_i or different by a quantity dependent on temperature alone. Hence the ξ_i in the phase equilibrium conditions (1.90) can be replaced by the ξ_{ic} for $i = 1, \ldots, n$. The probability distribution in the configurational subspace Λ_c is independent of kinetic terms by (2.14) and hence we can derive

$$U_c = \langle E_c(\sigma_c) \rangle = -\frac{\partial \ln Z_c}{\partial (T^{-1})} = \frac{\partial (A_c/T)}{\partial (T^{-1})} \tag{2.22}$$

by arguments similar to those used to obtain (2.5).

2.2 Distributions in General

Now consider probability distributions corresponding to the thermodynamic systems discussed in Sect. 1.7, where the independent variables are the temperature T, the η fields $\xi_1, \ldots, \xi_\eta$ and the $n - \eta$ dimensionless extensive variables $Q_{\eta+1}, \ldots, Q_n$. The space of microstates $\boldsymbol{\sigma}$ compatible with the prescribed values of $Q_{\eta+1}, \ldots, Q_n$ will be denoted by $\Lambda^{(\eta)}$. It includes Λ as a subspace. Each set of values of $\boldsymbol{\sigma}$ determines a value $E(\boldsymbol{\sigma})$ of the internal energy and values $\widehat{Q}_i(\boldsymbol{\sigma})$ of the dependent extensive variables $(1 \leq i \leq \eta)$. As $\boldsymbol{\sigma}$ varies over the space $\Lambda^{(\eta)}$ of permitted microstates, there will be fluctuations not only of the energy but also of $\widehat{Q}_1, \ldots, \widehat{Q}_\eta$. Note that the character of the distribution depends not only on the value of η but on the selection of the pairs of conjugate variables labelled $1, \ldots, \eta$ and $\eta+1, \ldots, n$ respectively. The case $\eta = 0$, however, is unambiguous. As for the canonical distribution, the general distribution can be regarded as giving microstate probabilities for the members of a collection or 'ensemble' of similar assemblies. The *Hamiltonian* $\widehat{H}_\eta(\boldsymbol{\sigma})$, which is defined by

$$\widehat{H}_\eta(\boldsymbol{\sigma}) = E(\boldsymbol{\sigma}) - \sum_{i=1}^{\eta} \xi_i \widehat{Q}_i(\boldsymbol{\sigma}) \, , \tag{2.23}$$

is a crucial quantity in the statistical mechanics of the system with independent thermodynamic variables T, $\xi_1, \ldots, \xi_\eta$, $Q_{\eta+1}, \ldots, Q_n$. It is the microscopic equivalent of the thermodynamic enthalpy defined by (1.59). With $\widehat{H}_\eta$ defined, the fundamental distribution law is that the probability of the microstate $\boldsymbol{\sigma}$ is given by

$$p_\eta(\sigma) = \frac{\exp[-\widehat{H}_\eta(\sigma)/T]}{Z_\eta} \tag{2.24}$$

with the generalized partition function

$$Z_\eta(T, \xi_1, \ldots, \xi_\eta, Q_{\eta+1}, \ldots, Q_n) = \sum_{\Lambda^{(\eta)}} \exp[-\widehat{H}_\eta(\sigma)/T]. \tag{2.25}$$

From (2.24) and (2.25) it can be seen that the probabilities satisfy a consistency condition similar to (2.3).

Now suppose that the thermodynamic enthalpy H_η is identified with the mean value of $\widehat{H}_\eta(\sigma)$. From (2.23), (2.24) and (2.25),

$$H_r = \langle \widehat{H}_\eta(\sigma) \rangle = \sum_{\Lambda^{(\eta)}} p_\eta(\sigma)\widehat{H}_\eta(\sigma) = U - \sum_{i=1}^{\eta} \xi_i Q_i = -\frac{\partial \ln Z_\eta}{\partial (T^{-1})}. \tag{2.26}$$

If the equilibrium value Q_j of a dependent extensive variable is identified with the mean value of $\widehat{Q}_j(\sigma)$ then, from (2.24),

$$Q_j = \langle \widehat{Q}_j(\sigma) \rangle = \sum_{\Lambda^{(\eta)}} p_\eta(\sigma)\widehat{Q}_j(\sigma) = T\frac{\partial \ln Z_\eta}{\partial \xi_j}, \qquad j = 1, \ldots, \eta. \tag{2.27}$$

In the microstate σ, the response of the system to variation of the independent extensive variable Q_j, $j = \eta + 1, \ldots, n$, is the partial derivative of $\widehat{H}_\eta(\sigma)$ with respect to Q_j and the mean value of this is identified with the thermodynamic value of the conjugate field ξ_j to give

$$\xi_j = \left\langle \frac{\partial \widehat{H}_\eta(\sigma)}{\partial Q_j} \right\rangle = \sum_{\Lambda^{(\eta)}} p_\eta(\sigma)\frac{\partial \widehat{H}_\eta(\sigma)}{\partial Q_j} = -T\frac{\partial \ln Z_\eta}{\partial Q_j}, \qquad j = \eta + 1, \ldots, n. \tag{2.28}$$

The 'averaging' relations (2.26), (2.27) and (2.28) become identical with the thermodynamic relations in (1.66) and (1.65), if the identification

$$F_\eta = -T \ln Z_\eta(T, \xi_1, \ldots, \xi_\eta, Q_{\eta+1}, \ldots, Q_n) \tag{2.29}$$

is made. System size has the same significance in the general distribution as in the canonical distribution. Scaling by Q_n, as in (1.76), a rigorously correct definition of the free energy density is

$$f_\eta = -T \lim_{Q_n \to \infty} \frac{\ln Z_\eta}{Q_n}. \tag{2.30}$$

However, an equivalent procedure is to retain the part of F_η, in (2.29), which is proportional to Q_n discarding terms of lower order.

There is a close connection between the general partition function Z_η and the canonical partition function Z, which corresponds to the case $\eta = 0$. This can be demonstrated by performing the summation of (2.25) in two stages, the first over the values of σ corresponding to fixed values of $\widehat{Q}_1, \ldots, \widehat{Q}_\eta$, the second over all values of the latter. However the first summation, for which all extensive variables are fixed, is precisely that required to obtain a canonical partition. Hence we have

$$Z_\eta(T, \xi_1, \dots, \xi_\eta, Q_{n-\eta+1}, \dots, Q_n) = \sum_{\{\widehat{Q}_1, \dots, \widehat{Q}_\eta\}} \left\{ \exp\left(\sum_{i=1}^{\eta} \xi_i \widehat{Q}_i / T \right) \times Z(T, \widehat{Q}_1, \dots, \widehat{Q}_n) \right\} . \quad (2.31)$$

Now suppose that $E = E_{\mathrm{kin}} + E_{\mathrm{c}}$, as assumed in Sect. 2.1.2. A configurational Hamiltonian and partition function are defined by

$$\begin{aligned} \widehat{H}_{\eta\mathrm{c}}(\boldsymbol{\sigma}_{\mathrm{c}}) &= E_{\mathrm{c}}(\boldsymbol{\sigma}_{\mathrm{c}}) - \sum_{i=1}^{\eta} \xi_{i\mathrm{c}} \widehat{Q}_i(\boldsymbol{\sigma}_{\mathrm{c}}) , \\ Z_{\eta\mathrm{c}} &= \sum_{\Lambda_{\mathrm{c}}^{(\eta)}} \exp[-\widehat{H}_{\eta\mathrm{c}}(\boldsymbol{\sigma}_{\mathrm{c}})/T] , \end{aligned} \quad (2.32)$$

where it is assumed that $\widehat{Q}_i$ are dependent on the configurational microstate. The $\xi_{i\mathrm{c}}$ $(i = 1, \dots, n)$ are defined by (2.18) and (2.19). By substituting $Z = Z_{\mathrm{kin}} Z_{\mathrm{c}}$ into (2.31) and using (2.17) for Z_{kin},

$$Z_\eta = \prod_{\{j\}}{}' [\psi_j(T)]^{M_j} Z_{r\mathrm{c}} , \quad (2.33)$$

where the prime indicates that only factors corresponding to M_j which are independent variables are included in the product. From (2.18), (2.19), (2.23), and (2.32)

$$\widehat{H}_\eta = E_{\mathrm{c}} + E_{\mathrm{kin}} - \sum_{i=1}^{\eta} \xi_i \widehat{Q}_i = \widehat{H}_{r\mathrm{c}} + E_{\mathrm{kin}} + T \sum_{\{j\}}{}'' \widehat{M}_j \ln[\psi_j(T)] , \quad (2.34)$$

where the double prime denotes a sum of factors corresponding to the $\widehat{M}_j$ which are dependent variables. Hence

$$\exp(-\widehat{H}_\eta/T) = \exp(-\widehat{H}_{\eta\mathrm{c}}/T) \exp(-E_{\mathrm{kin}}/T) \prod_{\{j\}}{}'' [\psi_j(T)]^{-\widehat{M}_j} . \quad (2.35)$$

Substitution of (2.33) and (2.35) into (2.24) followed by summation over the momentum microstates yields

$$p_{\eta\mathrm{c}}(\boldsymbol{\sigma}_{\mathrm{c}}) = \frac{\exp[-\widehat{H}_{\eta\mathrm{c}}(\boldsymbol{\sigma}_{\mathrm{c}})/T]}{Z_{\eta\mathrm{c}}} , \quad (2.36)$$

where $p_{\eta\mathrm{c}}(\boldsymbol{\sigma}_{\mathrm{c}})$ is the probability of the configurational microstate $\boldsymbol{\sigma}_{\mathrm{c}}$. The configurational free energy is defined by

$$F_{\eta c} = -T \ln Z_{\eta c}(T, \xi_{1c}, \dots, \xi_{\eta c}, Q_{\eta+1}, \dots, Q_n) \tag{2.37}$$

and then, from arguments similar to those used in deriving (2.27), (2.28) and (2.26),

$$Q_j = T \frac{\partial \ln Z_{\eta c}}{\partial \xi_{jc}} = -\frac{\partial F_{\eta c}}{\xi_{jc}}, \qquad j = 1, \dots, \eta, \tag{2.38}$$

$$\xi_{jc} = -T \frac{\partial \ln Z_{\eta c}}{\partial Q_j} = \frac{\partial F_{\eta c}}{\partial Q_j}, \qquad j = \eta + 1, \dots, n, \tag{2.39}$$

$$H_{\eta c} = U_c - \sum_{i=1}^{\eta} \xi_{ic} Q_i = -\frac{\partial \ln Z_{\eta c}}{\partial (T^{-1})} = F_{\eta c} + T S_{\eta c}, \tag{2.40}$$

where $S_{\eta c} = -\partial F_{\eta c}/\partial T$ is the configurational entropy and $H_{\eta c}$ is the configurational enthalpy. From (2.38) and (2.39)

$$\mathrm{d}F_{\eta c} = -S_{\eta c}\mathrm{d}T - \sum_{i=1}^{\eta} Q_i \mathrm{d}\xi_{ic} + \sum_{i=\eta+1}^{n} \xi_{ic} \mathrm{d}Q_i . \tag{2.41}$$

The configurational free energy density is defined by

$$f_{\eta c} = -T \lim_{Q_n \to \infty} \frac{\ln Z_{\eta c}}{Q_n} . \tag{2.42}$$

For large enough Q_n, we can put $F_{\eta c} = f_{\eta c} Q_n$ and, by (1.70), $Q_i = \rho_i Q_n$, $i = 1, \dots, n-1$. Allowing the extensive variables to change with all intensive variables fixed, (2.41) yields

$$f_{\eta c} \mathrm{d}Q_n = \left(\sum_{i=\eta+1}^{n-1} \xi_{ic} \rho_i + \xi_{nc} \right) \mathrm{d}Q_n , \tag{2.43}$$

which can be satisfied only if

$$f_{\eta c} = \sum_{i=\eta+1}^{n-1} \xi_{ic} \rho_i + \xi_{nc} . \tag{2.44}$$

A relation similar to the first member of (1.77) can be deduced by dividing (2.40) by Q_n. Again, by allowing the intensive variables to change at a fixed value of Q_n, (2.41) yields

$$\mathrm{d}f_{\eta c} = -s_c \mathrm{d}T - \sum_{i=1}^{\eta} \rho_i \mathrm{d}\xi_{ic} + \sum_{i=\eta+1}^{n-1} \xi_{ic} \mathrm{d}\rho_i . \tag{2.45}$$

From these results and those of Sect. 2.1 it can be deduced that all the relations of Sects. 1.7 and 1.8 are satisfied by the configurational thermodynamic variables. It follows that we can treat the statistical mechanics and thermodynamics of the configurational degrees of freedom as if the latter constituted a separate system. Henceforward we shall deal exclusively with configurational space and, accordingly, omit the index 'c'. Again, although the index 'η' in symbols like Z_η, F_η, $F_{\eta c}$ is useful in general theory, it will be omitted in particular cases where the value of η is well understood.

2.3 Particular Distributions

As remarked above, the canonical distribution is the case $\eta = 0$ of the general theory in Sect. 2.2. We now consider several other particular cases which are useful in practice. The distribution treated in Sect. 2.3.1 will be used in the next section.

2.3.1 The Constant Magnetic Field Distribution

This corresponds to the thermodynamic theory of Sect. 1.4 where T and the magnetic field $\mathcal{H}$, defined in (1.55), are the independent intensive variables. It can be obtained from the general theory of Sect. 2.2 by putting $\eta = 1$ and $Q_1 = \mathcal{M}$, the dimensionless magnetization defined in (1.54), so that $\xi_1 = \mathcal{H}$. We use a configurational partition function as defined by (2.32) and, omitting the indices c and $\eta = 1$ as indicated above, write

$$Z = \sum_{\Lambda} \exp[-\widehat{H}(\sigma)/T] ,$$
$$\widehat{H}(\sigma) = E(\sigma) - \mathcal{H}\mathcal{M}(\sigma) . \tag{2.46}$$

From (2.38), the equilibrium magnetization is given by

$$\mathcal{M} = T\frac{\partial \ln Z}{\partial \mathcal{H}} = -\frac{\partial F}{\partial \mathcal{H}} . \tag{2.47}$$

2.3.2 The Constant-Pressure (Isobaric) Distribution

Consider a liquid or isotropic solid where the pressure is the only mechanical field. We take T and P as the independent intensive variables so that, in the notation of Sect. 2.2, $\eta = 1$ and $Q_1 = V$. The independent extensive variables Q_i, $(i > 1)$ are the molecule numbers $M_1, \ldots, M_\kappa$, where there are κ components. There is a close analogy with the constant magnetic field distribution and relations can be translated from one distribution to the other by interchanging $-P$ and V with $\mathcal{H}$ and $\mathcal{M}$ respectively. Making the same assumption about separability as in Sect. 2.3.1, a configurational partition function can be defined by

$$Z = \sum_{\Lambda} \exp[-\widehat{H}(\sigma)/T] ,$$
$$\widehat{H}(\sigma) = E(\sigma) + P\widehat{V}(\sigma) . \tag{2.48}$$

The configurational *Gibbs potential* is then given by

$$G = -T \ln Z\,, \tag{2.49}$$

and, from (2.38) and (2.39), the equilibrium values of the volume V and the configurational chemical potentials μ_i are given by

$$\begin{aligned} V &= \langle \widehat{V} \rangle = -T \frac{\partial \ln Z}{\partial P} = \frac{\partial G}{\partial P}\,, \\ \mu_i &= -T \frac{\partial \ln Z}{\partial M_i} = \frac{\partial G}{\partial M_i}\,, \qquad i = 1, \ldots, \kappa\,. \end{aligned} \tag{2.50}$$

2.3.3 The Grand Distribution

This corresponds to a thermodynamic system with $\eta = \kappa$, T and the chemical potentials μ_i being independent intensive variables. A configurational grand partition function is defined by

$$\begin{aligned} Z &= \sum_{\Lambda} \exp[-\widehat{H}(\boldsymbol{\sigma})/T]\,, \\ \widehat{H}(\boldsymbol{\sigma}) &= E(\boldsymbol{\sigma}) - \sum_{i=1}^{\kappa} \mu_i \widehat{M}_i(\boldsymbol{\sigma}) \end{aligned} \tag{2.51}$$

with the configurational *grand potential* given by

$$D = -T \ln Z\,. \tag{2.52}$$

From (2.38), the equilibrium values of the molecule numbers are

$$M_i = T \frac{\partial \ln Z}{\partial \mu_i} = -\frac{\partial D}{\partial \mu_i}\,, \qquad i = 1, \ldots, \kappa\,, \tag{2.53}$$

and the fields which are not chemical potentials are

$$\xi_j = -T \frac{\partial \ln Z}{\partial Q_j} = \frac{\partial D}{\partial Q_j}\,, \qquad j = \kappa + 1, \ldots, n\,. \tag{2.54}$$

When $n \geq \kappa + 2$, a field which is not a chemical potential can be added to the set of independent intensive variables, so that $\eta = \kappa + 1$, giving a *constant-field grand distribution.*

2.3.4 Restricted Distributions for Lattice Models

Suppose that the centre of each molecule (or atom) occupies a lattice site and that the N sites are equivalent and form a regular lattice occupying a volume V. Consider the following conditions

(i) The volume per lattice site $v_0 = \widetilde{V}/N$ is constant.

(ii) Every lattice site is occupied by the centre of a molecule of one of the κ components.

Condition (i) will be assumed for all the lattice models discussed in this book, and v_0 can be used as the standard volume in (1.54) to define the reduced volume

$$V = \widetilde{V}/v_0 = N\,. \tag{2.55}$$

The volume v_0 must be replaced by a length per site and an area per site for one- and two-dimensional systems respectively. Condition (ii) is less general and we shall consider a number of models with vacant sites. However, when (ii) is assumed, it follows that

$$V = N = \sum_{j=1}^{\kappa} M_j = M\,, \tag{2.56}$$

so that the variables $M_1, \ldots, M_k, V$ are no longer independent, and n is reduced by one. Any partition function obtained under the condition $M = N$ is called 'restricted', and the same description can be applied to the model itself.

Using a transformation similar to that of (1.82), but incorporating a pressure term, it follows from (2.56) that

$$-P\mathrm{d}V + \sum_{j=1}^{\kappa} \mu_j \mathrm{d}M_j = \sum_{j=1}^{\kappa-1} (\mu_j - \mu_\kappa)\mathrm{d}M_j + (\mu_\kappa - P)\mathrm{d}N\,, \tag{2.57}$$

where the μ_j are now configurational chemical potentials. The set of $\kappa + 1$ fields μ_j, $(j = 1, \ldots, \kappa)$, $-P$ are thus replaced by the set of κ fields $\mu_j - \mu_\kappa$, $(j = 1, \ldots, \kappa - 1)$, $\mu_\kappa - P$. As we saw in Sect. 1.9 the conditions for phase equilibrium are that T, $\mu_j - \mu_\kappa$, $(j = 1, \ldots, \kappa - 1)$, $\mu_\kappa - P$ and any other relevant fields, such as $\mathcal{H}$, must be equal in the two phases. The new phase equilibrium relations can be explained by the fact that, under conditions (i) and (ii), a change in volume must be accompanied by a transfer of molecules, whereas an increase in the number of molecules of a given component at constant volume must be compensated for by a decrease in the number of molecules of another component.

Now consider the distribution in which $\xi_n = \mu_\kappa - P$ and $\xi_1, \ldots, \xi_{n-1}$, which include the fields $\mu_j - \mu_\kappa$, $(j = 1, \ldots, \kappa - 1)$ are taken as independent variables. The only independent extensive variable is $Q_n = N$ and the restricted partition function is given by

$$\begin{aligned} Z &= \sum_{\Lambda} \exp[-\widehat{H}(\sigma)/T]\,, \\ \widehat{H}(\sigma) &= E(\sigma) - \sum_{i=1}^{n-1} \xi_i \widehat{Q}_i(\sigma)\,. \end{aligned} \tag{2.58}$$

The associated free energy per site f is of type f_{n-1} in the nomenclature of Sect. 1.8. Using (2.42) and (2.44) we have, for large N,

$$f = -T\,\frac{\ln Z}{N} = u - Ts - \sum_{i=1}^{n-1} \rho_i \xi_i = \xi_n = \mu_k - P\,. \tag{2.59}$$

From (2.45)

$$\mathrm{d}f = -s\mathrm{d}T - \sum_{i=1}^{n-1} \rho_i \mathrm{d}\xi_i. \tag{2.60}$$

2.4 Magnetism and the Ising Model

Ferromagnetism and other forms of permanent magnetism are due to exchange interaction between unpaired spins occurring either directly or through conduction electrons or non-magnetic component atoms, (Martin 1967, Sinha and Kumar 1980, Mattis 1985, Izyumov and Skryabin 1988).

In the anisotropic Heisenberg model the energy eigenvalues for an assembly of spins on a regular lattice are derived from a Hamiltonian operator

$$\widehat{H} = -2 \sum_{\{i,j\}}^{(\mathrm{n.n.})} (I_x \hat{s}_{ix} \hat{s}_{jx} + I_y \hat{s}_{iy} \hat{s}_{jy} + I_z \hat{s}_{iz} \hat{s}_{jz})\,, \tag{2.61}$$

where $\hat{s}_{ix}$, $\hat{s}_{iy}$ and $\hat{s}_{iz}$ are the component operators associated with the spin on site i. The symbol n.n. stands for 'nearest neighbour', and the summation is over all nearest-neighbour pairs of sites (i,j), each pair occurring once only. The isotropic *Heisenberg model* is obtained when $I_x = I_y = I_z$ (Volume 2, Sect. 7.6).

In most magnetic materials with localized spins there is a direction of easy magnetization so that $|I_z|$ (say) is greater than $|I_x|$ and $|I_y|$ (Sinha and Kumar 1980). In the Ising model $I_x = I_y = 0$ and the Hamiltonian is now diagonal in the representation in which the commuting z-component spin operators $\hat{s}_{iz}$ are diagonal. It can thus be treated as the scalar configurational energy. The $\hat{s}_{iz}$ take the values $(s, s-1, \ldots, -s+1, -s)$, where s is the spin magnitude with possible values $\frac{1}{2}, 1, \frac{3}{2}, 2, \ldots$. For $s = \frac{1}{2}$ the microstates are more conveniently represented by $\sigma_i = 2\hat{s}_{iz}$ and each σ_i can take the values $\sigma_i = 1$ and $\sigma_i = -1$, referred to as 'up' and 'down' spins respectively. For a lattice of N sites the set of configurational internal variables is thus given by the vector

$$\boldsymbol{\sigma} = (\sigma_1, \sigma_2, \ldots, \sigma_{N-1}, \sigma_N)\,. \tag{2.62}$$

This has 2^N possible values and the configurational subspace has 2^N discrete points. For the ferromagnet with $I_z/2 = J > 0$ the configurational energy is

$$E(\sigma) = -J \sum_{\{i,j\}}^{(\text{n.n.})} \sigma_i \sigma_j \,. \tag{2.63}$$

Since the contribution of the nearest-neighbour site pair (i, j) is $-J$ if $\sigma_i = \sigma_j$ and $+J$ if $\sigma_i \neq \sigma_j$, this favours parallel alignment. For general spin magnitude s let $\sigma_i = \hat{s}_{iz}$, $J = 2I_z$ for integer s and $\sigma_i = 2\hat{s}_{iz}$, $J = I_z/2$ for half-integer s. This ensures that all σ_i are whole numbers and that (2.63) applies for all s. The $s = \frac{1}{2}$ case will be referred to simply as the Ising model and for any higher value of s we use the term *spin-s Ising model*, as in Example 2.4 for the spin-1 model. Since each site is occupied by a spin, Ising model partition functions are restricted in the sense of Sect. 2.3.4.

It is assumed that the magnetization of the Ising model is in the z-direction, so that the dimensionless magnetization associated with a given spin configuration on the lattice is

$$\widehat{\mathcal{M}} = \sum_{i=1}^{N} \sigma_i \,. \tag{2.64}$$

Now E, as given by (2.63), is invariant under the operation $\sigma_i \to -\sigma_i$ (all i), that is under reversal of all spins. This important symmetry property is true for all s and for the antiferromagnetic case, where $-J$ in (2.63) is replaced by $+J$ ($J > 0$). In the canonical distribution the magnetization $\mathcal{M}$ is a fixed independent variable and the summation over σ_i in the partition function Z is restricted to those lattice spin configurations for which $\sum_i \sigma_i = \mathcal{M}$. From the symmetry property of E and (2.64) it follows that Z is unchanged when $\mathcal{M} \to -\mathcal{M}$. Hence the Helmholtz free energy A satisfies the symmetry relation

$$A(N, \mathcal{M}, T) = A(N, -\mathcal{M}, T) \,. \tag{2.65}$$

In the constant magnetic field distribution it follows from (2.63), (2.64) and (2.46) that the configurational Hamiltonian for the Ising model ferromagnet is

$$\widehat{H} = -J \sum_{\{i,j\}}^{(\text{n.n.})} \sigma_i \sigma_j - \mathcal{H} \sum_{\{i\}} \sigma_i \,, \tag{2.66}$$

which is invariant under the operation $\sigma_i \to -\sigma_i$ (all i), $\mathcal{H} \to -\mathcal{H}$. The configurational constant-field partition function is thus

$$Z = \sum_{\Lambda} \exp\left(\frac{J}{T} \sum_{\{i,j\}}^{(\text{n.n.})} \sigma_i \sigma_j \right) \exp\left(\frac{\mathcal{H}}{T} \sum_{\{i\}} \sigma_i \right) . \tag{2.67}$$

This is a restricted partition function of the type defined by (2.58). Since the summation is over all spin configurations, Z is unchanged when $\mathcal{H} \to -\mathcal{H}$. It follows that the corresponding free energy satisfies the symmetry relation

$$F(N, T, \mathcal{H}) = F(N, T, -\mathcal{H}) \tag{2.68}$$

and then, from (2.47), the equilibrium magnetization obeys

$$\mathcal{M}(N,T,-\mathcal{H}) = -\mathcal{M}(N,T,\mathcal{H})\,. \tag{2.69}$$

A relative magnetization (or magnetization density) m is defined by

$$m = \mathcal{M}/N \tag{2.70}$$

and, since m is an intensive variable, (2.69) implies

$$m(T,-\mathcal{H}) = -m(T,\mathcal{H})\,. \tag{2.71}$$

Although certain cobalt salts are thought to have an Ising-type interaction (de Jongh and Miedema 1974, Wood 1975) the extreme degree of anisotropy in the Ising model makes it a rather artificial model for most real ferromagnets. Indeed any theory which, like the Ising or Heisenberg models, depends on localized spins is unrealistic for iron-group metals (Martin 1967, Mattis 1985). Nevertheless the Ising model ferromagnet in two or three dimensions does reproduce the most important property of a real ferromagnet in that it displays spontaneous magnetization below a critical (Curie) temperature, where there is a singularity in the heat capacity. Also the Ising model ferromagnet is mathematically equivalent to models representing a number of different physical phenomena, such as zero-field antiferromagnetism (Chap. 4), liquid–vapour phase separation (Chap. 5) and phase separation or sublattice ordering in mixtures (Chap. 6). When an accurate or approximate mathematical solution has been found for one model, a quantity of physical interest in another model can be obtained by identification with the appropriate quantity in the first model.

2.4.1 The One-Dimensional Ferromagnet in Zero Field

Suppose that N identical microsystems are situated at N equally spaced lattice points along a line and that each microsystem is in one of ν distinguishable states. Suppose also that in the distribution used the configurational Hamiltonian $\widehat{H}_c$, is the sum of contributions from nearest-neighbour pairs and that when the two microsystems of such a pair are in states α and β respectively, this contribution is $h_{\alpha\beta}$. The configurational partition function is

$$Z = \sum_{\{\alpha,\beta,\gamma,\ldots,\psi,\omega\}} \exp[-(h_{\alpha\beta} + h_{\beta\gamma} + \cdots + h_{\psi\omega})/T], \tag{2.72}$$

where the sum is over all the states $\alpha, \beta, \ldots, \omega$ on the N sites of the lattice. The variables

$$V_{\alpha\beta} = \exp(-h_{\alpha\beta}/T) \tag{2.73}$$

form the elements of the $\nu\times\nu$ *transfer matrix*[2] $\boldsymbol{V}$ and (2.72) can be expressed in the form

[2] A general discussion of the use of transfer matrices is given in Volume 2, Chap. 4.

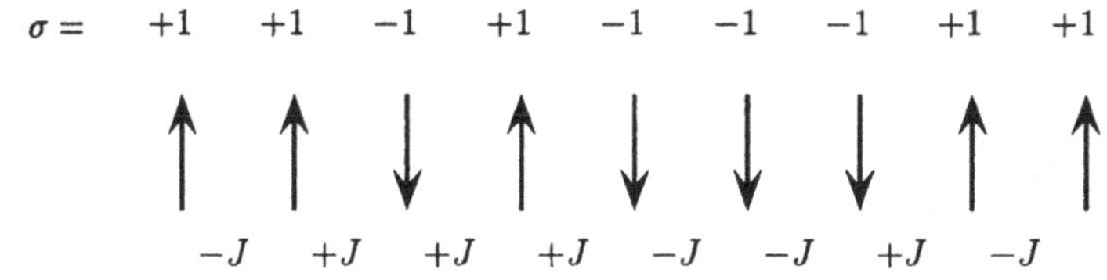

Fig. 2.1. One-dimensional Ising model ferromagnet; pair energies are shown underneath.

$$Z = \sum_{\{\alpha,\beta,\gamma,\ldots,\chi,\psi,\omega\}} V_{\alpha\beta} V_{\beta\gamma} \cdots V_{\chi\psi} V_{\psi\omega} \,. \tag{2.74}$$

We now consider the special case of the one-dimensional Ising model ferromagnet shown in Fig. 2.1 for which $\nu = 2$ (see also Thompson (1972a)). It may be thought that one-dimensional or linear lattice models are unrealistic. However, they provide examples of systems with interacting molecules ('many-body' systems in the terminology of theoretical physics) which are soluble and instructive. Moreover, we shall see in Sect. 2.7 that there is at least one case where the one-dimensional Ising model can be used as a simple model for certain aspects of a real physical system.

For the Ising model it is useful to define the Boltzmann factors

$$\mathfrak{X} = \exp(J/T) \,, \qquad \mathfrak{Z} = \exp(\mathcal{H}/T) \,. \tag{2.75}$$

For the one-dimensional Ising model with $\mathcal{H} = 0$ the Hamiltonian defined by (2.66) consists of a contribution $-J\sigma_k\sigma_{k+1}$ from each successive site pair $(k, k+1)$. The matrix $\boldsymbol{V}$, defined by (2.73), is given by

$$\boldsymbol{V} = \begin{pmatrix} \mathfrak{X} & \mathfrak{X}^{-1} \\ \mathfrak{X}^{-1} & \mathfrak{X} \end{pmatrix} \tag{2.76}$$

and has the property that

$$\sum_{\{\beta\}} V_{\alpha\beta} = \mathfrak{X} + \mathfrak{X}^{-1} \,, \tag{2.77}$$

is independent of α. Hence, summing over ω in (2.74), the constant magnetic field configurational partition function for $\mathcal{H} = 0$ is given by

$$Z = (\mathfrak{X} + \mathfrak{X}^{-1}) \sum_{\{\alpha,\beta,\gamma,\ldots,\chi,\psi\}} V_{\alpha\beta} V_{\beta\gamma} \cdots V_{\chi\psi} \,. \tag{2.78}$$

Repeating this operation over the $N - 2$ remaining factors gives

$$Z = 2(\mathfrak{X} + \mathfrak{X}^{-1})^{N-1} = 2^N [\cosh(J/T)]^{N-1} \tag{2.79}$$

and

$$\ln Z = N\{\ln 2 + \ln[\cosh(J/T)]\} - \ln[\cosh(J/T)] \,. \tag{2.80}$$

Since the last term divided by N goes to zero as N approaches infinity

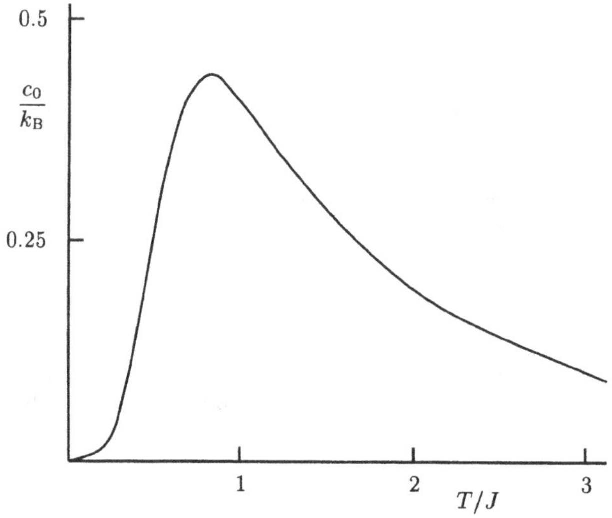

Fig. 2.2. Heat capacity plotted against temperature for the model of Fig. 2.1.

$$F = -NT\{\ln 2 + \ln[\cosh(J/T)]\} \tag{2.81}$$

in the thermodynamic limit.

With $\mathcal{H} = 0$ the enthalpy $H = U$, the configurational energy, and, by (2.40)

$$U = \frac{\partial(F/T)}{\partial(T^{-1})} = -NJ\tanh(J/T)\,. \tag{2.82}$$

Hence $U \to -NJ$ as $T/J \to 0$ and $U \to 0$ as $T/J \to \infty$. From (1.35) and (2.82) the constant-field configurational heat capacity per spin with $\mathcal{H} = 0$ is

$$c_0 = -N^{-1}k_B T\frac{\partial^2 F}{\partial T^2} = N^{-1}k_B\frac{\partial U}{\partial T} = k_B\left[\frac{J/T}{\cosh(J/T)}\right]^2\,. \tag{2.83}$$

From (2.83), $c_0 \to 0$ as $T/J \to 0$ and as $T/J \to \infty$. A plot against temperature is shown in Fig. 2.2. From the limiting value $-NJ$ of the configurational energy as $T/J \to 0$ it might be supposed that at low temperatures ($T/J \ll 1$) the majority of the dipoles are lined up in one of the two directions. However, this cannot be the case since, as will be shown in Sect. 2.4.2, the equilibrium magnetization $\mathcal{M}$ is zero for $\mathcal{H} = 0$ at all temperatures $T > 0$. To understand the behaviour of the model it is useful to derive the two-site correlation function $\langle\sigma_i\sigma_{i+k}\rangle$, which is positive if the k-th neighbour pair of dipoles tends to point in the same direction. First note that, since $\sigma_i^2 = 1$ for any site i,

$$\begin{aligned}\sigma_i\sigma_{i+k} &= \sigma_i\sigma_{i+1}^2\sigma_{i+2}^2\cdots\sigma_{i+k-1}^2\sigma_{i+k}\\ &= (\sigma_i\sigma_{i+1})(\sigma_{i+1}\sigma_{i+2})\cdots(\sigma_{i+k-1}\sigma_{i+k})\,.\end{aligned} \tag{2.84}$$

Now temporarily modify the model by supposing that the interaction parameter J has an individual value for each nearest-neighbour pair so that, in zero field,

$$\widehat{H} = -\sum_{k=1}^{N-1} J_k \sigma_k \sigma_{k+1} \,. \tag{2.85}$$

In exactly the same way as (2.79) was obtained, it can now be shown (Stanley 1987) that

$$Z = 2^N \prod_{k=1}^{N-1} \cosh(J_k/T) \,. \tag{2.86}$$

Using (2.84) and the expression (2.85) for $\widehat{H}$,

$$\begin{aligned}\langle \sigma_i \sigma_{i+k} \rangle &= Z^{-1} \sum_{\Lambda} (\sigma_i \sigma_{i+1})(\sigma_{i+1}\sigma_{i+2}) \cdots (\sigma_{i+k-1}\sigma_{i+k}) \exp(-\widehat{H}/T) \\ &= \frac{T^{k-1}}{Z} \frac{\partial}{\partial J_i} \frac{\partial}{\partial J_{i+1}} \cdots \frac{\partial Z}{\partial J_{i+k-1}} \,.\end{aligned} \tag{2.87}$$

Substituting from (2.86) and, finally, restoring the original model by putting $J_i = J$ (all i),

$$\langle \sigma_i \sigma_{i+k} \rangle = [\tanh(J/T)]^k . \tag{2.88}$$

The nearest-neighbour correlation function $\tanh(J/T)$ is obtained by putting $k = 1$ in (2.88): this provides an alternative way of deriving the expression (2.82) for U. From (2.88) the correlation function for an k-th neighbour pair is independent of its position on the lattice and is always positive but decreases as k increases. For any fixed value of k, $\langle \sigma_i \sigma_{i+k} \rangle \to 1$ as $T/J \to 0$ and to zero as $T/J \to \infty$. The picture which emerges is that, although the total numbers of up and down dipoles remain equal, blocks of similarly oriented dipoles form and steadily increase in length as T/J decreases. The approach of U to the limiting value $-NJ$ is explained by the number of unlike nearest-neighbour pairs, which represent block boundaries, becoming very small relative to the number of like pairs for $T/J \ll 1$. Although there is no *long-range order* in the sense of a majority of the dipoles becoming oriented in one direction, the properties of the assembly at low temperatures are dominated by the short-range order expressed by (2.88).

2.4.2 The One-Dimensional Ferromagnet in a Field

We now consider the behaviour of the model in non-zero external field ($\mathcal{H} \neq 0$). With the Hamiltonian given by (2.66) a term

$$-J\sigma_k \sigma_{k+1} - \frac{1}{2}\mathcal{H}(\sigma_k + \sigma_{k+1}) \tag{2.89}$$

can be attributed to each successive site pair and the transfer matrix (2.76) must be replaced by

$$V = \begin{pmatrix} \mathfrak{X}\mathfrak{Z} & \mathfrak{X}^{-1} \\ \mathfrak{X}^{-1} & \mathfrak{X}\mathfrak{Z}^{-1} \end{pmatrix}, \tag{2.90}$$

the Boltzmann factors $\mathfrak{X}$ and $\mathfrak{Z}$ being defined by (2.75). The property, expressed by (2.77), that the two row sums of $\boldsymbol{V}$ are equal, is no longer valid and the method of successive summation employed in Sect. 2.4.1 is inapplicable. The way out is to replace the open linear lattice by a ring, supposing that site N is a nearest neighbour of site 1 and adding a term $-J\sigma_N\sigma_1$ to the energy. Since every site now has two nearest neighbours the expression (2.89) summed over the N nearest-neighbour pairs gives the Hamiltonian exactly and there are no 'end-effect' terms. With the ring assumption an extra factor $V_{\omega\alpha}$ must be multiplied into each term in (2.74) to allow for the interaction between the microsystems on sites 1 and N and (2.74) can now be written

$$Z = \text{Trace}\{\boldsymbol{V}^N\}. \tag{2.91}$$

For the one-dimensional Ising model (2.91) yields

$$Z = \lambda_1^N + \lambda_2^N = \lambda_1^N[1 + (\lambda_2/\lambda_1)^N], \tag{2.92}$$

where λ_1 and λ_2 (with $\lambda_1 > \lambda_2$) are the eigenvalues of the symmetric matrix $\boldsymbol{V}$ given by (2.90). These eigenvalues can be obtained from

$$|\boldsymbol{V} - \lambda \boldsymbol{I}| = \lambda^2 - \mathfrak{X}[\mathfrak{Z} + \mathfrak{Z}^{-1}]\lambda + \mathfrak{X}^2 - \mathfrak{X}^{-2} = 0 \tag{2.93}$$

giving

$$\begin{aligned} \lambda_1 &= \frac{1}{2}\mathfrak{X}\{(\mathfrak{Z} + \mathfrak{Z}^{-1}) + [(\mathfrak{Z} - \mathfrak{Z}^{-1})^2 + 4\mathfrak{X}^{-4}]^{1/2}\}, \\ \lambda_2 &= \frac{1}{2}\mathfrak{X}\{(\mathfrak{Z} + \mathfrak{Z}^{-1}) - [(\mathfrak{Z} - \mathfrak{Z}^{-1})^2 + 4\mathfrak{X}^{-4}]^{1/2}\}. \end{aligned} \tag{2.94}$$

Since λ_1 and λ_2 are both positive and $\lambda_2/\lambda_1 < 1$,

$$0 < \ln[1 + (\lambda_2/\lambda_1)^N] < (\lambda_2/\lambda_1)^N \to 0 \tag{2.95}$$

as $N \to \infty$. Hence, in the thermodynamic limit,

$$F = -T \ln Z = -NT \ln \lambda_1 . \tag{2.96}$$

It is of interest to return briefly to the zero-field case $\mathcal{H} = 0$, where $\mathfrak{Z} = 1$. Then (2.94) reduces to

$$\begin{aligned} \lambda_1 &= \mathfrak{X} + \mathfrak{X}^{-1} = 2\cosh(J/T), \\ \lambda_2 &= \mathfrak{X} - \mathfrak{X}^{-1} = 2\sinh(J/T), \end{aligned} \tag{2.97}$$

and (2.92) can be written

$$Z = 2^N[\cosh(J/T)]^N\{1 + [\tanh(J/T)]^N\}. \tag{2.98}$$

The difference between this and (2.79) is due to the use of the ring lattice in one case and the open lattice in the other. However, substitution of λ_1 from (2.97) in (2.96) gives a thermodynamic limit expression identical to (2.81), which was derived for the open lattice. We have an example of a general rule that differences due to boundary conditions disappear in the thermodynamic limit. From (2.98) and (2.79) it can be seen that convergence is faster in the closed ring case, as we might intuitively expect. The transformation $\mathcal{H} \to -\mathcal{H}$ gives $\mathfrak{Z} \to 1/\mathfrak{Z}$ and it can be seen from (2.94) that the free energy given by (2.96) obeys the symmetry relation (2.68).

From (2.47), (2.96), (2.94) and (2.75) the relative magnetization is given by

$$m = \mathfrak{Z}\frac{\partial \ln \lambda_1}{\partial \mathfrak{Z}} = \frac{\sinh(\mathcal{H}/T)}{[\sinh^2(\mathcal{H}/T) + \exp(-4J/T)]^{1/2}} \,. \tag{2.99}$$

The symmetry condition (2.71) is satisfied and m is a continuous function of $\mathcal{H}$ with $m = 0$ at $\mathcal{H} = 0$ for any $T/J > 0$. For any fixed $T/J > 0$, $m \to \pm 1$ (depending on the sign of $\mathcal{H}$) as $|\mathcal{H}|/J \to \infty$ while, for any fixed $\mathcal{H}/J \neq 0$, $m \to \pm 1$ as $T/J \to 0$. For $T = 0$, m thus jumps at $\mathcal{H} = 0$ from the value -1, for $\mathcal{H} < 0$, to the value of $+1$, for $\mathcal{H} > 0$. Equation (2.71) is true for any lattice dimension. However, for two- or three-dimensional lattices, m is discontinuous at $\mathcal{H} = 0$ not only at $T = 0$ but for a range $0 \leq T < T_c$, where T_c is a critical temperature.

We now consider the isothermal susceptibility χ_T which was defined by (1.36). However, in the one-dimensional system we must replace the volume $\tilde{v}$ per atom by the lattice spacing, which is denoted by b. Then (2.99) yields

$$\frac{\mu_0 b \chi_T}{m_0^2} = \frac{\partial m}{\partial \mathcal{H}} = \frac{\exp(-4J/T)\cosh(\mathcal{H}/T)}{T[\sinh^2(\mathcal{H}/T) + \exp(-4J/T)]^{3/2}} \,, \tag{2.100}$$

m_0 being the moment of an Ising dipole and μ_0 the permeability of free space. On the $\mathcal{H} = 0$ axis (2.100) reduces to

$$\frac{\partial m}{\partial \mathcal{H}} = T^{-1}\exp(2J/T) \,. \tag{2.101}$$

For higher-dimensional lattices, χ_T has a singularity on the $\mathcal{H} = 0$ axis for a non-zero critical value of T, which is associated with the onset of long-range order. For the one-dimensional case, where long-range order is absent, (2.101) yields an essential singularity at $T/J = 0$ and $\partial m/\partial H \to \infty$ as $T/J \to 0$. For any $\mathcal{H}/J \neq 0$, it can be seen from (2.100) that $\partial m/\partial \mathcal{H} \to 0$ as $T/J \to 0$ and thus has no singularity, a situation which is paralleled in higher dimensions.

2.5 Fluctuations and Entropy

Thermodynamically a system with prescribed values of the temperature T and the extensive variables $Q, \ldots, Q_n$ has a determinate internal energy U, which we have identified with the mean value $\langle E(\sigma)\rangle$. However, the canonical distribution displays a range of energy values, with the probability of a microstate of energy $E(\sigma)$ being given by (2.1). For complete consistency with thermodynamics we have to show that for a large enough system the probability of a significant deviation of $E(\sigma)$ from U is small. We define the energy fluctuation by

$$\Delta E(\sigma) = E(\sigma) - \langle E(\sigma)\rangle = E(\sigma) - U\,, \tag{2.102}$$

and, to estimate the likely deviation of $E(\sigma)$ from U, we obtain the mean of $(\Delta E)^2$, which is the variance of E denoted by Var[E]. Thus

$$\mathrm{Var}[E] = \langle (E(\sigma) - U)^2\rangle = \langle E(\sigma)^2\rangle - U^2\,. \tag{2.103}$$

From (2.4) and (2.5),

$$\frac{\partial U}{\partial T} = \frac{\partial\langle E(\sigma)\rangle}{\partial T} = \frac{1}{T^2}\left[\langle E(\sigma)^2\rangle - U^2\right], \tag{2.104}$$

giving

$$\mathrm{Var}[E] = T^2\frac{\partial U}{\partial T}\,. \tag{2.105}$$

This is an example of a *fluctuation–response function relation.* In this case the fluctuating quantity is the energy and the response function is a heat capacity. The spread of energy values is best expressed by the relative fluctuation given by

$$\frac{(\mathrm{Var}[E])^{1/2}}{|U|} = \frac{T}{|U|}\left(\frac{\partial U}{\partial T}\right)^{1/2}. \tag{2.106}$$

Since U is proportional to M, the total number of molecules in the assembly, it follows from (2.106) that the relative fluctuation will normally be $\mathrm{O}(M^{-1/2})$ and hence very small if M is greater than 10^{20}, say. (Avogadro's number $M_{\mathrm{A}} = 6.022 \times 10^{23}$.) Hence observable deviations of the energy from the equilibrium value U are extremely unlikely. For instance, $U = \frac{3}{2}MT$ for a perfect monatomic gas and (2.106) gives the value $[2/(3M)]^{-1/2}$.

The mean or equilibrium value of the energy $E(\sigma)$ is determined by a balance between the probability of a single microstate, which by (2.1) decreases monotonically as $E(\sigma)$ increases, and the number of microstates with energy values near to $E(\sigma)$ which, in general, increases with $E(\sigma)$. From (2.106) there is an overwhelming probability that the system will be in one of a set of ν microstates, corresponding approximately to an energy range $U - \mathrm{Var}[E] < E(\sigma) < U + \mathrm{Var}[E]$. Since all these microstates have an energy near U, a reasonable estimate of ν can be obtained by supposing that the probability of each microstate is

$$p_U = \frac{\exp(-U/T)}{Z}\,. \tag{2.107}$$

Hence substituting for Z from (2.7) and using (1.61)

$$\nu = (p_U)^{-1} = \exp(U - A/T) = \exp(S) = \exp(\widetilde{S}/k_{\mathrm{B}}) . \tag{2.108}$$

The entropy is often described as a measure of disorder, that is of our lack of knowledge about which particular microstate is occupied.

For the general distribution discussed in Sect. 2.2, we can obtain a relation for the variance of the Hamiltonian $\widehat{H}_\eta$:

$$\mathrm{Var}[\widehat{H}_\eta] = \langle [\widehat{H}_\eta(\sigma) - H_\eta]^2 \rangle = T^2 \frac{\partial H_\eta}{\partial T} \tag{2.109}$$

and the relative fluctuation of $\widehat{H}_\eta$ is then given by

$$\frac{(\mathrm{Var}[\widehat{H}_\eta])^{1/2}}{|H_\eta|} = \frac{T}{|H_\eta|} \left(\frac{\partial H_\eta}{\partial T} \right)^{1/2} . \tag{2.110}$$

The dependent extensive variables also fluctuate, and it can be shown from (2.24) and (2.27) that

$$\mathrm{Var}[\widehat{Q}_j] = T \frac{\partial Q_j}{\partial \xi_j} , \qquad j = 1, \ldots, \eta . \tag{2.111}$$

Hence the relative fluctuation of Q_j is given by

$$\frac{(\mathrm{Var}[\widehat{Q}_j])^{1/2}}{|Q_j|} = \frac{1}{|Q_j|} \left(T \frac{\partial Q_j}{\partial \xi_j} \right)^{1/2} , \qquad j = 1, \ldots, \eta . \tag{2.112}$$

By scaling the extensive quantities with respect to M, the total number of molecules, it can be seen that the relative fluctuations given by (2.110) or (2.112) are of $\mathrm{O}(M^{-1/2})$, and observable deviations from equilibrium values are thus highly unlikely.

We showed in the context of the canonical distribution that in any factor space of microstates there is an independent probability distribution (2.14) of the same form as the general probability distribution. If separation between momentum and configurational terms occurs, giving a configurational partition function as defined by (2.32), then there is an independent probability distribution in configuration microspace. Thus in the zero-field Ising model, we can use (2.105) for the relative fluctuation of the configurational energy, giving, after substituting for U from (2.82),

$$\frac{(\mathrm{Var}[E])^{1/2}}{|U|} = \frac{T}{|U|} \left(\frac{\partial U}{\partial T} \right)^{1/2} = \frac{1}{\sqrt{N} \sinh(J/T)} . \tag{2.113}$$

This is of $\mathrm{O}(N^{-\frac{1}{2}})$ unless T is so large that T/J itself is of $\mathrm{O}(N^{\frac{1}{2}})$. For the Ising model in non-zero field $\mathcal{H}$, (2.112) gives the relative fluctuation of the magnetization as

$$\begin{aligned} \frac{(\mathrm{Var}[\widehat{\mathcal{M}}])^{1/2}}{|\mathcal{M}|} &= \frac{1}{\sqrt{N}|m|} \left(T \frac{\partial m}{\partial \mathcal{H}} \right)^{1/2} \\ &= \left[\frac{\cosh(\mathcal{H}/T)}{N \sinh^2(\mathcal{H}/T)} \right]^{1/2} \frac{\exp(-2J/T)}{[\sinh^2(\mathcal{H}/T) + \exp(-4J/T)]^{1/4}} . \end{aligned} \tag{2.114}$$

This is of $O(N^{-\frac{1}{2}})$ unless $\mathcal{H}/T$ is very small. The relative fluctuation compares the likely deviation of a variable from its equilibrium value with the equilibrium value itself and thus ceases to be useful when the equilibrium value approaches zero, as $\mathcal{M}$ does in the one-dimensional Ising model when $\mathcal{H} \to 0$. However, fluctuation–response function relations like (2.109) or (2.111) can still be used.

In the last paragraph, variances were calculated from known response functions. However in certain circumstances, response functions can be calculated from variances, using fluctuation–response function relations. For the zero-field Ising model, (2.111) yields the fluctuation–susceptibility relation

$$\left(\frac{\partial \mathcal{M}}{\partial \mathcal{H}}\right)_{\mathcal{H}=0} = T^{-1}\mathrm{Var}[\widehat{\mathcal{M}}]_{\mathcal{H}=0} = T^{-1}\langle \widehat{\mathcal{M}}^2 \rangle . \tag{2.115}$$

since $\mathcal{M} = 0$ at $\mathcal{H} = 0$. From (2.64)

$$\langle \widehat{\mathcal{M}}^2 \rangle = \left\langle \left(\sum_{i=1}^{N} \sigma_i \right)^2 \right\rangle = N + 2 \sum_{i,k;k>0} \langle \sigma_i \sigma_{i+k} \rangle . \tag{2.116}$$

For the open linear lattice, the correlation function $\langle \sigma_i \sigma_{i+k} \rangle$ is given by (2.88), and if $N - i$ is large we can write without appreciable error

$$\begin{aligned} \sum_{\{k>0\}} \langle \sigma_i \sigma_{i+k} \rangle &= \sum_{k=1}^{\infty} \tanh^k (J/T) \\ &= \tanh(J/T)[1 - \tanh(J/T)]^{-1}. \end{aligned} \tag{2.117}$$

In the thermodynamic limit it can be assumed that $N - i$ is large for all except a relatively negligible number of sites, and hence

$$\langle \widehat{\mathcal{M}}^2 \rangle = N \left[1 + \frac{2\tanh(J/T)}{1 - \tanh(J/T)} \right] = N \exp(2J/T) . \tag{2.118}$$

Substitution in (2.115) then reproduces equation (2.101) for $\partial m / \partial \mathcal{H}$ which we have thus succeeded in calculating from a fluctuation–susceptibility relation.

We now look at what happens when T is reduced to zero, with η independent fields and $n - \eta$ independent extensive variables kept constant. Suppose that there is a microstate $\boldsymbol{\sigma}_0$ such that $\widehat{H}_\eta(\boldsymbol{\sigma}_0) < \widehat{H}_\eta(\boldsymbol{\sigma})$ for all $\boldsymbol{\sigma} \neq \boldsymbol{\sigma}_0$. Then, by (2.24), the probability ratio $p(\boldsymbol{\sigma})/p(\boldsymbol{\sigma}_0) \to 0$ as $T \to 0$ for all $\boldsymbol{\sigma} \neq \boldsymbol{\sigma}_0$. Hence, at $T = 0$, $p(\boldsymbol{\sigma}_0) = 1$ and the microstate $\boldsymbol{\sigma}_0$ is called the *ground state*. Thus the equilibrium state at $T = 0$ can be identified by finding the least value of $\widehat{H}_\eta(\boldsymbol{\sigma})$. This remains true if Ω_0 microstates, with $\Omega_0 > 1$, have the same least Hamiltonian value $\widehat{H}_\eta(\boldsymbol{\sigma}_0)$. The ground state is now degenerate and each of the Ω_0 microstates composing the ground state has probability Ω_0^{-1} at $T = 0$. The one-dimensional ring Ising model ferromagnet has $\Omega_0 = 2$ at zero field, since the two states $\sigma_i = 1$ for all i and $\sigma_j = -1$ for all k both have the least Hamiltonian value $-NJ$. When

$\Omega_0 \to \infty$ in the thermodynamic limit $Q_n \to \infty$, the ground state is said to be *infinitely degenerate.*

By the first relation of (2.26), $H_\eta \to \widehat{H}_\eta(\sigma_0)$ as $T \to 0$, so that the ground state can be regarded thermodynamically as that of least enthalpy or of least enthalpy density $h_\eta = Q_n^{-1} H_\eta$. Since the occupation probability of microstates other than those composing the ground state tends to zero with T, we have $\nu \to \Omega_0$ as $T \to 0$, so that, by (2.108), $S = \ln(\Omega_0)$ at $T = 0$. Normally $Q_n^{-1}\ln(\Omega_0) \to 0$ as $Q_n \to \infty$ so that the entropy density $s = 0$ at $T = 0$. However, as will be seen in Sect. 8.12 and Chap. 10, there are exceptional cases where $s \neq 0$ at $T = 0$, and we then say that a *zero-point entropy* exists.

2.6 The Maximum-Term Method

In the canonical distribution, all the microstates having the same energy value E_ℓ can be regarded as grouped into an *energy level.* The number of microstates having energy E_ℓ is denoted by Ω_ℓ, call the *weight* or *degeneracy* of the level. The partition function, defined by (2.2), can be written as a sum over levels:

$$Z = \sum_{\{\ell\}} \Psi_\ell = \sum_{\{\ell\}} \Omega_\ell \exp(-E_\ell/T) , \tag{2.119}$$

where $E_{\ell+1} > E_\ell$. From (2.1) the probability of level ℓ is Ψ_ℓ/Z. For a system with a sufficient number of molecules (say, more than 10^{20}) there is, according to the discussion in Sect. 2.5, an overwhelming probability that the system will be in a level with E_ℓ close to the equilibrium internal energy value U. Thus a plot of Ψ_ℓ against E_ℓ will be very highly peaked near $E_\ell = U$ and we can identify U with $E_{\max}$, where $\Psi_{\max}$ is the largest value of Ψ_ℓ. If the gaps between successive levels are small in the vicinity of $E_{\max}$ we can regard E_ℓ as a continuous variable and write the maximization conditions as

$$\left(\frac{\mathrm{d}\Psi_\ell}{\mathrm{d}E_\ell}\right)_{E_\ell = E_{\max}} = 0 , \qquad \left(\frac{\mathrm{d}^2\Psi_\ell}{\mathrm{d}E^2}\right)_{E_\ell = E_{\max}} < 0 . \tag{2.120}$$

In view of the work on fluctuations we can readily accept the equality of $E_{\max}$ and U, but what is surprising is that we can put

$$\begin{aligned} S &= \ln(\Omega_{\max}) , \\ A &= E_{\max} - T\ln(\Omega_{\max}) = -T\ln(\Psi_{\max}) . \end{aligned} \tag{2.121}$$

Thus, in the thermodynamic limit, we are identifying $\ln(\Omega_{\max})$ with $\ln(\nu)$, ν being the measure of 'disorder'. However, it must be remembered that even if $\Omega_{\max}$, the degeneracy of the maximum probability level, is less than ν by a factor M^{-k}, for some $k > 0$, this will have no effect in the thermodynamic limit, since $M^{-1}\ln(M^k) \to 0$ as $M \to \infty$.

2.6.1 The One-Dimensional Ising Ferromagnet

The maximum-term method will now be applied to the Ising model ferromagnet on a one-dimensional ring lattice. Since the canonical distribution is used, the magnetization $\mathcal{M}$ is fixed and hence so are the numbers, denoted by N_1 and N_2 respectively, of $+1$ and -1 spins. We can write

$$N_1 = \frac{1}{2}\sum_{i=1}^{N}(1+\sigma_i)\,,\qquad N_2 = \frac{1}{2}\sum_{i=1}^{N}(1-\sigma_i)\,,\qquad \mathcal{M} = N_1 - N_2\,, \tag{2.122}$$

where the last relation follows from (2.64). Hence, on a lattice of N sites,

$$N_1 = \tfrac{1}{2}N(1+m)\,,\qquad N_2 = \tfrac{1}{2}N(1-m)\,, \tag{2.123}$$

where m is the relative magnetization, defined by (2.70). For any distribution on a ring lattice, the number of times a $+1$ spin on site i is followed by a -1 on site $i+1$ is equal to the number of times a -1 spin is followed by a $+1$. Denoting this number by X, there are

$$2X = \frac{1}{2}\sum_{i=1}^{N}(1-\sigma_i\sigma_{i+1})\,,\qquad (\sigma_{N+1}=\sigma_1)\,, \tag{2.124}$$

unlike nearest-neighbour pairs and hence the configurational energy is

$$E = -J\sum_{i=1}^{N}\sigma_i\sigma_{i+1} = -JN + 4JX\,. \tag{2.125}$$

The energy of a spin configuration on the ring lattice is determined by X. We need an expression for $\Omega(N_1,N_2,X)$, the number of distinct configurations for given N_1, N_2, and X, assuming that spins of the same sign are indistinguishable but lattice sites are distinguishable. It is useful to define a number $g(N_1,N_2,X)$ by

$$g(N_1,N_2,X) = \frac{N_1!N_2!}{(N_1-X)!(N_2-X)!(X!)^2} = g(N_2,N_1,X)\,. \tag{2.126}$$

Now arrange N_1 up spins in a row and distinguish X of them by a 'star'. Arrange N_2 down spins in another row, put a star on the last spin and distribute $X-1$ stars on the remaining -1 spins. The number of ways of assigning stars in this manner is

$$\frac{N_1!}{(N_1-X)!X!}\,\frac{(N_2-1)!}{(N_2-X)!(X-1)!} = g(N_1,N_2,X)\frac{X}{N_2}\,. \tag{2.127}$$

Put the first up spin on lattice site 1, the next up on lattice site 2 and continue until the first starred up has been placed. Since the star indicates an unlike pair, put the first down spin on the next lattice site and continue until the first starred down has been placed. Then resume putting the up spins on successive lattice sites and proceed, changing from spins of one sign to another after a starred spin has been placed. The proviso that the last of the row of down spins must carry a star is necessary because, whether placed on site N or a previous site, this spin is followed by an up spin. Each assignment of stars determines a distinct spin configuration on the lattice, and hence the number of configurations with site 1 occupied by an up spin is given by (2.127). Interchanging N_1 with N_2 in (2.127) gives the number of configurations with site 1 occupied by a down spin and hence, using the symmetry property of $g(N_1, N_2, X)$,

$$\Omega(N_1,N_2,X) = g(N_1,N_2,X)\left(\frac{X}{N_1}+\frac{X}{N_2}\right)$$

$$= g(N_1,N_2,X)\frac{NX}{N_1N_2}\,. \tag{2.128}$$

Since the logarithm of $NX/(N_1N_2)$ remains finite as $N \to \infty$, we can drop this factor from Ω. Hence a relation of form (2.119) for the configurational partition function can be written

$$\exp(JN/T)Z(N,\mathcal{M},T) = \sum_X \Psi(X)$$

$$= \sum_X g(N_1,N_2,X)\exp(-4JX/T)\,, \tag{2.129}$$

where the factor due to the constant energy term $-NJ$ has been taken over to the left-hand side. Using Stirling's formula for $\ln(n!)$, with $n \gg 1$,

$$\begin{aligned}\ln\Psi(X) = -4(J/T)X + [&N_1\ln(N_1) + N_2\ln(N_2) - 2X\ln(X)\\ &-(N_1-X)\ln(N_1-X) - (N_2-X)\ln(N_2-X)]\,.\end{aligned} \tag{2.130}$$

Hence the stationary point condition for the value $X_{\max}$ maximizing $\Psi(X)$ is

$$\begin{aligned}\left(\frac{\mathrm{d}\ln\Psi(X)}{\mathrm{d}X}\right)_{X=X_{\max}} &= \ln\left[\frac{(N_1-X_{\max})(N_2-X_{\max})}{X_{\max}^2}\right] - 4(J/T)\\ &= 0\,.\end{aligned} \tag{2.131}$$

It is easy to verify that $\mathrm{d}^2\ln\Psi(X)/\mathrm{d}X^2$ is negative for all X. From (2.131)

$$X_{\max}^2 = (N_1 - X_m)(N_2 - X_m)\mathfrak{X}^{-4}, \tag{2.132}$$

where $\mathfrak{X}$ is defined by (2.75) . From (2.121) we have

$$A(N,T,\mathcal{M}) = -NJ - T\ln\Psi(X_{\max}). \tag{2.133}$$

From (2.8), the magnetic field is given by

$$\mathcal{H} = \frac{\partial A}{\partial\mathcal{M}} = \frac{1}{N}\frac{\partial A}{\partial m} = \frac{1}{2}\ln\left(\frac{N_1 - X_{\max}}{N_2 - X_{\max}}\frac{N_2}{N_1}\right) \tag{2.134}$$

using (2.130), (2.131) and (2.123). Hence

$$(N_1 - X_{\max})N_2 = (N_2 - X_{\max})N_1 \mathfrak{Z}^2, \tag{2.135}$$

$\mathfrak{Z}$ being defined by (2.75). The reader will be asked in Example 2.5 to show that these results are equivalent to those of Sect. 2.4.2.

2.6.2 The General Distribution

The maximum-term method can be extended to the general distribution and to a set $X_1, X_2, \ldots, X_\nu$ of internal variables, of which one may be equal to or dependent on the energy. (The X_k are often called *order variables*.) We then have

$$Z_\eta = \sum_{\{X_1,\ldots,X_\nu\}} \Psi(X_k) = \sum_{\{X_1,\ldots,X_\nu\}} \Omega(X_k)\exp[-H_\eta(X_k)/T] \tag{2.136}$$

and the conditions for the maximizing values $X_{k\max}$ of X_k for $k = 1, \ldots, \nu$ are

$$\left(\frac{\partial \Psi}{\partial X_i}\right)_{X_k = X_{k\max}} = 0\,, \qquad i = 1, \ldots, \nu\,, \tag{2.137}$$

$$\sum_{\{i,j\}} \left(\frac{\partial^2 \Psi}{\partial X_i \partial X_j}\right)_{X_k = X_{k\max}} (X_i - X_{i\max})(X_j - X_{j\max}) < 0\,, \tag{2.138}$$

for any $X_i \neq X_{i\max}$. This means that the left-hand side of (2.138) is negative definite. If, for instance, the constant magnetic field distribution is used for the Ising model, X and m can be chosen as the two order variables while the Hamiltonian in an equation of type (2.136) for Z is

$$\widehat{H} = -NJ + 4JX - N\mathcal{H}m\,. \tag{2.139}$$

Since $\nu = 2$, (2.137) gives two relations, which are in fact equivalent to (2.131) and (2.134).

Analogously to (2.121), $\ln(Z_\eta)$ can be identified with the logarithm of the maximum term in the summation of (2.135) and put

$$F_\eta = -T \ln \Psi(X_{i\max}). \tag{2.140}$$

This relation can be given a thermodynamic interpretation. Suppose a non-equilibrium thermodynamic potential is defined for general X_i values by

$$F_\eta(X_i) = -T \ln \Psi(X_i)\,. \tag{2.141}$$

The conditions (2.137) and (2.138) which give a maximum value of Ψ clearly give a minimum value of $F_\eta(X_i)$, as T is one of the independent variables which are all kept at fixed values. Hence the equilibrium value of F_η, as given by (2.140), is identified as a minimum. This is in accordance with the thermodynamic principle that deviations from the equilibrium state at fixed values of $T, \xi_1, \ldots, \xi_\eta, Q_{\eta+1}, \ldots, Q_n$ cause F_η to increase from its equilibrium value. The conditions for a minimum value of F_η are

$$\left(\frac{\partial F_\eta}{\partial X_i}\right)_{X_k = X_{k\max}} = 0\,, \qquad i = 1, \ldots, \nu\,, \tag{2.142}$$

$$\sum_{i,j} \left(\frac{\partial^2 F_\eta}{\partial X_i \partial X_j}\right)_{X_k = X_{k\max}} (X_i - X_{i\max})(X_j - X_{j\max}) > 0\,, \tag{2.143}$$

which are equivalent to (2.137) and (2.138).

It is possible for two sets, $X_{i\max}$ and $X'_{i\max}$ respectively, of the order variables to satisfy conditions (2.142) and (2.143). Then the states defined by $X_{i\max}$ and $X'_{i\max}$ are both internally stable, that is stable with respect to small variations of X_i. Now suppose that $F_\eta(X'_{i\max}) > F_\eta(X_{i\max})$. The state $X'_{i\max}$ is thus internally stable but not stable with respect to the state $X'_{i\max}$. It is *metastable.* The true equilibrium or stable state is $X_{i\max}$ and is that given by averaging over the entire microstate space $\Lambda^{(\eta)}$. However, with constraints applying which exclude the true equilibrium range of the X_i, the metastable state can sometimes be obtained by averaging over a subspace $\Lambda^{(\eta)\prime}$. This can be described as 'broken ergodicity' (Palmer 1982), since, during a normal time span, the microstate can vary only over the subspace $\Lambda^{(\eta)\prime}$ rather than the entire space $\Lambda^{(\eta)}$.

Real substances can exist in states which are metastable with respect to ordering; for instance *quenched alloys* where the temperature has been lowered too rapidly for equilibrium rearrangements of the atoms to occur.[3] The persistence of these metastable states is determined by considerations similar to those discussed in Sect. 1.10.

2.7 A One-Dimensional Model for DNA Denaturation

In solution, the DNA molecule consists of two very long strands with chemical groups called *bases* at regular intervals along each strand. At low temperatures each base in one strand is connected to a base in the other strand by a link known as a *hydrogen bond* (H-bond). Successive bonds are slightly angled to each other, leading to the famous double-helix form of DNA. However, this is not important for the present work, and we can regard the molecule as a ladder with the H-bonds as rungs. As the temperature increases, the H-bonds tend to break, a process known as *denaturation, unzipping* or *the helix-coil transition.* Experimentally the fraction θ of bonds broken can be obtained from ultraviolet absorbance measurements and plotted against temperature.

[3] When, on the contrary, the temperature is lowered slowly enough for complete equilibrium to be maintained the alloy is said to be *annealed.*

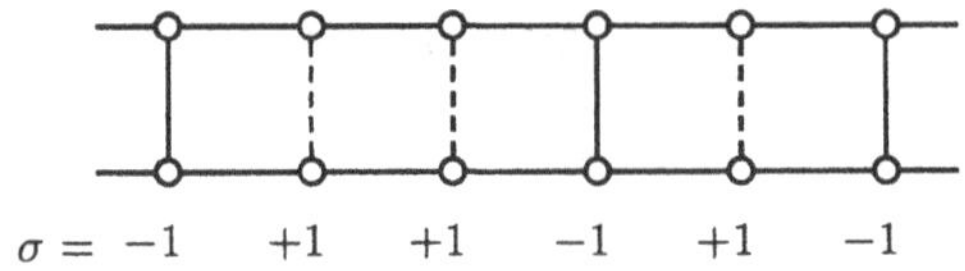

Fig. 2.3. The ladder model for DNA. The vertical segments represent intact and broken H-bonds.

Such a plot is known as a *melting curve*[4], and the melting temperature T_{m} is defined as that at which $\theta = \frac{1}{2}$. Observed melting curves are quite sharp, with a considerable change in θ occurring in an interval of a few degrees round T_{m}.

To develop a statistical mechanical theory we introduce a one-dimensional lattice, with each site corresponding to a rung of the ladder (see the reviews by Wartell and Montroll (1972), Gotoh (1983), Wartell and Benight (1985)). A variable σ_i is associated with each site i, where $\sigma_i = -1$ when the corresponding H-bond is intact and $\sigma_i = +1$ when it is broken (see Fig. 2.3). A partition function is constructed for the one-dimensional lattice, taking temperature, pressure and the concentrations in the ambient solution as independent variables, though only the first need be introduced explicitly. An Ising-type theory can then be constructed which, as Wartell and Benight (1985) remark, gives a semi-quantitative understanding of DNA melting.

It will be assumed that the number of sites is large enough to regard the last site as adjacent to the first and to use the thermodynamic limit without serious error. Suppose that the term in the partition function corresponding to all bonds intact is Ψ_0 and that each broken bond introduces a factor $\exp[2L(T)]$. Suppose also that it is easier for a bond to break if adjacent bonds are broken, and to allow for this cooperative effect a factor $\exp[-2K(T)]$, where $K > 0$, is introduced for each $+-$ nearest-neighbour pair. The partition function becomes

$$Z = \Psi_0 \sum_{\Lambda} \exp\left[L \sum_{i=1}^{N} (\sigma_i + 1) + K \sum_{k=1}^{N} (\sigma_i \sigma_{i+1} - 1)\right] , \tag{2.144}$$

where Λ is the space of the variables $\sigma_1, \ldots, \sigma_N$ and $\sigma_{N+1} = \sigma_1$. Disregarding constants it can be seen that for each nearest-neighbour pair there is a factor

$$\exp\left[K\sigma_i\sigma_{i+1} + \tfrac{1}{2}L(\sigma_i + \sigma_{i+1})\right] \tag{2.145}$$

in the general term in Z. Comparison with (2.89) shows that, apart from constant factors which have no effect on the bond configuration, the partition function is identical to that for a one-dimensional Ising model ferromagnet, with[5]

[4] The term 'melting', though sanctioned here by usage, is employed loosely. Properly speaking, melting is a first-order transition, not a continuous one.

[5] The variables K and L are then examples of *couplings*, which are introduced in the context of the coupling-density formulation of thermodynamics in Volume 2, Sect. 1.4.

$$K = J/T, \qquad L = \mathcal{H}/T. \tag{2.146}$$

The distribution of intact and broken bonds is thus the same as that of up and down spins in the Ising model. In particular, the equilibrium fraction of broken bonds

$$\theta = \tfrac{1}{2}\langle 1 + \sigma_i \rangle = \tfrac{1}{2} + \tfrac{1}{2}\langle \sigma_i \rangle = \tfrac{1}{2}(1 + m) \tag{2.147}$$

and then by (2.99), using the identifications (2.146),

$$\theta = \frac{1}{2}\left\{ 1 + \frac{\sinh L}{[\sinh^2(L) + \exp(-4K)]^{1/2}} \right\}. \tag{2.148}$$

Since $\exp[L(T)]$ represents the effect of a complicated chemical interaction between the polymer and the solution, it is not a simple Boltzmann factor and we treat it phenomenologically. First, note that, as T_m is defined as the temperature where $\theta = \tfrac{1}{2}$, then by (2.148), $L(T_m) = 0$. We may thus expand $L(T)$ in a Taylor series,

$$L(T) = a(T - T_m) + b(T - T_m)^2 + \cdots, \tag{2.149}$$

but for simplicity, only the first term is retained. Although the magnitude of K is important, its variation in the temperature range round T_m is not significant and we take K as constant. If K were zero, (2.148) would become

$$\theta = \tfrac{1}{2}\{1 + \tanh(L)\} = \tfrac{1}{2}\{1 + \tanh[a(T - T_m)]\}, \tag{2.150}$$

a result which could be obtained by deriving $\langle \sigma \rangle$ for a single isolated bond. For $K \neq 0$, (2.148) can be expressed in the form

$$\theta = \frac{1}{2}\left\{ 1 + \frac{\sinh\{a(T - T_m)\}}{[\sinh^2\{a(T - T_m)\} + \exp(-4K)]^{1/2}} \right\}. \tag{2.151}$$

The maximum slope of the plot of θ against T occurs at $T = T_m$ where

$$\left(\frac{d\theta}{dT} \right)_{T=T_m} = \tfrac{1}{2}a \exp(2K). \tag{2.152}$$

The cooperative effect is embodied in the parameter K, and the slope at T_m increases with K. In the limiting case $K = \infty$, θ becomes a Heaviside unit function equal to zero for $T < T_m$ and to one for $T > T_m$. However, the large experimental slope could, by (2.152), be attributed to either a large a or a large K. To obtain unequivocal evidence of the importance of the cooperative effect and to consider certain other effects, we have to examine the structure of the polymer in more detail.

The bases which form part of the DNA molecule are not identical and there are in fact four types, labelled A, T, G and C. However the only possible types of H-bond are AT or GC. In a natural DNA molecule the bases occur in an irregular but fixed order, and by the bonding rules just given the sequence of bases along one strand determines the sequence along the other strand. Hence if the strands separate, each can attract molecules from the medium to reproduce its complementary strand; this is the important replication property of DNA. Studies of the dependence of melting temperature on the relative proportions of AT and GC indicate that $T_{m1} < T_{m2}$, where T_{m1} and T_{m2} respectively denote the melting temperatures of pure AT and pure GC DNA molecules. The difference between these temperatures is between 40 and 50 degrees Celsius, depending to some extent on the medium (Wartell and Montroll 1972). Most early experiments on DNA gave smooth sigmoidal curves of θ against T, but this is now attributed (Wartell and Benight 1985) to the heterogeneity in base-pair sequence and molecular length of the samples. With homogeneous samples, the curves are less smooth and show several peaks of $d\theta/dT$ instead of a single peak in the centre of the melting range (Wada et al. 1980).

Consider a molecule in which a fraction x of the H-bonds is AT and a fraction $1 - x$ is GC. For a preliminary treatment, assume that there is no cooperative effect ($K = 0$) so that each bond is an isolated system and their order is immaterial. Supposing, for simplicity, that the parameter a is the same for both types of bond we define

$$L_1 = a(T - T_{m1}), \qquad L_2 = a(T - T_{m2}). \tag{2.153}$$

Equation(2.150) can then be applied, with $L = L_1$, to the fraction x of AT bonds and, with $L = L_2$, to the fraction $1 - x$ of GC bonds. This gives, for the overall fraction θ of broken bonds,

$$\theta - \frac{1}{2} = \frac{1}{2}[x \tanh(L_1) + (1 - x)\tanh(L_2)]. \tag{2.154}$$

Consider the case $T_{m1}/T_m = 0.928$ and $T_{m2}/T_m = 1.092$, T_m being the mean melting temperature defined in (2.162) below. This corresponds to melting temperatures of 48C and 98C for AT and GC respectively. For a reasonably narrow melting range in the cases $x = 1$ and $x = 0$ a large value of a is required so let $aT_m = 80$. With these values and $x = \frac{1}{2}$, (2.154) gives curve (i) in Fig. 2.4, which consists of sigmoidal portions near $T = T_{m1}$ and $T = T_{m2}$ separated by a long nearly horizontal stretch. Since no observed melting curve is like this, the assumption of non-interacting bonds is unsatisfactory.

The theory of an irregular sequence at AT and GC bonds with cooperative effects is difficult (see references above). However, the influence of the cooperative effect is clearly seen for a regular alternating sequence of AT and GC, and this is the case we shall discuss. For simplicity K is supposed to be the same for all nearest-neighbour pairs and we again assume (2.153). Consider a linear lattice with sites $1, 3, 5, \ldots$ corresponding to AT bonds, sites $2, 4, 6, \ldots$ corresponding to GC bonds. The number of sites N is taken to be even with sites N and 1 nearest neighbours. For the nearest-neighbour site pair $(i, i+1)$ there is a factor

$$\exp\left[K\sigma_i\sigma_{i+1} + \tfrac{1}{2}(L_1\sigma_i + L_2\sigma_{i+1})\right], \qquad \text{for } i \text{ odd}, \tag{2.155}$$

$$\exp\left[K\sigma_i\sigma_{i+1} + \tfrac{1}{2}(L_2\sigma_i + L_1\sigma_{i+1})\right], \qquad \text{for } i \text{ even}. \tag{2.156}$$

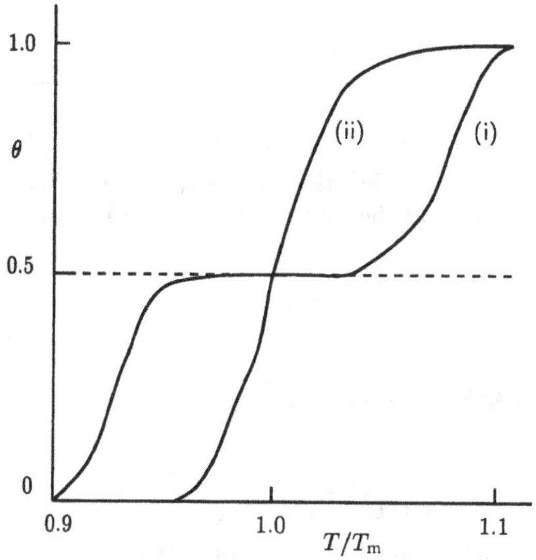

Fig. 2.4. Theoretical melting curves: (i) $\mathfrak{X} = 1$ for non-interacting bonds; (ii) $\mathfrak{X} = 0.0125$.

The configurational partition function can now be written as

$$Z = \sum_{\{\alpha,\beta,\gamma,\delta,\ldots,\chi,\psi,\omega\}} V_{\alpha\beta}V'_{\beta\gamma}\cdots V'_{\chi\psi}V_{\psi\omega}V'_{\omega\alpha}\,, \tag{2.157}$$

where $\alpha, \beta, \gamma, \delta, \ldots, \chi, \psi, \omega$ represent the states of the bonds on sites $1, 2, 3, 4, \ldots, N-2, N-1, N$ respectively. Here

$$\boldsymbol{V} = \begin{pmatrix} \exp[K + \frac{1}{2}(L_1 + L_2)] & \exp[-K + \frac{1}{2}(L_1 - L_2)] \\ \exp[-K + \frac{1}{2}(L_2 - L_1)] & \exp[K - \frac{1}{2}(L_1 + L_2)] \end{pmatrix}, \tag{2.158}$$

the matrix of the quantities given by (2.155), with $\boldsymbol{V}'$ corresponding similarly to (2.156). Thus $\boldsymbol{V}'$ is the transpose of $\boldsymbol{V}$. From (2.157),

$$Z = \mathrm{Trace}\{(\boldsymbol{V}\boldsymbol{V}')^{\frac{1}{2}N}\} = \lambda_1^{\frac{1}{2}N}\,, \tag{2.159}$$

where λ_1 is the largest eigenvalue of $\boldsymbol{V}\boldsymbol{V}'$. The last expression depends on assuming the thermodynamic limit. The fraction of broken bonds is given by

$$\theta = \frac{1}{4}\left[\langle 1 + \sigma_i\rangle_{i\ \mathrm{even}} + \langle 1 + \sigma_i\rangle_{i\ \mathrm{odd}}\right]$$

$$= \frac{1}{2} + \frac{1}{4}\left[\frac{\partial \ln(\lambda_1)}{\partial L_1} + \frac{\partial \ln(\lambda_1)}{\partial L_2}\right]. \tag{2.160}$$

After obtaining an expression for λ_1 (see Example 2.6) this yields

$$\theta - \frac{1}{2} = \frac{\mathfrak{X}^{-1}\sinh[2a(T - T_{\mathrm{m}})]}{2\left(\{\mathfrak{X}^{-1}\cosh[2a(T - T_{\mathrm{m}})] + \eta c_{\mathrm{m}}\}^2 - [\eta^{-1} - \eta]^2\right)^{\frac{1}{2}}}\,, \tag{2.161}$$

where

$$\mathfrak{X} = \exp(-2\kappa)\,,$$
$$T_{\rm m} = \frac{1}{2}(T_{\rm m1} + T_{\rm m2})\,, \qquad (2.162)$$
$$c_{\rm m} = \cosh[a(T_{\rm m2} - T_{\rm m1})]\,.$$

Thus $\theta = \frac{1}{2}$ for $T = T_{\rm m}$, and $\theta - \frac{1}{2}$ is an odd function of $T - T_{\rm m}$. At $T = T_{\rm m}$ the even derivatives of θ are thus zero and the first two odd derivatives are given by

$$\left(\frac{{\rm d}\theta}{{\rm d}T}\right)_{\rm m} = aQ(c_{\rm m}, \mathfrak{X})\,,$$
$$\left(\frac{{\rm d}^3\theta}{{\rm d}T^3}\right)_{\rm m} = 4a^3Q(c_{\rm m}, \mathfrak{X})\left\{1 - 3[Q(c_{\rm m}, \mathfrak{X})]^2(1 + c_{\rm m}\mathfrak{X}^2)\right\}, \qquad (2.163)$$

where

$$Q(c_{\rm m}, \mathfrak{X}) = \mathfrak{X}^{-1}(c_{\rm m} + 1)^{-\frac{1}{2}}[2 + (c_{\rm m} - 1)\mathfrak{X}^2]^{-\frac{1}{2}}\,. \qquad (2.164)$$

For the independent bond case when $\kappa = 0$ ($\mathfrak{X} = 1$), (2.161) reduces to (2.154) with $x = \frac{1}{2}$. At the opposite extreme, when $\kappa \to \infty$ ($\mathfrak{X} \to 0$), it follows from (2.161) that $\theta - \frac{1}{2} \to \frac{1}{2}{\rm sign}(T - T_{\rm m})$, so that the limiting form of θ is a Heaviside unit function. From (2.163), $({\rm d}\theta/{\rm d}T)_{\rm m}$ increases monotonically with κ and $({\rm d}\theta/{\rm d}T)_m \to \infty$ as $\kappa \to \infty$. When $({\rm d}^3\theta/{\rm d}T^3)_{\rm m} > 0$, ${\rm d}\theta/{\rm d}T$, which is an even function of $T - T_{\rm m}$, has a minimum at $T = T_{\rm m}$ and so must have a pair of maxima symmetrically placed about $T = T_{\rm m}$. For the values of $T_{\rm m1}/T_{\rm m}$, $T_{\rm m2}/T_{\rm m}$ and $aT_{\rm m}$ assumed above, $({\rm d}^3\theta/{\rm d}T^3)_{\rm m} > 0$ for $\mathfrak{X} > 0.006762$ ($\kappa < 2.498$), implying that the double maxima still exist at values of $\mathfrak{X}$ where there is no remaining vestige of the horizontal stretch around $T = T_{\rm m}$ and where the melting range has narrowed considerably. This is illustrated by curve (ii) in Fig. 2.4, which is drawn for $\mathfrak{X} = 0.0125$ ($\kappa = 2.1910$). The two peaks of ${\rm d}\theta/{\rm d}T$ are at $T/T_{\rm m} = 1 \pm 0.019$, approximately. This simple ${\rm d}\theta/{\rm d}T$ spectrum reflects the elementary periodic AT/GC bond distribution which we have assumed.

Although some important factors in denaturation have been identified, models of the type discussed above have the disadvantage that intact and broken bonds are treated in an equivalent way as states of linear lattice sites. However, where there are long sequences of broken bonds the two strands will not remain in a fixed relative position and will adopt an irregular 'loop' conformation. The additional entropy can be estimated by regarding the strands as two free polymers subject to the constraint that they possess common end-points. Various expressions are compared by Gotoh (1983). Under certain assumptions, the theory gives a first-order transition (see the review by Wiegel (1983)) which is surprising in a system where the interacting elements are ordered one-dimensionally, but presumably results from loop entropy terms acting as long-range interactions. However, the connection with experimental results is not clear.

For other physical realizations of one-dimensional lattice systems, see Ziman (1979), Chap. 2.

Examples

2.1 Show that in the one-dimensional zero-field Ising model the configurational entropy is given by

$$S = \frac{N(2J/T)}{1 + \exp(2J/T)} + N\ln[1 + \exp(-2J/T)].$$

Verify that $S/N \to \ln(2)$ as $T/J \to \infty$ and $S/N \to 0$ as $T/J \to 0$.

2.2 In the ν-state *Potts Model* each microsystem occupies one lattice site and can exist in ν distinct states. A 'like' nearest-neighbour pair, when both members of the pair are in same state, has interaction energy $-R$, whereas an 'unlike' nearest-neighbour pair has zero interaction energy. The *ferromagnetic Potts model* is when $R > 0$, which encourages nearest-neighbour pairs to be in the same state and the *antiferromagnetic Potts model* when $R < 0$, which encourages nearest-neighbour pairs to be in different states.
Assuming no interaction with any external field, construct the transfer matrix $\boldsymbol{V}$ for a one-dimensional Potts Model. Using the method of Sect. 2.4.1, obtain the partition function for an open linear lattice of N sites and show that in the thermodynamic limit, when $R > 0$, the configurational internal energy is

$$U = -\frac{NR\exp(R/T)}{\exp(R/T) + \nu - 1}\,.$$

Show that, apart from an additive constant term in the energy, the two-state Potts Model with $\nu = 2$ is equivalent to the Ising model with $J = \frac{1}{2}R$. Derive the heat capacity for general ν and $R > 0$ and show that the temperature where the heat capacity has a maximum decreases with ν.

2.3 In the Ising model ferromagnet with $\mathcal{H} \neq 0$ the internal energy $U = H + \mathcal{M}\mathcal{H}$ where H is the enthalpy derived in the constant magnetic field distribution. Given that λ_1 is the largest eigenvalue of the transfer matrix show that

$$H = -N\left[J\mathfrak{X}\frac{\partial\ln(\lambda_1)}{\partial\mathfrak{X}} + \mathcal{H}\mathfrak{Z}\frac{\partial\ln(\lambda_1)}{\partial\mathfrak{Z}}\right], \qquad U = -NJ\mathfrak{X}\frac{\partial\ln(\lambda_1)}{\partial\mathfrak{X}},$$

$\mathfrak{X}$ and $\mathfrak{Z}$ being defined by (2.75). Hence prove that

$$U = NJ\left\{\frac{4\mathfrak{X}^{-3}}{\lambda_1[(\mathfrak{X} - \mathfrak{Z}^{-1})^2 + 4\mathfrak{X}^{-4}]^{\frac{1}{2}}} - 1\right\}$$

and show that this expression reduces to $-NJ\tanh(J/T)$ when $\mathcal{H} = 0$.

2.4 In the *spin-1 Ising model* each dipole has three states, in which the spin variable $\sigma = -1, 0, 1$ respectively. The Hamiltonian for the constant magnetic field distribution has the same form as for the original Ising model being given by (2.66). Given $\mathcal{H} = 0$, construct the transfer matrix for the one-dimensional model and determine why the successive summation method of Sect. 2.4.1 is inapplicable. Again with $\mathcal{H} = 0$ show that the free energy for a ring lattice of N sites is

$$F = -NT\ln\left[\frac{2c+1+\sqrt{(2c-1)^2+8}}{2}\right], \qquad c = \cosh(J/T).$$

Obtain an expression for the configurational energy U and show that $U \to -NJ$ as $T/J \to 0$ and $U \to 0$ as $T \to \infty$.

2.5 In the maximum-term theory of the linear Ising model show, from (2.135), that

$$X_{\max} = \frac{N_1 N_2(\mathfrak{Z}^2 - 1)}{N_1\mathfrak{Z}^2 - N_2}.$$

By substitution into (2.132) and using (2.123) deduce that

$$m = \frac{\mathfrak{Z} - \mathfrak{Z}^{-1}}{\sqrt{(\mathfrak{Z} - \mathfrak{Z}^{-1})^2 + 4\mathfrak{X}^{-4}}}.$$

Show that this expression for m is equivalent to (2.99).

2.6 (i) Let $\boldsymbol{V}'$ be defined as the transpose of the matrix $\boldsymbol{V}$ given by (2.158). Show that the largest eigenvalue of the product matrix $\boldsymbol{V}\boldsymbol{V}'$ is

$$\lambda_1 = \phi + \sqrt{\phi^2 - 4\sinh(2K)},$$

where

$$\phi = \cosh(L_1 + L_2)\exp(2K) + \cosh(L_2 - L_1)\exp(-2K).$$

Deduce the expression (2.161) for the fraction θ of broken bonds, noting that L_1 and L_2 are defined by (2.153).

(ii) Show that, when $K = 0$, the expression (2.161) reduces to

$$\theta - \tfrac{1}{2} = \tfrac{1}{2}\{\tanh[a(T - T_{\mathrm{m}1})] + \tanh[a(T - T_{\mathrm{m}2})]\},$$

noting that this is (2.154) for the case $x = \frac{1}{2}$.

(iii) Show that, when $T_{\mathrm{m}1} = T_{\mathrm{m}2}$, (2.161) reduces to (2.151).

(iv) From (2.161) deduce that

$$\frac{\mathrm{d}\theta}{\mathrm{d}T} = a\exp(2K)[\phi^2 - 4\sinh^2(2K)]^{-\frac{3}{2}}$$
$$\times\{[\phi^2 - 4\sinh^2(2K)]\cosh(L_1 + L_2)$$
$$- \phi\sinh^2(L_1 + L_2)\exp(2K)\}\,.$$

Derive (2.163) for $T = T_{\mathrm{m}} = \frac{1}{2}(T_{\mathrm{m}1} + T_{\mathrm{m}2})$.

3. The Mean-Field Approximation, Scaling and Critical Exponents

3.1 The Ising Model Ferromagnet

We consider the Ising model on a regular lattice (see Appendix A.1) where each interior site has the same number of nearest-neighbour sites. This is called the *coordination number* of the lattice and will be denoted by z. We shall denote the dimension of the lattice by the symbol d. It is assumed that, in the thermodynamic limit, boundary sites can be disregarded and that, with N sites, the number of nearest-neighbour site pairs is $\frac{1}{2}zN$. Since the Ising model is restricted in the sense of Sect. 2.3.4, the Helmholtz free energy A depends on the two extensive variables $\mathcal{M}$ and N, the latter being equal to the number of spins and also to the reduced volume V if the volume v_0 per site is taken as the standard volume (see discussion after (2.55)). The configurational Helmholtz free energy per site a depends on T and the (relative) magnetization per site m, defined by (2.70), so that $n = 2$.

The value of m is prescribed for the canonical distribution, and we now consider spin configurations for given m. Since all sites are regarded as equivalent, (2.64) implies

$$m = \langle \sigma_i \rangle \,, \tag{3.1}$$

where the right-hand side is the mean of the spin variable on site i in the canonical distribution. Except at the critical point itself, the spins on widely separated sites i and j are independent and thus

$$\langle \sigma_i \sigma_j \rangle = \langle \sigma_i \rangle \langle \sigma_j \rangle = m^2 \,. \tag{3.2}$$

However, with i and j close together, the situation is different. For the Ising ferromagnet the spins tend to lie parallel and thus

$$\langle \sigma_i \sigma_j \rangle > m^2 \,. \tag{3.3}$$

This represents *short-range order*, and the difference $\langle \sigma_i \sigma_j \rangle - m^2$, which is a function of the separation of sites i and j, is a measure of the *correlation* between the spins (see Ziman 1979, Chap. 1). This quantity reduces to $\langle \sigma_i \sigma_j \rangle$ when $m = 0$, a result which holds for $d = 1$, $\mathcal{H} = 0$ at all temperatures. The correlation $\langle \sigma_i \sigma_{i+k} \rangle$ for k-th neighbour sites in this case was evaluated in Sect. 2.4. From (2.88), $\langle \sigma_i \sigma_{i+k} \rangle > 0$ for all k, but tends to zero as $k \to \infty$.

Now consider the mean $-J\langle\sigma_i\sigma_j\rangle$ of the interaction energy of a nearest-neighbour (i,j) pair. The contribution from a particular value of σ_i is $-J\sigma_i\langle\sigma_j\rangle_{\sigma_i}$ where $\langle\sigma_j\rangle_{\sigma_i}$ denotes the mean of σ_j for given σ_i and m. Thus spin σ_i can be regarded as acted on by a field $-J\langle\sigma_j\rangle_{\sigma_i}$ due to its neighbouring spin σ_j. The mean-field approximation consists in replacing $\langle\sigma_j\rangle_{\sigma_i}$ by m, which by (3.1) is the mean value of σ_j over all spin configurations compatible with the prescribed m, irrespective of the value of σ_i. Summing over σ_i and using (3.1) again gives $-Jm^2$ for $-J\langle\sigma_i\sigma_j\rangle$. This assumes that (3.2) applies even for nearest-neighbour pairs, and is equivalent to putting the correlation equal to zero, thereby disregarding short-range order. The energy E is the same for all terms in the canonical partition function and thus

$$E = -\tfrac{1}{2}zNJm^2 = U\,. \tag{3.4}$$

Equations (2.122) and (2.123) apply for a lattice of any dimension and hence the number of spin configurations compatible with a given value of m is

$$\Omega(m) = \frac{N!}{N_1!N_2!} = \frac{N!}{[\frac{1}{2}N(1+m)]![\frac{1}{2}N(1-m)]!}\,. \tag{3.5}$$

The canonical partition function in the mean-field approximation is thus

$$Z = \Omega(m)\exp(-E/T)\,. \tag{3.6}$$

Since (3.4) and (3.5) correspond to a random mixture of N_1 positive and N_2 negative spins on the lattice, the mean-field method can be termed the *randomized approximation.* Another term in frequent use is the *zeroth-order approximation.*

3.1.1 Free Energy and Magnetization

Using (3.4)–(3.6) and Stirling's formula the Helmholtz free energy density is given by

$$\begin{aligned} a &= N^{-1}T\ln Z \\ &= \tfrac{1}{2}T[(1+m)\ln(1+m) + (1-m)\ln(1-m) - 2\ln 2] - \tfrac{1}{2}zJm^2\,. \end{aligned} \tag{3.7}$$

It can be seen that a is an even function of m, thus satisfying the symmetry relation (2.65). The second derivative of a with respect to m is

$$\frac{\partial^2 a}{\partial m^2} = \frac{T}{1-m^2} - zJ \tag{3.8}$$

and we define the critical temperature T_c by

$$T_c = zJ\,. \tag{3.9}$$

From (3.8) the curve of a against m is convex downwards for all m in the relevant range $(-1,1)$ when $T > T_c$, as shown by the broken line in Fig. 3.1. At $T = T_c$ the curvature at $m = 0$ changes sign and for $T < T_c$ the form of a

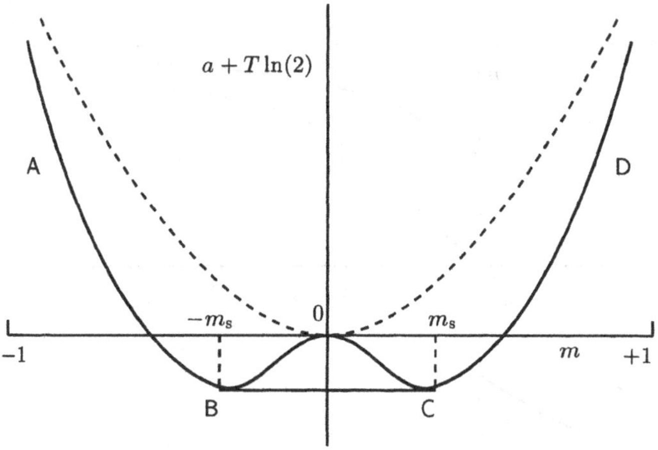

Fig. 3.1. Free energy–magnetization isotherms: broken curve, $T > T_c$; full curve, $T < T_c$.

is shown by the full line ABOCD in Fig. 3.1. Since $n = 2$, a is a free energy density of type f_{n-2} as discussed in Sect. 1.10 (iii), and the points B and C where the curve of a is touched by the double tangent BC represent conjugate phases. The loop BOC, lying above BC, is typical of classical theories and represents unstable states, the stable equilibrium a being given by the convex envelope ABCD. The situation is similar to that shown on Fig. 1.2 for the van der Waals gas. However, because of the symmetry of the a curves about $m = 0$ the double tangent BC is horizontal and the magnetizations corresponding to the points B and C are equal and opposite. They are denoted by $\pm m_s$ ($m_s > 0$), where m_s is a function of $T < T_c$.

The field $\mathcal{H}$ is conjugate to the density m, and either from the second relation of (1.34), or from the more general relation (1.78), (with $\eta = 0$) it follows that

$$\mathcal{H} = \frac{\partial a}{\partial m} = \frac{1}{2}T\ln\left[\frac{1+m)}{1-m}\right] - zJm\,. \tag{3.10}$$

The last expression derives from (3.7) and, since it is an odd function of m, the symmetry relation (2.71) is satisfied. From the broken curve in Fig. 3.1 the slope $\partial a/\partial m$ is a continuous function of m for $T > T_c$. Hence the corresponding plot of m against $\mathcal{H}$ is continuous, as shown by the broken line in Fig. 3.2. The situation is quite different for $T < T_c$. The slope $\partial a/\partial m$ and hence $\mathcal{H}$ are zero at the end-points B and C of the common tangent (Fig. 3.1). Hence the magnetizations $\pm m_s$ correspond to $\mathcal{H} = 0$; m_s is thus termed the *spontaneous magnetization.* The slopes of a against m on the convex envelope approach zero continuously as the ends B and C of the common tangent are approached from the left to right respectively. The form of the plot of m against $\mathcal{H}$ for $T < T_c$ is thus given by the full line in Fig. 3.2 and

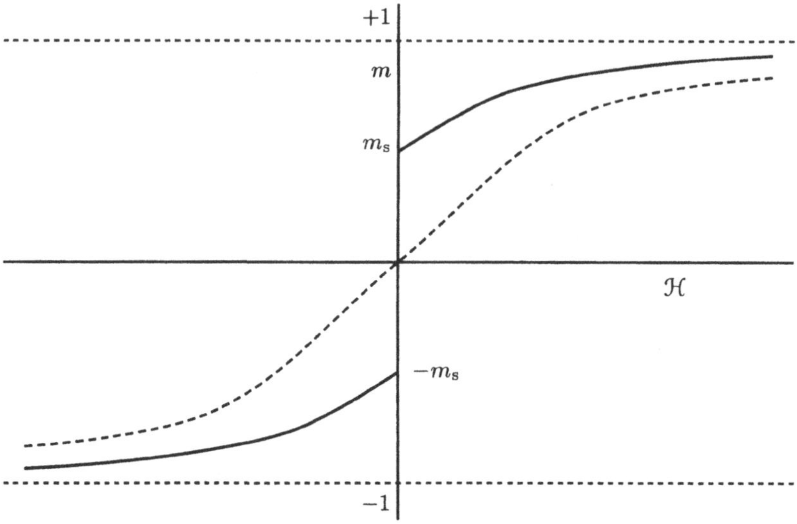

Fig. 3.2. Magnetization–field isotherms, as in Fig. 3.1.

$$\lim_{\mathcal{H}\to\pm 0} m = \pm m_{\mathrm{s}} \,. \tag{3.11}$$

There is a discontinuity in the density m as $\mathcal{H}$ passes through the value zero at constant $T < T_{\mathrm{c}}$ and this constitutes a first-order transition. Since the internal energy U is an even function of m it remains continuous as $\mathcal{H}$ passes through zero.

It has been shown that m is a smoothly varying function of $\mathcal{H}$ and T except for the discontinuity (3.11). If states of the system are represented by points in the $(T, \mathcal{H})$ plane then there is a segment of discontinuity (first-order transition line) occupying the interval $(0, T_{\mathrm{c}})$ on the T-axis, as shown in Fig. 3.3, and all thermodynamic functions are smooth except on this interval. Although it has been derived by the rather crude mean-field approximation,

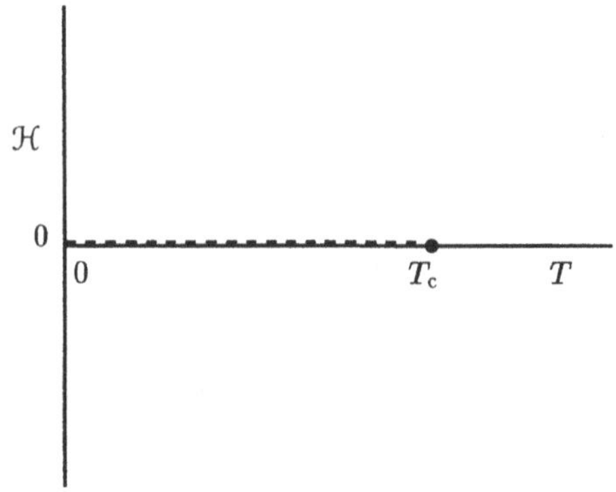

Fig. 3.3. Phase diagram in the temperature–field plane.

this simple form of phase diagram is believed to be qualitatively correct for the Ising model ferromagnet with $d = 2$ or 3. The symmetry about the T-axis is of course a consequence of the $\mathcal{H}$-reversal symmetry expressed by the relations in Sect. 2.4. For both theoretical and experimental ferromagnets the critical temperature T_c is called the *Curie temperature.*

3.1.2 Fluctuations in Zero Field

The zero-field magnetization is obtained by putting $\mathcal{H} = 0$ and $m = m_s$ in (3.10) to give

$$\frac{1}{2}\ln\frac{1+m_s}{1-m_s} - \frac{zJ}{T}m_s = m_s\left(1 - \frac{T_c}{T} + \frac{1}{3}m_s^2 + \frac{1}{5}m_s^4 + \cdots\right) = 0\,. \qquad (3.12)$$

For $T > T_c$ the only solution is $m_s = 0$. For $T < T_c$, $m_s = 0$ is still a solution of (3.12) but now corresponds to a non-equilibrium state. The equilibrium value of m_s is given by

$$\frac{1}{3}m_s^2 + \frac{1}{5}m_s^4 + \cdots = \frac{T_c - T}{T}\,. \qquad (3.13)$$

It follows that $m_s \to 0$ as $T \to T_c - 0$ so that the magnetization is continuous at $T = T_c$. As $T \to 0$ the series on the left-hand side must diverge and so $m_s \to 1$ (see Fig. 3.4). From (3.4) the zero-field configurational internal energy per spin is given by

$$u = \begin{cases} 0\,, & T > T_c\,, \\ -\frac{1}{2}zJm_s^2\,, & T < T_c\,. \end{cases} \qquad (3.14)$$

The result for $T > T_c$ occurs because short-range ordering is neglected in the mean-field approximation. The configurational heat capacity per spin, when $\mathcal{H} = 0$ and $T < T_c$, is given by

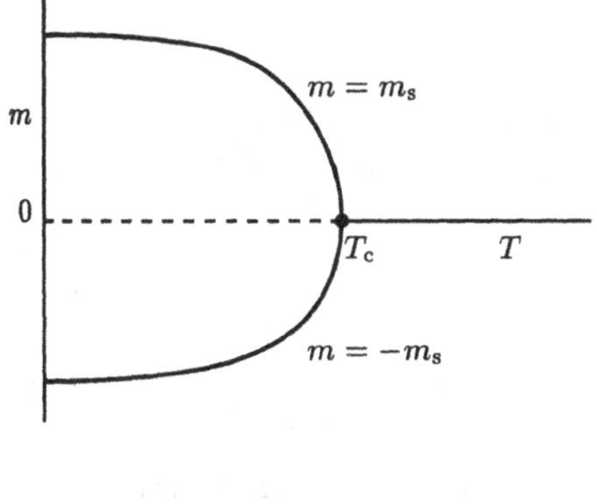

Fig. 3.4. The spontaneous magnetization plotted against temperature.

$$
\begin{aligned}
c_0 = k_{\mathrm{B}} \left(\frac{\partial u}{\partial T}\right)_{\mathcal{H}=0} &= -\frac{1}{2} z k_{\mathrm{B}} J \frac{\partial (m_{\mathrm{s}}^2)}{\partial T} \\
&= \frac{1}{2} k_{\mathrm{B}} \left(\frac{T_{\mathrm{c}}}{T}\right)^2 \left[\frac{1}{3} + \frac{2}{5} m_{\mathrm{s}}^2 + \cdots\right]^{-1} ,
\end{aligned} \tag{3.15}
$$

where the last form follows from (3.13). From (3.15) and (3.14)

$$
\begin{aligned}
\lim_{T \to T_{\mathrm{c}} - 0} c_0 &= \frac{3}{2} k_{\mathrm{B}} , \\
c_0 = 0 , \qquad & T > T_{\mathrm{c}} .
\end{aligned} \tag{3.16}
$$

Hence there is a finite discontinuity in c_0 at $T = T_{\mathrm{c}}$, as shown in Fig. 3.5. As described in Sect. 1.9 and illustrated by (3.11), a first-order transition occurs where there is discontinuous change in one or more of the thermodynamic densities. We have just seen that in mean-field theory the Ising model ferromagnet undergoes a transition from a disordered (i.e. unmagnetized) state to a long-range ordered (i.e. spontaneously magnetized) state when T decreases through the value T_{c} with $\mathcal{H} = 0$. Although a change of symmetry has thus occurred, the energy, entropy and magnetization densities are continuous at $T = T_{\mathrm{c}}$ so that the transition cannot be first-order. In Ehrenfest's classification (Ehrenfest 1933, Pippard 1967) a pth-order transition, for $p > 1$, is one in which there are finite discontinuities in the $(p-1)$th-order derivatives of the densities with respect to fields, but where all lower order derivatives and the densities themselves are continuous. In this scheme the Ising model transition at $T = T_{\mathrm{c}}$ is second-order since, by (3.15) and (3.16), there is a discontinuity in $\partial u/\partial T$. However, in an exact theory of the $d = 2$ Ising model (see Sect. 8.9), the finite discontinuity is replaced by a singularity where $\partial u/\partial T \to \infty$ as the critical temperature is approached from either above or below. Even in mean-field theory there is a symmetrical infinity at

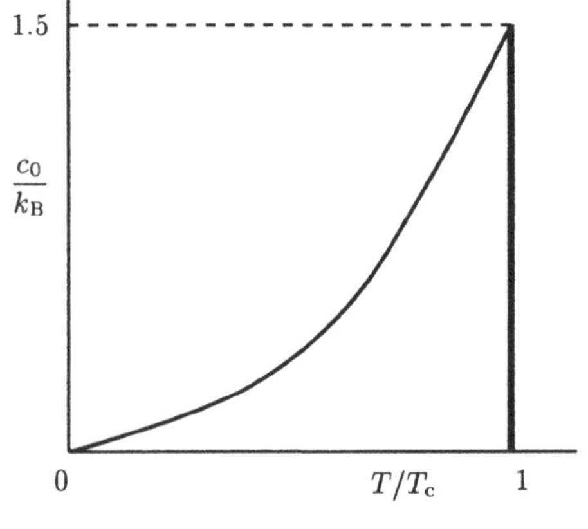

Fig. 3.5. The zero-field heat capacity from mean-field theory.

$T = T_c$ in another first derivative of density, as is shown by (3.18) below, and Example 3.3. As an alternative to the Ehrenfest scheme, Fisher (1967) proposed a classification of transitions into two categories, 'first-order' and 'continuous' (i.e. with continuous densities). The term 'higher-order' may be used instead of 'continuous'. However, in our view there is no ambiguity in describing transitions where the densities are continuous, but where there are discontinuities or infinities in some of their first derivatives with respect to fields, as 'second-order', and we shall use this term where appropriate.

We now consider the zero-field susceptibility for $T > T_c$. Using (3.9) and a power series expansion of the logarithmic term, (3.10) can be expressed in the form

$$m\left(1 - \frac{T_c}{T} + \frac{1}{3}m^2 + \frac{1}{5}m^4 + \cdots\right) - \frac{\mathcal{H}}{T} = 0\,, \tag{3.17}$$

from which it follows that $m \to 0$ as $\mathcal{H} \to 0$ for $T > T_c$. Hence

$$\left(\frac{\partial m}{\partial \mathcal{H}}\right)_{\mathcal{H}=0} = \lim_{\mathcal{H}\to 0} \frac{m}{\mathcal{H}} = \frac{1}{T - T_c}\,, \tag{3.18}$$

which diverges as $T \to T_c + 0$.

To discuss fluctuations in m it is necessary to use the constant magnetic field distribution. From (2.46), (2.65) and (3.4)

$$\widehat{H} = -\frac{1}{2}zNJ\hat{m}^2 - N\mathcal{H}\hat{m}\,, \tag{3.19}$$

where the 'hat' notation for $\hat{m}$ indicates that the magnetization is no longer an independent variable and must now be summed over in the construction of the partition function. Thus

$$\begin{aligned} Z(N,T,\mathcal{H}) &= \sum_{\{\hat{m}\}} \exp(N\mathcal{H}\hat{m}/T)\Psi(\hat{m})\,, \\ \Psi(\hat{m}) &= \Omega(\hat{m})\exp\left(\tfrac{1}{2}zNJ\hat{m}^2/T\right) = \exp[-Na(\hat{m})/T]\,, \end{aligned} \tag{3.20}$$

where $\Omega(\hat{m})$ is given by (3.5) and a by (3.7). Since Ψ is identical in form to the canonical partition function, (3.20) is an example of (2.31). If the method of the maximum term is used to obtain the equilibrium value m then, from (3.20), the equation (3.10) is recovered. Since $m = 0$ for $T > T_c$ we have

$$\mathrm{Var}[\hat{m}] = \frac{\displaystyle\sum_{\{\hat{m}\}} \hat{m}^2\Psi(\hat{m})}{\displaystyle\sum_{\{\hat{m}\}} \Psi(\hat{m})}\,. \tag{3.21}$$

Only values of $\hat{m}$ close to the equilibrium value zero will contribute significantly to the sums in (3.21), and hence, using (3.7),

$$\ln \Psi(\hat{m}) = -NT^{-1}a(\hat{m}) = N\ln(2) - N\,\frac{T - T_c}{2T}\,\hat{m}^2\,. \tag{3.22}$$

Since $a(\hat{m})$ is an even function of $\hat{m}$, only even powers appear in its Maclaurin expansion which we have truncated after the squared term. The term $N\ln(2)$ can be omitted since it cancels in (3.21). The sums in this equation can be replaced by integrals since, by (2.123), the interval between successive values of $\hat{m}$ is only $2/N$. Again, as only the neighbourhood of $\hat{m} = 0$ is significant, the limits ± 1 for $\hat{m}$ can be replaced by $\pm\infty$ in the integrals. Finally, after making the substitution

$$x = \hat{m}\sqrt{\frac{N(T - T_c)}{2T}} \tag{3.23}$$

we have

$$\mathrm{Var}[\hat{m}] = \frac{2T}{N(T - T_c)} \frac{\int_{-\infty}^{\infty} x^2 \exp(-x^2)\mathrm{d}x}{\int_{-\infty}^{\infty} \exp(-x^2)\mathrm{d}x} = \frac{T}{N(T - T_c)}. \tag{3.24}$$

Now the fluctuation–susceptibility equation (2.115) applies for any d, provided the equilibrium value of $\mathcal{M}$ is zero. Using (2.70), it can be expressed in the form

$$\left(\frac{\partial m}{\partial \mathcal{H}}\right)_{\mathcal{H}=0} = NT^{-1}\mathrm{Var}[\hat{m}] \tag{3.25}$$

and, combining with (3.24), we recover (3.18).

It is important to consider the microscopic behaviour of the Ising model just above a critical point where the susceptibility diverges. By (2.101) the exact zero-field susceptibility of the $d = 1$ Ising model ferromagnet tends to infinity as $T \to 0$. From Sect. 2.5, this results from the k-th neighbour correlation $[\tanh(J/T)]^k$ tending to unity as $T \to 0$. The susceptibility divergence is thus related to the divergence in the size of the region of local order, that is the length of the blocks of parallel spins. An exact treatment would reveal similar behaviour for $d = 2$ or 3 just above the critical temperature T_c. Although the overall numbers of up and down spins remain equal there are large regions of local order or 'patches' of parallel spins. As $T \to T_c + 0$ the patches become infinitely large which means that a quantity called the *correlation length* is approaching infinity. This is analogous to critical opalescence in a fluid, referred to in Chap. 1, where the local fluctuations (i.e. regions where the density differs appreciably from the mean) become large just above the critical point. In the mean-field method, only overall fluctuations, which change the relative magnetization $\hat{m}$ from its equilibrium value, are recognized, although (3.24) shows that these are sufficient to produce a susceptibility divergence as $T \to T_c + 0$. In an accurate treatment, which must include the effect of local fluctuations, the susceptibility divergence is sharper, especially for $d = 2$ (see the discussion of critical indices in Sect. 3.5 and the exact solution for $d = 2$ in Chap. 8).

For $T < T_c$ there is what at first sight seems to be a paradox in relation to the properties of the zero-field Ising model ferromagnet for $d > 1$. With $\mathcal{H} = 0$, $\widehat{H}$ as given by (2.66) is equal to E, which is invariant under a change of sign of the spin variables. Hence $\langle\sigma_i\rangle = -\langle\sigma_i\rangle = 0$ for all sites i, which by (2.64) implies that the mean $\langle\hat{m}\rangle = 0$ in the constant magnetic field distribution with $\mathcal{H} = 0$. The same result follows from mean-field theory since $\Psi(\hat{m})$, as given by (3.20), is an even function of $\hat{m}$ and hence $\sum_{\hat{m}} \hat{m}\Psi(\hat{m}) = 0$. Since this is true for all T it appears to be inconsistent with long-range order (i.e. a spontaneous magnetization $m_s \neq 0$) for $T < T_c$ and $\mathcal{H} = 0$. However, although $\langle\hat{m}\rangle = 0$, there is no concentration of probability density near $\hat{m} = 0$ in either accurate or mean-field theory. There is instead a symmetrically bifurcated probability distribution with density concentrated near $\hat{m} = m_s$ and $\hat{m} = -m_s$. In mean-field theory, $\Psi(\hat{m})$ is given by the last relation of (3.20) and from the curve ABOCD in Fig. 3.1 there are equal maxima of $\Psi(\hat{m})$ at $\hat{m} = \pm m_s$ with a minimum at $\hat{m} = 0$, for $T < T_c$. For N of the order of Avogadro's number M_A the maxima are very sharp. However, for any non-zero $\mathcal{H}$, however small, one of the two probability maxima becomes dominant. Suppose that, with $T < T_c$, the state $m = m_s > 0$ is reached by letting $\mathcal{H} \to +0$. Then, with any realistic assumptions about the dynamics of the system, the waiting time for a 'flip' to the state $m = -m_s < 0$ will be very large compared to normal observation times. The same will apply when the initial state is $m = -m_s < 0$. This situation for $T < T_c$ is similar to that in the metastable states discussed near the end of Sect. 2.6, but the microstate now varies over one of the two equal subspaces into which the space of microstates is divided, corresponding respectively to positive and negative magnetization. For either possibility the spatial symmetry is reduced as compared with the $T > T_c$ state $m = 0$, since the system is no longer invariant with respect to rotations or reflections which reverse the magnetization axis. As Palmer (1982) remarks: 'Broken symmetry is a particular case of ... broken ergodicity'.

The large-scale properties of the Ising model ferromagnet are expressed quantitatively by *critical parameters* such as $T_c/(zJ)$, which are discussed in Sect. 8.9. (Behaviour very near the critical point is characterized by *critical exponents*; see Sect. 3.4.) The mean-field method ignores the correlation between spins on nearest-neighbour sites. It might be expected that this correlation would weaken as the coordination number z becomes larger, and it is found that accurate critical parameters do become closer to mean-field values as z increases. However, the dimension d of the lattice is more important. For $d = 1$ the mean-field method gives $T_c = zJ = 2J$, but this is spurious since, from the exact results of Chap. 2 there is no non-zero critical temperature. For $d = 2$ and $d = 3$, accurate methods do predict a non-zero critical temperature, but the accurate parameters for $d = 3$ are considerably closer to mean-field values than those for $d = 2$ (see Chap. 8, Table 8.1, and Domb 1974, Table 13).

3.2 Interpretations of the Mean-Field Method

The mean-field method was developed independently by a number of authors. As will be seen below, van der Waals' treatment of attractive forces was mean-field in character. The Weiss approach (Fowler and Guggenheim 1949), initiated before the development of the quantum theory of magnetism, consisted in adding a term λm to the field $\mathcal{H}$ acting on the dipoles. With $\lambda = zJ$ this is equivalent to the mean-field method, (Example 3.1).

There are two limiting cases in which the mean-field results can be considered accurate, both involving long-distance interactions. The rest of this section will be devoted to these.

3.2.1 Many-Neighbour Interactions and the Lebowitz–Penrose Theorem

The Ising interaction can be extended to n-th neighbour distance by generalizing (2.63) to

$$E = -\sum_{k=1}^{n} J_k \sum_{\{i,j\}}^{(k)} \sigma_i \sigma_j \,, \tag{3.26}$$

where the summation labelled (k) is over all k-th neighbour pairs of sites, each pair being counted only once. In the mean-field approximation all pairs $\sigma_i \sigma_j$ are replaced by m^2 so that

$$E = -\frac{1}{2} N \left(\sum_{k=1}^{n} z_k J_k \right) m^2 \,, \tag{3.27}$$

z_k denoting the number of kth neighbours of any site. All results derived in Sect. 3.1 are thus still applicable, the only change being the replacement of zJ by $\sum_{k=1}^{n} z_k J_k$. Suppose that now J_k is set equal to J for all $k = 1, \ldots, n$. Since the correlation between the two spins of a pair will diminish as the distance between them increases it might be expected that accurate critical parameters will become closer to mean-field values as n increases, and this is found to be the case (Wood 1975, Table 14.) However, the weight of numerical evidence is that the critical exponents for small finite n are the same as those for the nearest neighbour model ($n = 1$).

On the other hand, it is possible that mean-field theory may apply for interactions of very long range. Consider, for instance, a fluid where the range is so large that any density fluctuations may be neglected. The energy of interaction of the given molecule with other molecules is then proportional to the overall density and can be written as $-2\kappa M/V$ where V is the reduced volume defined in (1.54). Adding over all molecules and dividing by two to avoid double counting, the long-range interactions contribute $-\kappa M^2/V$ to the internal energy U and the Helmholtz free energy A. By (1.9) or (1.65), this gives a term $-\kappa\rho^2$ in the pressure field P and $-2\kappa\rho$ in the chemical potential μ, where $\rho = M/V$. The pressure contribution is identical in form to the attractive pressure term in van der Waals theory, described in Sect. 1.6. The contribution to the free energy density a, taken with respect to V, is also $-\kappa\rho^2$. Similar reasoning for a lattice magnetic model gives a term proportional to $-m^2$ in the energy per site, as in mean-field theory. This intuitive idea about a long-range interaction was made precise by Lebowitz and Penrose (1966) and extended to the quantum mechanical case by Lieb (1966) (see also Hemmer and Lebowitz 1976). Take a fluid model of dimension d, called the 'reference model', with free energy $a_0(T, \rho)$ per unit volume and introduce an additional interaction energy $w(r)$ between any two molecules at distance r where

$$w(r) = \gamma^d u(\gamma r) , \tag{3.28}$$

$u(x)$ being a monotonic function of x, decreasing faster than x^{-d} at large x. For fixed r, $w(r)$ can thus be made as small as we like by increasing γ, while on the other hand the space integral of $w(r)$, which is proportional to

$$\int_0^\infty \gamma^d u(\gamma r) r^{d-1} \mathrm{d}r , \tag{3.29}$$

is independent of γ. Lebowitz and Penrose showed that, as $\gamma \to \infty$, the limiting form of the free energy per unit volume is the convex envelope of

$$a(T, \rho) = a_0(T, \rho) - \kappa\rho^2 , \tag{3.30}$$

where κ is proportional to the space integral. Thompson (1972b) has discussed the application of this result to lattice models. The mean-field Ising model theory can thus be regarded as a reference model of independent lattice spins with a Lebowitz–Penrose Ising-type limiting interaction added.

3.2.2 A Distance-Independent Interaction

Consider a spin-$\frac{1}{2}$ model on N lattice sites with an equal Ising-type interaction between each pair of spins, irrespective of distance. This *distance-independent model*, which was first proposed by Husimi (1953) and Temperley (1954) and further discussed by Kac (1968) and Brankov and Zagrebnov (1983), can easily be shown to give a mean-field form of the configurational energy and the free energy. Extending the summation in (2.63) to every pair of spins and using $\sigma_i^2 = 1$

$$\begin{aligned} E &= -\varepsilon \sum_{\{i \neq j\}}^{\text{(all pairs)}} \sigma_i \sigma_j = -\frac{1}{2}\varepsilon \sum_{\{i\}} \sigma_i \sum_{\{j \neq i\}} \sigma_j \\ &= -\frac{1}{2}\varepsilon \Big(\sum_{\{i\}} \sigma_i \Big)^2 + \frac{1}{2}\varepsilon N . \end{aligned} \tag{3.31}$$

For given relative magnetization m this gives

$$E = -\frac{1}{2}\varepsilon N^2 m^2 + \frac{1}{2}\varepsilon N\,. \tag{3.32}$$

Given that, with constant ε, E/N would diverge in the thermodynamic limit $N \to \infty$, we choose $\varepsilon = zJ/N$ to obtain

$$E = -\frac{1}{2}zNJm^2 + \frac{1}{2}zJ\,. \tag{3.33}$$

Apart from the last term, which can be disregarded, (3.33) is identical to (3.4).

3.3 The Mean-Field Method for a More General Model

Suppose that the lattice of N sites is occupied by members of species $1, \dots, \kappa$, with one member to each site. One species may be 'holes' or vacant sites. The number of members of species j on the lattice is denoted by N_j, $j = 1, \dots, \kappa$ and the fraction of sites occupied by species j is $x_j = N_j/N$ where

$$\sum_{j=1}^{\kappa} N_j = N\,, \qquad \sum_{j=1}^{\kappa} x_j = 1\,. \tag{3.34}$$

Each (j, ℓ) nearest neighbour pair makes a contribution $\varepsilon_{j\ell}$ to the configurational energy, which may therefore be written as

$$E = \sum_{\{j,\ell\}} \varepsilon_{j\ell} N_{j\ell}\,, \tag{3.35}$$

where $N_{j\ell}$ is the number of (j, ℓ) nearest neighbour pairs. Now a pair of widely separated sites can be regarded as uncorrelated, and hence the probability of both sites being occupied by species j is x_j^2, whereas the probability of one being occupied by a j and the other by an ℓ,for $j \neq \ell$, is $2x_j x_\ell$. The mean-field approximation results from the assumption that these probabilities apply for all site pairs, including nearest neighbours. Hence

$$E = \frac{1}{2}zN \sum_{\{j\}} \varepsilon_{jj} x_j^2 + zN \sum_{\{j<\ell\}} \varepsilon_{j\ell} x_j x_\ell \tag{3.36}$$

since, disregarding boundary effects, the total number of nearest neighbour site pairs is $\frac{1}{2}zN$. The Ising model ferromagnet is an example of a model with $\kappa = 2$, species 1 and 2 being up and down spins respectively. Then $\varepsilon_{11} = \varepsilon_{22} = -J$, $\varepsilon_{12} = J$ and, from (3.36),

$$E = -\frac{1}{2}zNJ(x_1^2 + x_2^2) + zNJx_1x_2 = -\frac{1}{2}zNJ(x_1 - x_2)^2\,, \tag{3.37}$$

which reproduces (3.4), since, from (2.122), $m = x_1 - x_2$. The number of configurations for given values of the N_j is

$$\Omega(N_j) = \frac{N!}{\prod_{j=1}^{\kappa} N_j!} \tag{3.38}$$

and hence the corresponding contribution to the canonical partition function is

$$\Psi(N_j) = \Omega(N_j)\exp(-E/T), \tag{3.39}$$

with E given by (3.36). If all the N_j are thermodynamic extensive variables or determined by them, then Ψ itself is the canonical partition function, and the Helmholtz free energy per site is

$$a = -N^{-1}T\ln\Psi = T\sum_{j=1}^{\kappa} x_j \ln x_j + \frac{1}{2}z\sum_{j=1}^{\kappa} \varepsilon_{jj}x_j^2 + z\sum_{\{j<\ell\}} \varepsilon_{j\ell}x_j x_\ell, \tag{3.40}$$

where Stirling's formula has been applied to $\ln\Omega(N_j)$. This is applicable to the Ising model where N_1 and N_2 are determined by $\mathcal{M}$ and N so that (3.7) is a particular case of (3.40).

If not all the N_j are determined by thermodynamic extensive variables then the canonical partition function is a sum over the $\Psi(N_j)$. However, the expression (3.40) for a may be regarded as a non-equilibrium free energy as defined in (2.141), while some of the x_j are then order variables with respect to which a must be minimized so as to obtain equilibrium conditions. As discussed in Sect. 2.6, this is equivalent to applying a maximum-term procedure to the partition function. It is easy to construct a non-equilibrium free energy for an alternative distribution. For instance, if some or all of the species are magnetic spins in various orientational states then a constant magnetic field free energy per site is

$$f = a - \mathcal{H}m, \tag{3.41}$$

where m may be expressed in terms of the x_j (see Example 3.4 and Chap. 4).

3.4 Critical Points and Critical Exponents

In this section we return to the subject of critical phenomena in general. This is a continuation of the Sect. 1.9, but we now have the advantage of having studied in detail a magnetic as well as a fluid system. It is supposed that $n = 2$ in the sense of Sect. 1.7 so that there are three extensive variables S, Q_1 and Q_2 which are conjugate to T and the fields ξ_1 and ξ_2 respectively. The field-density formulation of Sect. 1.8 will be used with densities $s = S/Q_2$ and $\rho = Q_1/Q_2$. For convenience we put $\xi_1 = \xi$, $\xi_2 = \phi$. Then, from (1.77) and (1.78) with $\eta = 0$, the Helmholtz free energy density $a(\rho, T)$ is given by

$$a = \phi + \xi\rho, \tag{3.42}$$

$$\mathrm{d}a = -s\mathrm{d}T + \xi\mathrm{d}\rho. \tag{3.43}$$

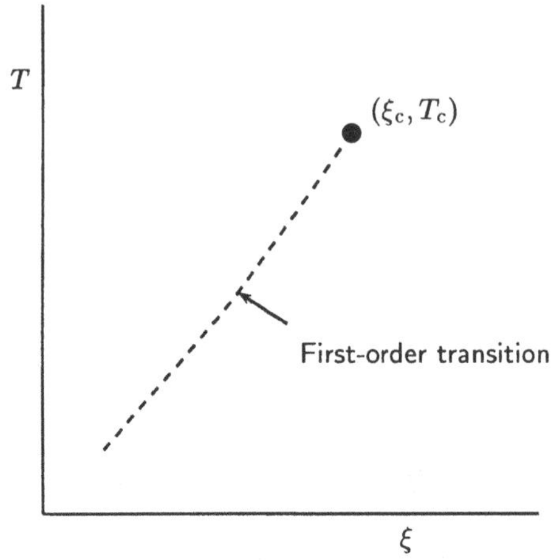

Fig. 3.6. Transition curve in the temperature–field plane.

In the most common type of critical behaviour, a line of first-order phase transitions in the (ξ, T) plane terminates at a critical point (ξ_c, T_c). The situation is shown in Fig. 3.6. The corresponding picture in the (ρ, T) plane is shown in Fig. 3.7. The liquid–vapour equilibrium line in the (T, P) plane of Fig. 1.4 and the interval of discontinuity $(0, T_c)$ on the T-axis of the $(T, \mathcal{H})$ plane of Fig. 3.3 are examples of first-order transition lines terminating at a critical point. The coexistence curve in Fig. 1.3 and the spontaneous magnetization curves in Fig. 3.4 are of the type shown in Fig. 3.7.

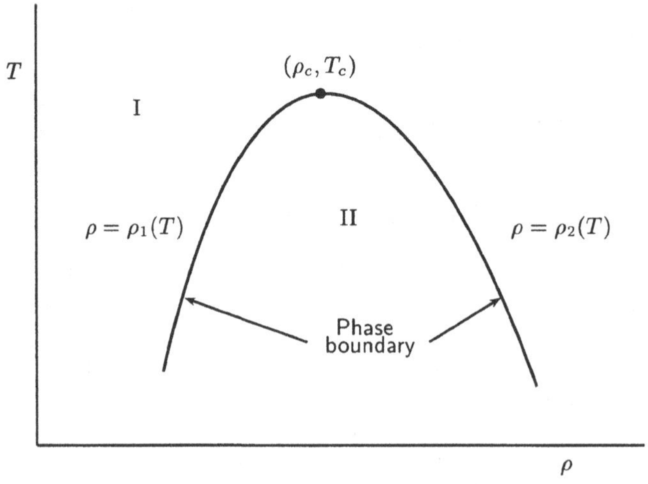

Fig. 3.7. Phase boundary dividing the coexistence region from the single-phase region in the temperature–density plane.

The (ρ, T) plane in Fig. 3.7 is divided into two regions, I and II, separated by a phase boundary which can be expressed either in terms of the single equation

$$T = T^*(\rho) \tag{3.44}$$

or, in terms of the two equations,

$$\rho = \rho_1(T) \qquad \text{and} \qquad \rho = \rho_2(T)\,, \tag{3.45}$$

with $\rho_1(T) < \rho_2(T)$, for $T < T_c$, and $\rho_1(T_c) = \rho_2(T_c) = \rho_c$. Region II is a region of coexistence between two phases. We can define the diameter $\rho_d(T)$ of the coexistence region by

$$\rho_d(T) = \frac{1}{2}[\rho_2(T) + \rho_1(T)] \tag{3.46}$$

and the width $\rho_w(T)$ of the coexistence region by

$$\rho_w(T) = \rho_2(T) - \rho_1(T)\,. \tag{3.47}$$

In order to bring some precision to this rather informal discussion of critical behaviour we need to make a number of assumptions. These are based on Griffiths (1967a). Although they are not universally valid, they apply to a wide class of systems.

(i) $a(\rho, T)$, $s(\rho, T)$ and the field $\xi(\rho, T)$ are continuous functions of ρ and T in a neighbourhood around the critical point (ρ_c, T_c). They are regular in that neighbourhood except possibly on the phase boundary.

(ii) The phase boundary $T = T^*(\rho)$ is a convex upward function of ρ which is regular except possibly at the critical point where it attains its unique maximum.

(iii) The diameter and width of the coexistence region are regular functions of T except possibly at the critical point where $\rho_w(T)$ is more singular than $\rho_d(T)$.

(iv) For fixed T, $a(\rho, T)$ is a convex downward function of ρ, linear in region II.

(v) For fixed ρ, $a(\rho, T)$ is a convex upward function of T.

(vi) The curve

$$\xi = \xi^*(T) = \xi(\rho_1(T), T) = \xi(\rho_2(T), T)\,, \qquad T \le T_c \tag{3.48}$$

in the (ξ, T) plane is a regular function of T, even at the critical point, which can be analytically continued into the region $T > T_c$.

In the simple lattice gas transcription of the Ising model described in Sects. 5.3 and 5.4 the coexistence diameter coincides with the critical isochore $\rho = \rho_c$. This is a special case of the *rectilinear diameter law* which states that $\rho_{\rm d}(T)$ is an asymptotically linear function of T in a neighbourhood of the critical point. It is now known that this law is not generally true. There are a number of models (Hemmer and Stell 1970, Widom and Rowlinson 1970, Mermin 1971a, 1971b, Wheeler 1977) in which the coexistence diameter is singular at the critical point. One type of such model is treated in Sect. 9.12. In fact, it may be shown (Mermin and Rehr 1971), on the basis of some reasonable assumptions, that, unless the model has some special symmetry, as is the case for the Ising model, $\mathrm{d}\rho_{\rm d}/\mathrm{d}T$ is at least as singular as the constant-volume heat capacity c_ρ.

Assumption (iii) would appear to be valid for a wide class of systems, and since

$$\begin{aligned} \rho_1(T) &= \rho_{\rm d}(T) - \frac{1}{2}\rho_{\rm w}(T)\,, \\ \rho_2(T) &= \rho_{\rm d}(T) + \frac{1}{2}\rho_{\rm w}(T)\,, \end{aligned} \tag{3.49}$$

it follows from (iii) that each branch of the phase boundary and functions calculated from them have the same singular behaviour as $T \to T_{\rm c} - 0$. From assumption (iv) it follows that $\xi(\rho, T)$ is a monotonic non-decreasing function of ρ along an isotherm. In region II, ξ has a constant value along an isotherm given, from (vi), by $\xi = \xi^*(T)$. The response function $(\partial\rho/\partial\xi)_T$ is thus non-negative in region I and infinite in region II (but not on the phase boundary, where it is defined by its limiting value when approaching from region I). From assumption (v), $s(\rho, T)$ is a monotonically non-decreasing function of T along an isochore, and the heat capacity $c_\rho = T(\partial s/\partial T)_\rho$ is non-negative.

We now choose the scaled variables

$$\begin{aligned} t &= \left(\frac{T}{T_{\rm c}} - 1\right)\,, \\ \theta &= \rho - \rho_{\rm c} \end{aligned} \tag{3.50}$$

and expand the free energy density in the form

$$a(\rho, T) = a(\rho_{\rm c}, T_{\rm c}) + \theta\xi^*(t) + T_{\rm c}g(\theta, t)\,. \tag{3.51}$$

From this it follows that $g(0, 0) = 0$ and

$$\zeta(\theta, t) = \frac{\partial g}{\partial \theta} = \frac{1}{T_{\rm c}}[\xi(\theta, t) - \xi^*(t)] \tag{3.52}$$

is zero throughout region II and on the phase boundary.

Since we have assumed in (vi) that ξ^* is a regular function of T, and hence of t, it follows that any non-regular behaviour of $a(\rho, T)$ is contained in $g(\theta, t)$. In particular the non-regular behaviour of the entropy s is contained in the function

$$\chi(\theta, t) = -\frac{\partial g}{\partial t} = s + \frac{\theta}{T_{\rm c}}\frac{\mathrm{d}\xi^*}{\mathrm{d}t}\,. \tag{3.53}$$

The modern theory of critical phenomena focuses to a substantial extent on a set of indices called *critical exponents* which describe the behaviour near the critical point (see, for example, Stanley 1987 or Volume 2, Chap. 2). There is now a set of agreed symbols for most of these exponents, and for the system under discussion they are defined in the following way:[1]

(a) On the phase boundary, as $t \to -0$,

$$\rho_{\mathrm{w}} \simeq B_1(-t)^{\beta}. \tag{3.54}$$

In view of assumption (iii) and equations (3.49), an alternative expression, valid on both branches of the phase boundary, is

$$|\theta| \simeq B|t|^{\beta}. \tag{3.55}$$

(b) Along the critical isotherm $t = 0$, as $|\theta| \to 0$,

$$\xi - \xi_{\mathrm{c}} = \xi(\theta, 0) - \xi^*(0) = T_{\mathrm{c}}\zeta(\theta, 0) \simeq C\theta|\theta|^{\delta-1}. \tag{3.56}$$

(c) The limiting properties of the compressibility of a fluid or the susceptibility of a magnet are those of

$$\varphi_T = T\left(\frac{\partial\rho}{\partial\xi}\right)_T = \left[\frac{T_{\mathrm{c}}}{T}\left(\frac{\partial\zeta}{\partial\theta}\right)_t\right]^{-1}. \tag{3.57}$$

On the critical isochore $\theta = 0$ ($\rho = \rho_{\mathrm{c}}$), as $t \to +0$,

$$\varphi_T \simeq D|t|^{-\gamma} \tag{3.58}$$

and on the phase boundary, as $t \to -0$,

$$\varphi_T \simeq \begin{cases} D_1|t|^{-\gamma'} & \text{along } \rho = \rho_1(T)\ , \\ D_2|t|^{-\gamma'} & \text{along } \rho = \rho_2(T)\ . \end{cases} \tag{3.59}$$

(d) For the heat capacity at constant density

$$c_\rho = k_{\mathrm{B}}T\left(\frac{\partial s}{\partial T}\right)_\rho \simeq k_{\mathrm{B}}T\left(\frac{\partial\chi}{\partial t}\right)_\theta. \tag{3.60}$$

On the critical isochore $\theta = 0$, as $t \to +0$,

$$c_\rho \simeq K|t|^{-\alpha} \tag{3.61}$$

and on the phase boundary, as $t \to -0$,

$$c_\rho \simeq \begin{cases} K_1|t|^{-\alpha'} & \text{along } \rho = \rho_1(T)\ , \\ K_2|t|^{-\alpha'} & \text{along } \rho = \rho_2(T)\ . \end{cases} \tag{3.62}$$

[1] The asymptotic relation symbols $\simeq$ and $\sim$ are defined as follows:
$f(x) \simeq g(x)$ as $x \to x_0$ means that $f(x)/g(x) \to 1$ as $x \to x_0$,
$f(x) \sim g(x)$ as $x \to x_0$ means that $f(x)/g(x) \to$ some constant $\neq 0$ as $x \to x_0$.

From equation (3.61)

$$\lim_{|t|\to 0} |t|^\alpha c_\rho = K\,, \tag{3.63}$$

where K is finite and non-zero. Thus if c_ρ approaches a finite limit as $T \to T_c + 0$ the critical exponent $\alpha = 0$. A finite discontinuity in c_ρ as the path in the (ρ, T) plane passes through the point (ρ_c, T_c) onto either the $\rho = \rho_1$, or the $\rho = \rho_2$ branch thus implies $\alpha = \alpha' = 0$. Now suppose that c_ρ behaves as $\ln|t|$, as in the exact theory of the $d = 2$ Ising model (see Chap. 8). Then, since $|t|^n \ln|t| \to 0$ as $|t| \to 0$ for any $n > 0$, it follows again that $\alpha = \alpha' = 0$. It thus appears that zero values for α (or any other critical exponent) can correspond to quite different types of singularity. A more detailed theory of critical exponents and scaling is needed to distinguish these cases and this is presented in Volume 2, Chap. 2.

3.5 Scaling and Exponent Relations

Suppose that in a neighbourhood of the critical point in region I we have the asymptotic form

$$\zeta \simeq \zeta(\theta, t) \tag{3.64}$$

for the dependence of ζ on θ and t. It can be seen that the asymptotic forms (3.55) and (3.56) are preserved under the transformation

$$\begin{aligned} &\theta \to \lambda\theta\,, \\ &t \to \lambda^{y_t} t\,, \\ &\zeta \to \lambda^{y_\zeta}\zeta\,, \end{aligned} \tag{3.65}$$

for all $\lambda > 0$, if

$$y_t = 1/\beta\,, \qquad y_\zeta = \delta\,. \tag{3.66}$$

So we now suppose that (3.64) can also be transformed in the same way. This means that

$$\lambda^\delta \zeta(\theta, t) \simeq \zeta(\lambda\theta, \lambda^{1/\beta} t)\,. \tag{3.67}$$

Setting $\lambda = |\theta|^{-1}$ we now have

$$\zeta \simeq |\theta|^\delta \zeta\left(\pm 1, \frac{t}{|\theta|^{1/\beta}}\right)\,, \tag{3.68}$$

where $+1$ and -1 correspond respectively to those parts of the neighbourhood in which $\theta > 0$ and $\theta < 0$. Given that ξ is a regular monotonic non-decreasing function of ρ along an isotherm in region I it follows that

$$\zeta\left(\pm 1, \frac{t}{|\theta|^{1/\beta}}\right) = \pm\zeta_1\left(\frac{t}{|\theta|^{1/\beta}}\right) \tag{3.69}$$

and we have

$$\zeta \simeq \theta|\theta|^{\delta-1}\zeta_1\left(\frac{t}{|\theta|^{1/\beta}}\right) . \tag{3.70}$$

Comparing (3.56) and (3.70) we see that $C = \zeta_1(0)$. Since $\zeta = 0$ on the phase boundary, if $\zeta_1(x)$ has the root x_0, then the amplitude B of (3.55) is given by $B = -x_0$.

We now differentiate (3.67) with respect to θ to give

$$\lambda^{1-\delta}\varphi_T \sim \varphi_T(\lambda\theta, \lambda^{1/\beta}t) . \tag{3.71}$$

Setting $\lambda = |t|^{-\beta}$ we have

$$\varphi_T \sim |t|^{\beta(1-\delta)}\varphi_T\left(\frac{\theta}{|t|^{\beta}}, \pm 1\right) , \tag{3.72}$$

where $+1$ and -1 now correspond respectively to $t > 0$ and $t < 0$. On the critical isochore for $t > 0$

$$\varphi_T \simeq |t|^{\beta(1-\delta)}\varphi_T(0,1) . \tag{3.73}$$

On the phase boundary, for $\theta < 0$,

$$\varphi_T \simeq |t|^{\beta(1-\delta)}\varphi_T(-B,-1) \tag{3.74}$$

and on the phase boundary, for $\theta > 0$,

$$\varphi_T \simeq |t|^{\beta(1-\delta)}\varphi_T(B,-1) . \tag{3.75}$$

Comparing equations (3.73)–(3.75) with (3.58)–(3.59) we see that

$$\gamma = \gamma' \tag{3.76}$$

and $D = \varphi_T(0,1)$, $D_1 = \varphi_T(-B,-1)$ and $D_2 = \varphi_T(B,-1)$. We further see that

$$\gamma = \beta(\delta - 1) , \tag{3.77}$$

which is the *Widom scaling law*, (Widom 1964). To obtain the heat capacity exponents, defined in (3.61)–(3.62) we need the asymptotic scaling relationship for the free energy. From (3.52)

$$g(\theta,t) = \int \zeta(\theta,t)\mathrm{d}\theta + w(t) . \tag{3.78}$$

We shall assume that $w(t)$ is a regular function of t in a neighbourhood of the critical point. Then, from (3.67), for the singular part g_{sing} of g, obtained by subtracting off w, we have

$$\lambda^{\delta+1}g_{\mathrm{sing}} \simeq g_{\mathrm{sing}}(\lambda\theta, \lambda^{1/\beta}t) . \tag{3.79}$$

Differentiating twice with respect to t we have

$$\lambda^{\delta+1-2/\beta}c_p \sim c_p(\lambda\theta, \lambda^{1/\beta}t). \tag{3.80}$$

Setting $\lambda = |t|^{-\beta}$ we have

$$c_\rho \simeq |t|^{\beta(\delta+1)-2} c_\rho \left(\frac{\theta}{|t|^\beta}, \pm 1 \right) , \tag{3.81}$$

where $+1$ and -1 again correspond to $t > 0$ and $t < 0$ respectively. On the critical isochore, for $t > 0$,

$$c_\rho \simeq |t|^{\beta(\delta+1)-2} c_\rho(0,1) , \tag{3.82}$$

on the phase boundary, for $\theta < 0$,

$$c_\rho \simeq |t|^{\beta(\delta+1)-2} c_\rho(-B,-1) \tag{3.83}$$

and on the phase boundary, for $\theta > 0$,

$$c_\rho \simeq |t|^{\beta(\delta+1)-2} c_\rho(B,-1) . \tag{3.84}$$

Comparing equations (3.82)–(3.84) with (3.61)–(3.62) we see that

$$\alpha = \alpha' \tag{3.85}$$

and $E = c_\rho(0,1)$, $E_1 = c_\rho(-B,-1)$ and $E_2 = c_\rho(B,-1)$. We further see that

$$\alpha = 2 - \beta(\delta + 1) . \tag{3.86}$$

From (3.77) and (3.86) we have

$$\alpha + 2\beta + \gamma = 2 , \tag{3.87}$$

which is the *Essam–Fisher scaling law* (Essam and Fisher 1963).

There is one difficulty arising in connection with the Ising model ferromagnet which can now be resolved. Here $\rho = m$ so the constant-density heat capacity is c_m. However, the heat capacity usually obtained is c_0 (i.e. c_ξ for $\xi = \mathcal{H} = 0$). For $T > T_c$ the 'critical isochore' $m = 0$ corresponds to $\mathcal{H} = 0$ and hence $c_0 = c_m$. For $T < T_c$, it follows from the result in Example 1.3 that

$$c_0 - c_m = k_B T \left(\frac{\partial m}{\partial T} \right)^2_{\mathcal{H}=0} \left(\frac{\partial m}{\partial \mathcal{H}} \right)^{-1} , \tag{3.88}$$

where c_0 and c_m are 'per molecule' heat capacities. However $\mathcal{H} = 0$ corresponds to the phase boundary where $\partial m / \partial T$ varies as $|t|^{\beta-1}$ by (3.55) and $\partial m / \partial \mathcal{H}$ as $|t|^{-\gamma}$ by (3.59) and (3.76). Hence $c_0 - c_m$ varies as $|t|^{2\beta+\gamma-1} = |t|^{-\alpha}$ by the Essam–Fisher relation (3.87). It follows that c_0, like c_m, varies as $|t|^{-\alpha}$.

It should be emphasized that, in general, the exponent of c_ξ is not equal to α. In fact (see Volume 2, Chap. 2), it is equal to γ, defined by (3.58) unless certain symmetry relations are satisfied. If they are, then the exponents for c_ρ and c_ξ are equal, as has just been shown for the Ising model.

3.6 Classical Critical Exponents

We now apply our analysis to the two classical models of a phase transition considered in Sects. 3.1 and 1.6 respectively. We shall see that these produce the same *classical* (or *mean-field*) values for the critical exponents.

3.6.1 The Ising Model Ferromagnet: Mean-Field Approximation

Here the treatment is simplified by the $m \to -m$, $\mathcal{H} \to -\mathcal{H}$ symmetry expressed in the relations of Sect. 2.4 and illustrated by Figs. 3.1–3.4. For the Ising model $Q_1 = \mathcal{M}$ and $Q_2 = N$ so that, in the notation introduced at the beginning of Sect. 3.4, $\rho = m$ and its conjugate field $\xi = \mathcal{H}$. From (3.43),

$$\phi = a(T, m) - m\mathcal{H} = f\,, \tag{3.89}$$

where, provided that m has its equilibrium value for the given $\mathcal{H}$, f is the constant magnetic field free energy density. In Sect. 3.1 the phase transition conditions were deduced geometrically and it was found that the transition line is the interval $(0, T_c)$ of the $\mathcal{H} = 0$ axis so that the critical point field $\mathcal{H}_c = 0$. The coexistence densities are $\rho_1 = m_s$ and $\rho_2 = -m_s$, and the critical density $m_c = 0$. Since $a(T, -m_s) = a(T, m_s)$ and $\mathcal{H} = 0$ on the transition line it follows from (3.89) that f is continuous across the transition line. Hence the analytic conditions (1.90) for phase equilibrium are satisfied.

Since $\mathcal{H}_c = 0$ and $m_c = 0$, the scaled variables θ and ζ, defined by (3.50) and (3.52), are given by $\theta = m$ and $\zeta = \mathcal{H}/T_c$. The critical exponents introduced in (3.55)–(3.62) can be obtained from relations given in Sect. 3.1. From (3.17)

$$\zeta \simeq tm + \frac{1}{3}m^3\,. \tag{3.90}$$

As this is the asymptotic form corresponding to (3.64) we apply the transformation

$$\begin{aligned} m &\to \lambda m\,,\\ t &\to \lambda^{1/\beta} t\,,\\ \zeta &\to \lambda^{\delta}\zeta \end{aligned} \tag{3.91}$$

to give

$$\zeta \simeq \lambda^{1-\delta+\beta^{-1}} tm + \frac{1}{3}\lambda^{3-\delta} m^3\,. \tag{3.92}$$

Since this must be equivalent to (3.90) for all $\lambda > 0$ it follows that

$$\beta = \tfrac{1}{2}\,, \tag{3.93}$$

$$\delta = 3\,. \tag{3.94}$$

Alternatively, these values for β and δ follow from (3.13) and (3.17) respectively with $T = T_c$. The relation (3.18) gives

$$\gamma = 1 \tag{3.95}$$

and hence the Widom scaling law (3.77) is obeyed. Since $m_s \to 0$ as $T \to T_c - 0$, it follows from the expression given in Example 3.3 for $(\partial m/\partial\mathcal{H})_{\mathcal{H}=0}$ where $T < T_c$ that $\gamma' = 1$. Hence the exponent relation (3.76) is satisfied.

From the arguments given at the end of Sect. 3.5, the zero-field heat capacity c_0 may be used instead of c_m on both sides of the transition point. Since c_0 remains finite (in fact zero) as $T \to T_c + 0$,

$$\alpha = 0 , \tag{3.96}$$

which, with (3.93) and (3.95), satisfies the Essam–Fisher scaling law (3.87). Since, by (3.16), c_0 also remains finite as $T \to T_c - 0$, $\alpha' = 0$, satisfying the exponent relation (3.85).

In contrast to these mean-field results the accurate critical exponents for the Ising model, change according to dimension d of the system.[2] For $d = 2$, the exact solution (Sect. 8.9) gives $\alpha = 0$, $\beta = \frac{1}{8}$, $\gamma = \frac{7}{4}$, with $\delta = 15$ following from the Widom scaling law (3.77) and the value for α in this case corresponding to a logarithmic singularity. For $d = 3$ Le Guillou and Zinn-Justin (1980) used field-theory to predict that $\alpha = \alpha' = 0.110(5)$, $\beta = 0.325(2)$ and $\gamma = \gamma' = 1.241(2)$, with $\delta = 4.80(2)$ given by the Widom scaling law (3.77). High and low-temperature series expansions give good agreement with these values (see, for example, Volume 2, Chap. 7). It was suggested by Domb and Sykes (1961) that $\gamma = \frac{5}{4}$ and by Essam and Fisher (1963) that $\beta = \frac{5}{16}$. This set of attractively simple values is completed, using the Essam–Fisher and Widom scaling laws, with $\alpha = \frac{1}{8}$, $\delta = 5$.

3.6.2 The Van der Waals Gas

Here the deduction of the critical exponents for $T < T_c$ is harder than for the Ising model because of the lack of symmetry in the (T, ρ) or (T, ξ) plane. We now use the parameter b as the standard volume v_0 of (1.54) and the critical parameters of (1.52) have the form

$$T_c = \tfrac{8}{27}\varepsilon , \qquad v_c = 3 , \qquad P_c = \tfrac{1}{27}\varepsilon , \tag{3.97}$$

where $\varepsilon = c/b$ is of the dimensions of energy. Using 'per molecule' densities as in Sect. 1.6, $Q_1 = V$, $Q_2 = M$, $\rho = v$ and the conjugate field $\xi = -P$ and from (3.43) and (1.47),

$$\phi = a + Pv = \mu . \tag{3.98}$$

In terms of the scaled variables t and θ defined in (3.50) the van der Waals equation (1.43) takes the form

$$\frac{P}{P_c} = \frac{8P}{T_c} = \frac{8(1+t)}{\theta + 2} - \frac{27}{(\theta + 3)^2} \tag{3.99}$$

and along the critical isochore

$$-\left(\frac{\partial P}{\partial v}\right)_{v=3} = -\left(\frac{\partial P}{\partial \theta}\right)_{\theta=0} = \frac{1}{4}(T - T_c) . \tag{3.100}$$

[2] The factors which affect the values of critical exponents for any model and the related idea of *universality classes* is discussed in Volume 2, Chap. 2.

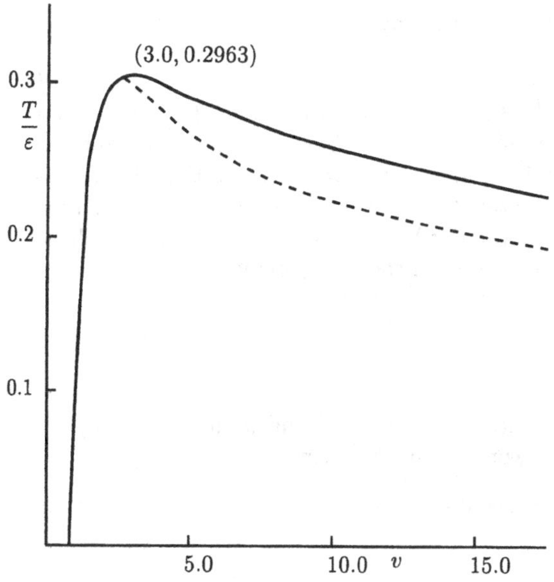

Fig. 3.8. The phase boundary and the coexistence diameter for the van der Waals fluid.

With (3.57) and (3.58) this gives $\gamma = 1$. Using (3.99) we define

$$\frac{\Delta P}{T_c} = \frac{P(t,\theta) - P(t,0)}{T_c} = -\frac{\theta}{2(2+\theta)}\left[t + \frac{\theta^2}{4(\theta+3)^2}\right] \tag{3.101}$$

and, on the critical isotherm $T = T_c$ $(t = 0)$,

$$\zeta = \frac{P(0,\theta) - P_c}{T_c} = -\frac{1}{144}\theta^3 + \mathrm{O}(\theta^4), \tag{3.102}$$

giving with (3.56), $\delta = 3$. Now, from (1.44) and (3.99),

$$\mu = T\frac{v}{v-1} - \frac{2\varepsilon}{v} - T\ln(v-1) + \psi(T), \tag{3.103}$$

where the last term depends on T alone. This then gives

$$\begin{aligned}\frac{\Delta\mu}{T_c} &= \frac{\mu(t,\theta) - \mu(t,0)}{T_c}\\ &= -(1+t)\left\{\frac{\theta}{2(\theta+2)} + \ln\left[\tfrac{1}{2}(\theta+2)\right]\right\} + \frac{9\theta}{4(\theta+3)}\\ &= -\tfrac{3}{4}\theta\left[t\left(1 - \tfrac{1}{3}\theta + \tfrac{5}{36}\theta^2 - \tfrac{1}{16}\theta^3 + \cdots\right) + \tfrac{1}{36}\theta^2 - \tfrac{11}{432}\theta^3 \cdots\right].\end{aligned} \tag{3.104}$$

To satisfy the phase equilibrium conditions (1.41) for $T < T_c$ $(t < 0)$, values of θ_1 and θ_2 must be found such that

$$\begin{aligned}\Delta P(t,\theta_1) &= \Delta P(t,\theta_2),\\ \Delta\mu(t,\theta_1) &= \Delta\mu(t,\theta_2).\end{aligned} \tag{3.105}$$

To order $|t|^{\frac{1}{2}}$ this can be achieved by putting

$$\theta_1 = -\theta_2 = 6|t|^{1/2} \tag{3.106}$$

and hence, from (3.55), $\beta = \frac{1}{2}$. Thus the Widom relation (3.77) is obeyed and (3.106) shows that both branches of the coexistence curve exhibit the same singular behaviour. This is in accordance with assumption (iii) above concerning the coexistence diameter. It can, in fact, be proved directly that this assumption is correct for the van der Waals gas. Fig. 3.8 shows the phase boundary and the coexistence diameter in the (v, T) plane.

Examples

3.1 Show that the mean-field relation between m and $\mathcal{H}$ for the Ising model ferromagnet can be expressed in the form

$$m = \tanh h', \qquad h' = (\mathcal{H} + zJm)/T.$$

(This is a *Weiss field equation.*)

3.2 For the Ising model ferromagnet in the mean-field approximation, show that the constant magnetic field free energy per site f and the reduced entropy per site s can be expressed as

$$\begin{aligned} f &= -T\ln(2\cosh h') + \frac{1}{2}zJ\tanh^2 h', \\ s &= \ln(2\cosh h') - h'\tanh h', \end{aligned}$$

where h' is defined as in question 1.

3.3 For the Ising model ferromagnet in the mean-field approximation, show that for $\mathcal{H} = 0$, $T < T_c$,

$$\frac{\partial m}{\partial \mathcal{H}} = \frac{1 - m_s^2}{T - T_c(1 - m_s^2)} \, .$$

Show that for $T \ll T_c$ the expression on the right-hand side is approximately $(4/T)\exp(-2T_c/T)$ and thus tends to zero with T.

3.4 For the spin-1 Ising model ferromagnet in the mean-field approximation, show that a non-equilibrium constant magnetic field free energy per site can be written in the form

$$\begin{aligned} f = T[x_1\ln(x_1) + x_2\ln(x_2) + (1 - x_1 - x_2)\ln(1 - x_1 - x_2)] \\ -\frac{1}{2}zJ(x_1 - x_2)^2 - (x_1 - x_2)\mathcal{H} \, . \end{aligned}$$

Here x_1 and x_2 are fractions of spins with $\sigma = +1$ and $\sigma = -1$ respectively and have the character of order variables. From the conditions for f to have a stationary value with respect to x_1 and x_2, show that the spontaneous relative magnetization is given by

$$m_s = (x_1 - x_2)_{\text{equilibrium}} = \frac{y - y^{-1}}{1 + y + y^{-1}},$$

where $y = \exp(zJm_s/T)$. Show that the critical temperature $T_c = 2zJ/3$.

4. Antiferromagnets and Other Magnetic Systems

4.1 The One-Dimensional Antiferromagnet

The Ising model ferromagnet can be transformed to an antiferromagnet by changing J to $-J$ ($J > 0$) in the configurational energy expression (2.63). The analysis of the one-dimensional case in Chap. 2 remains valid, and replacing J by $-J$ the zero-field correlation relation (2.88) becomes

$$\langle \sigma_i \sigma_{i+k} \rangle = (-1)^k [\tanh(J/T)]^k, \tag{4.1}$$

which is the same as in the ferromagnet for even k, but is reversed in sign for odd k. At low temperatures the regions of local order are blocks of alternate spins, and nearest-neighbour pairs of parallel spins are the local fluctuations which form the boundaries between these regions. As will be shown in Sect. 4.2, the equilibrium state at $\mathcal{H} = 0$ for the antiferromagnet can be obtained from that of the ferromagnet by reversing every other spin, (see Fig. 4.1). The values of U, F, and hence S, at $\mathcal{H} = 0$ are unaltered by the change $J \to -J$, as can be seen from (2.81) and (2.82). The zero-field heat capacity given by (2.83) is also unaffected.

For $\mathcal{H} \neq 0$, there is no simple connection between the states of the ferromagnet and the antiferromagnet although the basic symmetry relations of Sect. 2.4 remain valid. In the ferromagnet the field reinforces the interaction in promoting parallel spins, but in the antiferromagnet the field tends to upset the antiparallel nearest-neighbour spin pairs favoured by the interaction. The latter effect can be clearly seen at $T = 0$, and to investigate this we replace J by $-J$ in (2.99) to obtain the relation

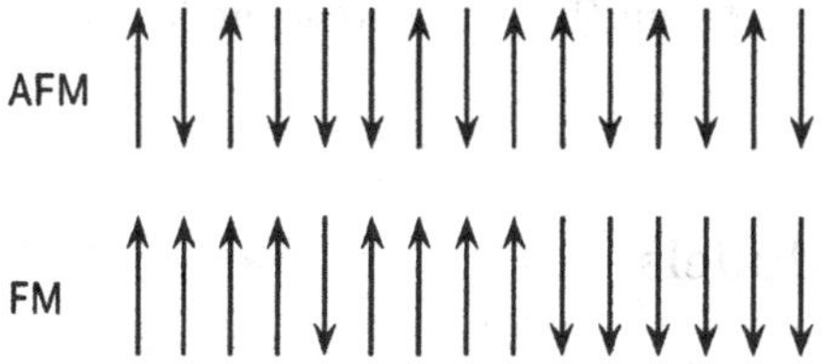

Fig. 4.1. One-dimensional antiferromagnet (*top row*) and ferromagnet (*lower row*).

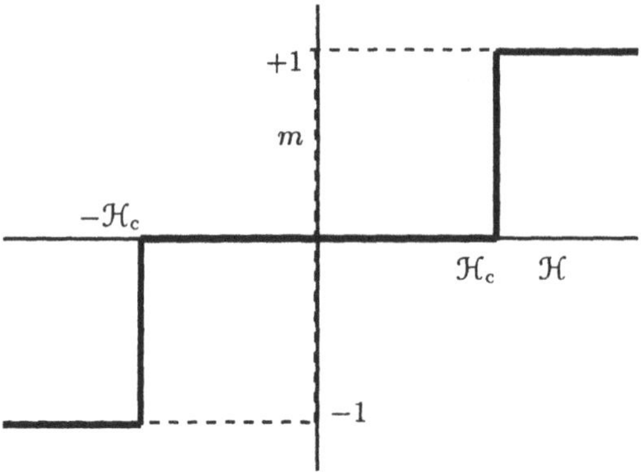

Fig. 4.2. Magnetization as a function of field at zero temperature: full lines the antiferromagnet, broken lines the ferromagnet.

$$m = \frac{\sinh(\mathcal{H}/T)}{[\sinh^2(\mathcal{H}/T) + \exp(4J/T)]^{1/2}} \tag{4.2}$$

for the relative magnetization of the $d = 1$ antiferromagnet. When $T \ll |\mathcal{H}|$,

$$\sinh^2(\mathcal{H}/T) \simeq \frac{1}{4}\exp(2|\mathcal{H}|/T) \tag{4.3}$$

and thus, from (4.2), as $T/J \to 0$,

$$m \to \begin{cases} -1, & \mathcal{H} < -\mathcal{H}_c, \\ 0, & -\mathcal{H}_c < \mathcal{H} < \mathcal{H}_c, \\ 1, & \mathcal{H} > \mathcal{H}_c, \end{cases} \tag{4.4}$$

where $\mathcal{H}_c = 2J$ is the *critical field.* The behaviour of the antiferromagnet on the $T = 0$ axis is thus quite different from that of the ferromagnet (see Fig. 4.2). Since all thermodynamic functions are analytic for $T > 0$ in the $d = 1$ case, the transition points $\pm\mathcal{H}_c$ on the $T = 0$ axis are isolated singularities. The zero-field susceptibility of the one-dimensional antiferromagnet also contrasts with that of the ferromagnet. Changing J to $-J$ $(J > 0)$ in (2.101) yields

$$\frac{\partial m}{\partial \mathcal{H}} = \frac{1}{T\exp(2J/T)} \tag{4.5}$$

for the antiferromagnet. Thus $\partial m/\partial\mathcal{H} \to 0$ as $T/J \to 0$ or $T/J \to \infty$ and has a maximum at $T = 2J$. The low susceptibility values thus found for $T \ll J$ are another consequence of the 'conflict' between the antiferromagnetic spin interaction and the magnetic field.

4.2 Antiferromagnetic Ising Models

So far it has been assumed that, if boundary effects are disregarded, all lattice sites in the Ising model are 'equivalent'. We must now look more carefully at what this term implies. Firstly, it means that, as for all regular lattices listed in Appendix A.1, the unoccupied lattice looks exactly the same when viewed from any site. That is each site has the same number z_p of p-th neighbour sites for any p. It also means that the interaction of a member of any species j with an external field is the same for all sites and the interaction energy of a member of species j and a member of species ℓ depends only on the relative position of the lattice sites they occupy.[1] This condition could easily be extended from two-body (pairwise) interactions to n-body interactions.

The equivalence of the lattice sites means that it may be possible to divide the lattice into two equal sublattices, labelled $\mathfrak{a}$ and $\mathfrak{b}$. The sublattices are *interpenetrating* if all nearest-neighbour sites of any $\mathfrak{a}$-site are $\mathfrak{b}$-sites and vice versa. A lattice which can be divided into two equal interpenetrating sublattices is called *loose-packed* whereas a lattice for which division of this kind is not possible is *close-packed*. The $d = 1$ lattice is loose-packed since it can be divided into interpenetrating sublattices simply by labelling alternate sites $\mathfrak{a}$ and $\mathfrak{b}$. The $\mathfrak{a}$ and $\mathfrak{b}$ sublattices for the loose-packed square, honeycomb, simple cubic, body-centred cubic and diamond lattices are shown in Appendix A.1.

We shall use the constant magnetic field distribution exclusively in this chapter. The antiferromagnet with nearest-neighbour interactions only on a loose-packed lattice will be called 'simple', and by adapting (2.66) its Hamiltonian may be written as

$$\widehat{H} = J \sum_{\{i,j\}}^{(\text{n.n.})} \sigma_i^{(\mathfrak{a})} \sigma_j^{(\mathfrak{b})} - \mathcal{H}\Big(\sum_{\{i\}} \sigma_i^{(\mathfrak{a})} + \sum_{\{j\}} \sigma_j^{(\mathfrak{b})}\Big), \tag{4.6}$$

$\sigma_i^{(\mathfrak{a})}$ and $\sigma_j^{(\mathfrak{b})}$ respectively denoting the spin variables for a site i of sublattice $\mathfrak{a}$ and a site j of sublattice $\mathfrak{b}$. The parameter J of (2.66)has been changed to $-J$ $(J > O)$ and the fact that all nearest-neighbour pairs consist of an $\mathfrak{a}$-site and a $\mathfrak{b}$-site has been used. For $\mathcal{H} = 0$ (4.6) can be written

$$\widehat{H} = E = -J \sum_{\{i,j\}}^{(\text{n.n.})} \sigma_i^{(\mathfrak{a})} (-\sigma_j^{(\mathfrak{b})}) . \tag{4.7}$$

This is the Hamiltonian of a ferromagnet with spin variables $\sigma_i^{(\mathfrak{a})}$ and $-\sigma_j^{(\mathfrak{b})}$. There is a one-to-one correspondence between microstates of equal energy in the zero-field ferromagnet and antiferromagnet. These microstates are connected by the operation of reversing the spins on one sublattice, leaving those on the other sublattice unchanged, while changing the sign of the interaction parameter. It follows that in zero field the partition function for the antiferromagnet is equal to that for the ferromagnet. Hence, as we have seen for the one-dimensional case, the free energy, entropy, internal energy and heat

[1] 'Relative position' involves the direction as well as the length of the lattice vector connecting the two sites.

capacity are the same. The critical temperature T_c, where there is a heat capacity singularity at $\mathcal{H} = 0$, is the Curie point for the ferromagnet, and is termed the *Néel point* for the antiferromagnet. Experimentally, the phenomenon of antiferromagnetism is commoner than that of ferromagnetism (Martin 1967).

The type of long-range order in the antiferromagnet below the critical temperature is quite different from that in the ferromagnet. We define a relative magnetization for each sublattice by the relations

$$m_{\mathfrak{a}} = (2/N) \sum_{\{i\}} \sigma_i^{(\mathfrak{a})}, \qquad m_{\mathfrak{b}} = (2/N) \sum_{\{j\}} \sigma_j^{(\mathfrak{b})}, \tag{4.8}$$

where the sums are taken over all sites of sublattice $\mathfrak{a}$ and sublattice $\mathfrak{b}$ respectively. The overall relative magnetization m is then given by

$$m = N^{-1}\Big(\sum_{\{i\}} \sigma_i^{(\mathfrak{a})} + \sum_{\{j\}} \sigma_j^{(\mathfrak{b})}\Big) = \frac{1}{2}(m_{\mathfrak{a}} + m_{\mathfrak{b}}). \tag{4.9}$$

For the zero-field ferromagnet above the critical temperature T_c, $m_{\mathfrak{a}} = m_{\mathfrak{b}} = m = 0$ whereas below T_c, $m_{\mathfrak{a}} = m_{\mathfrak{b}} = \pm m_s$. Now, from (4.7), the zero-field equilibrium state of the antiferromagnet can be derived from that of the ferromagnet by reversing the magnetization of one of the sublattices. Hence above the Néel point ($T > T_c$) the magnetizations of both sublattices are zero, but below it ($T < T_c$) we have either

$$\left.\begin{array}{ll} m_{\mathfrak{a}} = m_s\,, & m_{\mathfrak{b}} = -m_s\,, \\ & \text{or} \\ m_{\mathfrak{a}} = -m_s\,, & m_{\mathfrak{b}} = m_s\,, \end{array}\right\} \quad m = 0\,. \tag{4.10}$$

The magnetizations of the two sublattices are equal and opposite, giving zero overall magnetization. This is the first example we meet of the important co-operative phenomenon of *sublattice ordering*. The degeneracy displayed in (4.10) for $\mathcal{H} = 0$, $T < T_c$, parallels the degeneracy $m = \pm m_s$ of the ferromagnet. As $T/J \to 0$, the Ising model antiferromagnet approaches the state $m_{\mathfrak{a}} = \pm 1$, $m_{\mathfrak{b}} = \mp 1$ where all nearest-neighbour spin pairs are antiparallel (see Fig. 4.3). Since, from (4.10), m is a continuous function of $\mathcal{H}$ as the interval $(0, T_c)$ of the $\mathcal{H} = 0$ axis is crossed, this interval is no longer a first-order transition line, as it is in the ferromagnet.

For $\mathcal{H} \neq 0$ there is no correspondence between the states of the antiferromagnet and those of the ferromagnet. From the equivalence of sites in the antiferromagnetic model it may be expected that, if either the magnetic field or thermal disordering is strong enough to offset the antiferromagnetic spin interaction, the equilibrium state will be one in which $m_{\mathfrak{a}} = m_{\mathfrak{b}} = m$. Such a state is called *paramagnetic* as opposed to the antiferromagnetic state in which the spins on sublattices $\mathfrak{a}$ and $\mathfrak{b}$ are oppositely oriented. The transition curve in the $(T, \mathcal{H})$ plane on which the change of symmetry from the paramagnetic state to the antiferromagnetic state takes place for $d > 1$ has to be investigated by approximation or series methods even for $d = 2$, since no exact results are available for either the ferromagnet or antiferromagnet with $\mathcal{H} \neq 0$ and $T > 0$. However, from the zero-field results above, the transition curve meets the $\mathcal{H} = 0$ axis at $T = T_\mathrm{c}$ and from the symmetry relations of Sect. 2.4 it must be symmetrical about $\mathcal{H} = 0$. Also, the points where it meets the $T = 0$ axis can be determined quite easily. At $T = 0$, the equilibrium state is one of minimum enthalpy, as shown at the end of Sect. 2.5. In the $T = 0$ paramagnetic state, all $\sigma_i^{(\mathrm{a})} = \sigma_j^{(\mathrm{b})} = \pm 1$, according as $\mathcal{H} \gtrless 0$ and the configurational energy is $\frac{1}{2}zNJ$. In the $T = 0$ antiferromagnetic state, on the other hand, all $\sigma_i^{(\mathrm{a})} = 1$ and all $\sigma_j^{(\mathrm{b})} = -1$ (or vice versa); the configurational energy is $-\frac{1}{2}zNJ$ and the magnetization is zero. Thus, if H_P and H_A denote the enthalpies in the paramagnetic and antiferromagnetic states respectively, by (1.32),

$$H_\mathrm{P} = \frac{1}{2}zNJ - N|\mathcal{H}|\,, \qquad H_\mathrm{A} = -\frac{1}{2}zNJ\,, \tag{4.11}$$

and $H_\mathrm{A} < H_\mathrm{P}$, making the antiferromagnetic state stable if

$$-\mathcal{H}_\mathrm{c} < \mathcal{H} < \mathcal{H}_\mathrm{c} = zJ. \tag{4.12}$$

The last relation defines the *critical field* $\mathcal{H}_\mathrm{c}$ and the zero-temperature values for the magnetization are the same as those for the one-dimensional case given in (4.4) and Fig. 4.2 with the coordination number z replacing the $d = 1$ value $z = 2$. However, for $d > 1$, the transition points $\mathcal{H} = \pm\mathcal{H}_\mathrm{c}$ at $T = 0$ are no longer isolated singularities.

To sum up our deductions from the basic symmetries of the antiferromagnetic model on a loose-packed lattice:

(i) At $\mathcal{H} = 0$ and $T < T_\mathrm{c}$, by simultaneous reversal of the signs of all spins on one sublattice and of the sign of the a–b interaction, a correspondence is established between the states of the antiferromagnet and those of a ferromagnet in which $\langle\sigma_i\rangle$ has the same value for all lattice sites i.

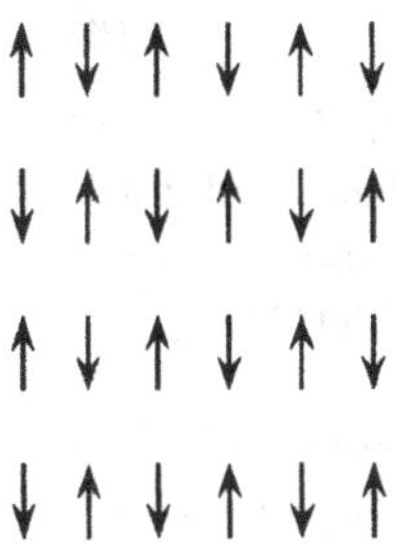

Fig. 4.3. The antiferromagnetic ground state for the square lattice in zero field.

(ii) For sufficiently large $|\mathcal{H}|$ or T the antiferromagnet may go into a paramagnetic state in which $\langle\sigma_i\rangle$ has the same value for all lattice sites i.

It is useful to generalize the simple antiferromagnetic model by introducing a ferromagnetic interaction (i.e. tending to spin parallelism) between spins on the same sublattice. In a loose-packed lattice (see Appendix A.1) both sites of any second-neighbour pair lie on the same sublattice, and we introduce an energy term $-J_2\sigma_i\sigma_j$ for each such pair. The Hamiltonian (4.6) is modified to

$$\begin{aligned}\widehat{H} = J_1 \sum_{\{i,j\}}^{(\text{n.n.})} \sigma_i^{(\text{a})}\sigma_j^{(\text{b})} - J_2 \sum_{\{i,j\}}^{(\text{s.n.})} \sigma_i^{(\text{a})}\sigma_j^{(\text{b})} - J_2 \sum_{\{i,j\}}^{(\text{s.n.})} \sigma_i^{(\text{b})}\sigma_j^{(\text{b})} \\ - \mathcal{H}\Big(\sum_{\{i\}} \sigma_i^{(\text{a})} + \sum_{\{j\}} \sigma_j^{(\text{b})}\Big),\end{aligned} \tag{4.13}$$

'(s.n.)' denoting the sum over second-neighbour pairs and both J_1 and J_2 are taken to be positive. At $\mathcal{H} = 0$ we thus have

$$\begin{aligned}\widehat{H} &= J_1 \sum_{\{i,j\}}^{(\text{n.n.})} \sigma_i^{(\text{a})}\sigma_j^{(\text{b})} - J_2 \sum_{\{i,j\}}^{(\text{s.n.})} \sigma_i^{(\text{a})}\sigma_j^{(\text{a})} - J_2 \sum_{\{i,j\}}^{(\text{s.n.})} \sigma_i^{(\text{b})}\sigma_j^{(\text{b})} \\ &= - J_1 \sum_{\{i,j\}}^{(\text{n.n.})} \sigma_i^{(\text{a})}(-\sigma_j^{(\text{b})}) - J_2 \sum_{\{i,j\}}^{(s.n.)} \sigma_i^{(\text{a})}\sigma_j^{(a)} - J_2 \sum_{\{i,j\}}^{(\text{s.n.})} (-\sigma_i^{(\text{b})})(-\sigma_j^{(\text{b})}) \,.\end{aligned} \tag{4.14}$$

Property (i) above is thus still applicable, but the corresponding ferromagnet has second as well as nearest-neighbour interactions, its configurational energy being given by (3.26) with $n = 2$. Since all sites remain equivalent, property (ii) is still applicable. The enthalpies for the paramagnetic and antiferromagnetic states at $T = 0$ are

$$\begin{aligned}H_{\text{P}} &= \tfrac{1}{2}\text{z}_1 N J_1 - \tfrac{1}{2}\text{z}_2 N J_2 - N|\mathcal{H}|\,, \\ H_{\text{A}} &= -\tfrac{1}{2}\text{z}_1 N J_1 - \tfrac{1}{2}\text{z}_2 N J_2\end{aligned} \tag{4.15}$$

and the critical field is

$$\mathcal{H}_{\text{c}} = \text{z}_1 J_1 \,. \tag{4.16}$$

The simple antiferromagnet is recovered by putting $J_2 = 0$, $J_1 = J$ and $\text{z}_1 = \text{z}$. It can be seen that the critical field is not altered by the introduction of the second-neighbour interaction. This is because the second-neighbour energy contribution is the same for the paramagnetic and antiferromagnetic ground states.

We now consider briefly, the thermodynamics of the system, which is restricted in the sense of Sect. 2.3.4, since every lattice site is occupied by a spin. As there is only one component we can put $\xi_1 = \mathcal{H}$, $\xi_2 = \xi_n = \mu - P$. The use of the constant-field Hamiltonian implies that $\xi_1 = \mathcal{H}$ is an independent variable so that the partition function is of type (2.58), and by (2.59) the free energy per site $f = \mu - P$. It follows that magnetized phases are in equilibrium if T, $\mathcal{H}$ and f have the same values in each phase. For the Ising model ferromagnet these conditions apply by symmetry to the oppositely magnetized phases on the segment $0 < T < T_c$ of the $\mathcal{H} = 0$ axis in the $(T, \mathcal{H})$ plane. For the models considered here with $J_2 \neq 0$, phase transition lines can occur away from the $\mathcal{H} = 0$ axis, and to investigate them the relation

$$df = -m d\mathcal{H} \tag{4.17}$$

at constant T, derived from (1.34) or (2.60), is essential.

4.3 Mean-Field Theory

A mean-field theory of the model with Hamiltonian (4.13) will be developed by extending the procedure of Sect. 3.1. The nearest-neighbour $\langle \sigma_i^{(a)} \sigma_j^{(b)} \rangle$ is replaced by $m_a m_b$ and the second-neighbour $\langle \sigma_i^{(a)} \sigma_j^{(a)} \rangle$ and $\langle \sigma_i^{(b)} \sigma_j^{(b)} \rangle$ are replaced by m_a^2 and m_b^2 respectively. The number of configurations for given values of m_a and m_b is the product $\Omega(m_a)\Omega(m_b)$ with Ω defined by (3.5) except that N is replaced by $\frac{1}{2}N$, the number of sites in each sublattice. Using Stirling's formula, (4.9) and (3.41) a constant-field configurational free energy per site is

$$\begin{aligned} f = \frac{1}{4}T\,[&(1+m_a)\ln(1+m_a) + (1-m_a)\ln(1-m_a) \\ &+ (1+m_b)\ln(1+m_b) + (1-m_b)\ln(1-m_b) - 4\ln(2)] \\ &+ \frac{1}{2}z_1 J_1 m_a m_b - \frac{1}{4}z_2 J_2 (m_a^2 + m_b^2) - \frac{1}{2}\mathcal{H}(m_a + m_b)\,. \end{aligned} \tag{4.18}$$

Since, in the constant magnetic field distribution, the sublattice magnetizations m_a and m_b are functions of the independent variables T and $\mathcal{H}$, this is an example of a non-equilibrium thermodynamic potential which, as in the general equation (2.141), depends on several order variables. The equilibrium state corresponds to the minimum of f with respect to the order variables m_a and m_b. Before deriving the conditions for this minimum it is useful to define a function

$$g(m) = \frac{1}{2}\ln\left[\frac{1+m}{1-m}\right] - \tau^{-1}(1+\lambda)m = -g(-m)\,. \tag{4.19}$$

The two new parameters occurring here and one for later use are defined by

$$\lambda = \frac{z_2 J_2}{z_1 J_1}\,, \qquad \tau = \frac{T}{z_1 J_1}\,, \qquad h = \frac{\mathcal{H}}{z_1 J_1}\,. \tag{4.20}$$

Here the parameter λ expresses the magnitude of the second-neighbour interaction relative to the first, and τ and h are respectively a reduced temperature and field. The first derivative of $g(m)$ is given by

$$g'(m) = \frac{1}{1-m^2} - \tau^{-1}(1+\lambda). \tag{4.21}$$

The conditions for a stationary value of f can now be expressed in terms of the function $g(m)$ as

$$\begin{aligned} \frac{2}{T}\frac{\partial f}{\partial m_{\mathfrak{a}}} &= g(m_{\mathfrak{a}}) + \tau^{-1}(m_{\mathfrak{a}} + m_{\mathfrak{b}}) - \tau^{-1}h = 0, \\ \frac{2}{T}\frac{\partial f}{\partial m_{\mathfrak{b}}} &= g(m_{\mathfrak{b}}) + \tau^{-1}(m_{\mathfrak{a}} + m_{\mathfrak{b}}) - \tau^{-1}h = 0. \end{aligned} \tag{4.22}$$

The stationary value is a minimum if

$$\begin{aligned} \Delta &= \frac{4}{T^2}\left[\frac{\partial^2 f}{\partial m_{\mathfrak{a}}^2}\frac{\partial^2 f}{\partial m_{\mathfrak{b}}^2} - \left(\frac{\partial^2 f}{\partial m_{\mathfrak{a}}\partial m_{\mathfrak{b}}}\right)^2\right] \\ &= g'(m_{\mathfrak{a}})g'(m_{\mathfrak{b}}) + \tau^{-1}\{g'(m_{\mathfrak{a}}) + g'(m_{\mathfrak{b}})\} > 0. \end{aligned} \tag{4.23}$$

By substituting conditions (4.22) in (4.18), it can be shown that, at equilibrium,

$$\frac{4f}{T} = \ln\left[\frac{(1-m_{\mathfrak{a}}^2)(1-m_{\mathfrak{b}}^2)}{16}\right] - \tau^{-1}[2m_{\mathfrak{a}}m_{\mathfrak{b}} - \lambda(m_{\mathfrak{a}}^2 + m_{\mathfrak{b}}^2)]. \tag{4.24}$$

4.3.1 The Paramagnetic State

The paramagnetic state, where $m_{\mathfrak{a}} = m_{\mathfrak{b}}$, is stable for large values of τ or h. From (4.9) and (4.22)

$$\begin{aligned} m_{\mathfrak{a}} &= m_{\mathfrak{b}} = m, \\ g(m) + 2m\tau^{-1} &= h\tau^{-1}. \end{aligned} \tag{4.25}$$

From (4.23) the paramagnetic state is internally stable when

$$\tau[g'(m)]^2 + 2g'(m) = \tau g'(m)[g'(m) + 2\tau^{-1}] > 0. \tag{4.26}$$

As h or τ decreases, $g'(m)$ will approach zero before $g'(m) + 2\tau^{-1}$, the latter expression being in fact always positive for $\lambda \le 1$ by (4.21). Hence the paramagnetic state will become unstable with respect to sublattice ordering when

$$g'(m) = 0. \tag{4.27}$$

By eliminating m between (4.27) and (4.25) we obtain the equation $h = h^*(\tau)$ of the *critical curve* in the (τ, h) plane (see Example 4.2), outside which the paramagnetic state is stable. Any second-order transitions, that is ones in which m is continuous, from the paramagnetic to the antiferromagnetic state occur on this curve where we shall show that there is a discontinuity in $(\partial m/\partial h)_\tau$. From (4.27), on the critical curve

$$m = m^*(\tau) = \left(\frac{1+\lambda-\tau}{1+\lambda}\right)^{\frac{1}{2}} . \tag{4.28}$$

From (4.25)

$$\left(\frac{\partial m}{\partial h}\right)_\tau = \frac{1}{\tau g'(m)+2} , \tag{4.29}$$

so that $(\partial m/\partial h)_\tau \to \frac{1}{2}$ as $m \to m^* + 0$ (i.e. from the paramagnetic side). The critical curve cuts the $h = 0$ axis orthogonally where $g'(0) = 0$, giving a critical (Néel) temperature

$$\tau_c = 1 + \lambda , \qquad T_c = z_1 J_1 + z_2 J_2 . \tag{4.30}$$

This is the point where the paramagnetic state becomes unstable when temperature is reduced at zero field.

4.3.2 The Antiferromagnetic State

Here there is sublattice order and $m_a \neq m_b$. At $h = 0$, the equilibrium equations (4.22) are satisfied by $m_a = \pm m_s$, $m_b = \mp m_s$ provided that

$$g(m_s) = 0 . \tag{4.31}$$

Consider a ferromagnet with nearest and second-neighbour interactions J_1 and J_2 respectively. From (3.27) all relations of Sect. 3.1 are satisfied, with zJ replaced by $z_1 J_1 + z_2 J_2$. Hence (4.31) is the relation giving the spontaneous magnetization m_s for such a ferromagnetic model. In particular, this implies that the Curie temperature for the ferromagnet is equal to the Néel temperature given by (4.30) for the antiferromagnet. Mean-field theory thus yields the ferromagnet–antiferromagnet equivalence at $\mathcal{H} = 0$ deduced above from the Hamiltonian (4.13).

In the antiferromagnetic state the degree of sublattice order can be expressed by the parameter

$$\psi = \frac{1}{2}(m_a - m_b) , \tag{4.32}$$

which, from the previous paragraph, is equal to m_s in zero field. However, for $h \neq 0$ the simple relationship between antiferromagnet and ferromagnet no longer exists. Some of the most interesting properties of the model depend on behaviour near the critical curve. Accordingly, we define

$$\begin{aligned} \delta &= m - m^* , \\ \varepsilon &= h - h^* , \end{aligned} \tag{4.33}$$

which are the deviations of the reduced magnetization and field respectively from their values on the critical curve. From (4.9), (4.32) and (4.33) the sublattice magnetizations are

$$m_{\mathfrak{a}} = m^* + \delta + \psi\,, \qquad m_{\mathfrak{b}} = m^* + \delta - \psi\,. \tag{4.34}$$

A relation between ψ and δ is obtained from the difference of the two equations (4.22). After substitution from (4.34),

$$g(m^* + \delta + \psi) - g(m^* + \delta - \psi) = 0\,. \tag{4.35}$$

Note that the field terms have been eliminated and that the left-hand side is an odd function of ψ, which can be written in the form $\psi f(\psi^2, \delta)$. Hence, after Taylor expansions of $g(m^* + \delta \pm \psi)$ in powers of $\delta \pm \psi$ and some manipulation, we obtain

$$\psi^2 = A_1(\tau)\delta + A_2(\tau)\delta^2 + \mathrm{O}(\delta^3)\,, \tag{4.36}$$

where

$$A_1(\tau) = 2m^*(1 - 4\alpha)\,, \qquad A_2(\tau) = \frac{1}{5}(5 - 20\alpha + 64\alpha^2 - 64\alpha^3) \tag{4.37}$$

and

$$\alpha = \frac{1}{1 + 3m^{*2}} = \frac{1 + \lambda}{4 + 4\lambda - 3\tau}\,. \tag{4.38}$$

Substitution of $m = m^*$ and $h = h^*$ in (4.25) and combination with (4.22) yields

$$\frac{1}{2}[g(m^* + \delta + \psi) + g(m^* + \delta - \psi)] - g(m^*) = \tau^{-1}(\varepsilon - 2\delta)\,. \tag{4.39}$$

The left-hand side is an even function of ψ. Hence, using Taylor expansions of $g(m^* + \delta \pm \psi)$ as above, and rearranging, only even powers of ψ are left. We can then substitute for ψ^2 from (4.36) and, after a good deal of algebra, obtain the relation

$$\varepsilon = B_1(\tau)\delta + B_2(\tau)\delta^2 + \mathrm{O}(\delta^3)\,, \tag{4.40}$$

for $\delta < 0$, where

$$B_1(\tau) = 2(\alpha\lambda + \alpha - \lambda)\,, \qquad B_2(\tau) = (8m^*/5\tau)\alpha^2(1 + \lambda)^2(3 - 8\alpha)\,. \tag{4.41}$$

It is now easy to derive an expression for $\partial m/\partial h$ on the antiferromagnetic side of the critical curve. When a higher-order transition occurs on this curve, the magnetizations $m_{\mathfrak{a}}$ and $m_{\mathfrak{b}}$ are continuous and both approach m^* as $h \to h^* - 0$. Thus the equilibrium value of $\delta \to 0$ as $\varepsilon \to -0$. Substituting from (4.41) and (4.38),

$$\left(\frac{\partial m}{\partial h}\right)_{h\to h^*-0} = \lim_{\varepsilon\to -0}\left(\frac{\delta}{\varepsilon}\right) = \frac{1}{B_1(\tau)}$$

$$= \frac{4\lambda + 4 - 3\tau}{2[3\lambda\tau - (\lambda+1)(3\lambda-1)]}. \tag{4.42}$$

When the critical curve meets the τ-axis, at $\tau = \tau_c = 1 + \lambda$, the last expression reduces to $\frac{1}{2}$, and it increases from $\frac{1}{2}$ as τ decreases from τ_c. Since $\partial m/\partial h \to \frac{1}{2}$, for all τ, as $h \to h^* + 0$ (i.e. when the critical curve is approached from the paramagnetic side) there is a discontinuity in $\partial m/\partial h$ on the critical curve, confirming that the transition there is second-order in the mean-field approximation. This assumes that $B_1(\tau) > 0$; we consider the situation $B_1(\tau) < 0$ in Sect. 4.4.

4.3.3 The Simple Antiferromagnet

We can now apply this analysis to the simple antiferromagnet, for which $\lambda = 0$ since there are no second-neighbour interactions. Now, by (4.16) and (4.20), the critical field values for the antiferromagnetic–paramagnetic transition on the $\tau = 0$ axis correspond to $h = \pm 1$. It can be shown (see Example 4.2) that the critical curve $h = h^*(\tau)$ meets the $\tau = 0$ axis at the points $h = \pm(1-\lambda)$, which coincide with the transition points when $\lambda = 0$. Thus, when $\lambda = 0$, the antiferromagnetic–paramagnetic transitions can be second-order and therefore occur on the critical curve in the (τ, h) plane for the entire range $0 < \tau < \tau_c$. However, there is a stability condition which must be satisfied. Since $\mathrm{d}\mathcal{H} = z_1 J_1 \mathrm{d}h$ by (4.20), it follows from (4.17) and (1.107) that a homogeneous phase is stable only if

$$\left(\frac{\partial h}{\partial m}\right)_\tau > 0 \tag{4.43}$$

at all points. From (4.29) the condition (4.43) is satisfied everywhere in the paramagnetic region. Just on the antiferromagnetic side of the critical curve (i.e. as $m \to m^* - 0$) we have, from (4.40),

$$\left(\frac{\partial h}{\partial m}\right)_\tau = B_1(\tau) \tag{4.44}$$

and, for $\lambda = 0$, $B_1(\tau) > 0$ everywhere on the critical curve. We shall show below that (4.43) is satisfied on the $h = 0$ axis, and it can be confirmed numerically that this is true in the rest of the antiferromagnetic region. Hence, for the simple antiferromagnet, the transition line is identical to the critical curve which is the $\lambda = 0$ line in Fig. 4.4. Although the finite discontinuity in $(\partial m/\partial h)_\tau$ at the transition may be an artifact of mean-field theory, the general picture of the phase diagram for the simple antiferromagnet, with the antiferromagnetic and paramagnetic regions separated by a higher-order transition line, is confirmed by accurate series work (Domb 1974).

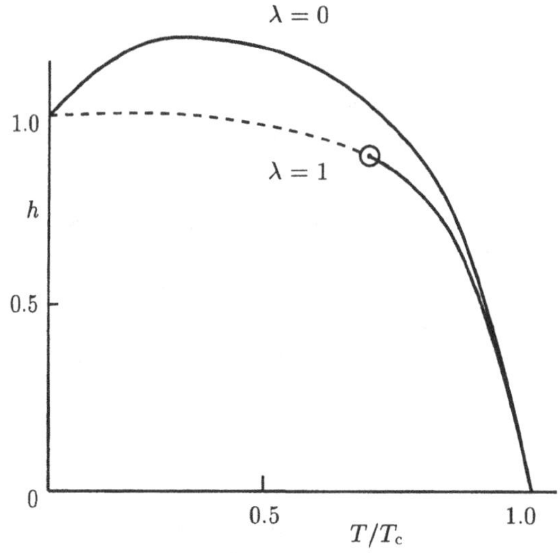

Fig. 4.4. Second-order (*full lines*) and first-order (*broken line*) transition curves in temperature–field plane for the antiferromagnet ($\lambda = 0$) and metamagnet ($\lambda = 1$); the tricritical point is circled.

We now consider the zero-field susceptibility. For $T > T_c$, $m_a = m_b = m = 0$ and, from (4.21) and (4.29) with $\lambda = 0$,

$$\left(\frac{\partial m}{\partial h}\right)_T = \frac{1}{\tau + 1} = \frac{1}{\tau + \tau_c} . \tag{4.45}$$

Hence, from (1.36) and (4.20),

$$\chi_T = \frac{C}{z_1 J_1}\left(\frac{\partial m}{\partial h}\right)_T = \frac{C}{T + T_c} , \qquad C = \frac{m_0^2}{v\mu_0} . \tag{4.46}$$

It can be shown from (4.9), (4.22) and (4.23) that, for general values of m_a, m_b, λ and τ,

$$\left(\frac{\partial m}{\partial h}\right)_T = \frac{g'(m_a) + g'(m_b)}{2\tau\Delta} . \tag{4.47}$$

For $\mathcal{H} = 0$ and $T < T_c$, equation (4.10) applies and then (4.47), (4.21) and (4.23) give, for $\lambda = 0$,

$$\chi_T = \frac{C}{T(1 - m_s^2)^{-1} + T_c} . \tag{4.48}$$

From (4.48), (3.12) and (3.13), χ_T and $d\chi_T/dT \to 0$, as $T \to 0$, whereas $d\chi_T/dT > 0$, for $T \le T_c$. From (4.46), $d\chi_T/dT < 0$, for $T > T_c$. Hence, while χ_T is continuous at $T = T_c$, $d\chi_T/dT$ has a finite discontinuity, giving a cusp in the (χ_T, T) curve. This contrasts with accurate series results (Fisher and Sykes 1962, Domb 1974) which indicate that $d\chi_T/dT$ is continuous at all T and that χ_T has a maximum at a temperature which is about $1.5T_c$ for $d = 2$ and $1.08T_c$ for $d = 3$.

When an experimental plot of χ_T^{-1} against T is asymptotic at high temperatures to a line $A(T - T_p)$, for some constant A, a *Curie–Weiss law* is said to be obeyed, and T_p is called a *paramagnetic Curie temperature.* From (3.18), $T_p = T_c$ in the mean-field approximation for the Ising model ferromagnet. In a real ferromagnet, the paramagnetic Curie temperature is 10 or 20K above the ferromagnetic Curie temperature (Martin 1967). From (4.46), $T_p = -T_c$ for the mean-field antiferromagnet, and a negative paramagnetic Curie temperature is a well-known characteristic of real antiferromagnetic substances for which the susceptibility below the Néel point depends on the direction in which the external magnetic field is applied. The curve (4.48) corresponds qualitatively to the response to a field along the zero-field magnetization axis of the sublattices. For responses to fields perpendicular to the axis of zero-field magnetization see Fisher (1963), Martin (1967) and Mattis (1985).

4.4 Metamagnetism: Tricritical Points and Critical End-Points

In a metamagnetic material the array of sites occupied by the magnetic dipoles is again divided into equivalent sublattices $\mathfrak{a}$ and $\mathfrak{b}$, whose zero-field magnetization is equal and opposite at low temperatures. However, the $\mathfrak{a}$ and $\mathfrak{b}$ sites now lie in alternate planes; there is a ferromagnetic interaction between nearest-neighbour $\mathfrak{a}$–$\mathfrak{a}$ or $\mathfrak{b}$–$\mathfrak{b}$ pairs of spins and an antiferromagnetic interaction between the closest $\mathfrak{a}$–$\mathfrak{b}$ pairs of spins (Martin 1967). The reason for this difference in the interactions lies in the structure of the crystal and particularly in non-magnetic atoms situated between the magnetic sublattice planes. A two-dimensional version of the metamagnetic ground state is shown in Fig. 4.5. For an Ising model metamagnet the Hamiltonian is identical to (4.13), except that the 'second neighbour' designation should be replaced by 'nearest neighbour in the same sublattice'. The symmetry properties (i) and (ii) are unaffected. The zero-field metamagnet is equivalent to a non-

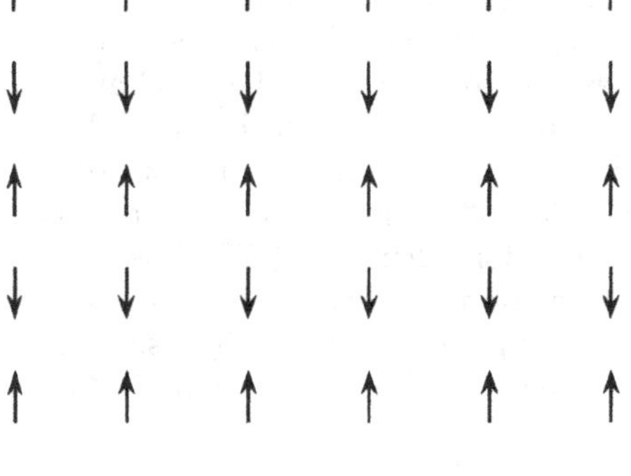

Fig. 4.5. Ground state for square lattice metamagnet in zero field.

isotropic ferromagnet with interaction parameters J_1 and J_2, and a stable paramagnetic state exists at large $|\mathcal{H}|$ and T.

The mean-field approximation for the free energy per site is again (4.18), z_1 and z_2 now respectively denoting the numbers of nearest neighbours of a given site in its own and in the other sublattice. A positive λ is now an essential characteristic of the model. Hence, on the $\tau = 0$ axis, the end points $h = \pm(1-\lambda)$ of the critical curves are not coincident with the transition points $h = \pm 1$. This indicates that, at low temperatures, there cannot be a second-order transition from the antiferromagnetic state, where $m_\mathfrak{a}$ and $m_\mathfrak{b}$ have opposite signs, to the paramagnetic state where $m_\mathfrak{a} = m_\mathfrak{b} = m$. If $m_\mathfrak{a} = -m_\mathfrak{b} = m_\mathrm{s} > 0$ at $h = 0$, and h is then increased continuously at a constant value of τ, there comes a point where the equilibrium equations (4.22) yield $m_\mathfrak{b} = 0$, and $m_\mathfrak{b}$ changes sign from negative to positive. For $0 < \tau \ll \tau_\mathrm{c}$ it can be shown (see Example 4.3) that $(\partial h/\partial m)_\tau < 0$ when $m_\mathfrak{b} = 0$. There is thus an instability loop in the plot of h against m at constant τ resulting (see Sect. 1.10 (ii)) in phase separation, with the point where $m_\mathfrak{b} = 0$ lying in the region of intrinsic instability. Hence a line of first-order transitions exists in the upper half of the (τ, h) plane, terminating at $\tau = 0$, where we would expect it to meet the h-axis at the transition point $h = 1$. This is confirmed (see Example 4.3) by the fact that, if the values of h and τ where $m_\mathfrak{b} = 0$ are plotted in the (τ, h) plane, this curve meets the h-axis at $h = 1$. The remainder of this section will he concerned with how this first-order transition line terminates at its upper end. There is also a mirror-image transition line in the $h < 0$ half-plane, but we need consider only the $h > 0$ line explicitly.

Since $(\partial h/\partial m)_\tau = B_1(\tau)$ at $m = m^* - 0$, the stability condition (4.43) is not satisfied on the antiferromagnetic side of the critical curve if $B_1(\tau) < 0$. From (4.42), $B_1(\tau) = 2$ at $\tau = \tau_\mathrm{c} = (1+\lambda)$ and decreases steadily with τ, passing through the value zero at $\tau = \tau_\mathrm{t}$ where

$$\tau_\mathrm{t} = \frac{(3\lambda - 1)(\lambda + 1)}{3\lambda} = \frac{(3\lambda - 1)\tau_\mathrm{c}}{3\lambda} . \tag{4.49}$$

For $\tau_\mathrm{c} > \tau > \tau_\mathrm{t}$ there is a second-order transition on the critical curve, as discussed at the end of Sect. 4.3. For $\tau_\mathrm{t} > \tau > 0$ and λ sufficiently large (in fact, $\lambda > \frac{3}{5}$, as shown below), the plot of h against m is similar to that in Fig. 4.6 where the loop implies a first-order transition. The magnetizations m_1 and m_2 in the conjugate phases correspond to the ends of the equal-areas tie line shown, and the slope discontinuity at $m = m^*$ falls in the instability interval (m_1, m_2). Hence, for the range $\tau_\mathrm{t} > \tau > 0$, there is a line of first-order transitions, as shown for the $\lambda = 1$ case in Fig. 4.4. (The method of Sect. 1.10 (i), with $f_{n-1} = f$ and $\xi_i = z_1 J_1 h$, was used for calculation.) We thus have a new type of critical phenomenon. The line of first-order transitions in the (τ, h) plane ends at the point $\tau = \tau_\mathrm{t}$, $h = h^*(\tau_\mathrm{t})$ but is continued by a second-order transition line. Such a point is called *tricritical*, and τ_t, given by (4.49), is a *tricritical temperature*.

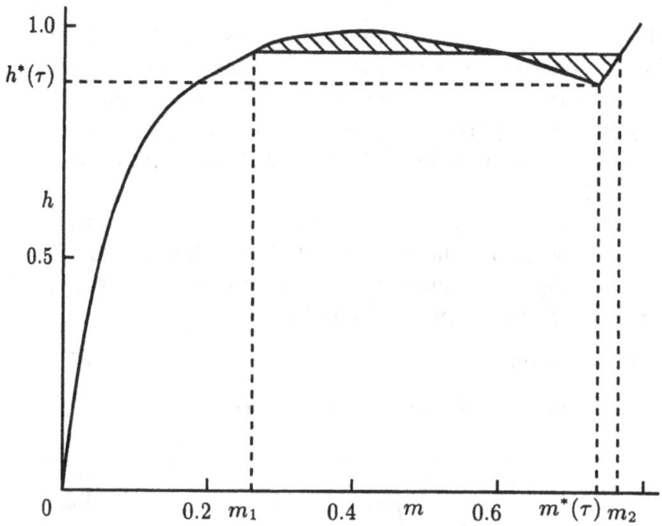

Fig. 4.6. Field-magnetization plot for $\lambda = 1$, $\tau = \frac{1}{2}\tau_c$, showing an instability loop and the equal-areas tie line.

For small values of λ the first-order transition line starting at $h = 1$ on the $\tau = 0$ axis cannot end at a tricritical point; in fact from (4.49), $\tau_t < 0$ for $\lambda < 1/3$. Hence there is an ordinary critical point where, by (1.102),

$$\left(\frac{\partial h}{\partial m}\right)_\tau = 0\,, \qquad \left(\frac{\partial^2 h}{\partial m^2}\right)_\tau = 0\,. \tag{4.50}$$

Since the condition (4.43) is satisfied everywhere in the paramagnetic region the critical point must lie in the antiferromagnetic region, that is inside the critical curve in the (τ, h) plane. Fig. 4.7 shows part of the (τ, h) plane for $\lambda = 0.3$, the first-order transition line again being obtained by the method of Sect. 1.10 (i). For any τ between the values corresponding to points Q and C, the plot of h against m will show a loop, but the point $m = m^*$, where a slope discontinuity occurs, lies outside the instability interval associated with the loop. Hence, when h is increased at constant τ, there is a first-order transition on the segment QC, and then an antiferromagnetic–paramagnetic second-order transition at $h = h^*$ on the critical line. At the temperature corresponding to Q the point $m = m^*$ enters the instability interval of the first-order transition, and the line of second-order transitions on the critical curve terminates. We have, in fact, another new critical phenomenon and a point like Q where a line of second-order transitions terminates on a first-order transition line is known as a *critical end-point*.

It is now easy to find the value $\lambda = \lambda_1$ where the low-λ critical end-point phase pattern goes over to the high-λ tricritical point phase pattern. Consider Fig. 4.7 and suppose that λ steadily increases. The segment QC will shrink until eventually the critical end-point Q coalesces with the critical point C at $\lambda = \lambda_1$. When coalescence occurs, the conditions (4.50) defining Q are satisfied on the antiferromagnetic side of the critical curve so that, by (4.40), $B_1(\tau) = B_2(\tau) = 0$. Comparison of the expressions in (4.41) then yields $\lambda_1 = \frac{3}{5}$. (This result was derived by Bidaux et al. (1967), who give a comprehensive set of diagrams.)

We now consider the tricritical regime again and in particular the situation just below the tricritical temperature where $\tau = \tau_t - \theta$, $\theta \ll 1$. The conjugate antiferromagnetic and paramagnetic phases existing for $\tau < \tau_t$, are labelled by the indices 1 and 2. Noting that, from (4.17) and (4.20),

$$(z_1 J_1)^{-1} \mathrm{d}f = -\mathrm{d}(hm) + h\mathrm{d}m \tag{4.51}$$

the phase equilibrium conditions can be written in the form

$$\begin{aligned} &h_2 - h_1 = 0\,, \\ &(z_1 J_1)^{-1}[f_2 - f_1] = -h_2 m_2 + h_1 m_1 + \int_{m_1}^{m_2} h\mathrm{d}m = 0\,. \end{aligned} \tag{4.52}$$

Using the quantities ε and δ, defined by (4.33), the equilibrium conditions (4.52) can be expressed as

$$\begin{aligned} &\varepsilon_1 = \varepsilon_2\,, \\ &\varepsilon_1\delta_1 - \int_0^{\delta_1} \varepsilon\mathrm{d}\delta = \varepsilon_2\delta_2 - \int_0^{\delta_2} \varepsilon\mathrm{d}\delta\,, \end{aligned} \tag{4.53}$$

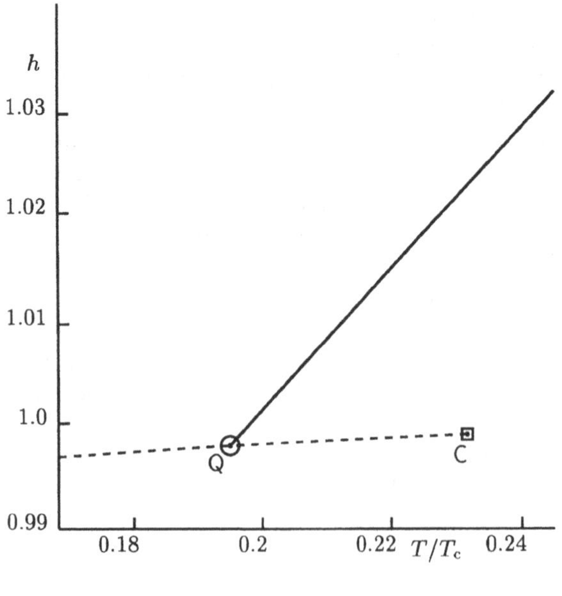

Fig. 4.7. As in Fig. 4.4 for $\lambda = 0.3$, with critical end-point Q and critical point C.

with $\delta_1 < 0$ and $\delta_2 > 0$. On the antiferromagnetic side of the critical curve where $\delta < 0$ we use (4.40). Since $\theta \ll 1$ and $B_1(\tau_t) = 0$ we can, if only leading terms in θ are required, write

$$B_1(\tau) = -\mathfrak{A}\,\theta\,, \qquad B_2(\tau) = B_2(\tau_t) = \mathfrak{B}\,, \tag{4.54}$$

where

$$\mathfrak{A} = \frac{6\lambda^2}{(1+\lambda)^2}\,, \qquad \mathfrak{B} = \frac{8\lambda^2(3\lambda)^{\frac{1}{2}}(3-5\lambda)}{5(1+\lambda)^2(3\lambda-1)}\,. \tag{4.55}$$

For $\delta > 0$, using (4.29) with $g'(m) = 0$,

$$\varepsilon = \left(\frac{\partial h}{\partial m}\right)_{m=m^*+0} \delta = 2\delta\,. \tag{4.56}$$

For the tricritical range $\lambda > \frac{3}{5}$, $\mathfrak{B} < 0$ so that (4.40) and (4.48)–(4.56) yield a plot of the same general form as Fig. 4.6. Substitution in the two relations of (4.53) yields

$$-\mathfrak{A}\theta\delta_1 + \mathfrak{B}\delta_1^2 = 2\delta_2\,, \qquad -\frac{1}{2}\mathfrak{A}\theta\delta_1^2 + \frac{2}{3}\mathfrak{B}\delta_1^3 = \delta_2^2\,. \tag{4.57}$$

After a little algebra we then have, for the magnetizations in the conjugate phases and the first-order transition value of the field,

$$m_1(\tau) = m^*(\tau) + \delta_1(\theta) = m^*(\tau) - \frac{3\mathfrak{A}}{4|\mathfrak{B}|}\theta + \mathrm{O}(\theta^2)\,,$$
$$m_2(\tau) = m^*(\tau) + \delta_2(\theta) = m^*(\tau) - \frac{3\mathfrak{A}^2}{32|\mathfrak{B}|}\theta^2 + \mathrm{O}(\theta^3)\,, \tag{4.58}$$

$$h(\tau) = h^*(\tau) + \frac{3\mathfrak{A}^2}{16|\mathfrak{B}|}\theta^2 + \mathrm{O}(\theta^3)\,. \tag{4.59}$$

From (4.59) the first-order transition line and the critical curve have a common tangent at the tricritical point ($\tau = \tau_t$, $\theta = 0$) as shown for $\lambda = 1$ in Fig. 4.4. From (4.59) the paramagnetic branch $m = m_2(\tau)$ of the coexistence curve in the (τ, m) plane has a common tangent with the critical magnetization curve $m = m^*(\tau)$ at $\tau = \tau_t$, but the antiferromagnetic branch $m = m_1(\tau)$ meets it at a non-zero angle. Hence the coexistence curve has a cusp at the tricritical point, in contrast to the continuous tangent at a critical point. (For an example in another system, see Fig. 6.10.)

At the tricritical point $h - h^* \simeq \mathfrak{B}(m^* - m)^2$ for $m < m^*$ (i.e. on the antiferromagnetic side of the critical curve). Now, from (4.55), $\mathfrak{B} < 0$ for $\lambda < \frac{3}{5}$, giving $h - h^* > 0$. This confirms that, for $\lambda < \frac{3}{5}$, the tricritical point falls in the instability range of the first-order transition. As we have seen,

τ_t becomes negative for $\lambda < \frac{1}{3}$. The subject of tricritical points and critical end-points is taken up again in Sects. 6.8–6.10 and general references are given.

4.5 Ferrimagnetism: Compensation Points

Like an antiferromagnet, a ferrimagnet in zero field has oppositely magnetized sublattices below a critical temperature. However, the magnetizations of the sublattices do not completely cancel and so there is still a spontaneous magnetization. The crystal structure of a real ferrimagnet can be extremely complicated (Martin 1967), but we shall introduce ferrimagnetism through a generalization of the Ising model antiferromagnet considered in Sects. 4.2 and 4.3. Suppose that a lattice of N sites is divided into sublattices $\mathfrak{a}$ and $\mathfrak{b}$ of $N_\mathfrak{a}$ and $N_\mathfrak{b}$ sites respectively, where $N_\mathfrak{a}$ and $N_\mathfrak{b}$ may be unequal. Each $\mathfrak{a}$–site has $z_{\mathfrak{aa}}$ nearest-neighbour $\mathfrak{a}$–sites and $z_{\mathfrak{ab}}$ nearest-neighbour $\mathfrak{b}$–sites; $z_{\mathfrak{bb}}$ and $z_{\mathfrak{ba}}$ are similarly defined. Then

$$
\begin{aligned}
&z_{\mathfrak{aa}} + z_{\mathfrak{ab}} = z\,, \\
&\qquad\qquad\qquad N_\mathfrak{a} z_{\mathfrak{ab}} = N_\mathfrak{b} z_{\mathfrak{ba}}\,, \\
&z_{\mathfrak{bb}} + z_{\mathfrak{ba}} = z\,,
\end{aligned}
\tag{4.60}
$$

where z is the coordination number of the lattice, and the final relation follows from counting the number of $\mathfrak{a}$–$\mathfrak{b}$ nearest-neighbour pairs in two different ways. A two-dimensional example from Bell (1974a) and based on the triangular lattice is shown in Fig. 4.8. Here the $\mathfrak{b}$-sites form a Kagomé sublattice, $N_\mathfrak{b} = 3N_\mathfrak{a}$, $z_{\mathfrak{aa}} = 0$, $z_{\mathfrak{ab}} = 6$, $z_{\mathfrak{bb}} = 4$, $z_{\mathfrak{ba}} = 2$. We assume ferromagnetic interaction energies $-J_{\mathfrak{aa}}\sigma_i^{(\mathfrak{a})}\sigma_j^{(\mathfrak{a})}$ and $-J_{\mathfrak{bb}}\sigma_i^{(\mathfrak{b})}\sigma_j^{(\mathfrak{b})}$ for $\mathfrak{a}$–$\mathfrak{a}$ and $\mathfrak{b}$–$\mathfrak{b}$ nearest-neighbour pairs respectively, and an antiferromagnetic interaction energy $J_{\mathfrak{ab}}\sigma_i^{(\mathfrak{a})}\sigma_j^{(\mathfrak{b})}$ for $\mathfrak{a}$–$\mathfrak{b}$ nearest-neighbour pairs, $J_{\mathfrak{aa}}$, $J_{\mathfrak{bb}}$ and $J_{\mathfrak{ab}}$ being

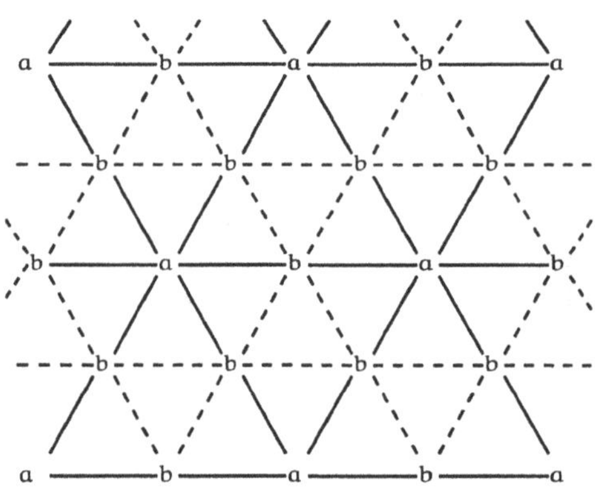

Fig. 4.8. Plane triangular lattice ferrimagnet. (Reprinted from Bell (1974a), by permission of the publisher IOP Publishing Ltd.)

all positive by convention. The spins on b-sites have magnetic moment m_0 and those on a-sites magnetic moment rm_0. We define a parameter s by

$$s = N_{\mathfrak{b}}/N_{\mathfrak{a}} = z_{\mathfrak{ab}}/z_{\mathfrak{ba}} \tag{4.61}$$

and no generality is lost by taking $s \geq 1$. Sublattice relative magnetizations, $m_{\mathfrak{a}}$ and $m_{\mathfrak{b}}$, and the overall relative magnetization m are defined by generalizing (4.8) and (4.9):

$$m_{\mathfrak{a}} = N_{\mathfrak{a}}^{-1} \sum_{\{i\}} \sigma_i^{(\mathfrak{a})}, \qquad m_{\mathfrak{b}} = N_{\mathfrak{b}}^{-1} \sum_{\{j\}} \sigma_j^{(\mathfrak{b})},$$
$$m = (rN_{\mathfrak{a}} + N_{\mathfrak{b}})^{-1} \left(r \sum_{\{i\}} \sigma_i^{(\mathfrak{a})} + \sum_{\{j\}} \sigma_j^{(\mathfrak{b})} \right) = \frac{rm_{\mathfrak{a}} + sm_{\mathfrak{b}}}{r+s}. \tag{4.62}$$

With all spins in sublattice a completely aligned, its magnetic moment is of magnitude $rN_{\mathfrak{a}}m_0$, whereas the corresponding magnitude for sublattice b is $N_{\mathfrak{b}}m_0 = sN_{\mathfrak{a}}m_0$. The ratio of the 'magnetic weights' of sublattices a and b is thus r/s. Because of the antiferromagnetic interaction between a and b, $m_{\mathfrak{a}}$ and $m_{\mathfrak{b}}$ will have opposite signs in small magnetic fields at $T = 0$. Hence for small positive $\mathcal{H}$ ($\mathcal{H}$ still defined by (1.55)), $m_{\mathfrak{a}} = -m_{\mathfrak{b}} = 1$ for $r > s$ and $m_{\mathfrak{b}} = -m_{\mathfrak{a}} = 1$ for $r < s$. The change in interaction energy involved in the transition to the state where $m_{\mathfrak{a}} = m_{\mathfrak{b}} = 1$ at $T = 0$ is $2N_{\mathfrak{b}}z_{\mathfrak{a}}J_{\mathfrak{ab}} = 2N_{\mathfrak{a}}z_{\mathfrak{ab}}J_{\mathfrak{ab}}$. For $r > s$ the corresponding change in field energy is $-2N_{\mathfrak{b}}\mathcal{H}$, and for $r < s$ we have $-2N_{\mathfrak{a}}r\mathcal{H}$. The critical fields at $T = 0$ are

$$\mathcal{H}_{\mathrm{c}} = \begin{cases} z_{\mathfrak{ba}}J_{\mathfrak{ab}}, & r > s, \\ (s/r)z_{\mathfrak{ba}}J_{\mathfrak{ab}}, & s > r. \end{cases} \tag{4.63}$$

Equation (4.18) for free energy in the mean-field approximation is easily generalized to

$$\begin{aligned} Nf(m_{\mathfrak{a}}, m_{\mathfrak{b}}) = \tfrac{1}{2}T\,\{&N_{\mathfrak{a}}[(1+m_{\mathfrak{a}})\ln(1+m_{\mathfrak{a}}) + (1-m_{\mathfrak{a}})\ln(1-m_{\mathfrak{a}})] \\ &+ N_{\mathfrak{b}}[(1+m_{\mathfrak{b}})\ln(1+m_{\mathfrak{b}}) + (1-m_{\mathfrak{b}})\ln(1-m_{\mathfrak{b}})]\} \\ &- NT\ln(2) + N_{\mathfrak{b}}z_{\mathfrak{ba}}J_{\mathfrak{ab}}m_{\mathfrak{a}}m_{\mathfrak{b}} \\ &- \tfrac{1}{2}N_{\mathfrak{a}}z_{\mathfrak{aa}}J_{\mathfrak{aa}}m_{\mathfrak{a}}^2 - \tfrac{1}{2}N_{\mathfrak{b}}z_{\mathfrak{bb}}J_{\mathfrak{bb}}m_{\mathfrak{b}}^2 \\ &- (N_{\mathfrak{a}}rm_{\mathfrak{a}} + N_{\mathfrak{b}}m_{\mathfrak{b}})\mathcal{H}. \end{aligned} \tag{4.64}$$

To express the equilibrium conditions as succinctly as possible we define the parameters

$$\lambda_{\mathfrak{a}} = \frac{z_{\mathfrak{aa}}J_{\mathfrak{aa}}}{z_{\mathfrak{ba}}J_{\mathfrak{ab}}}, \qquad \lambda_{\mathfrak{b}} = \frac{z_{\mathfrak{bb}}J_{\mathfrak{bb}}}{z_{\mathfrak{ba}}J_{\mathfrak{ab}}},$$
$$h = \frac{\mathcal{H}}{z_{\mathfrak{ba}}J_{\mathfrak{ab}}}, \qquad \tau = \frac{T}{z_{\mathfrak{ba}}J_{\mathfrak{ab}}}, \tag{4.65}$$

and then

$$\frac{N}{N_{\mathfrak{a}}T}\frac{\partial f}{\partial m_{\mathfrak{a}}}=\frac{1}{2}\ln\left[\frac{1+m_{\mathfrak{a}}}{1-m_{\mathfrak{a}}}\right]+\frac{sm_{\mathfrak{b}}-\lambda_{\mathfrak{a}}m_{\mathfrak{a}}-rh}{\tau}=0\,,$$
$$\frac{N}{N_{\mathfrak{b}}T}\frac{\partial f}{\partial m_{\mathfrak{b}}}=\frac{1}{2}\ln\left[\frac{1+m_{\mathfrak{b}}}{1-m_{\mathfrak{b}}}\right]+\frac{m_{\mathfrak{a}}-\lambda_{\mathfrak{b}}m_{\mathfrak{b}}-h}{\tau}=0\,. \tag{4.66}$$

For any solution $(m_{\mathfrak{a}}, m_{\mathfrak{b}})$ at given positive h, there is a solution $(-m_{\mathfrak{a}}, -m_{\mathfrak{b}})$ at $-h$. A stationary point given by (4.66) is a minimum if

$$\Delta=\frac{N^2T^2}{N_{\mathfrak{a}}N_{\mathfrak{b}}}\left[\frac{\partial^2 f}{\partial m_{\mathfrak{a}}^2}\frac{\partial^2 f}{\partial m_{\mathfrak{b}}^2}-\left(\frac{\partial^2 f}{\partial m_{\mathfrak{a}}\partial m_{\mathfrak{b}}}\right)^2\right]$$
$$=\left(\frac{1}{1-m_{\mathfrak{a}}^2}-\frac{\lambda_{\mathfrak{a}}}{\tau}\right)\left(\frac{1}{1-m_{\mathfrak{b}}^2}-\frac{\lambda_{\mathfrak{b}}}{\tau}\right)-\frac{s}{\tau^2}>0\,. \tag{4.67}$$

The high-temperature zero-field state $m_{\mathfrak{a}} = m_{\mathfrak{b}} = 0$ becomes unstable when $\Delta = 0$, giving the critical temperature

$$\tau_{\mathrm{c}}=\frac{1}{2}\{\lambda_{\mathfrak{a}}+\lambda_{\mathfrak{b}}+[(\lambda_{\mathfrak{a}}-\lambda_{\mathfrak{b}})^2+4s]^{\frac{1}{2}}\}\,. \tag{4.68}$$

The model treated in Sect. 4.3 is recovered by putting $\lambda_{\mathfrak{a}} = \lambda_{\mathfrak{b}} = \lambda$, $s = 1$, $r = 1$. Since, for $\tau < \tau_{\mathrm{c}}$, there is both spontaneous magnetization and oppositely magnetized sublattices, the terms 'Curie' and 'Néel' are equally appropriate for τ_{c}. However the former is usually chosen.

Since the lattice sites are no longer equivalent, the symmetry properties (i) and (ii), discussed in Sect. 4.2, do not apply for the Ising model ferrimagnet. At $\mathcal{H} = 0$ it is still, of course, possible to find an equivalent model by simultaneously changing all $\sigma_i^{(\mathfrak{b})}$ to $-\sigma_i^{(\mathfrak{b})}$ and $J_{\mathfrak{ab}}$ to $-J_{\mathfrak{ab}}$. However, though all interactions are then ferromagnetic, $\langle\sigma_i^{(\mathfrak{a})}\rangle \neq \langle\sigma_i^{(\mathfrak{b})}\rangle$ and the useful relation (4.10) is not satisfied. Property (i) and equation (4.10) do still apply, however, in the particular case

$$J_{\mathfrak{aa}} = J_{\mathfrak{bb}} = J_{\mathfrak{ab}} = J\,, \tag{4.69}$$

which will be termed the 'simple ferrimagnet'. Here the zero-field 'reversal' operation produces an Ising model ferromagnet with the same interaction parameter for all nearest-neighbour pairs; in fact, it makes all sites equivalent. For the ferrimagnet there is no paramagnetic state with all $\langle\sigma_i\rangle$ equal except at $\mathcal{H} = 0$, $\tau > \tau_{\mathrm{c}}$ when all $\langle\sigma_i\rangle = 0$ and $\tau = 0$, $\mathcal{H} < -\mathcal{H}_{\mathrm{c}}$ or $\mathcal{H} > \mathcal{H}_{\mathrm{c}}$. Second-order transitions, which correspond to a change of the symmetry of the state, thus no longer occur for $\mathcal{H} \neq 0$. Since there is a spontaneous magnetization which has opposite signs for $\mathcal{H} = +0$ and $\mathcal{H} = -0$ the interval $(0, \tau_{\mathrm{c}})$ of the τ-axis is a first-order transition line. There may also be first-order transition lines starting from $\mathcal{H} = \pm\mathcal{H}_{\mathrm{c}}$ on the $\mathcal{H}$-axis.

In some real ferrimagnets there is a *compensation temperature* between $T = 0$ and $T = T_{\mathrm{c}}$ where the spontaneous magnetization is zero. The curve of spontaneous magnetization against T has a downward cusp at this temperature.

4.5.1 Zero Field

We now investigate the conditions for the occurrence of a compensation point in the mean-field theory of the Ising model ferrimagnet. For definiteness we put $h = +0$, that is we consider limiting values as $h \to 0$ from above, so that the spontaneous magnetization is non-negative and hence, by (4.62), $rm_{\mathfrak{a}} + sm_{\mathfrak{b}} \geq 0$. A compensation point occurs when $rm_{\mathfrak{a}} + sm_{\mathfrak{b}} = 0$ which implies $rm_{\mathfrak{a}} = -sm_{\mathfrak{b}}$ and

$$|rm_{\mathfrak{a}}|/|sm_{\mathfrak{b}}| = 1\,. \tag{4.70}$$

The signs of $m_{\mathfrak{a}}$ and $m_{\mathfrak{b}}$ both reverse at the compensation point so that one zero-field solution $(m_{\mathfrak{a}}, m_{\mathfrak{b}})$ of the equilibrium equations (4.66) is exchanged for the alternative $(-m_{\mathfrak{a}}, -m_{\mathfrak{b}})$. Although $m_{\mathfrak{a}}$ and $m_{\mathfrak{b}}$ are discontinuous, their magnitudes $|m_{\mathfrak{a}}|$ and $|m_{\mathfrak{b}}|$ and the ratio $|rm_{\mathfrak{a}}|/|sm_{\mathfrak{b}}|$ are continuous. Just below the Néel temperature τ_c, $m_{\mathfrak{a}}$ and $m_{\mathfrak{b}}$ are small so that (4.66) with $h = 0$ can be linearized to give

$$\frac{m_{\mathfrak{a}}}{m_{\mathfrak{b}}} = \frac{s}{\lambda_{\mathfrak{a}} - \tau} = \lambda_{\mathfrak{b}} - \tau\,. \tag{4.71}$$

Equating the last two expressions gives (4.68) for τ_c, again, as would be expected, and hence

$$\lim_{T \to T_c - 0} \frac{|rm_{\mathfrak{a}}|}{|sm_{\mathfrak{b}}|} = \frac{r}{r_0}\,, \tag{4.72}$$

where

$$r_0 = \tau_c - \lambda_{\mathfrak{a}} = \frac{1}{2}\{\lambda_{\mathfrak{b}} - \lambda_{\mathfrak{a}} + [(\lambda_{\mathfrak{b}} - \lambda_{\mathfrak{a}})^2 + 4s]^{\frac{1}{2}}\}\,. \tag{4.73}$$

It can be verified that $r_0 > s$ if and only if $s < \lambda_{\mathfrak{b}} - \lambda_{\mathfrak{a}} + 1$. At $\tau = 0$, $|rm_{\mathfrak{a}}|/|sm_{\mathfrak{b}}| = r/s$. Hence, if r lies between r_0 and s, $|rm_{\mathfrak{a}}|/|sm_{\mathfrak{b}}|$ is greater than 1 at one end of the range $[0, \tau_c]$ and less than 1 at the other. So (4.70) will be satisfied at some temperature between 0 and τ_c, giving a compensation point. Given that we take $s \geq 1$, the conditions for the existence of such a point are therefore

$$\begin{gathered} s < \lambda_{\mathfrak{b}} - \lambda_{\mathfrak{a}} + 1, \quad \text{and} \quad r_0 > r > s, \\ \text{or} \\ s > \lambda_{\mathfrak{b}} - \lambda_{\mathfrak{a}} + 1, \quad \text{and} \quad s > r > r_0. \end{gathered} \tag{4.74}$$

With r and s fixed, it can be seen that, when $r > s$, a compensation point can be made to occur by increasing $\lambda_{\mathfrak{b}}$ relative to $\lambda_{\mathfrak{a}}$, and vice versa. Compensation points thus exist because stronger ferromagnetic interactions in one sublattice counteract its lower magnetic weight in an interval below τ_c, but cannot do so for lower values of τ where $|m_{\mathfrak{a}}|$ and $|m_{\mathfrak{b}}|$ are nearer to unity. In the simple Ising model ferrimagnet, defined by (4.69), the spontaneous magnetization is $|r-s|m_s^{(\text{I.M.})}/(r+s)$, where $m_s^{(\text{I.M.})}$ is the spontaneous magnetization of the equivalent ferromagnet. The sublattice magnetizations are too closely locked together to allow a compensation point. It is easily shown that in this case $s = \lambda_b - \lambda_a + 1$ and $r_0 = s$. For the general case, given that the necessary inequality is satisfied, the compensation temperature can be found by substituting $h = 0$ and $rm_a + sm_b = 0$ into (4.66) (see Example 4.5). Spontaneous magnetization curves are shown in Fig. 4.9, for the model of Fig. 4.8 with $\lambda_a = 0$, $\lambda_b = 1$. Here $r_0 = \frac{1}{2}(1+\sqrt{13}) = 2.3028 < s = 3$.

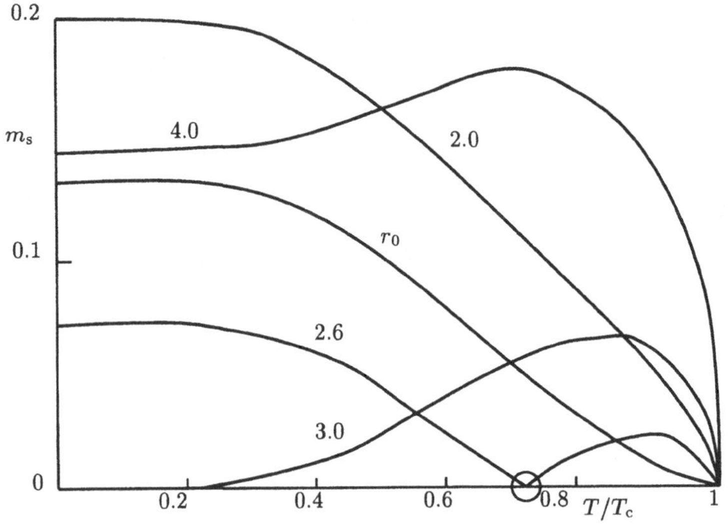

Fig. 4.9. Spontaneous magnetization–temperature curves for the model of Fig. 4.8. Each curve is labelled with the value of r.

4.5.2 Non-Zero Field

Suppose first that $s > r$. Then, at $h = +0$, $m_b > 0$ and $m_a < 0$ for small τ. However, for large positive values of h, m_a and m_b, though still unequal, must both be positive. Thus, as h is increased at constant τ, there must be a point where $m_a = 0$. Substitution in (4.66) yields an $m_a = 0$ curve defined by

$$\begin{aligned} m_b &= \frac{rh}{s}\,, \\ \frac{1}{2}\ln\left[\frac{1+m_b}{1-m_b}\right] &- \frac{r_0}{\tau}m_b = 0\,, \end{aligned} \tag{4.75}$$

where

$$\tau_0 = \lambda_b + (s/r)\,. \tag{4.76}$$

This curve meets the $\tau = 0$ axis at the critical h values of $\pm s/r$. The second equation of (4.75) is similar in form to (3.12) and has solutions $m_b \neq 0$ when $\tau < \tau_0$. At such a solution the gradient of the expression on the left-hand side is positive so that

$$\frac{1}{1-m_b^2} - \frac{\tau_0}{\tau} > 0\,. \tag{4.77}$$

If $m_a = 0$ is substituted in (4.67) we obtain

$$\Delta = \left(\frac{1}{1-m_b^2} - \frac{\lambda_b}{\tau}\right)\left(1 - \frac{\lambda_a}{\tau}\right) - \frac{s}{\tau^2}\,. \tag{4.78}$$

From (4.77) the first bracket is positive and so $\Delta < 0$ for a temperature range which includes all $\tau \leq \lambda_a$ and an interval above λ_a. If h is plotted against m for any τ in this range there will be an interval where $(\partial h/\partial m)_\tau < 0$, giving an instability loop. Thus, in the (τ, h) plane, there are two first-order transition lines, symmetrical about the $h = 0$ axis, which start at the critical field values $h = \pm s/r$ on the $\tau = 0$ axis. There are discontinuities in m_a and m_b, involving a change in the sign of m_a, at the transition lines.

We next investigate how these transition lines end. The curve defined by (4.75) cuts the $h = 0$ axis orthogonally at $\tau = \tau_0$ and from (4.68), (4.73) and (4.76), $\tau_0 > \tau_c$ if and only if $r < r_0$. Since $r < s$ there is therefore no compensation point if the curve (4.75) intersects the $h = 0$ axis at a point outside the interval $(0, \tau_c)$. Noting that, by definition, $|m_a| \leq 1$ and $|m_b| \leq 1$, the stability condition $\Delta > 0$ is satisfied for all m_a and m_b when $\tau > \tau_c$. Hence when $\tau_0 > \tau_c$ the instability regions on the $m = m(\tau, h)$ surface associated with the first-order transition lines starting at $\tau = 0$, $h = \pm s/r$ do not reach the $h = 0$ plane. These transition lines are therefore terminated in the (τ, h) plane by critical points at which $\tau < \tau_c$, $h \neq 0$, a situation illustrated by the $r = 1$ curve in Fig. 9.5 of Chap. 9. Now suppose that $\tau_0 < \tau_c$ which implies $r_0 < r < s$ and hence the existence of a compensation point. With $\tau_0 < \tau_c$, $\Delta < 0$ at every point on the curve (4.75) so that the regions of instability on the $m = m(\tau, h)$ surface do reach the $h = 0$ plane. The transition lines starting at $\tau = 0$, $h = \pm s/r$ in the (τ, h) plane meet the $h = 0$ axis orthogonally. For $h = +0$ (or $h = -0$) there are discontinuities in m_a and m_b at the compensation temperature and hence the crossing point must be identified with the compensation point. The situation is illustrated by the $r = 2.7$ curve in Fig. 9.5 of Chap. 9. The discontinuities in m are sketched in Fig. 4.10. The case $\lambda_a = 0$ needs separate consideration. When $m_a = 0$

$$\Delta = \frac{1}{1-m_b^2} - \frac{\lambda_b}{\tau} - \frac{s}{\tau^2}\,. \tag{4.79}$$

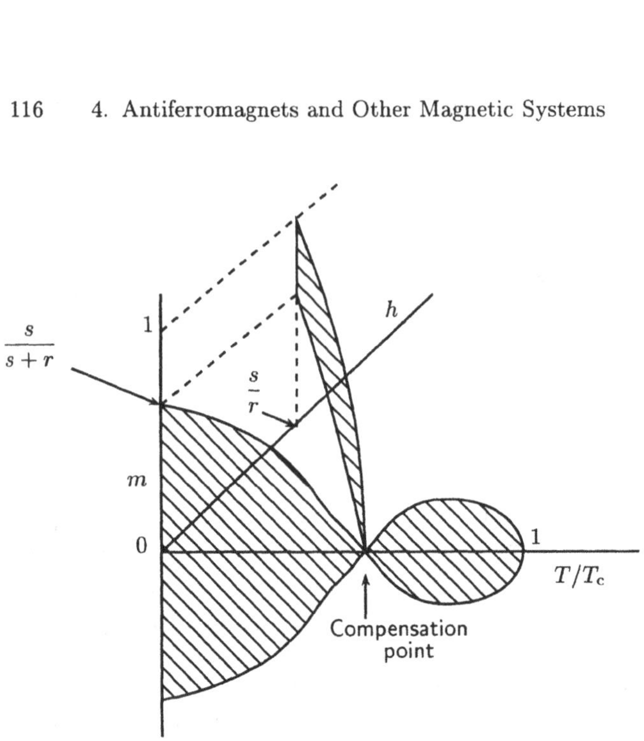

Fig. 4.10. Discontinuities in the magnetization surface for a Ising model ferrimagnet with a compensation point. The hatched lines are parallel to the m-axis.

Since $m_{\mathfrak{b}}$ is a solution of the second of equations (4.75) it can be shown that, as $\tau \to 0$, $(1 - m_{\mathfrak{b}}^2)^{-1}$ approaches infinity faster than τ^{-n} for any positive integer n. It follows that $\Delta > 0$ for small τ and that critical fields $h = \pm s/r$ on the $\tau = 0$ axis represent isolated singularities. However if $\tau_0 < \tau_{\mathrm{c}}$, implying the existence of a compensation point, then $\Delta < 0$ on the curve (4.75) in a temperature interval below τ_0. At the compensation point the interval $(0, \tau_{\mathrm{c}})$ on the $h = 0$ axis is crossed orthogonally by a first-order transition line which terminates at critical points in the (τ, h) plane for which $\tau > 0$. This situation is illustrated by Fig. 9.6 of Chap. 9. So far we have assumed $s > r$ in considering first-order transitions off the $h = 0$ axis, but the theory for $r > s$ runs in parallel, with an $m_{\mathfrak{b}} = 0$ curve replacing the $m_{\mathfrak{a}} = 0$ curve defined by (4.75). Finally it may be noted that when the Ising model ferrimagnet was introduced, all interactions were, for simplicity, taken as nearest neighbour. However, interactions of any range can be incorporated into the mean-field theory by replacing such quantities as $z_{\mathfrak{a}\mathfrak{a}} J_{\mathfrak{a}\mathfrak{a}}$ by the appropriate sums.

Although these results for the Ising model ferrimagnet have been obtained from the mean-field approximation, they are borne out qualitatively by results obtained by exact transformation theory for particular cases in Chap. 9 and by the real-space renormalization group methods described in Volume 2, Chap. 6. Of course, values for such parameters as r_0 differ considerably from mean-field ones. The most interesting new cooperative phenomenon is the compensation point formed by the orthogonal intersection of first-order transition lines and illustrated in Fig. 4.10.

Examples

4.1 For the one-dimensional antiferromagnet considered in Sect. 4.1 show that $\partial^2 m/\partial \mathcal{H}^2 = 0$ where

$$\exp(4J/T) - 2 = \cosh(2\mathcal{H}/T)\,.$$

Show that the curve defined by this relation meets the $T = 0$ axis at $\mathcal{H} = \mathcal{H}_c$ and the $\mathcal{H} = 0$ axis at $T = 4J/\ln(3)$.

4.2 For the antiferromagnet or metamagnet in the mean-field approximation, show from (4.25) and (4.27) that the critical curve (i.e. the curve where the paramagnetic state becomes internally unstable) in the (τ, h) plane is given by

$$h^* = \frac{1}{2}\tau \ln\left[\frac{1+m^*}{1-m^*}\right] + (1-\lambda)m^*\,, \qquad m^* = \pm\left(\frac{1+\lambda-\tau}{1+\lambda}\right)^{\frac{1}{2}}.$$

Hence show that, on the $h > 0$ branch of this curve, as $\tau \to 0$, $h^* \to 1 - \lambda$ and as $\tau \to 1 + \lambda$, $h^* \to 0$.

4.3 For the antiferromagnet or metamagnet in the mean-field approximation, show that, when $m_{\mathfrak{b}} = 0$, $m_{\mathfrak{a}} = h = m_{\mathrm{s}}$ where $m_{\mathrm{s}} > 0$ is the zero-field sublattice magnetization at the same temperature. Show that the curve thus defined in the (τ, h) plane meets the h-axis at $h = 1$ and the τ-axis at $\tau = \tau_{\mathrm{c}}$.
From (4.47) and (4.23) show that

$$\left(\frac{\partial h}{\partial m}\right)_\tau = \frac{[\tau g'(m_{\mathfrak{a}}) + 1][\tau g'(m_{\mathfrak{b}}) + 1] - 1}{\frac{1}{2}\tau[g'(m_{\mathfrak{a}}) + g'(m_{\mathfrak{b}})]}\,.$$

Hence show that, on the $m_{\mathfrak{b}} = 0$ curve,

$$\left(\frac{\partial h}{\partial m}\right)_\tau = \frac{[\tau(1-m_{\mathrm{s}}^2)^{-1} - \lambda](\tau - \lambda) - 1}{\frac{1}{2}[\tau(1-m_{\mathrm{s}}^2)^{-1} + \tau - 2(1+\lambda)]}\,.$$

Using (4.31) and a relation similar to (3.13), show that, on the $m_{\mathfrak{b}} = 0$ curve, $(\partial h/\partial m)_\tau < 0$ for $\tau \le \lambda$.

4.4 For the Ising model ferrimagnet in the mean-field approximation, show that the zero-field susceptibility above the critical temperature τ_{c} is given by

$$\chi_T^{-1} = K\frac{(\tau - \lambda_{\mathfrak{a}})(\tau - \lambda_{\mathfrak{b}}) - s}{r^2(\tau - \lambda_{\mathfrak{b}}) + s(\tau - \lambda_{\mathfrak{a}}) - 2rs}\,,$$

where K is a constant. Hence show that $\chi_T^{-1} \to 0$ as $\tau \to \tau_c + 0$. Show that this expression can be transformed to

$$\chi_T^{-1} = \frac{K}{r^2+s}\left\{\tau + \frac{2rs - \lambda_b s - \lambda_a r^2}{r^2+s} - \frac{s}{r^2+s}\frac{[r(\lambda_a - \lambda_b) + r^2 - s]^2}{(r^2+s)\tau - (\lambda_a s + \lambda_b r^2 + 2rs)}\right\}.$$

Hence show that the paramagnetic Curie temperature

4.5 Curie temperature!paramagnetic is lower than τ_c and sketch the form of χ_T^{-1} above the critical temperature.

4.6 For the Ising model ferrimagnet in the mean-field approximation show, by substituting $h = 0$ and $sm_b = -rm_a$ into the equilibrium equations (4.66), that at a compensation point

$$\ln\left[\frac{s + rm_a}{s - rm_a}\right] - \frac{s + \lambda_b r}{s(r + \lambda_a)}\ln\left[\frac{1 + m_a}{1 - m_a}\right] = 0,$$

$$\tau^{-1} = \frac{1}{m_a(r + \lambda_a)}\ln\left[\frac{1 + m_a}{1 - m_a}\right].$$

If $r < s$, show that there is a non-zero solution for m_a if $r_0 < r$, where r_0 is defined by (4.73). If $r > s$, show that the corresponding condition is $r < r_0$.

5. Lattice Gases

5.1 Introduction

In a classical fluid composed of M similar spherical molecules contained in a volume $\widetilde{V}$, the molecules are regarded as centres of force situated at points $\boldsymbol{r}_1, \boldsymbol{r}_2, \ldots, \boldsymbol{r}_M$ in $\widetilde{V}$. The configurational energy is the sum of $\frac{1}{2}M(M-1)$ terms, each representing the interaction energy of a pair of centres and dependent on the distance between them. Thus, the configurational energy is

$$E(\boldsymbol{r}_1, \boldsymbol{r}_2, \ldots, \boldsymbol{r}_M) = \sum_{\{i<j\}} u(|\boldsymbol{r}_i - \boldsymbol{r}_j|). \tag{5.1}$$

The pair interaction energy $u(r)$ is positive for small r and $u(r) \to \infty$ as $r \to 0$. For larger r values, $u(r)$ becomes negative with a minimum at $r = r_0$ and, finally, $u(r) \to 0$ as $r \to \infty$ (see Fig. 5.1). The interaction is, therefore, strongly repulsive at small r, but becomes attractive for $r > r_0$. In Chap. 1 we saw how van der Waals incorporated the essential elements of a hard core repulsion and an attractive interaction into a simple theory.[1] Now suppose that the approximation is made of replacing the continuous volume by a regular lattice of N sites and confining the centres of force to lattice sites with not more than one centre on each site. The M sites coinciding with a centre of force are said to be 'occupied' by molecules and the remaining $N - M$ sites are 'vacancies' or 'holes'. A model of this type is called a *lattice gas*.[2] The volume per site is

$$v_0 = \widetilde{V}/N\,, \tag{5.2}$$

which is regarded as constant. For $d = 1$ and $d = 2$, 'volume' must be replaced by 'length' and 'area' respectively. An interaction energy ε_k is associated with each k-th neighbour pair of occupied sites, where $\varepsilon_k = u(r_k)$, r_k being the k-th neighbour pair distance on the lattice. An alternative is to put $\varepsilon_k = \bar{u}(r_k)$, where $\bar{u}$ is an average of $u(r)$ in a suitable neighbourhood of the k-th neighbour site.

[1] For an introduction to continuous fluid models, see Rowlinson and Swinton (1982), Chap. 7.

[2] This is accepted terminology although the term 'gas' is not entirely appropriate since phase separation occurs for $d = 2, 3$. It would be better to use the name 'lattice fluid'.

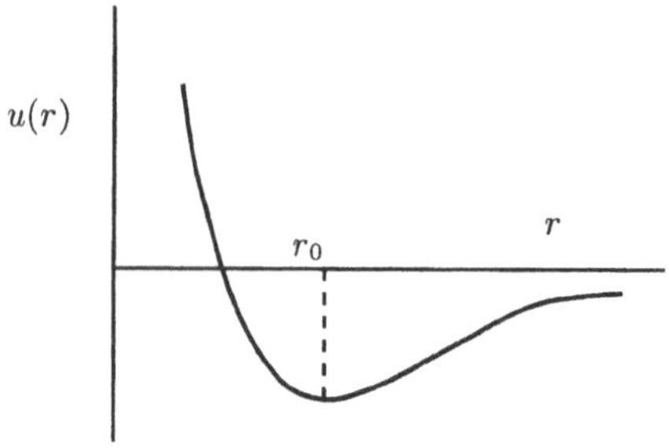

Fig. 5.1. The interaction energy for two molecules with centres at a distance r.

If the mesh is so coarse that $r_1 = r_0$ (see Fig. 5.1) then $\varepsilon_1 = u(r_0) < 0$. The *simple lattice gas* is obtained if we then put $\varepsilon_k = 0$ for $k > 1$, so that the interaction is confined to nearest-neighbour molecules. The hard core and attractive elements are still present, the hard core because two molecules cannot approach nearer than r_1 and the attraction because $\varepsilon_1 < 0$. The simple lattice gas is mathematically equivalent to the Ising model ferromagnet as is shown in Sect. 5.3.

The hard core can be extended by putting $\varepsilon_1 = \infty$ so that no two nearest-neighbour sites can both be occupied; this is a *nearest-neighbour exclusion model.* With $\varepsilon_1 = \varepsilon_2 = \cdots = \varepsilon_m = \infty$ we have an m-th neighbour exclusion model. This type of model can undergo a solid–fluid (melting) transition (Runnels 1972). Generally, the chief advantage of lattice gas models is their versatility as they can be modified to allow for non-isotropic and many-body interactions (see Chaps. 7 and 9).

If the standard volume in (1.54) is taken to be the constant volume per site v_0, then the dimensionless volume is $V = N$, the number of lattice sites, and the conjugate field is $-P = -\widetilde{P}v_0$. For a $d = 3$ lattice $\widetilde{P}$ has the familiar dimension of energy/volume=force/area, but for $d = 2$ and $d = 1$ it has respectively the dimensions of energy/area=force/length and energy/length=force. A two-dimensional or surface pressure can be measured experimentally for monolayers and the one-dimensional pressure is just a compression applied to the lattice regarded as a rod. The number density ρ is defined so that $\rho = 1$ at closest packing. Where two molecules can occupy nearest-neighbour sites,

$$\rho = M/N \,. \tag{5.3}$$

If we define the Helmholtz free energy density as 'per site', that is per unit of reduced volume, then (1.77) and (1.78) with $n = 2$, $\eta = 0$ give

$$\begin{aligned} a(T,\rho) &= \rho\mu - P\,, \\ \mathrm{d}a &= \mu \mathrm{d}\rho\,, \end{aligned} \tag{5.4}$$

where the second relation applies at constant T. Hence

$$P = \rho\left(\frac{\partial a}{\partial \rho}\right)_T - a\,. \tag{5.5}$$

An alternative approach is to think of the volume as divided into N equal 'cells'. It is assumed that:

(i) there is not more than one molecule in each cell;

(ii) the interaction energy between two molecules depends only on the relative position of their cells and is independent of their location inside the cells.

If a lattice site is associated with each cell, the corresponding 'lattice gas' and 'cell' models are completely equivalent.

5.2 The One-Dimensional Lattice Gas and the Continuous Limit

Critical phenomena will not occur at $T > 0$ in a $d = 1$ model with interactions of finite range because any long-range order can be destroyed by a small number of local fluctuations. (We have seen in the $d = 1$ Ising model that with a few 'wrong' nearest-neighbour pairs a configurational energy near the ground-state energy can be attained without a majority of spins pointing in one direction.) Hence a $d = 1$ lattice gas will not display phase separation. However it can, as we shall see in Sect. 5.5, illustrate cooperative phenomena based on short-range order. Also the relation between lattice and continuous models is particularly clear in one dimension.

We consider lattice approximants to a *Takahashi model*, which is a one-dimensional continuous model in which M hard-rod molecules of length a lie on a line of total length L. The hard-rod property is introduced by putting $u(r) = \infty$ for $r < a$, where r is the distance between the centres of a pair of molecules. For $r \geq a$, $u(r)$ is piecewise continuous and $u(r) = 0$ for $r \geq 2a$. The last condition ensures that interaction occurs only between successive molecules on the line. It is convenient to introduce a function $v(R) = u(r)$ where $R = r - a$ is the length of unoccupied line between the two rods (see Fig. 5.2(a)). Clearly $v(R) = 0$ for $R \geq a$.

5.2.1 The Lattice Gas

Now take a $d = 1$ lattice of N sites with nearest-neighbour site distance b where $b = a/n$, n being a positive integer and a measure of the 'fineness' of the mesh. Place M molecules on the lattice with k-th neighbour interactions $\varepsilon_k = u(kb+0)$ where $u(r)$ is the interaction energy function for the continuous model.[3] Thus $\varepsilon_1 = \cdots = \varepsilon_{n-1} = \infty$, which ensures that the hard core of each molecule is a rod occupying n sites. Clearly we have an $(n-1)$-th neighbour

[3] The +0 symbol indicates that if $r = kb$ is a point where $u(r)$ has a discontinuity we take the limiting value of $u(r)$ as r approaches kb from above.

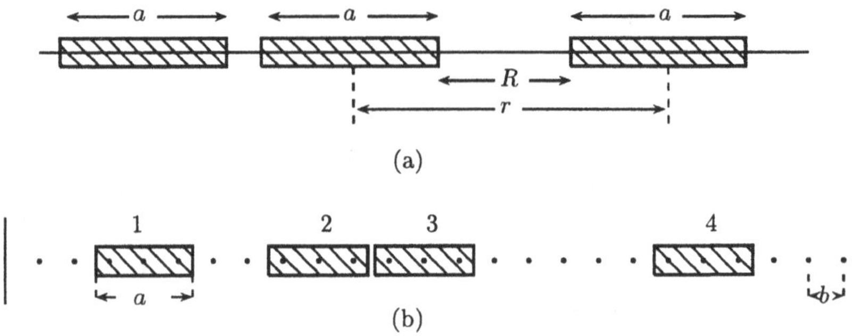

Fig. 5.2. (a) The continuous one-dimensional model. **(b)** The one-dimensional lattice model with each molecule occupying three sites. The left-hand boundary is shown with $q_0 = q_1 = 2$, $q_2 = 0$, $q_3 = 4$.

exclusion model in the sense of Sect. 5.1. Also, $\varepsilon_k = 0$ for $k \geq 2n$. It is convenient to introduce new interaction parameters ε'_q where $\varepsilon'_q = \varepsilon_k$ for $q = k - n$. Here q is the number of vacant sites between the two hard rod molecules and corresponds to the continuous variable R introduced above. In fact, $\varepsilon'_q = v(qb + 0)$ and $\varepsilon'_q = 0$ for $q \geq n$. The microstate is completely specified by the set of integers $q_0, q_1, \ldots, q_{M-1}, q_M$ where q_j is the number of vacant sites between the jth molecule from the left and the $(j+1)$th molecule. The integers q_0 and q_M are the numbers of vacant sites between the first and last molecules respectively and the boundaries (see Fig. 5.2(b)). It is assumed that there is no interaction between the boundaries and the adjacent molecules. Then the number of sites N and the configurational energy E are given in terms of the q_j by

$$\begin{aligned} N &= Mn + \sum_{j=0}^{M} q_j\,, \\ E &= \sum_{j=1}^{M-1} \varepsilon'_{q_j}\,. \end{aligned} \tag{5.6}$$

The constant-pressure distribution of Sect. 2.3.2, will be used, in which the one-dimensional pressure $P = \widetilde{P}b$, with M and T independent variables. The number of lattice sites is now a fluctuating quantity and (according to the convention prescribed in Sect. 2.2) will be specified as $\widehat{N}$. Then from (2.48),

$$\widehat{H} = E + P\widehat{N}\,. \tag{5.7}$$

Since there is only one component, (2.48), (2.49) and (1.12) yield

$$Z = \exp(M\mu/T) = \sum_{\widehat{N}=Mn}^{\infty} \xi^{\widehat{N}} \sum_{q_j} \exp(-E/T)\,, \tag{5.8}$$

where the inner summation is over all values of q_j compatible with a given $\widehat{N}$ and

$$\xi = \exp(-\widetilde{P}b/T) = \exp(-P/T)\,. \tag{5.9}$$

The two stage summation of (5.8) is equivalent to summing all q_j from 0 to ∞. For j in the range 1 to $M-1$ each q_j yields an identical factor. After substituting the expressions for N and E given by (5.6) into (5.8) and using the condition $\varepsilon'_q = 0$ for $q \geq n$ this has the form

$$\sum_{q=0}^{\infty} \xi^q \exp(-\varepsilon'_q/T) = \sum_{q=0}^{n-1} \xi^q \exp(-\varepsilon'_q/T) + \frac{\xi^n}{1-\xi}\,. \tag{5.10}$$

The partition function contains a product of $M-1$ of these factors together with factors for q_0 and q_M, the numbers of vacant sites between the end rods and the boundaries. These each give the geometrical progression

$$\sum_{i=0}^{\infty} \xi^i = \frac{1}{1-\xi}\,. \tag{5.11}$$

Finally, the Mn term in the first of equations (5.6) gives the factor ξ^{Mn} and thus

$$Z = \frac{\xi^n}{(1-\xi)^2}\Psi^{M-1},$$

$$\Psi = \xi^n \left[\sum_{q=0}^{n-1} \xi^q \exp(-\varepsilon'_q/T) + \frac{\xi^n}{1-\xi}\right]. \tag{5.12}$$

Since $M^{-1}\ln[\xi^n(1-\xi)^{-2}\Psi^{-1}] \to 0$, as $M \to \infty$, at constant P and T,

$$G = M\mu = -T\ln Z = -MT\ln\Psi\,, \tag{5.13}$$

in the thermodynamic limit. An alternative form for Ψ is

$$\Psi = \xi^n \left\{\sum_{q=0}^{n-1} \xi^q[\exp(-\varepsilon'_q/T) - 1] + \frac{1}{1-\xi}\right\}, \tag{5.14}$$

where the Boltzmann factors $\exp(-\varepsilon'_q/T)$ have been replaced by *Mayer factors* $\exp(-\varepsilon'_q/T) - 1$. Using (5.14) we can average $\widehat{N}$ over the constant-pressure distribution to obtain ℓ_n, the mean length per molecule (for a particular value of n). In this case the equilibrium value of $\widehat{N}$ can be obtained from (2.50). With (5.13) and (5.9) this gives

$$\ell_n = \frac{\widehat{N}b}{M} = -Tb\frac{\partial \ln\Psi}{\partial P} = b\xi\frac{\partial \ln\Psi}{\partial \xi}\,. \tag{5.15}$$

From (5.14) and (5.15), and noting that the constant $a = bn$,

$$\ell_n = a + \frac{b\xi}{1-\xi} + \frac{b\sum_{q=0}^{n-1}\xi^q[q-(q+1)\xi][\exp(-\varepsilon'_q/T)-1]}{1+(1-\xi)\sum_{q=0}^{n-1}\xi^q[\exp(-\varepsilon'_q/T)-1]} . \tag{5.16}$$

As $P \to \infty$ at constant T the parameter $\xi \to 0$ and, from (5.16), $\ell_n \to a$, which means that the hard-core rods are tightly packed on the lattice. For an assembly with hard-core interactions only, all $\varepsilon'_q = 0$ and the last term of (5.16) disappears. For the general case the first and second terms on the right-hand side of (5.16) represent hard-core effects, whereas the last term results from interactions outside the hard core.

5.2.2 The Continuum Limit

The continuum limit is attained by taking $n \to \infty$ and $b = a/n \to 0$ with a being kept constant. For this b must be shown explicitly, by using $\widetilde{P}$ rather than P and, from (5.9), for small b

$$1 - \xi = \widetilde{P}b/T + \mathrm{O}(b^2) . \tag{5.17}$$

The sums in (5.16) become Riemann integrals as $b \to 0$ and

$$\lim_{n\to\infty} \ell_n = \ell_c , \tag{5.18}$$

where

$$\ell_c = a + \frac{T}{\widetilde{P}} + \frac{\int_0^a (\widetilde{P}R/T - 1)\exp(-\widetilde{P}R/T)[\exp(-v(R)/T)-1]\mathrm{d}R}{1 + (\widetilde{P}/T)\int_0^a \exp(-\widetilde{P}R/T)[\exp(-v(R)/T)-1]\mathrm{d}R} . \tag{5.19}$$

This is a form of the equation of state for the Takahashi model described at the beginning of this section (Lieb and Mattis 1966, Mattis 1993). Thus it has been shown that, for rods of fixed length, the equation of state for the lattice approximation goes over to that of the continuous model when the lattice mesh becomes infinitely fine. It can be shown that the same limiting form is attained if, instead of using $\varepsilon'_q = v(qb+0)$, the averaged value

$$\varepsilon'_q = \frac{1}{b}\int_{qb}^{(q+1)b} v(R)\mathrm{d}R \tag{5.20}$$

is used. In fact, the convergence is then frequently faster (Bell 1980).

For a *Tonks gas*, which is the case where there is the hard-core interaction only, $v(R) = 0$, and the final term in (5.19) disappears giving

$$\widetilde{P} = \frac{T}{\ell_c - a} . \tag{5.21}$$

If the attractive term $c/\tilde{v}^2$ is omitted from the van der Waals equation (1.43) then it becomes identical in form to (5.21), with the volumes $\tilde{v}$ and b taking the place of the lengths ℓ_c and a respectively. It follows that the van der Waals treatment of hard-core terms is accurate in one dimension. A density ρ can be defined by

$$\rho = \frac{nM}{N} = \frac{a}{\ell_n} \to \frac{a}{\ell_c} \qquad \text{as } n \to \infty , \tag{5.22}$$

with $\rho = 1$, its maximum value, at closest packing. From Sect. 3.2 we can treat the Tonks gas as a reference model and incorporate a Lebowitz–Penrose-type long-range interaction by adding a term $-c/\ell_c^2$ to the right-hand side of (5.21). The resulting equation is then identical in form to that of van der Waals.

5.3 The Simple Lattice Gas and the Ising Model

Consider the simple lattice gas, described in Sect. 5.1, putting $\varepsilon_1 = -\varepsilon$, ($\varepsilon > 0$). A microstate can be specified by the N quantities σ_i, where $\sigma_i = 1$ or -1 according to whether site i is occupied or vacant. Using the grand distribution of Sect. 2.3.3, the configurational Hamiltonian for this one-component assembly is, from (2.51),

$$\widehat{H} = -\frac{1}{4}\varepsilon \sum_{\{i,j\}}^{(\text{n.n.})} (\sigma_i + 1)(\sigma_j + 1) - \frac{1}{2}\mu \sum_{\{i\}} (\sigma_i + 1) . \tag{5.23}$$

It will be seen that there is a contribution to the first summation only from terms where $\sigma_i = \sigma_j = 1$. To establish the equivalence with the ferromagnetic Ising model, use the fact that there are in all $\frac{1}{2}zN$ nearest-neighbour site pairs and rewrite (5.23) as

$$\widehat{H} = -\frac{1}{8}N(4\mu + z\varepsilon) - \frac{1}{4}\varepsilon \sum_{\{i,j\}}^{(\text{n.n.})} \sigma_i\sigma_j - \frac{1}{4}(2\mu + z\varepsilon) \sum_{\{i\}} \sigma_i . \tag{5.24}$$

Apart from the constant term this is the Hamiltonian (2.66) for a ferromagnetic Ising model, the positive and negative spins respectively corresponding to the molecules and holes in the lattice gas. Formally we can write, for the equivalent Ising model,

$$J_e = \frac{1}{4}\varepsilon , \qquad \mathcal{H}_e = \frac{1}{4}(2\mu + z\varepsilon) , \tag{5.25}$$

$$\widehat{H} = -\frac{1}{8}N(4\mu + z\varepsilon) + \widehat{H}^{(\text{I.M.})} , \tag{5.26}$$

where the last term in (5.26) denotes the Ising model Hamiltonian with $J = J_e$ and $\mathcal{H} = \mathcal{H}_e$, as given by (5.25). It follows from (2.51) that the grand partition function for the lattice gas is related to the constant magnetic field partition function for the ferromagnetic Ising model by

$$Z(N,T,\mu) = \exp\left[\frac{N(4\mu + z\varepsilon)}{8T}\right] Z^{(\text{I.M.})}(N,T,\mathcal{H}). \tag{5.27}$$

Unlike the ferromagnet–antiferromagnet equivalence, which is confined to $\mathcal{H} = 0$, this applies for all $\mathcal{H}$.

Thermodynamic functions for the lattice gas can now be obtained from those for the Ising model. From (1.19), (2.52), (5.27) and (5.2)

$$\begin{aligned} P &= \tfrac{1}{8}(4\mu + z\varepsilon) + N^{-1}T\ln Z^{(\text{I.M.})}(N,T,\mathcal{H}_\text{e}) \\ &= \tfrac{1}{8}(4\mu + z\varepsilon) - f^{(\text{I.M.})}(T,\mathcal{H}_\text{e}), \end{aligned} \tag{5.28}$$

where $f^{(\text{I.M.})}$ is the constant magnetic field free energy per site for the ferromagnetic Ising model. The fluid number density ρ satisfies the relations

$$\rho = M/N = \tfrac{1}{2}(1 + \langle\sigma_i\rangle) = \tfrac{1}{2}(1 + m), \tag{5.29}$$

where M is the equilibrium number of lattice gas molecules at the given values of T and μ, and m is the relative magnetization in the equivalent Ising model.

As an illustration, we show that (5.28) is satisfied by the explicit expressions previously obtained for the one-dimensional Ising model and lattice gas respectively. Using the notation of Sect. 5.2, the simple lattice gas case corresponds to putting $n = 1$ ($b = a$) and $\varepsilon_0' = \varepsilon_1 = -\varepsilon$. Equations (5.9), (5.12) and (5.13) then yield a relation between μ, P and T in the form

$$\Psi = \exp(-\mu/T) = \xi\left[\exp(\varepsilon/T) + \frac{\xi}{1-\xi}\right], \tag{5.30}$$

$$\xi = \exp(-P/T). \tag{5.31}$$

To obtain P in terms of μ and T, (5.30) is first expressed as a quadratic equation

$$\Psi\xi^{-2} - [\Psi + \exp(\varepsilon/T)]\xi^{-1} + [\exp(\varepsilon/T) - 1] = 0 \tag{5.32}$$

for ξ^{-1}. Solving and choosing the larger root to ensure that $P > 0$ gives

$$\exp(P/T) = \frac{1}{2\Psi}[\exp(\varepsilon/T) + \Psi + \{[\exp(\varepsilon/T) - \Psi]^2 + 4\Psi\}^{\frac{1}{2}}]. \tag{5.33}$$

With $z = 2$, (5.25) gives

$$\Psi = \mathfrak{X}^4/\mathfrak{Z}^2, \qquad \exp(\varepsilon/T) = \mathfrak{X}^4, \tag{5.34}$$

where, as in (2.75), $\mathfrak{X} = \exp(J_\text{e}/T)$, $\mathfrak{Z} = \exp(\mathcal{H}_\text{e}/T)$. Substitution into (5.33) then gives

$$\exp(P/T) = \mathfrak{Z}\mathfrak{X}^{-1}\lambda_1 = \exp\left[\frac{2\mu + \varepsilon}{4T}\right]\lambda_1, \tag{5.35}$$

where λ_1 is defined by (2.94). By (2.96) this is equivalent to the relation (5.28) in the $d = 1$ case. In Example 5.3 the reader is asked to verify (5.29) for $d = 1$.

5.4 Phase Separation in the Simple Lattice Gas

We now consider the simple lattice gas for $d = 2, 3$ and in particular the significance of the first-order transition line on the $\mathcal{H} = 0$ axis in the $(T, \mathcal{H})$ plane of the equivalent Ising model (see Fig. 3.3). Defining

$$\Delta\mu = \mu + \frac{1}{2}z\varepsilon = 2\mathcal{H}_c\,, \tag{5.36}$$

the internal state of the fluid depends on the values of T and $\Delta\mu$, which may be regarded as defining a point in a $(T, \Delta\mu)$ plane. By (5.36) the $\Delta\mu = 0$ axis is equivalent to the $\mathcal{H} = 0$ axis in the $(T, \mathcal{H})$ plane of the Ising model. From (5.29)

$$\lim_{\Delta\mu \to \pm 0} \rho = \frac{1}{2}(1 \pm m_s)\,, \tag{5.37}$$

for $T < T_c$, where m_s is the Ising model spontaneous magnetization defined by (3.11). There is thus a discontinuous density increase, equal to m_s, as the $\Delta\mu = 0$ axis is crossed from below to above. From the Ising model symmetry

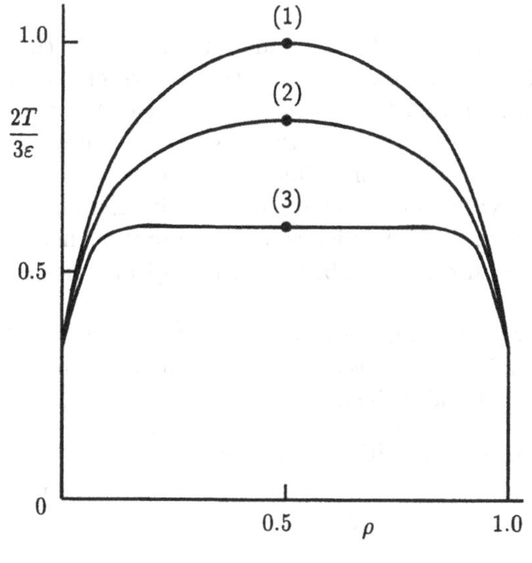

Fig. 5.3. Plane triangular lattice gas vapour–liquid coexistence curves: (1) mean-field (2) first-order pair (3) exact.

relation (2.68) and from (5.28) a pressure symmetry relation for the fluid can be deduced in the form

$$P(T, \Delta\mu) = P(T, -\Delta\mu)\,. \tag{5.38}$$

From this relation

$$\lim_{\Delta\mu \to +0} P(T, \Delta\mu) = \lim_{\Delta\mu \to -0} P(T, \Delta\mu)\,, \tag{5.39}$$

so that the phase equilibrium conditions (1.41) are satisfied along the interval $(0, T_c)$ of the $\Delta\mu = 0$ axis, where ρ is discontinuous, T_c being the Ising model critical temperature with $J = \frac{1}{4}\varepsilon$. For $T \geq T_c$, $\rho = \frac{1}{2}$ on the $\Delta\mu = 0$ axis by (5.29) and all thermodynamic quantities are continuous. It follows that T_c is the critical temperature for separation into a (high-density) liquid and a (low-density) vapour phase and the interval $(0, T_c)$ on the $\Delta\mu = 0$ axis is a first-order transition line. The condition $\Delta\mu = 0$ defines the critical isochore and the critical density is $\rho_c = \frac{1}{2}$. In Fig. 5.3 the conjugate phase densities for the triangular lattice (z = 6) are shown as functions of T using the mean-field approximation, the first-order pair approximation (see Chap. 7) and the exact solution (see Chap. 8) respectively. From (5.37) the coexistence curves in the (ρ, T) plane are symmetrical about the line $\rho = \frac{1}{2}$. Since $\Delta\mu = 0$ at phase equilibrium the corresponding pressure, by (5.28), is

$$P = -\frac{1}{8}z\varepsilon - f^{(\mathrm{I.M.})}(T, 0)\,. \tag{5.40}$$

As $T \to 0$ the Ising model ferromagnet tends to complete spin alignment and hence $f^{(\mathrm{I.M.})} \to -\frac{1}{2}zJ$. Since $J_c = \frac{1}{4}\varepsilon$, this implies $P \to 0$, as should be the case for a physically reasonable model.

5.5 A One-Dimensional Water-Like Model

The existence of a density maximum at 4 °C in liquid water is due to the persistence above the melting point of regions of open structure resembling that of ice I (see Appendix A.3). As this is a short-range order effect it is possible to simulate it in one dimension. Although the directional character of hydrogen bonding cannot be built into a one-dimensional model there is an open structure stable at low pressures if the configuration with minimum enthalpy has a density ρ (defined by (5.22)) which is less than unity. This is sufficient to give density maxima on certain isobars, and it is of interest to compare results for lattice and continuous models.

Several 'open structure' models were considered by Bell (1969, 1980). The one discussed here is the ***displaced parabolic well model*** with the continuous potential function shown in Fig. 5.4 and defined by

$$v(R) = \begin{cases} 0\,, & 0 \leq R \leq \frac{1}{2}a,\; a \leq R\,, \\ -\dfrac{8\mathcal{V}_0}{a^2}(2R - a)(a - R)\,, & \frac{1}{2}a \leq R \leq a\,. \end{cases} \tag{5.41}$$

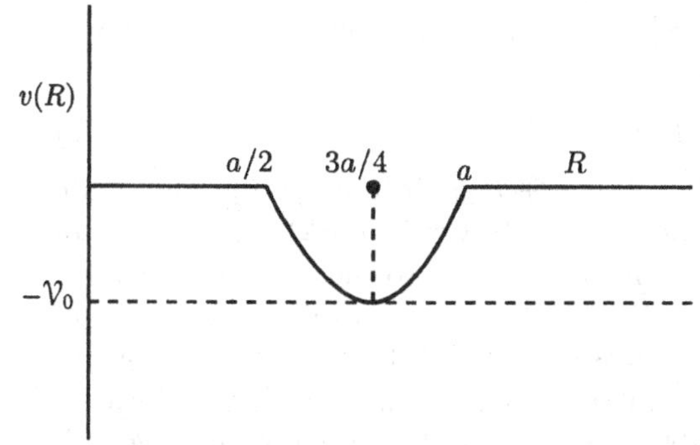

Fig. 5.4. Displaced parabolic well potential.

The minimum energy is $-\mathcal{V}_0$ at $R = \frac{3}{4}a$. Now consider the ground state, which is the configuration of least enthalpy, towards which the system tends as $T \to 0$. If the separation between successive centres in the ground state is $a+R$ then, since the constant-pressure distribution is used, the enthalpy per molecule is

$$h(R) = (a+R)\widetilde{P} + v(R)\,. \tag{5.42}$$

The least value of $h(R)$ for given $\widetilde{P}$ may occur either at $R = 0$, which gives a close-packed ground state, or at some point $R = R_{\text{min}}$, where there is a minimum defined by

$$\frac{\mathrm{d}h}{\mathrm{d}R} = 0, \qquad \frac{\mathrm{d}^2 h}{\mathrm{d}R^2} > 0\,. \tag{5.43}$$

From (5.41) a minimum can occur only in the range $\frac{1}{2}a < R < a$ where, by (5.42),

$$\begin{aligned} \frac{\mathrm{d}h}{\mathrm{d}R} &= \widetilde{P} + \frac{8\mathcal{V}_0}{a^2}(4R - 3a)\,, \\ \frac{\mathrm{d}^2 h}{\mathrm{d}R^2} &= \frac{32\mathcal{V}_0}{a^2}\,. \end{aligned} \tag{5.44}$$

Substitution of (5.44) into (5.43) yields

$$\begin{aligned} R_{\text{min}} &= \frac{3}{4}a - \frac{\widetilde{P}a^2}{32\mathcal{V}_0}\,, \\ h(R_{\text{min}}) &= \frac{8\mathcal{V}_0}{a^2}(a^2 - 2R_{\text{min}}^2) + \widetilde{P}a\,. \end{aligned} \tag{5.45}$$

Since, for consistency, $R_{\text{min}} > \frac{1}{2}a$, a minimum exists only for $\widetilde{P}$ in the range $(0, 8\mathcal{V}_0/a)$. However, the pressure range where $h(R_{\text{min}})$ is the least value of $h(R)$ is much more restricted. From the second relation of (5.45), $h(R_{\text{min}}) < h(0) = \widetilde{P}a$, when $R_{\text{min}} > a/\sqrt{2}$. There will be an open ground state with $R =$

$R_{\min}$ for the pressure range $(0, \widetilde{P}_0)$ where $\widetilde{P}_0$ corresponds to $R_{\min} = a/\sqrt{2}$. From the first relation of (5.45),

$$\widetilde{P}_0 = 8(\mathcal{V}_0/a)(3 - 2\sqrt{2}) = 1.3726(\mathcal{V}_0/a)\,. \tag{5.46}$$

The length per molecule ℓ at $T = 0$ is $a + R_{\min}$ in the open ground state and thus decreases from $\ell = 1.75a$ at $\widetilde{P} = 0$ to $\ell = (1 + \sqrt{2})a/\sqrt{2} = 1.7071a$ at $\widetilde{P} = \widetilde{P}_0$. Then there is a discontinuous decrease to the close-packed value $\ell = a$, which applies for all $\widetilde{P} > \widetilde{P}_0$. For $T > 0$, values of ℓ, and hence of the density ρ, can be obtained by substituting (5.41) into (5.18) and (5.19). It can be shown that, for any $\widetilde{P}$ in the range $(0, \widetilde{P}_0)$, ℓ decreases initially as T increases from zero (Bell 1969). Since ℓ must eventually increase with T, there is a maximum on any density isobar in the open structure pressure range $(0, \widetilde{P}_0)$.

In view of the behaviour of $v(R)$ in the two halves of the range $0 < R < a$, an even value of n is necessary for lattice approximations to the displaced well model. The simplest case is $n = 2$ and, using the averaging procedure of equation (5.20),

$$\varepsilon'_0 = 0\,, \qquad \varepsilon'_1 = -\frac{2}{3}\mathcal{V}_0\,. \tag{5.47}$$

There are only two possibilities for the ground state. One is close packing with $q = 0$ for all successive pairs of molecules, and the other is the open minimum energy configuration with $q = 1$. The molecular enthalpies for the two states are respectively

$$\begin{aligned} h(0) &= \widetilde{P}a\,, \\ h(1) &= \widetilde{P}(a + b) + \varepsilon'_1 = \frac{3}{2}\widetilde{P}a - \frac{2}{3}\mathcal{V}_0\,. \end{aligned} \tag{5.48}$$

For $\widetilde{P}a < \frac{4}{3}\mathcal{V}_0$, $h(0) > h(1)$ and the open structure with $\ell = \frac{3}{2}a$ is stable at $T = 0$. On the other hand, for $\widetilde{P}a > \frac{4}{3}\mathcal{V}_0$, $h(1) > h(0)$ and the close-packed state with $\ell = a$ is stable at $T = 0$. Hence there is a discontinuous decrease from $\ell_2 = \frac{3}{2}a$ to $\ell_2 = a$ at $\widetilde{P} = \frac{4}{3}\mathcal{V}_0/a$, which corresponds to $\widetilde{P}_0$ in the continuous model.

For $T > 0$, putting $n = 2$ and, correspondingly, $b = \frac{1}{2}a$, (5.16) yields, after substitution from (5.47),

$$\ell_2 = a + \frac{a\xi}{2(1 - \xi)} + \frac{a(1 - 2\xi)\xi[\exp(2\mathcal{V}_0/3T) - 1]}{2\{1 + \xi(1 - \xi)[\exp(2\mathcal{V}_0/3T) - 1]\}}\,, \tag{5.49}$$

where

$$\xi = \exp(-\widetilde{P}a/2T)\,. \tag{5.50}$$

Now $\xi \to 0$ as $T \to 0$ at fixed $\widetilde{P}$, whereas $\xi \exp(\frac{2}{3}\mathcal{V}_0/T) \to 0$ or $\to \infty$ according to whether $\widetilde{P}a > \frac{4}{3}\mathcal{V}_0$ or $\widetilde{P}a < \frac{4}{3}\mathcal{V}_0$. Hence $\ell_2 \to \frac{3}{2}a$ as $T \to 0$ for $\widetilde{P}a < \frac{4}{3}\mathcal{V}_0$ and $\ell_2 \to a$ as $T \to 0$ for $\widetilde{P}a > \frac{4}{3}\mathcal{V}_0$, agreeing with the ground-state analysis. After a little manipulation, (5.49) becomes

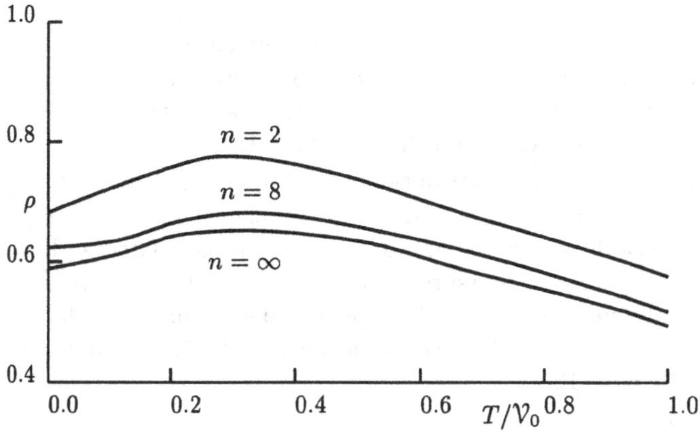

Fig. 5.5. Temperature–density isobars for $\widetilde{P} = \mathcal{V}_0/a$ with the potential of Fig. 5.4. n denotes number of sites occupied by hard core of one molecule. (Reprinted from Bell (1980), by permission of the publisher IOP Publishing Ltd.)

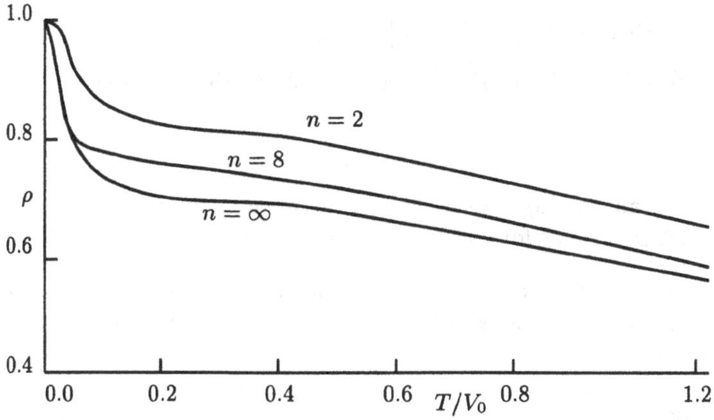

Fig. 5.6. As Fig. 5.5 with $\widetilde{P} = \frac{3}{2}\mathcal{V}_0/a$. (Reprinted from Bell (1980), by permission of the publisher IOP Publishing Ltd.)

$$\ell_2 = \frac{3}{2}a - \frac{a(1-2\xi)}{2[1-\xi]\{1+\xi(1-\xi)[\exp(2\mathcal{V}_0/3T)-1]\}} \,. \tag{5.51}$$

Hence, when $\ell_2 = \frac{3}{2}a$ at $T = 0$, there is a temperature range starting at $T = 0$ where $\ell_2 < \frac{3}{2}a$. Thus the density ρ initially increases as T is increased from 0 for constant $\widetilde{P} < \frac{4}{3}\mathcal{V}_0/a$. From (5.51), ℓ_2 regains its $T = 0$ value of $\frac{3}{2}a$ when $\xi = 1$, which corresponds to $T = \widetilde{P}a/\ln(4)$. For a pressure $\widetilde{P}$ in the open structure range $(0, \frac{4}{3}\mathcal{V}_0)$, the density has a maximum at some temperature $T < \widetilde{P}a/\ln(4)$.

Lattice approximations can also be obtained for higher n values. In Figs. 5.5 and 5.6 density isobars calculated for $n = 8$, using the averaging relation (5.20) for ε'_q, are compared with density isobars calculated for the extreme cases $n = 2$ and $n = \infty$ (continuous fluid). The curves in Fig. 5.5 are for $\widetilde{P} = \mathcal{V}_0/a$ which is in the open structure range in all three cases whereas those in Fig. 5.6 are for $\widetilde{P} = \frac{3}{2}\mathcal{V}_0/a$ which is outside the open structure range. It can be seen that a density maximum on an isobar is associated with an open ground state at the same pressure. The curves for lattice approximations are qualitatively similar to those for the continuous fluid, and for $n = 8$ are quite close quantitatively. Lattice gas water models for $d = 3$ are considered in Sects. 7.5 and 9.12.

Examples

5.1 Using the methods of Sect. 3.3, show that the mean-field Helmholtz free energy per site for the simple lattice gas is

$$a(T,\rho) = T[\rho \ln \rho + (1-\rho)\ln(1-\rho)] - \frac{1}{2}z\varepsilon\rho^2 .$$

By (5.5) or otherwise, find the critical point by solving the equations

$$\frac{\partial P}{\partial \rho} = 0 , \qquad \frac{\partial^2 P}{\partial \rho^2} = 0 .$$

Show that (5.25) and (5.28) are satisfied in the mean-field approximation and that $\mu = -\frac{1}{2}z\varepsilon$ for conjugate phases.

5.2 A continuous one-dimensional square well attractive potential model is defined, using the terminology of Sect. 5.2 and with $\varepsilon > 0$, by the relations

$$v(R) = \begin{cases} -\varepsilon , & 0 \le R < a , \\ 0 , & a < R . \end{cases}$$

Show that the equation of state for the nth lattice approximant is

$$\ell_n = a + \frac{a}{n\{\exp[\widetilde{P}a/(nT)] - 1\}} - \frac{a[\exp(\varepsilon/T) - 1]\exp(-\widetilde{P}a/T)}{1 + [\exp(\varepsilon/T) - 1][1 - \exp(-\widetilde{P}a/T)]} .$$

Given that ℓ_c denotes the length per molecule in the original continuous model, show that

$$\ell_c - \ell_n = \frac{T}{\widetilde{P}} - \frac{a}{n\{\exp[\widetilde{P}a/(nT)] - 1\}} .$$

Hence prove that $\ell_c > \ell_n$ for all n.

5.3 Using the Gibbs-Duhem relation (1.16), show that, for a $d = 1$ lattice gas at constant temperature,

$$\rho = a\left(\frac{\partial \widetilde{P}}{\partial \mu}\right)_T,$$

where the density ρ is defined by (5.22). Using (5.33) or otherwise, show that for the simple $d = 1$ lattice gas

$$\rho = \frac{1}{2} + \frac{\exp(\varepsilon/T) - \exp(-\mu/T)}{2\{[\exp(\varepsilon/T) - \exp(-\mu/T)]^2 + 4\exp(-\mu/T)\}^{1/2}}.$$

Hence show that $\rho = \frac{1}{2}(1+m)$, where m is the relative magnetization of the equivalent ferromagnetic Ising model.

5.4 A continuous *hard shoulder model* is like the model of question 2, except that $-\varepsilon$ is changed to ε ($\varepsilon > 0$) so that the interaction energy of rods with a gap $R < a$ between them is repulsive. Write down an expression for ℓ_n and show that, as $T/\varepsilon \to 0$ at fixed $\widetilde{P}$, $\ell_n \to a$ for $\widetilde{P}a > \varepsilon$ and $\ell_n \to 2a$ for $\widetilde{P}a < \varepsilon$. Explain this behaviour in terms of the enthalpy of the ground state.

6. Solid Mixtures and the Dilute Ising Model

6.1 The Restricted Grand Partition Function

We consider an isotropic solid mixture of M_{A} atoms of component A and M_{B} of component B in volume $\widetilde{V}$. This is modelled by a regular lattice of N sites, the volume v_0 per lattice site being as usual taken as constant and used as the standard volume in (1.54). Each site is occupied by an atom so that

$$M = M_{\mathrm{A}} + M_{\mathrm{B}} = \widetilde{V}/v_0 = V = N\,. \tag{6.1}$$

This is a restricted system in the sense of Sect. 2.3.4 with $n = 2$ and $Q_1 = M_{\mathrm{A}}$, $\xi_1 = \mu_{\mathrm{A}} - \mu_{\mathrm{B}}$, $Q_2 = N$ and $\xi_2 = \mu_{\mathrm{B}} - P$. A partition function of type (2.58) is obtained by taking N, T and $\mu_{\mathrm{B}} - \mu_{\mathrm{B}}$ as independent variables with

$$\begin{aligned} Z &= \sum_{\{M_{\mathrm{A}}+M_{\mathrm{B}}=M\}} \exp(-\widehat{H}/T)\,, \\ \widehat{H} &= E - (\mu_{\mathrm{A}} - \mu_{\mathrm{B}})M_{\mathrm{A}}\,, \end{aligned} \tag{6.2}$$

where E is the configurational energy. This is known as the *restricted grand partition function.* The free energy f per site is equal to $-T \ln Z/N$ for large N and, from (2.59),

$$f = a - (\mu_{\mathrm{A}} - \mu_{\mathrm{B}})x_{\mathrm{A}} = \mu_{\mathrm{B}} - P\,, \tag{6.3}$$

where the mole fraction $x_{\mathrm{A}} = M_{\mathrm{A}}/M = M_{\mathrm{A}}/N$ is the density conjugate to $\mu_{\mathrm{A}} - \mu_{\mathrm{B}}$ and $a = u - Ts$ is the Helmholtz free energy per site. From (2.60),

$$\mathrm{d}f = -s\mathrm{d}T - x_{\mathrm{A}}\mathrm{d}(\mu_{\mathrm{A}} - \mu_{\mathrm{B}})\,. \tag{6.4}$$

The condition for phase equilibrium is that T, $\mu_{\mathrm{A}} - \mu_{\mathrm{B}}$ and f have the same values in the conjugate phases. It follows from (6.4) that, on a phase equilibrium line in the $(\mu_A - \mu_B, T)$ plane,

$$\Delta s\mathrm{d}T + \Delta x_{\mathrm{A}}\mathrm{d}(\mu_{\mathrm{A}} - \mu_{\mathrm{B}}) = 0\,, \tag{6.5}$$

where Δs and Δx_{A} denote differences between the two phases. Hence, at the point where the composition of the two phases is the same ($\Delta x_{\mathrm{A}} = 0$) $\mathrm{d}T/\mathrm{d}(\mu_{\mathrm{A}} - \mu_{\mathrm{B}}) = 0$, so that the phase equilibrium curve has a stationary value. By a similar argument to that in Sect. 1.9.2 the conjugate phase curves in the (x_{A}, T) plane touch at the azeotropic point where $\Delta x_{\mathrm{A}} = 0$ and their common tangent there is horizontal, as in Fig. 1.7. Such behaviour is known both experimentally and theoretically when first-order transitions to a sublattice ordered state occur, although the term 'azeotrope' is not usually used for solid mixtures.

6.2 Binary Mixtures

In the lattice mixture we associate a variable σ_i with each site i where $\sigma_i = 1$ or -1 respectively according to whether the site is occupied by an A or a B atom. For a nearest-neighbour site pair (i, j) the expression $\frac{1}{4}(1+\sigma_i)(1+\sigma_j)$ is equal to 1 if both sites are occupied by A atoms and is zero otherwise. With similar expressions for B–B or A–B occupation,

$$
\begin{aligned}
M_{\mathrm{A}} &= \frac{1}{2}\sum_{\{i\}}(1+\sigma_i)\,, \qquad M_{\mathrm{B}} = \frac{1}{2}\sum_{\{i\}}(1-\sigma_i)\,,\\
N_{\mathrm{AA}} &= \frac{1}{4}\sum_{\{i,j\}}^{(\mathrm{n.n.})}(1+\sigma_i)(1+\sigma_j)\,,\\
N_{\mathrm{BB}} &= \frac{1}{4}\sum_{\{i,j\}}^{(\mathrm{n.n.})}(1-\sigma_i)(1-\sigma_j)\,,\\
N_{\mathrm{AB}} &= \frac{1}{4}\sum_{\{i,j\}}^{(\mathrm{n.n.})}[(1+\sigma_i)(1-\sigma_j)+(1-\sigma_i)(1+\sigma_j)]\\
&= \frac{1}{4}zN - \frac{1}{2}\sum_{\{i,j\}}^{(\mathrm{n.n.})}\sigma_i\sigma_j\,,
\end{aligned}
\tag{6.6}
$$

where N_{AA}, N_{BB} and N_{AB} respectively denote the numbers of AA, BB and AB nearest-neighbour pairs. Taking the configurational energy as a sum of nearest-neighbour pair interaction terms gives

$$
\begin{aligned}
E &= N_{\mathrm{AA}}\varepsilon_{\mathrm{AA}} + N_{\mathrm{BB}}\varepsilon_{\mathrm{BB}} + N_{\mathrm{AB}}\varepsilon_{\mathrm{AB}}\\
&= \frac{1}{8}zN\varepsilon_0 + \frac{1}{4}z(\varepsilon_{\mathrm{AA}}-\varepsilon_{\mathrm{BB}})\sum_{\{i\}}\sigma_i - \frac{1}{4}\varepsilon\sum_{\{i,j\}}^{(\mathrm{n.n.})}\sigma_i\sigma_j\,,
\end{aligned}
\tag{6.7}
$$

where

$$
\begin{aligned}
\varepsilon_0 &= 2\varepsilon_{\mathrm{AB}} + \varepsilon_{\mathrm{AA}} + \varepsilon_{\mathrm{BB}}\,,\\
\varepsilon &= 2\varepsilon_{\mathrm{AB}} - \varepsilon_{\mathrm{AA}} - \varepsilon_{\mathrm{BB}}\,.
\end{aligned}
\tag{6.8}
$$

From (6.6), (6.7) can be expressed in the alternative forms

$$E = \frac{1}{2}z(\varepsilon_{AA}M_A + \varepsilon_{BB}M_B) + \frac{1}{2}\varepsilon N_{AB}\,, \tag{6.9}$$

$$E = \frac{1}{2}z\varepsilon_{BB}N + z(\varepsilon_{AB} - \varepsilon_{BB})M_A - \varepsilon N_{AA}\,. \tag{6.10}$$

6.2.1 The Equivalence to the Ising Model

Substituting (6.7) into the expression (6.2) for the restricted grand Hamiltonian $\widehat{H}$ and comparison with (2.66) yields

$$\widehat{H} = \frac{1}{8}zN\varepsilon_0 - \frac{1}{2}N(\mu_A - \mu_B) + \widehat{H}^{(\text{I.M.})}, \tag{6.11}$$

where $\widehat{H}^{(\text{I.M.})}$ is the Hamiltonian of the equivalent Ising model which depends on parameters J_e and $\mathcal{H}_e$ defined by

$$J_e = \pm\frac{1}{4}\varepsilon\,, \qquad \mathcal{H}_e = \frac{1}{2}\Delta\mu\,, \tag{6.12}$$

$$\Delta\mu = \mu_A - \mu_B - \frac{1}{2}z(\varepsilon_{AA} - \varepsilon_{BB})\,. \tag{6.13}$$

If ε is positive then the + sign applies and the equivalent Ising model is ferromagnetic, whereas if ε is negative the − sign applies and the equivalent Ising model is antiferromagnetic. From (6.11) the restricted grand partition function satisfies the relation

$$Z(N,T,\Delta\mu) = \exp\{N[z\varepsilon_0 - 4(\mu_A - \mu_B)]/8T\}\, Z^{(\text{I.M.})}(N,T,\mathcal{H}_e) \tag{6.14}$$

and the free energy per site is given in terms of the constant magnetic field free energy per site of the Ising model by

$$f(T,\Delta\mu) = \frac{1}{8}z\varepsilon_0 - \frac{1}{2}(\mu_A - \mu_B) + f^{(\text{I.M.})}(T,\mathcal{H}_e)\,. \tag{6.15}$$

Since A and B atoms correspond respectively to $\sigma_i = 1$ and $\sigma_i = -1$ the mole fractions in the solid mixture are given in terms of the relative magnetization m of the equivalent Ising model by

$$x_A = \frac{1+m}{2}\,, \qquad x_B = \frac{1-m}{2}\,. \tag{6.16}$$

Now assume that

$$\varepsilon_{AB} > \frac{1}{2}(\varepsilon_{AA} + \varepsilon_{BB}) \quad \text{or equivalently} \quad \varepsilon > 0 \tag{6.17}$$

implying that AA or BB nearest-neighbour pairs are increasingly preferred to AB pairs as the temperature is reduced. From (6.12) the solid mixture is equivalent to an Ising model ferromagnet at the same temperature with $J = J_e = \frac{1}{4}\varepsilon$ and $\mathcal{H} = \mathcal{H}_e = \frac{1}{2}\Delta\mu$. From (6.13) and (6.15) two solid mixture phases have the same values of $\mu_A - \mu_B$ and f when the corresponding Ising model phases have the same values of $f^{(\text{I.M.})}$ and $\mathcal{H}_e$. Thus, if T_c is the Ising ferromagnet critical temperature when $J = \frac{1}{4}\varepsilon$, the interval $0 \le T < T_c$ on the T-axis in the $(T, \Delta\mu)$ plane for the solid mixture is a first-order transition line. It follows from (6.16) that at each temperature in this interval an A–rich phase with $x_A = \frac{1}{2}(1+m_s)$ is conjugate to a B–rich phase with $x_A = \frac{1}{2}(1-m_s)$, where m_s is the Ising ferromagnet spontaneous magnetization.

6.2.2 The Equivalence to a Lattice Gas

Comparing (6.12) and (6.13) with (5.25), a solid mixture and a lattice gas at the same temperature are equivalent when ε has the same value for both models and

$$\mu = \mu_A - \mu_B - z(\varepsilon_{AB} - \varepsilon_{BB}), \tag{6.18}$$

provided (6.17) applies. Equating the expressions for $f^{(\text{I.M.})}(T, \mathcal{H})$ given by (6.15) and (5.28) respectively and, using (6.18),

$$-P_g = f(T, \Delta\mu) - \frac{1}{2} z\varepsilon_{BB}, \tag{6.19}$$

where f is the free energy per site in the solid mixture and P_g the pressure field in the lattice gas. (The subscript g distinguishes P_g from the pressure field in the solid mixture.) Again, from (6.16) and (5.29) the number density in the lattice gas and the mole fractions in the solid mixture are connected by

$$x_A = \rho, \qquad x_B = 1 - \rho. \tag{6.20}$$

Hence the conjugate phase curves shown in Fig. 5.3 apply for the solid mixture if ρ is replaced by x_A.

6.3 Order–Disorder on Loose-Packed Lattices

Now consider a solid mixture on a loose-packed lattice (see Sect. 4.2 and Appendix A.1) with the inequality (6.17) reversed so that

$$\varepsilon_{AB} < \frac{1}{2}(\varepsilon_{AA} + \varepsilon_{BB}) \quad \text{or equivalently} \quad \varepsilon < 0. \tag{6.21}$$

There is now a tendency to the formation of AB pairs which, at low temperatures, results in sublattice ordering with A atoms preferentially occupying one sublattice and the B atoms the other. This is observed in certain real systems such as body-centred cubic equimolar ($x_A = x_B = \frac{1}{2}$) FeAl where, below a critical temperature, the Fe tends to concentrate on one simple cubic sublattice and the Al on the other (see Appendix A.1). Similar behaviour is found in body-centred cubic CuZn (β brass). For an order-disorder transition to occur, the energy of cohesion of the lattice must be significantly larger than the energy of sublattice ordering. Otherwise the latter will persist up to the melting point of the solid, as in intermetallic compounds. In terms of our parameters this means that $|\varepsilon_0|$ must be large compared to $|\varepsilon|$. The first quantitative theory of sublattice ordering was due to Bragg and Williams (1934) and is equivalent to the mean-field treatment of Sect. 6.4 for the particular case of a loose-packed lattice with $x_\mathrm{A} = x_\mathrm{B}$.

When (6.21) applies, the equivalent Ising model is the simple antiferromagnet, with $J_\mathrm{e} = \frac{1}{4}|\varepsilon| > 0$. Denote the fractions of A and B on sublattice $\mathfrak{a}$ in the solid mixture by $x_{\mathrm{A}\mathfrak{a}}$ and $x_{\mathrm{B}\mathfrak{a}}$ respectively and similarly define $x_{\mathrm{A}\mathfrak{b}}$ and $x_{\mathrm{B}\mathfrak{b}}$. Then since A and B correspond respectively to positive and negative spins in the antiferromagnet,

$$\begin{aligned} x_{\mathrm{A}\mathfrak{a}} &= \frac{1+m_\mathfrak{a}}{2}, \qquad & x_{\mathrm{B}\mathfrak{a}} &= 1 - x_{\mathrm{A}\mathfrak{a}} = \frac{1-m_\mathfrak{a}}{2}, \\ x_{\mathrm{A}\mathfrak{b}} &= \frac{1+m_\mathfrak{b}}{2}, \qquad & x_{\mathrm{B}\mathfrak{b}} &= 1 - x_{\mathrm{A}\mathfrak{b}} = \frac{1-m_\mathfrak{b}}{2}, \end{aligned} \tag{6.22}$$

where $m_\mathfrak{a}$ and $m_\mathfrak{b}$ are the sublattice relative magnetizations. For the antiferromagnet we know from Sect. 4.2 that the overall relative magnetization $m = 0$, for all T, when $\mathcal{H} = 0$. Hence by (6.16) the corresponding condition $\Delta\mu = 0$ gives $x_\mathrm{A} = x_\mathrm{B} = \frac{1}{2}$ for all T. For $T > T_\mathrm{c}$, $m_\mathfrak{a} = m_\mathfrak{b} = 0$ in the antiferromagnet, giving the disordered state $x_{\mathrm{A}\mathfrak{a}} = x_{\mathrm{A}\mathfrak{b}} = x_{\mathrm{B}\mathfrak{a}} = x_{\mathrm{B}\mathfrak{b}} = \frac{1}{2}$. For $T < T_\mathrm{c}$, (6.22) and for definiteness, the first set of relations of (4.10) yield the ordered state

$$\begin{aligned} x_{\mathrm{A}\mathfrak{a}} = x_{\mathrm{B}\mathfrak{b}} &= \frac{1+m_\mathrm{s}}{2}, \\ x_{\mathrm{A}\mathfrak{b}} = x_{\mathrm{B}\mathfrak{a}} &= \frac{1-m_\mathrm{s}}{2}. \end{aligned} \tag{6.23}$$

The change of sign $m_\mathrm{s} \to -m_\mathrm{s}$ corresponds to an interchange of the equal sublattices $\mathfrak{a}$ and $\mathfrak{b}$. As $T \to 0$ the ferromagnet spontaneous magnetization $m_\mathrm{s} \to 1$ and the solid solution approaches a perfectly ordered ground state with all $\mathfrak{a}$ sites occupied by A atoms and all $\mathfrak{b}$ sites by B atoms. One of the best-known properties of real alloys is the heat capacity anomaly at the order-disorder transition temperature. From the ferromagnet–antiferromagnet equivalence at zero field and the antiferromagnet–solid solution equivalence, the heat capacity for this model with $x_\mathrm{A} = x_\mathrm{B} = \frac{1}{2}$ is equal to that of the Ising ferromagnet at $\mathcal{H} = 0$. Exact results for the $d = 2$ case of the latter are shown in Fig. 8.6 and a mean-field plot is given in Fig. 3.5.

In the context of binary alloys the parameter $s = N_\mathfrak{b}/N_\mathfrak{a}$ introduced in (4.61) is termed the *stoichiometric ratio* and a mixture for which $x_\mathrm{B}/x_\mathrm{A} = s$ is called *stoichiometric*. For $x_\mathrm{B}/x_\mathrm{A} = s$, a state of perfect order with all A atoms on sublattice $\mathfrak{a}$ and all B atoms on sublattice $\mathfrak{b}$ is possible. For equal sublattices, $s = 1$ and $x_\mathrm{B}/x_\mathrm{A} = 1$ in a stoichiometric mixture, which thus corresponds to $\Delta\mu = 0$ or $\mathcal{H}_\mathrm{e} = 0$ in the equivalent Ising model. Non-stoichiometric mixtures ($x_\mathrm{B}/x_\mathrm{A} \neq 1$) correspond to $\frac{1}{2}\Delta\mu = \mathcal{H}_\mathrm{e} \neq 0$, and no

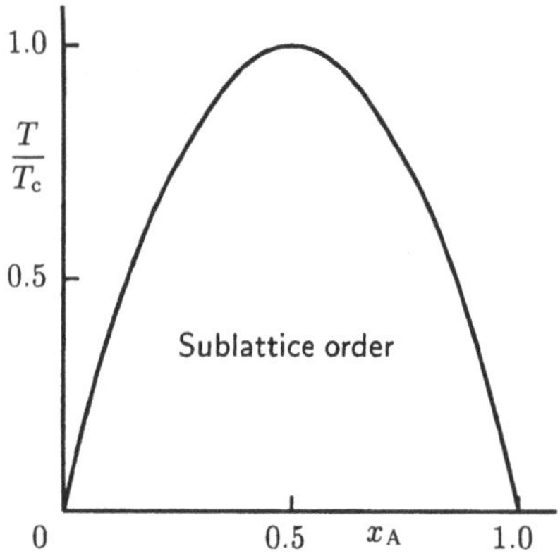

Fig. 6.1. The critical curve for sublattice ordering on a loose-packed lattice in the mean-field approximation.

exact results are available even for $d = 2$, so that approximation or series methods must be used.

From (6.22) the antiferromagnetic state $m_{\mathfrak{a}} \neq m_{\mathfrak{b}}$ corresponds to sublattice order in the solid solution and the paramagnetic state $m_{\mathfrak{a}} = m_{\mathfrak{b}} = m$ to disorder. Since the equivalent Ising model is the simple antiferromagnet with $\lambda = 0$ the order-disorder transition is second-order over its whole temperature range and corresponds to the critical curve for the antiferromagnet. A mean-field approximation to the order-disorder transition curve in the (x_{A}, T) plane can be obtained from (6.16) and the second relation of Example 4.2, with $\lambda = 0$,

$$x_{\mathrm{A}} = \frac{1}{2}\left\{1 \pm \sqrt{1-\tau}\right\} ,$$
$$\tau = \frac{T}{T_{\mathrm{c}}} = \frac{T}{zJ_{\mathrm{e}}} = \frac{4T}{z|\varepsilon|} . \tag{6.24}$$

This curve, which is clearly symmetrical about $x_{\mathrm{A}} = \frac{1}{2}$, is shown in Fig. 6.1. Every point in the (x_{A}, τ) plane corresponds to an equilibrium state, and the map of the antiferromagnet $\mathcal{H} = 0$ axis is the entire $x_{\mathrm{A}} = \frac{1}{2}$ line from $T = \infty$ to $T = 0$. This contrasts with the superficially similar Fig. 5.3 where the map of the ferromagnetic $\mathcal{H} = 0$ curve bifurcates at the critical point into the right- and left-hand branches of the coexistence curve, and the entire area inside the latter represents non-equilibrium states.

We can ask if other forms of sublattice order are possible when $x_{\mathrm{A}}/x_{\mathrm{B}}$ is far from 1. The body-centred cubic lattice can, for instance, be split into four equal sublattices by dividing the simple cubic sublattice $\mathfrak{a}$ into two face-centred cubic arrays $\mathfrak{a}_1$ and $\mathfrak{a}_2$ and similarly dividing $\mathfrak{b}$ into $\mathfrak{b}_1$ and $\mathfrak{b}_2$ (see Fig. 6.2). There may thus be a type of order with stoichiometric ratio $s = 3$ in which the $\mathfrak{a}_1$ sites (say) are preferentially occupied by A and the rest of the body-centred cubic lattice by B. However, this will not occur if there are nearest-neighbour interactions only. Suppose that $x_{\mathrm{A}} < \frac{1}{2}$ and that all A

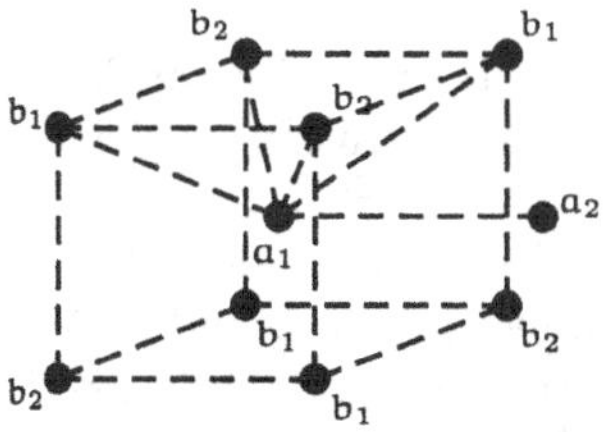

Fig. 6.2. Division of the body-centred cubic lattice into four face-centred cubic sublattices.

atoms are on $\mathfrak{a}$ sites. Then all the nearest-neighbour sites to any $\mathfrak{a}$ site are occupied by B atoms and $N_{AB} = zM_A = 8x_A N$. By (6.9) the configurational energy is thus the same for all distributions of the A atoms on the $\mathfrak{a}$ sites and there is no advantage in the A atoms concentrating on the $\mathfrak{a}_1$ face-centred cubic sublattice. The degeneracy is removed if there are second-neighbour interactions obeying an inequality similar to (6.21). Experimentally, an ordered state with the Al atoms occupying one face-centred cubic sublattice is found near the composition Fe_3Al for the body-centred cubic Fe–Al alloy. However, this is not so for other body-centred cubic alloys such as Fe–Co and Fe–Cr, the difference presumably depending on the strength of the effective second-neighbour interaction. A mean-field theory incorporating second-neighbour interactions on the body-centred cubic lattice and leading to 1:3 as well as 1:1 ordering has been developed (Elcock 1958). Other cases when $x_A/x_B \neq 1$ are considered in subsequent sections.

6.4 The Order Parameter and Landau Expansion

We now consider the possibility of unequal sublattices and intra-sublattice nearest-neighbour pairs using the nomenclature of Sect. 4.5, so that (4.60) and (4.61) apply. As in Sect. 4.5 we assume $s \geq 1$. An order parameter θ is defined by

$$\begin{aligned} x_{A\mathfrak{a}} &= x_A(1+s\theta)\,, \qquad & x_{B\mathfrak{a}} &= 1 - x_{A\mathfrak{a}} = x_B - x_A s\theta\,, \\ x_{A\mathfrak{b}} &= x_A(1-\theta)\,, \qquad & x_{B\mathfrak{b}} &= 1 - x_{A\mathfrak{b}} = x_B + x_A\theta\,. \end{aligned} \tag{6.25}$$

The necessary conditions

$$x_A = \frac{x_{A\mathfrak{a}} + s x_{A\mathfrak{b}}}{1+s}, \qquad x_B = \frac{x_{B\mathfrak{a}} + s x_{B\mathfrak{b}}}{1+s} \tag{6.26}$$

are clearly satisfied. In a stoichiometric mixture, $x_B/x_A = s$, giving

$$x_A = \frac{1}{1+s}\,, \qquad x_B = \frac{s}{1+s}\,. \tag{6.27}$$

For $x_A < (1+s)^{-1}$ the range of possible θ values is $(-s^{-1}, 1)$ and for $s(1+s)^{-1} > x_A > (1+s)^{-1}$ it is $(-s^{-1}, s^{-1}x_B/x_A)$. The disordered state $x_{A\mathfrak{a}} = x_{A\mathfrak{b}} = x_A$, $x_{B\mathfrak{a}} = x_{B\mathfrak{b}} = x_B$ corresponds to $\theta = 0$. For values of x_B/x_A in the neighbourhood of the stoichiometric ratio the equilibrium value of θ can be expected to be positive at low temperatures, giving a concentration of A on $\mathfrak{a}$ sites and B on $\mathfrak{b}$ sites.

A mean-field theory will now be developed. From Sect. 3.3 the probability of a pair of nearest-neighbour $\mathfrak{a}$ sites being occupied by an A and a B is $2x_{\mathrm{A}\mathfrak{a}}x_{\mathrm{B}\mathfrak{a}}$ and similarly for a nearest-neighbour pair of $\mathfrak{b}$ sites. Slightly extending the theory, the probability of this occupation for a nearest-neighbour site pair consisting of one $\mathfrak{a}$ site and one $\mathfrak{b}$ site is $x_{\mathrm{A}\mathfrak{a}}x_{\mathrm{B}\mathfrak{b}} + x_{\mathrm{A}\mathfrak{b}}x_{\mathrm{B}\mathfrak{a}}$. Hence the expected total number of AB nearest-neighbour pairs is, using the last relation of (4.60),

$$N_{\mathrm{AB}} = \mathrm{z}_{\mathfrak{aa}}N_{\mathfrak{a}}x_{\mathrm{A}\mathfrak{a}}x_{\mathrm{B}\mathfrak{a}} + \mathrm{z}_{\mathfrak{bb}}N_{\mathfrak{b}}x_{\mathrm{A}\mathfrak{b}}x_{\mathrm{B}\mathfrak{b}} + \frac{1}{2}(\mathrm{z}_{\mathfrak{ab}}N_{\mathfrak{a}} + \mathrm{z}_{\mathfrak{ba}}N_{\mathfrak{b}})(x_{\mathrm{A}\mathfrak{a}}x_{\mathrm{B}\mathfrak{b}} + x_{\mathrm{A}\mathfrak{b}}x_{\mathrm{B}\mathfrak{a}})\,. \tag{6.28}$$

The number of configurations is the product of expressions like (3.38) for the $\mathfrak{a}$ and $\mathfrak{b}$ sublattices respectively, with N replaced by $N_{\mathfrak{a}}$ in one expression and $N_{\mathfrak{b}}$ in the other. In a similar way, from (3.40), the Helmholtz free energy per site is given by

$$a = N^{-1}T\{N_{\mathfrak{a}}[x_{\mathrm{A}\mathfrak{a}}\ln x_{\mathrm{A}\mathfrak{a}} + x_{\mathrm{B}\mathfrak{a}}\ln x_{\mathrm{B}\mathfrak{a}}] + N_{\mathfrak{b}}[x_{\mathrm{A}\mathfrak{b}}\ln x_{\mathrm{A}\mathfrak{b}} + x_{\mathrm{B}\mathfrak{b}}\ln x_{\mathrm{B}\mathfrak{b}}]\} + \frac{1}{2}\mathrm{z}(\varepsilon_{\mathrm{AA}}x_{\mathrm{A}} + \varepsilon_{\mathrm{BB}}x_{\mathrm{B}}) - \frac{1}{2}|\varepsilon|N^{-1}N_{\mathrm{AB}}\,, \tag{6.29}$$

where the expression (6.9) for the configurational energy is used.

Since θ is an order variable, a is a non-equilibrium free energy which must be minimized with respect to θ. Using (4.60) and (4.61) a can be expressed in the form

$$a = a_0 + \Delta a(\theta)\,, \tag{6.30}$$

where

$$a_0 = T(x_{\mathrm{A}}\ln x_{\mathrm{A}} + x_{\mathrm{B}}\ln x_{\mathrm{B}}) + \frac{1}{2}\mathrm{z}(\varepsilon_{\mathrm{AA}}x_{\mathrm{A}} + \varepsilon_{\mathrm{BB}}x_{\mathrm{B}}) - \frac{1}{2}\mathrm{z}|\varepsilon|x_{\mathrm{A}}x_{\mathrm{B}}\,, \tag{6.31}$$

$$\Delta a(\theta) = \frac{T}{1+s}\{x_{\mathrm{A}}(1+s\theta)\ln(1+s\theta) + x_{\mathrm{B}}(1-ys\theta)\ln(1-ys\theta) + s[x_{\mathrm{A}}(1-\theta)\ln(1-\theta) + x_{\mathrm{B}}(1+y\theta)\ln(1+y\theta)]\} - \frac{1}{2}|\varepsilon|sx_{\mathrm{A}}^2(\mathrm{z} - \mathrm{z}_{\mathfrak{aa}} - \mathrm{z}_{\mathfrak{bb}})\theta^2 \tag{6.32}$$

with $\Delta a(0) = 0$ and the *mole ratio parameter*

$$y = x_{\mathrm{A}}/x_{\mathrm{B}}. \tag{6.33}$$

(For a stoichiometric mixture, $y = 1/s$.) The equilibrium value of θ satisfies the relation

$$\begin{aligned}\frac{\partial a}{\partial \theta} &= \frac{\partial \Delta a(\theta)}{\partial \theta} \\ &= \frac{Tsx_{\mathrm{A}}}{1+s} \ln\left[\frac{(1+s\theta)(1+y\theta)}{(1-\theta)(1-ys\theta)}\right] - |\varepsilon| s x_{\mathrm{A}}^2 (\mathrm{z} - \mathrm{z}_{aa} - \mathrm{z}_{bb})\theta \\ &= 0 \,. \end{aligned} \tag{6.34}$$

One solution is always $\theta = 0$ and other solutions with $\theta \neq 0$ appear at low temperatures if $\mathrm{z} > \mathrm{z}_{aa} + \mathrm{z}_{bb}$.

Landau expansions are described in detail in Volume 2, Chap. 3. In the case considered here the Landau expansion for the Helmholtz free energy density a is obtained by expanding $\Delta a(\theta)$ in powers of the order parameter θ. From (6.32),

$$\begin{aligned}\Delta a(\theta) = {} & \frac{sx_{\mathrm{A}}}{2x_{\mathrm{B}}}[T - |\varepsilon|(\mathrm{z} - \mathrm{z}_{aa} - \mathrm{z}_{bb})x_{\mathrm{A}}x_{\mathrm{B}}]\theta^2 \\ & + \frac{Tsx_{\mathrm{A}}}{1+s}\sum_{n=3}^{\infty} \frac{[1 + y(-y)^{n-2}][1 + s(-s)^{n-2}]\theta^n}{n(n-1)} \,. \end{aligned} \tag{6.35}$$

The essence of the Landau method is the deduction of critical properties from the behaviour of the first few coefficients, and this can be illustrated by the loose-packed lattice case. With equal sublattices, $s = 1$ and the odd powers of θ disappear in (6.35). This is to be expected since, when $s = 1$, the expression on the right of (6.32) becomes an even function of θ. For loose-packed lattices we have, in addition, $\mathrm{z}_{aa} = \mathrm{z}_{bb} = 0$ and (6.35) becomes

$$\Delta a(\theta) = B_2(T)\theta^2 + B_4(T)\theta^4 + B_6(T)\theta^6 + \cdots \,, \tag{6.36}$$

where

$$B_{2n}(T) = \begin{cases} \dfrac{x_{\mathrm{A}}}{2x_{\mathrm{B}}}(T - \mathrm{z}|\varepsilon|x_{\mathrm{A}}x_{\mathrm{B}}) \,, & n = 1 \,, \\[2ex] \dfrac{Tx_{\mathrm{A}}(1 + y^{2n-1})}{2n(2n-1)} \,, & n > 1 \,. \end{cases} \tag{6.37}$$

When, as here, the coefficients $B_{2n}(T)$ for $n > 1$ are intrinsically positive there is a second-order transition at the temperature where $B_2(T) = 0$. This is because Δa has a unique minimum at $\theta = 0$ for $B_2(T) > 0$, but for $B_2(T) < 0$ there is a maximum at $\theta = 0$ and symmetrically placed minima on either side. For the latter case the plot of Δa against θ is similar to the curve ABOCD in Fig. 3.1. From (6.37), the condition $B_2(T) = 0$ yields the critical curve relation

$$T = \mathrm{z}|\varepsilon|x_{\mathrm{A}}x_{\mathrm{B}} = \mathrm{z}|\varepsilon|x_{\mathrm{A}}(1 - x_{\mathrm{A}}) \,, \tag{6.38}$$

which is equivalent to (6.24). Comparison of (6.25) with $s = y = 1$ and (6.23) shows that for an equimolar mixture ($x_A = x_B = \frac{1}{2}$) the order parameter $\theta = m_s$. It follows that where $\Delta\mu$, as defined by (6.13), is zero then

$$\theta = \Delta x \,, \tag{6.39}$$

where $\Delta x = |x_A - x_B|$ for either of the conjugate phases in the solid mixture model of Sect. 6.2 with the same values of $|\varepsilon|$ and T. A case where the sublattices are unsymmetrical so that $s \neq 1$ and the odd powers of θ in (6.35) do not disappear is discussed in the next section.

6.5 First-Order Sublattice Transitions

We now consider the face-centred cubic lattice (see Appendix A.1). This divides into four equal simple cubic sublattices, any site in one of the latter having four nearest-neighbour sites in each of the other three which are thus completely equivalent with respect to the first sublattice.[1] Now attach the label $\mathfrak{a}$ to the sites in one simple cubic sublattice and the label $\mathfrak{b}$ to all other sites in the face-centred cubic. Hence

$$\begin{aligned} &N_{\mathfrak{b}} = 3N_{\mathfrak{a}} \,, && s = N_{\mathfrak{b}}/N_{\mathfrak{a}} = 3 \,, \\ &z_{\mathfrak{aa}} = 0 \,, && z_{\mathfrak{ab}} = 12 \,, \\ &z_{\mathfrak{bb}} = 8 \,, && z_{\mathfrak{ba}} = 4 \,. \end{aligned} \tag{6.40}$$

Figure 6.3 shows an $\mathfrak{a}$ site at the origin O and four nearest-neighbour $\mathfrak{b}$ sites in the plane Oxz. There are similar sets of four nearest-neighbour $\mathfrak{b}$ sites in the planes Oxy and Oyz. We shall study the transition from the disordered state $\theta = 0$ to the ordered states with $\theta > 0$. Substituting (6.40) into (6.35),

$$\Delta a(\theta) = B_2(T)\theta^2 + B_3(T)\theta^3 + B_4(T)\theta^4 + \cdots \,, \tag{6.41}$$

where

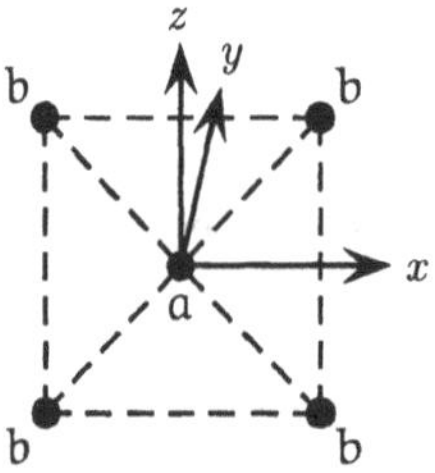

Fig. 6.3. Site of face-centred cubic sublattice $\mathfrak{a}$ and four of its twelve nearest neighbours in sublattice $\mathfrak{b}$.

[1] This equivalence does not apply when a loose-packed lattice like the body-centred cubic is divided into four equal sublattices. From Fig. 6.2 it can be seen that an $\mathfrak{a}_1$ site has four nearest neighbours in $\mathfrak{b}_1$ and four in $\mathfrak{b}_2$ but none in $\mathfrak{a}_2$.

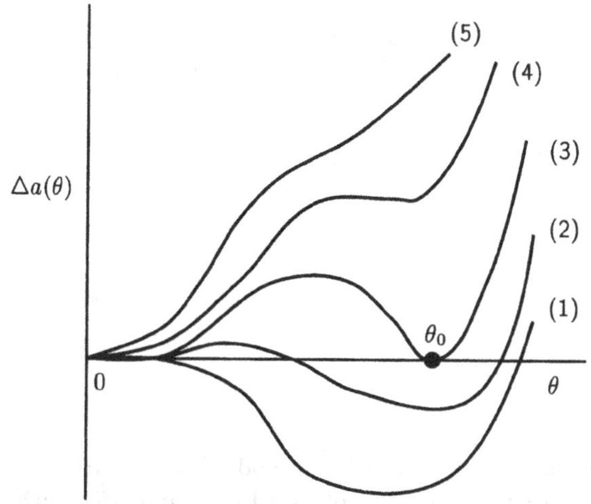

Fig. 6.4. Free energy isotherms plotted against the order parameter for the unsymmetrical case.

$$B_2(T) = \frac{3}{2}y[T - 4|\varepsilon|x_A x_B]\,,$$
$$B_3(T) = -x_A T(1-y^2)\,, \qquad (6.42)$$
$$B_4(T) = \frac{14}{8}x_A T(1+y^3)\,.$$

At the temperature $T = 4|\varepsilon|x_A x_B$, where $B_2(T) = 0$, the leading term is $B_3(T)\theta^3$, where $B_3(T) < 0$ for $y < 1$, and $\Delta a(\theta)$ has the form of curve (1) on Fig. 6.4. The equilibrium value of θ corresponds to the minimum of $\Delta a(\theta)$, and the transition occurs at a higher temperature. Curves (2), (3), (4) and (5) are for progressively increasing temperatures above $4|\varepsilon|x_A x_B$. On curve (2) the minimum at $\theta = 0$ represents a metastable state whereas on curve (4) the minimum at $\theta > 0$ represents a metastable state. The intermediate curve (3) corresponds to the transition temperature T_0. Here the minima at $\theta = 0$ and $\theta = \theta_0$ both lie on the θ-axis, and the transition is a first-order one from $\theta = 0$ to $\theta = \theta_0$, with a discontinuous change $-6|\varepsilon|x_A^2\theta_0^2$ in the configurational energy per site, by (6.32) and (6.40). For a given x_A the values of θ_0 and T_0 are determined by the relations

$$\Delta a(\theta_0) = 0\,, \qquad \frac{\partial \Delta a(\theta_0)}{\partial \theta_0} = 0\,. \qquad (6.43)$$

For the stoichiometric ratio $y = \frac{1}{3}$ ($x_A = \frac{1}{4}$, $x_B = \frac{3}{4}$), $B_2(T) = 0$ at $T = \frac{3}{4}|\varepsilon|$. Equation (6.43) yields $T_0 = 0.822|\varepsilon|$ and $\theta_0 = 0.463$, (Fowler and Guggenheim 1949, Sect. 1325).

We have discussed the behaviour of the system as T varies at constant x_A. Now, the equilibrium Helmholtz free energy per site $a(T, x_A)$ depends on the single density variable x_A (since $x_B = 1 - x_A$) and hence is a free energy density f_{n-2} in the sense of Chap. 1. By considering the variation of $a(T, x_A)$

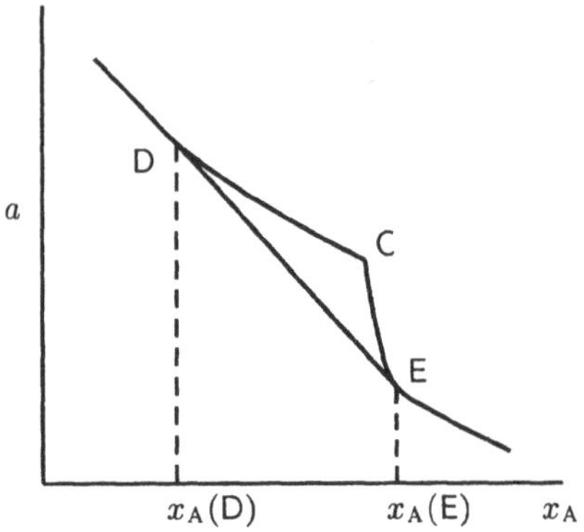

Fig. 6.5. Free energy-mole fraction isotherm showing convex envelope.

with x_{A} at constant T we shall show, using the method of Sect. 1.10 (iii), that the first-order transition occurs in a range of x_{A} which is unstable with respect to phase separation. We can express the equilibrium $\Delta a(T, x_{\mathrm{A}})$ as $\Delta a(T, x_{\mathrm{A}}, \theta_{\mathrm{e}})$, where $\theta = \theta_{\mathrm{e}}(T, x_{\mathrm{A}})$ is the equilibrium value of θ, that is the value of θ at the lowest minimum of the plot of $\Delta a(T, x_{\mathrm{A}}, \theta)$ against θ. Hence, at constant T,

$$\frac{\partial \Delta a}{\partial x_{\mathrm{A}}} = \left(\frac{\partial \Delta a}{\partial x_{\mathrm{A}}}\right)_{\theta_{\mathrm{e}}} + \left(\frac{\partial \Delta a}{\partial x_{\mathrm{A}}}\right) \frac{\partial \theta}{\partial x_{\mathrm{A}}} = \left(\frac{\partial \Delta a}{\partial x_{\mathrm{A}}}\right)_{\theta_{\mathrm{e}}}, \tag{6.44}$$

since $\partial(\Delta a)/\partial\theta = 0$ at equilibrium. It follows that

$$\frac{\partial a}{\partial x_{\mathrm{A}}} = \frac{\partial a_0}{\partial x_{\mathrm{A}}} + \left(\frac{\partial \Delta a}{\partial x_{\mathrm{A}}}\right)_{\theta_{\mathrm{e}}}, \tag{6.45}$$

where a_0 is given by (6.31). Suppose that x_{A} increases from zero at the fixed temperature $T = T_0$ given by (6.43). For small values of x_{A}, $\theta_{\mathrm{e}} = 0$ and hence $\Delta a(\theta) = 0$ so that $\partial a/\partial x_{\mathrm{A}} = \partial a_0/\partial x_{\mathrm{A}}$. At the particular value of x_{A} implied in (6.43), θ_{e} jumps from zero to θ_0 and $\partial a/\partial x_{\mathrm{A}}$ undergoes a discontinuous change equal to $(\partial \Delta a/\partial x_{\mathrm{A}})_{\theta=\theta_0}$. This quantity is negative, since the effect of ordering is to reduce a, and hence the plot of a against x_{A} is as shown in Fig. 6.5, where the first-order transition occurs at point C (see also Example 6.3). To complete the convex envelope, a double tangent DE must be drawn, and for values of x_{A} between $x_{\mathrm{A}}(\mathsf{D})$ and $x_{\mathrm{A}}(\mathsf{E})$ there is separation into a disordered phase where $x_{\mathrm{A}} = x_{\mathrm{A}}(\mathsf{D})$ and an ordered phase where $x_{\mathrm{A}} = x_{\mathrm{A}}(\mathsf{E})$. Conjugate phases in the (x_{A}, T) plane are shown in Fig. 6.6. Suppose that the temperature of a mixture with $x_{\mathrm{A}} = x_{\mathrm{A}}^*$ is reduced slowly enough for equilibrium to be maintained.[2] Then at $T = T_1$ an ordered phase starts to separate out from the disordered mixture, and for $T_1 > T > T_2$ the system consists of ordered and disordered phases in equilibrium. For $T < T_2$ the equilibrium system is a homogeneous ordered phase. The temperature T_0 at

[2] In metallurgical terminology this is called *annealed.*

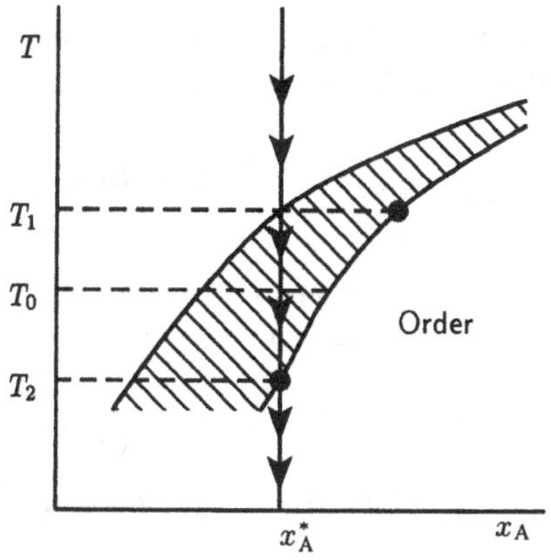

Fig. 6.6. Conjugate phases in the mole fraction–temperature plane.

which the (metastable) homogeneous mixture would undergo a first-order transition lies between T_1 and T_2.

Behaviour of this type is well-known experimentally. In fact, Irani (1972) remarks that 'order–disorder transformations in almost every system, except perhaps the $L2_0$ structures[3] are first-order transitions' and 'most systems exhibit a two-phase field away from the stoichiometric composition'. There is a mean-field theory (Shockley 1938) for the whole composition range on the face-centred cubic lattice, and our treatment of 1:3 ordering is part of this. However, there are discrepancies between experimental observations for the well-known Au–Cu alloys and mean-field results. In particular, at the stoichiometric composition $AuCu_3$ ($x_A = \frac{1}{4}$) the observed phase separation curves have a common maximum or azeotropic point, resembling point P in Fig. 1.7. In mean-field theory, on the other hand, the phase separation curves have a common maximum only at $x_A = \frac{1}{2}$. These discrepancies probably arise both from the simplicity of the model and from the crudity of the mean-field approximation. Ordering on the face-centred cubic lattice is discussed in terms of the first-order approximation in Sect. 7.6.

6.6 The Equilibrium Dilute Ising Model and Equivalent Models

We consider an important new set of mathematically equivalent models.

6.6.1 Model I: The Equilibrium Dilute Ising Model

Suppose that in a lattice A–B mixture like that of Sect. 6.1 the A atoms carry Ising spins with ferromagnetic interaction. We then have an Ising model

[3] Body-centred cubic lattice divided into equal interpenetrating sublattices.

ferromagnet diluted by a non-magnetic component B. Associating a spin variable $\tau_i = \pm 1$ with each A atom[4] the reduced magnetic moment

$$\widehat{\mathcal{M}} = \sum_{\{i\}}^{(\mathrm{A})} \tau_i \,, \tag{6.46}$$

where the summation is over all sites occupied by A atoms and $\mathcal{M} = \langle \widehat{\mathcal{M}} \rangle$ is conjugate to the applied magnetic field $\mathcal{H}$. The system is a restricted one, with $n = 3$, and we put

$$\xi_1 = \mu_\mathrm{A} - \mu_\mathrm{B} \,, \qquad \xi_2 = \mathcal{H} \,, \qquad \xi_3 = \mu_\mathrm{B} - P \,. \tag{6.47}$$

A partition function of type (2.58) is obtained by taking N, T, $\mu_\mathrm{A} - \mu_\mathrm{B}$ and $\mathcal{H}$ as independent variables with

$$\widehat{H} = -\varepsilon N_\mathrm{AA} + z(\varepsilon_\mathrm{AB} - \varepsilon_\mathrm{BB}) M_\mathrm{A} - (\mu_\mathrm{A} - \mu_\mathrm{B}) M_\mathrm{A} - J \sum_{\{i,j\}}^{(\mathrm{A,n.n.})} \tau_i \tau_j - \mathcal{H} \sum_{\{i\}}^{(\mathrm{A})} \tau_i \,. \tag{6.48}$$

Here the non-magnetic part of the configurational energy is expressed in the form (6.10), with the constant term $\frac{1}{2} z \varepsilon_\mathrm{BB} N$ omitted, and the sums are respectively over nearest-neighbour pairs and single sites occupied by A atoms. It is assumed that $J > 0$ and that (6.17) applies. The partition function can be called *constant magnetic field restricted grand* and reduces to (6.2) when $J = \mathcal{H} = 0$. A simpler alternative form for the Hamiltonian is obtained by associating the spin component value $\tau_i = 0$ with each site occupied by a B atom and by *defining* parameter μ by (6.18). Then (6.48) becomes

$$\widehat{H} = -\varepsilon \sum_{\{i,j\}}^{(\mathrm{n.n.})} \tau_i^2 \tau_j^2 - \mu \sum_{\{i\}} \tau_i^2 - J \sum_{\{i,j\}}^{(\mathrm{n.n.})} \tau_i \tau_j - \mathcal{H} \sum_{\{i\}} \tau_i \,. \tag{6.49}$$

We define a relative magnetization m with range $(-1, 1)$ by

$$m = \langle \tau_i \rangle_\mathrm{A} = \frac{\mathcal{M}}{M_\mathrm{A}} = \frac{\mathcal{M}}{x_\mathrm{A} N} \,. \tag{6.50}$$

The density per site conjugate to $\mathcal{H}$ is then $\mathcal{M}/N = m x_\mathrm{A}$. From (2.59) the free energy per site is

$$f = a - (\mu_\mathrm{A} - \mu_\mathrm{B}) x_\mathrm{A} - x_\mathrm{A} m \mathcal{H} = \mu_\mathrm{B} - P \,, \tag{6.51}$$

and, from (2.60), since $\mathrm{d}(\mu_\mathrm{A} - \mu_\mathrm{B}) = \mathrm{d}\mu$,

$$\mathrm{d}f = -s\mathrm{d}T - x_\mathrm{A} \mathrm{d}\mu - (x_\mathrm{A} m) \mathrm{d}\mathcal{H} \,. \tag{6.52}$$

The condition for phase equilibrium is that T, μ (or $\mu_\mathrm{A} - \mu_\mathrm{B}$), $\mathcal{H}$ and f take the same values in the conjugate phases.

---

[4] Using τ_i avoids confusion with the σ_i of Sect. 6.2, which has a different meaning.

Like the original Ising model Hamiltonian (2.66), $\widehat{H}$ is invariant under a simultaneous change of sign of $\mathcal{H}$ and all spin variables. Hence the symmetry relations (2.68), (2.69) and (2.71) are still applicable provided that the values of μ are the same on both sides. It also follows that

$$x_{\mathrm{A}}(T,\mu,\mathcal{H}) = x_{\mathrm{A}}(T,\mu,-\mathcal{H})\,. \tag{6.53}$$

If $\varepsilon = \mu = 0$, then (6.49) gives the Hamiltonian of the spin-1 Ising model with $\tau_i = \pm 1, 0$ (see Sect. 2.4 or Example 2.4).

6.6.2 Model II: The Ising Model Lattice Gas

This can be obtained from the results of Sect. 6.6.1 by replacing the B atoms by vacant sites or from the simple lattice gas of Chap. 5 by letting the molecules carry Ising spins. Since $M = M_{\mathrm{A}} \neq N = V$ the system is not a restricted one. We have $n = 3$ with

$$\xi_1 = \mu\,, \qquad \xi_2 = \mathcal{H}\,, \qquad \xi_3 = -P_{\mathrm{g}}\,, \tag{6.54}$$

where μ is the chemical potential and P_{g} the lattice gas pressure field. Taking N, T, μ and $\mathcal{H}$ as independent variables, the Hamiltonian $\widehat{H}$ has the form (6.49) where $-\varepsilon$ is the non-magnetic part of the interaction energy of a nearest-neighbour pair of molecules and $\tau_i = 0$ now represents an empty site. This Hamiltonian gives a constant magnetic field grand partition function. Let

$$\rho = \frac{M}{N}\,, \qquad m = \frac{\mathcal{M}}{M} = \frac{\mathcal{M}}{\rho N}\,. \tag{6.55}$$

Then, from (1.79),

$$-P_{\mathrm{g}} = a - \rho\mu - m\rho\mathcal{H}\,, \tag{6.56}$$

$$\mathrm{d}P_{\mathrm{g}} = s\mathrm{d}T + \rho\mathrm{d}\mu + \rho m\mathrm{d}\mathcal{H}\,. \tag{6.57}$$

From the similarity in form of the Hamiltonians the dependent intensive variables for models I and II are related by

$$f = -P_{\mathrm{g}}\,, \qquad x_{\mathrm{A}} = \rho\,, \qquad m = m\,, \tag{6.58}$$

with the same values of μ, $\mathcal{H}$ and T on both sides.

6.6.3 Model III: The Symmetrical Ternary Solid Mixture

Suppose that each lattice site is occupied by an atom of component B, C or D. The restricted grand Hamilton of (6.2) can be generalized to

$$\widehat{H} = E - (\mu_{\mathrm{C}} - \mu_{\mathrm{B}})M_{\mathrm{C}} - (\mu_{\mathrm{D}} - \mu_{\mathrm{B}})M_{\mathrm{D}}\,, \tag{6.59}$$

the independent thermodynamic variables being N, T, $\mu_{\mathrm{C}} - \mu_{\mathrm{B}}$ and $\mu_{\mathrm{D}} - \mu_{\mathrm{B}}$. There are six nearest-neighbour pair interaction energies $\varepsilon_{\mathrm{BB}}$, $\varepsilon_{\mathrm{CC}}$, $\varepsilon_{\mathrm{DD}}$, $\varepsilon_{\mathrm{BC}}$,

ε_{BD} and ε_{CD} but this model is equivalent to model I only if the C-D symmetry conditions

$$\varepsilon_{CC} = \varepsilon_{DD}, \qquad \varepsilon_{BC} = \varepsilon_{BD}, \tag{6.60}$$

are satisfied. If C, D and B are identified with $\tau_i = 1, -1$ and 0 respectively and (6.60) is assumed

$$\begin{aligned} &\varepsilon_{CC} = \varepsilon_{AA} - J, \qquad \varepsilon_{CD} = \varepsilon_{AA} + J, \qquad \varepsilon_{BC} = \varepsilon_{AB}, \\ &\mu_C = \mu_A + \mathcal{H}, \qquad \mu_D = \mu_A - \mathcal{H}. \end{aligned} \tag{6.61}$$

Then, using (6.18), the Hamiltonian (6.59) can be written in the form (6.49) with

$$\begin{aligned} &\varepsilon = 2\varepsilon_{BC} - \frac{1}{2}(\varepsilon_{CC} + \varepsilon_{CD}) - \varepsilon_{BB}, \qquad J = \frac{1}{2}(\varepsilon_{CD} - \varepsilon_{CC}), \\ &\mu = \frac{1}{2}(\mu_C + \mu_D) - \mu_B - z(\varepsilon_{BC} - \varepsilon_{BB}), \quad \mathcal{H} = \frac{1}{2}(\mu_C - \mu_D). \end{aligned} \tag{6.62}$$

The dependent intensive variables for models I and III are related by

$$f = f, \qquad \frac{1}{2}x_A(1+m) = x_C, \qquad \frac{1}{2}x_A(1-m) = x_D. \tag{6.63}$$

6.6.4 Model IV: The Symmetrical Lattice Gas Mixture

This model can be derived from model III by replacing each B atom by a vacant site or from that of Sect. 6.6.2 by identifying $\tau_i = 1$ and $\tau_i = -1$ respectively with occupation by a C or D atom. Taking N, T, μ_C and μ_B as independent variables, the grand (*not* restricted grand) Hamiltonian is

$$\widehat{H} = E - M_C\mu_C - M_D\mu_D. \tag{6.64}$$

There are three nearest-neighbour interaction energies ε_{CC}, ε_{DD} and ε_{CD}, but the model is equivalent to the previous ones only if the symmetry condition

$$\varepsilon_{CC} = \varepsilon_{DD} \tag{6.65}$$

applies. Then with

$$\begin{aligned} &\varepsilon_{CC} = -\varepsilon - J, \qquad \varepsilon_{CD} = -\varepsilon + J, \\ &\mu_C = \mu + \mathcal{H}, \qquad \mu_D = \mu - \mathcal{H}, \end{aligned} \tag{6.66}$$

(6.64) can then be written in the form (6.49) with

$$\begin{aligned} &\varepsilon = -\frac{1}{2}(\varepsilon_{CC} + \varepsilon_{CD}), \qquad J = \frac{1}{2}(\varepsilon_{CD} - \varepsilon_{CC}), \\ &\mathcal{H} = \frac{1}{2}(\mu_C - \mu_D), \qquad \mu = \frac{1}{2}(\mu_C + \mu_D). \end{aligned} \tag{6.67}$$

The dependent intensive variables for models II and IV are related by

$$P_g = P_g, \quad \frac{1}{2}\rho(1+m) = \frac{M_C}{N} = \rho_C, \quad \frac{1}{2}\rho(1-m) = \frac{M_D}{N} = \rho_D. \tag{6.68}$$

6.6.5 Other Models

Replacement of J by $-J$ ($J > 0$) yields another four equivalent models. Model I is changed to a dilute Ising antiferromagnet and model II to an antiferromagnetic lattice gas. On a loose-packed lattice with $\mathcal{H} = 0$, the Hamiltonian (6.49) is invariant to a change from J to $-J$ ($J > 0$) coupled with a reversal of sign for all τ_i on one of the equal sublattices. Hence the zero-field ferromagnet–antiferromagnet equivalence of Sect. 4.2 is maintained for models I and II. For models III and IV, separation into C-rich and D-rich phases is replaced by sublattice segregation of C and D.

6.6.6 Applications

We now give a brief account of the development and application of the *dilute Ising models.* Meijering (1950, 1951) developed a theory of ternary mixtures in which the symmetrical case was equivalent to model III in the mean-field approximation. Bell (1953) treated model III using the mean-field approximation and the first-order pair method (see Chap. 7). He pointed out the equivalence of models III to I and also considered the antiferromagnetic case, deriving a relation between the sublattice order parameter and conjugate phase composition which corresponds to the binary mixture relation (6.39). Wheeler and Widom (1970) also used a first-order method for model III. Blume (1966) and Capel (1966) developed mean-field theories for model I in the restricted case $\varepsilon = 0$. This work was extended by Blume, Emery and Griffiths (1971) to model I with $\varepsilon > 0$, which is hence sometimes called the *Blume–Emery–Griffiths model.* These authors pointed out that the phase diagram for the model can contain tricritical points, in the sense defined by Griffiths (1970) (see Sect. 6.9 below). They were concerned with modelling low-temperature liquid mixtures of the helium isotopes ^{4}He and ^{3}He, identifying them with A and B respectively. ^{4}He undergoes a second-order transition to a superfluid state characterized by a two-dimensional vector order variable. The representation by model I, with $\mathcal{H} = 0$, involves replacing the real ordering process by a simpler one in the hope, which is largely justified, that the general phase characteristics will not be affected (see Sect. 6.8 below). In a rather similar way, monolayers at the air–water or oil–water interface, which display quite complicated cooperative phenomena, can be represented by model II with $\mathcal{H} = 0$ (see Sect. 7.4 below).

6.7 Mean-Field Theory and the Dilute Ising Model

We develop a mean-field theory for model I of Sect. 6.6.1, which can easily be transcribed to the models II, III or IV of Sects. 6.6.2–6.6.4. To simplify the notation, let $x_{\mathrm{A}} = x$, $x_{\mathrm{B}} = 1 - x$. By the generalized mean-field theory of Sect. 3.3 the number N_{AA} of AA nearest-neighbour pairs can be replaced by

$\frac{1}{2}zNx^2$. From (6.50) the fractions of sites occupied by A atoms with $\tau_i = 1$ and $\tau_i = -1$ respectively are $\frac{1}{2}x(1+m)$ and $\frac{1}{2}x(1-m)$ so that, by an argument similar to that following (3.36), the magnetic energy can be replaced by $-\frac{1}{2}Jzx^2m^2$. Then using (3.38), (6.10), (6.18) and (6.51), the free energy density

$$f = T\left[x\ln x + (1-x)\ln(1-x)\right] - \frac{1}{2}z\varepsilon x^2 - x\bar{\mu}$$
$$+ \frac{1}{2}x\left\{T\left[(1+m)\ln(1+m) + (1-m)\ln(1-m)\right]\right.$$
$$\left. - zJxm^2 - 2m\mathcal{H}\right\}, \tag{6.69}$$

where

$$\bar{\mu} = \mu + T\ln(2). \tag{6.70}$$

The equilibrium free energy density f is a function of T, μ and $\mathcal{H}$ so that the densities x and m must be regarded as order variables with respect to which the expression (6.69) has to be minimized. Equating to zero the partial derivatives of f with respect to m and x yield respectively

$$\mathcal{H} = \frac{1}{2}T\ln\left(\frac{1+m}{1-m}\right) - Jxm, \tag{6.71}$$

$$\bar{\mu} = T\ln\left(\frac{x}{1-x}\right) - z(\varepsilon + Jm^2)x - m\mathcal{H}$$
$$+ \frac{1}{2}T\left\{(1+m)\ln(1+m) + (1-m)\ln(1-m)\right\}. \tag{6.72}$$

We shall need the first and second derivatives of $\bar{\mu}$ with respect to x at constant T and $\mathcal{H}$, treating m as a function of x given implicitly by (6.71). From (6.72) and (6.71),

$$\left(\frac{\partial\bar{\mu}}{\partial x}\right)_{T,\mathcal{H}} = \frac{T}{x(1-x)} - \frac{1}{2}zJx\frac{\partial(m^2)}{\partial x} - z(\varepsilon + Jm^2), \tag{6.73}$$

$$\left(\frac{\partial^2\bar{\mu}}{\partial x^2}\right)_{T,\mathcal{H}} = T\left[\frac{1}{(1-x)^2} - \frac{1}{x^2}\right] - \frac{3}{2}zJ\frac{\partial(m^2)}{\partial x} - \frac{1}{2}zJx\frac{\partial^2(m^2)}{\partial x^2}. \tag{6.74}$$

With $\mathcal{H} = 0$ (6.71) is identical to (3.10) except that zJ is replaced by xzJ and it follows that, at equilibrium, $m = 0$ for $T > xzJ$ and $m = \pm m_s$ ($m_s > 0$) for $T < xzJ$. The critical curve on which second-order transitions from the disordered ($m = 0$) to be ordered ($m \neq 0$) state take place is therefore represented in the (x, T) plane by the line

$$T = xzJ, \tag{6.75}$$

which is shown in Fig. 6.7. The critical curve $\bar{\mu} = \bar{\mu}^*(T)$ in the $(\bar{\mu}, T)$ plane

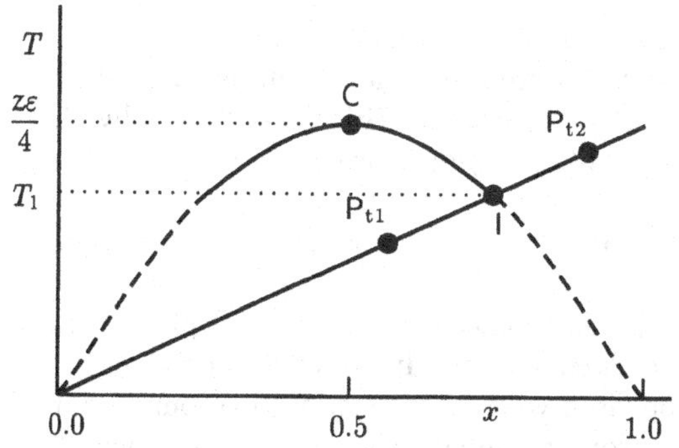

Fig. 6.7. Critical line and conjugate disordered phase curve in the mole fraction–temperature plane for dilute Ising model.

can be obtained by putting $m = 0$ and $x = T/(zJ)$ in (6.72).

Now consider the disordered state. With $m = 0$, (6.69) and (6.72) are similar to the corresponding mean-field relations for the binary solid mixture of Sect. 6.2 or, putting $x = \rho$, for the simple lattice gas. The only difference is that $\bar{\mu}$ replaces μ. Hence, from (5.25) and Sect. 5.4 (see also Example 5.1), a first-order transition line in the $\mathcal{H} = 0$ plane is specified by

$$\bar{\mu} = -\frac{1}{2}z\varepsilon\,, \qquad T < T_c\,. \tag{6.76}$$

The critical value of x is $\frac{1}{2}$ and, in the mean-field approximation, the critical temperature $T_c = \frac{1}{4}z\varepsilon$. The first-order transition is from a B-rich disordered phase (Δ_1) to an A-rich disordered phase (Δ_2). For the equivalent lattice gas model II of Sect. 6.6.2, Δ_1 and Δ_2 are respectively vapour and liquid.

These results can easily be confirmed from the formalism of this section. With T and $\mathcal{H}$ constant, (6.52) gives

$$\mathrm{d}f = -x\mathrm{d}\mu = -x\mathrm{d}\bar{\mu}\,. \tag{6.77}$$

From Sect. 1.10(ii), phase separation is associated with an instability loop including a range of x where

$$\frac{\partial \bar{\mu}}{\partial x} < 0 \tag{6.78}$$

and the critical point is given by

$$\frac{\partial \bar{\mu}}{\partial x} = 0\,, \qquad \frac{\partial^2 \bar{\mu}}{\partial x^2} = 0\,. \tag{6.79}$$

These last relations are satisfied by (6.73) and (6.74), with m^2 and its derivatives put equal to zero, if $\frac{1}{2}$ is substituted for x and $\frac{1}{4}z\varepsilon$ for T. Substituting $\bar{\mu} = -\frac{1}{2}z\varepsilon$ into (6.72), with $m = 0$, gives

$$T\ln\left(\frac{x}{1-x}\right) - \frac{1}{2}z\varepsilon(2x-1) = 0\,. \tag{6.80}$$

For $T < \frac{1}{4}z\varepsilon$, (6.80) has pairs of solutions $(x, 1-x)$. If we put $m = 0$, $\bar{\mu} = -\frac{1}{2}z\varepsilon$ in (6.69) then we obtain an expression such that $f(T,x) = f(T,1-x)$.

Hence (6.80) gives the Δ_1–Δ_2 coexistence curve, symmetrical about $x = \frac{1}{2}$, in the (x, T) plane, as shown in Fig. 6.7. Here the critical point is labelled by C and the intersection of the curve (6.80) with the critical line (6.75) by Q. From (6.80) and (6.75) the coordinates $(x_\mathrm{q}, T_\mathrm{q})$ of Q are given by

$$x_\mathrm{q} \ln\left(\frac{x_\mathrm{q}}{1 - x_\mathrm{q}}\right) - \frac{\varepsilon}{2J}(2x_\mathrm{q} - 1) = 0\,, \qquad T_\mathrm{q} = x_\mathrm{q} zJ\,. \tag{6.81}$$

For $T < T_\mathrm{q}$, the conjugate phase Δ_2 is replaced by an ordered phase and (6.80) no longer represents the coexistence curve. If the effect of the short-range magnetic ordering in component A were fully taken into account, as it would be in an exact theory, the disordered phase coexistence curve would no longer be symmetric about $x = \frac{1}{2}$. The only exception is the case $J = 0$, discussed in Sect. 9.9.

6.8 Multicritical Points in the Dilute Ising Model

It has been found that there is a critical curve in the $(\bar{\mu}, T)$ plane for the dilute Ising model just as there is in the $(\mathcal{H}, T)$ plane for the metamagnet (see Sect. 4.4).[5] To look for phase separation instability on the critical curve we need an expression for m at given T valid for values of x just greater than x^*, where x^* is the critical value of x for magnetic ordering, defined by (6.75). From (6.71), with $\mathcal{H} = 0$, and an expansion similar to that of (3.13) we have, for constant T and $x > x^*$,

$$m^2 = \frac{3(x - x^*)}{x^*} - \frac{27(x - x^*)^2}{5(x^*)^2} + \cdots \tag{6.82}$$

and hence

$$\left(\frac{\partial(m^2)}{\partial x}\right)_{x=x^*+0} = \frac{3}{x^*}\,, \qquad \left(\frac{\partial^2(m^2)}{\partial x^2}\right)_{x=x^*+0} = -\frac{54}{5(x^*)^2}\,. \tag{6.83}$$

Now, using (6.73) and (6.83), we define

$$\begin{aligned} B_1^{(+)}(T) &= \left(\frac{\partial\bar{\mu}}{\partial x}\right)_{x=x^*+0} = \frac{zJ}{1 - x^*} - z\varepsilon - \frac{3}{2}zJ\,, \\ B_1^{(-)}(T) &= \left(\frac{\partial\bar{\mu}}{\partial x}\right)_{x=x^*-0} = \frac{zJ}{1 - x^*} - z\varepsilon\,, \\ B_2(T) &= \frac{1}{2}\left(\frac{\partial^2\bar{\mu}}{\partial x^2}\right)_{x=x^*+0}\,. \end{aligned} \tag{6.84}$$

[5] Note that x here corresponds to m in the theory of the metamagnet whereas m here corresponds to ψ, given by (4.32).

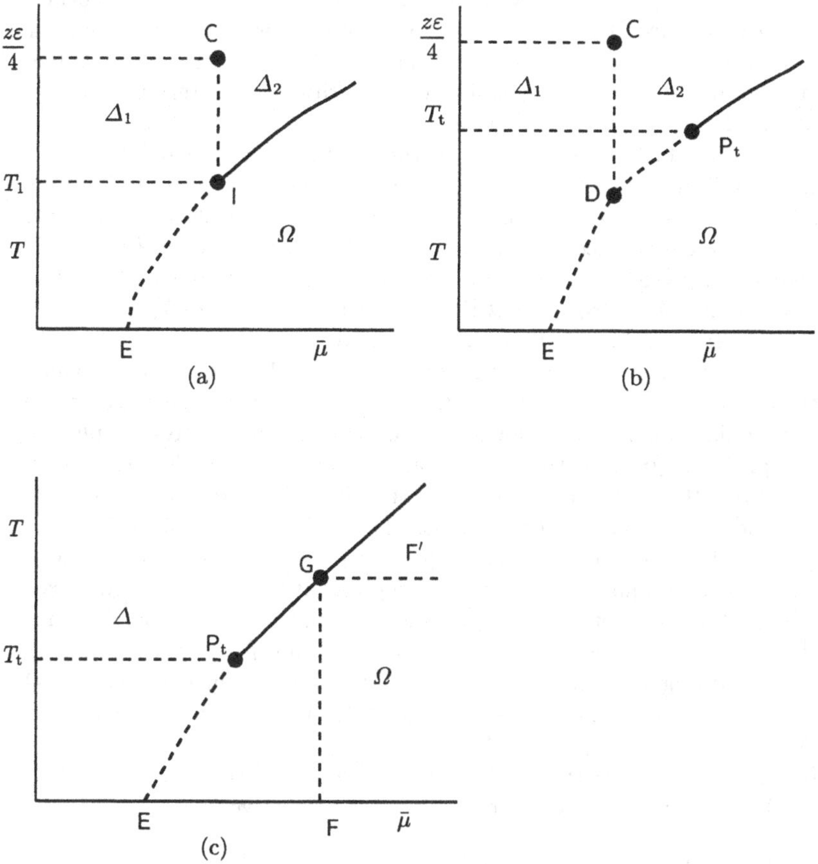

Fig. 6.8. Phase diagrams in the $\mathcal{H} = 0$ plane as ε/J decreases. Full lines are second-order transition curves (critical curves) and broken lines first-order transition curves. **(a)** $\varepsilon/J > 3.8019$, **(b)** $2.7747 < \varepsilon/J < 3.8019$, **(c)** $\varepsilon/J < 2.7747$.

Thus $B_1^{(-)}(T) > B_1^{(+)}(T)$, and instability will appear on the ordered side of the critical curve when $B_1^{(+)}(T) = 0$. From (6.84) the coordinates of the instability point P_t in the (x, T) plane are

$$x_\mathrm{t} = \frac{2\varepsilon + J}{2\varepsilon + 3J}\,, \qquad T_\mathrm{t} = zJx_\mathrm{t}\,. \tag{6.85}$$

A little numerical work shows that P_t and Q coincide (i.e. $x_\mathrm{t} = x_\mathrm{q}$) when $\varepsilon = 3.8019J$.

For $\varepsilon > 3.8019J$, $x_\mathrm{t} < x_\mathrm{q}$ and, in the (x, T) plane, the point P_t, like $P_{\mathrm{t}1}$ in Fig. 6.7, lies inside the Δ_1–Δ_2 coexistence curve. The phase diagram in the $(\bar{\mu}, T)$ plane is sketched in Fig. 6.8(a), where CQ is the line of Δ_1–Δ_2 first-order transitions and QE is the line of Δ_1–Ω first-order transitions, the ordered phase being labelled by Ω. The point E represents the equilibrium at $T = 0$ of a pure B phase with a completely ordered ($m = 1$) pure A phase, and it is easily verified that at this point $\bar{\mu} = -\frac{1}{2}\mathrm{z}(\varepsilon + J)$. The line of equilibrium second-order transitions on the critical curve terminates at Q which is therefore a *critical end-point.*

For $\varepsilon < 3.8019J$, $x_\mathrm{q} < x_\mathrm{t}$ and, in the (x, T) plane, the point P_t, like $\mathsf{P}_{\mathrm{t}2}$ in Fig. 6.7, lies outside the Δ_1–Δ_2 coexistence curve. Thus, as T passes below T_t, Δ_2–Ω phase separation begins at the point P_t on the critical curve. P_t is therefore a *tricritical point.* The phase diagram in the $(\bar{\mu}, T)$ plane is sketched in Fig. 6.8(b). The Δ_2–Ω first-order transition line which starts at P_t terminates at a *triple point* D where it meets the Δ_1–Δ_2 and Δ_1–Ω first-order transition lines CD and DE respectively.

When the ratio ε/J is further reduced, the Δ_1–Δ_2 critical point, which occurs at $x = \frac{1}{2}$ and $T = \frac{1}{4}\mathrm{z}\varepsilon$, falls into the instability range initiated at the tricritical point. There are no longer two distinguishable disordered phases, and the phase diagram in the $(\bar{\mu}, T)$ plane assumes the simple form shown in Fig. 6.8(c). Here, the Δ–Ω first-order transition line EP_t meets the Δ–Ω second-order transition line at the tricritical point P_t. The transition between the types of phase diagram shown in Figs. 6.8(b) and (c) occurs at the value of ε/J where there is equilibrium between the phase at C and an ordered phase. This can be found by setting $T = \frac{1}{4}\mathrm{z}\varepsilon$, and equating f and $\bar{\mu}$ respectively for $x = \frac{1}{2}$ and $m = 0$ to the appropriate expressions for f and $\bar{\mu}$ with $m > 0$. The two resulting relations together with the equilibrium equation (6.71) with $\mathcal{H} = 0$ determine the values of x and m for the ordered phase and the ratio ε/J. The solution gives $\varepsilon = 2.7747J$.

Now consider the tricritical phase separation for $T = T_\mathrm{t} - \theta$ where $0 < \theta \ll T_\mathrm{t}$. For the ordered and disordered phases respectively

$$
\begin{aligned}
\bar{\mu} - \bar{\mu}^* &= B_1^{(+)}(T_\mathrm{t} - \theta)(x - x^*) + B_2(T_\mathrm{t} - \theta)(x - x^*)^2 + \cdots, \\
\bar{\mu} - \bar{\mu}^* &= B_1^{(-)}(T_\mathrm{t} - \theta)(x - x^*) + \cdots,
\end{aligned}
\qquad (6.86)
$$

where $B_1^{(\pm)}(T)$ and $B_2(T)$ are defined by (6.84) and x^* by (6.75). Recalling that $B_1^{(+)}(T_\mathrm{q}) = 0$ we can, if we require only leading terms in θ, write, using (6.84) and (6.85),

$$
B_1^{(+)}(T_\mathrm{t} - \theta) = -A\theta , \qquad A = \left(\frac{3J + 2\varepsilon}{2J}\right)^2 . \qquad (6.87)
$$

To the same order of accuracy, $B_1^{(-)}(T)$ and $B_2(T)$ can be replaced by their $B_1^{(-)}(T_\mathrm{t})$ and $B_2(T_\mathrm{t})$ forms. From (6.74), (6.83) and (6.84),

$$
B_1^{(-)}(T_\mathrm{t}) = \frac{3}{2}\mathrm{z}J , \qquad B_2(T_\mathrm{t}) = \frac{\mathrm{z}J(3J + 2\varepsilon)(6J^2 + 40J\varepsilon + 40\varepsilon^2)}{80J^2(J + 2\varepsilon)} . \qquad (6.88)
$$

The plot of $\bar{\mu}$ against x has the form shown in Fig. 6.9. Applying analysis similar to that at the end of Sect. 4.4, the compositions of the ordered and disordered conjugate phases are given respectively by

$$x_\Omega = x^* + \frac{3}{4}\left[\frac{A}{B_2(T_\mathrm{t})}\right]\theta + \mathrm{O}(\theta^2)\,,$$
$$x_\Delta = x^* - \left[\frac{A^2}{8zJB_2(T_\mathrm{t})}\right]\theta^2 + \mathrm{O}(\theta^3) \tag{6.89}$$

and the value of $\bar{\mu}$ at phase equilibrium by

$$\bar{\mu} = \bar{\mu}^* - \left[\frac{3A^2}{16B_2(T_\mathrm{t})}\right]\theta^2 + \mathrm{O}(\theta^3)\,. \tag{6.90}$$

From (6.90), the first-order and second-order transition lines in the $(\bar{\mu}, T)$ plane which meet at the tricritical point have a common tangent there. From (6.89) the tricritical point is the apex of a cusp in a coexistence curve in the (x, T) plane, as shown in Fig. 6.10 which corresponds to the $(\bar{\mu}, T)$ plane diagram of Fig. 6.8(c). A diagram like that of Fig. 6.10 qualitatively resembles that observed for ^{4}He–^{3}He mixtures (Blume et al. 1971). The mean-field critical curve relation (6.75) can be expressed as $x^* = T/T_\mathrm{c}$, where T_c is the critical temperature for $x = 1$ (pure A), but the observed values at the tricritical point for the ^{4}He–^{3}He mixture are $T_\mathrm{t}/T_\mathrm{c} = 0.4$, $x_\mathrm{t} = 0.331$. When $\varepsilon > 0$ and short-range ordering is taken into account, $T/T_\mathrm{c} > x^*$ on critical curves (Bell 1953). However, in view of the schematic nature of I as a model for the ^{4}He–^{3}He mixture, close quantitative agreement between even exact results for model I and experiment is unlikely. For instance, the mole fraction x of A in the A-rich conjugate phase $\to 1$ as $T \to 0$, but in the ^{4}He–^{3}He systems the ^{4}He mole fraction $\to 0.94$, owing to quantum statistical effects.

Finally, note that, in spite of similar behaviour near the tricritical point, there are global differences between the dilute Ising model and the metamagnet of Sect. 4.4. For the latter, there is a critical end-point for $\lambda < \frac{3}{5}$,

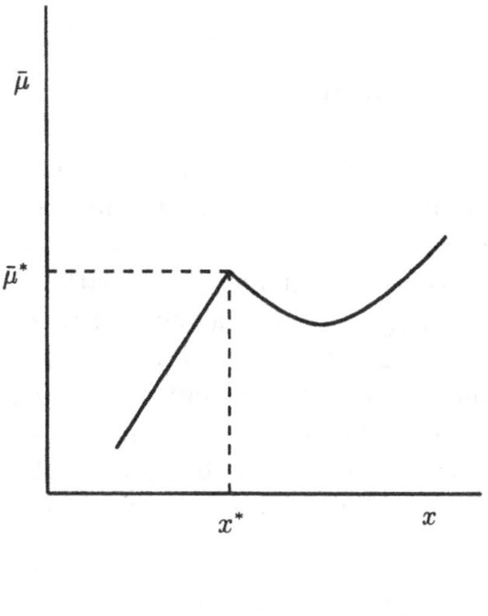

Fig. 6.9. Chemical potential–mole fraction isotherm for $T < T_\mathrm{t}$ showing instability on the critical curve.

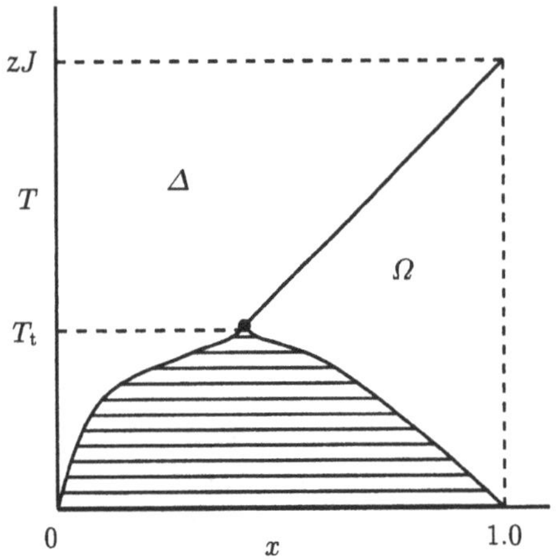

Fig. 6.10. Phase diagram in mole fraction–temperature plane corresponding to Fig. 6.8(c). The horizontal lines are phase separation tie lines.

and when λ increases through this value, a tricritical point is formed by the coalescence of the critical end-point and the 'ordinary' critical point (i.e. Q and C in Fig. 4.7). For the dilute Ising model there is a critical end-point when $\varepsilon/J > 3.8019$, but when ε/J decreases through this value, the critical end-point bifurcates into a tricritical point and a triple point. This leads to the intermediate phase pattern of Fig. 6.8(b), which is absent in the metamagnetic case. It is significant that the length of the first-order transition line above the critical end-point lies in the ordered region for the metamagnet but in the disordered region for the dilute Ising model. Hence, for the latter, coalescence of the critical point C with a point on the critical curve would require $B_1^{(-)}(T) = 0$, where $B_1^{(-)}(T)$ is given by (6.84). Since $B_1^{(-)}(T) = 0$ implies $B_1^{(+)}(T) < 0$, such a coalescence could not occur in an equilibrium state in mean-field theory.

6.9 Multicritical Phenomena with Additional Thermodynamic Dimension

By the results derived at the end of the account of Sect. 6.6.1, $\mathcal{H} = 0$ defines a plane of symmetry in the three-dimensional $(\bar{\mu}, \mathcal{H}, T)$ space of fields. So far, phase diagrams have been obtained only for this plane, but we now consider the extension into three dimensions of the 'simple' phase pattern of Fig. 6.8(c). A spontaneous magnetization $m_s(\bar{\mu}, T) > 0$ is associated with any point $(\bar{\mu}, T)$ in the ordered region of the plane $\mathcal{H} > 0$. From (6.71), $m \to m_s$ if this point is approached from the $\mathcal{H} > 0$ half-space but $m \to -m_s$ if it is approached from the $\mathcal{H} < 0$ half-space (cf.(3.11)). Hence there is a discontinuity in m of magnitude $2m_s$ at $\mathcal{H} = 0$, although by (6.53) the mole fraction x is continuous. It is necessary to distinguish two ordered phases labelled Ω_+ ($\mathcal{H}, m > 0$) and Ω_- ($\mathcal{H}, m < 0$) respectively and in the $(\bar{\mu}, T, \mathcal{H})$ space the ordered region of the $\mathcal{H} = 0$ plane becomes a *surface of first-order transitions* from Ω_+ to Ω_-. One boundary of the ordered region in the $\mathcal{H} = 0$ plane is the first-order transition line $\mathsf{P_tE}$. At any point on $\mathsf{P_tE}$ we now recognize two phases Ω_+ and Ω_-, with equal and opposite magnetizations, in equilibrium with a Δ ($m = 0$)–phase, where the value of x is lower. Hence $\mathsf{P_tE}$ is a *line of triple points* in $(\bar{\mu}, \mathcal{H}, T)$ space. These results are in accordance with the generalized phase rule (1.92) which yields $\mathcal{F}(3,p) = 4 - p$. Here p is the number of phases in equilibrium and $\mathcal{F}(3,p)$ the number of degrees of freedom. Thus, $\mathcal{F}(3,2) = 2$, giving a surface in field space whereas $\mathcal{F}(3,3) = 1$, giving a line.

Now consider a vertical ($\bar{\mu}$ = constant) segment in the ordered region of the $\mathcal{H} = 0$ plane, like FG in Fig. 6.8(c). The segment FG is a line of first-order Ω_+–Ω_- transitions in the $\bar{\mu}$ = constant plane, and hence the point G where this line terminates is a critical point. Since G is an arbitrary point on the critical curve, the latter becomes a *line of critical points* in $(\bar{\mu}, \mathcal{H}, T)$ space.[6]

As remarked above, three distinct phases, designated Ω_+, Ω_- and Δ, are in equilibrium at any point on the line $\mathsf{P_tE}$, and $\mathsf{P_tE}$ is a boundary of the Ω_+–Ω_- transition surface. It may therefore be supposed that there are two other first-order transition surfaces which meet on $\mathsf{P_tE}$. These are respectively a Δ–Ω_+ transition surface in the $\mathcal{H} > 0$ half-space, and a Δ–Ω_- surface in the $\mathcal{H} < 0$ half-space. By field-reversal symmetry these surfaces will be mirror images in the $\mathcal{H} = 0$ plane and they can be regarded as a pair of 'wings', as shown in Fig. 6.11. This picture is confirmed by the proof given below of the existence of the critical curves $\mathsf{P_tK}$ and $\mathsf{P_tL}$, which form the upper boundaries of the 'wings'. Although $m \neq 0$ whenever $\mathcal{H} \neq 0$, the Δ-phase region in $(\bar{\mu}, T, \mathcal{H})$ space can be defined as that which is continuous with the Δ ($m = 0$) area on the $\mathcal{H} = 0$ plane. At a Δ–$\Omega_\pm$ transition surface, $|m|$ and x increase discontinuously. However, for very large positive $\mathcal{H}$, $\tau_i = 1$ for nearly all sites occupied by an A. Thus component A effectively consists of a single species and the system is a simple binary mixture with the parameter ε replaced by $\varepsilon + J$. The situation is similar for large negative $\mathcal{H}$. Hence on the critical curves $\mathsf{P_tK}$ or $\mathsf{P_tL}$, $T \to \frac{1}{4}\mathrm{z}(\varepsilon + J)$ and $x \to \frac{1}{2}$ as $|\mathcal{H}| \to \infty$.

At constant $\bar{\mu}$ and T, from (6.52), $\mathrm{d}f = -xm\mathrm{d}\mathcal{H}$. From (1.102), critical curves in the $\mathcal{H} \neq 0$ regions are defined by the relations

$$\left(\frac{\partial \mathcal{H}}{\partial (xm)}\right)_{\bar{\mu},T} = 0\,, \qquad \left(\frac{\partial^2 \mathcal{H}}{\partial^2 (xm)}\right)_{\bar{\mu},T} = 0\,. \tag{6.91}$$

Substitution of (6.71) into (6.72) gives

$$\bar{\mu} = T \ln\left(\frac{x}{1-x}\right) - \mathrm{z}\varepsilon x + \frac{1}{2} T \ln(1 - m^2). \tag{6.92}$$

[6] We now restrict the term 'critical curve' to the equilibrium part only.

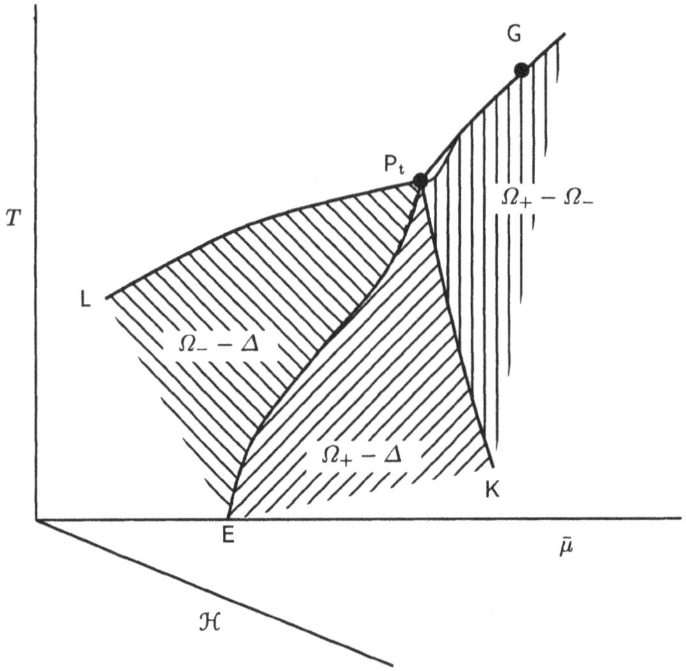

Fig. 6.11. Phase diagram in chemical potential–magnetic field–temperature space corresponding to Fig. 6.8(c). The broken curve $\mathsf{P_tE}$ is a line of triple points and bounds the vertically hatched Ω_+–Ω_- transition surface in the symmetry plane $\mathcal{H} = 0$. The 'wings' bounded by $\mathsf{P_tE}$, $\mathsf{P_tK}$ and $\mathsf{P_tL}$ respectively lie outside the $\mathcal{H} = 0$ plane.

Now (6.91) implies that $\mathcal{H}$ has been replaced as independent variable by the density xm. However, since, by (6.92), x and hence xm depend only on m at fixed $\bar{\mu}$ and T, it follows that (6.91) is equivalent to

$$\left(\frac{\partial \mathcal{H}}{\partial m}\right)_{\bar{\mu},T} = 0\,, \qquad \left(\frac{\partial^2 \mathcal{H}}{\partial m^2}\right)_{\bar{\mu},T} = 0\,, \tag{6.93}$$

or alternatively, regarding m as a function of x, to similar relations with x replacing m. Thus to demonstrate the presence of critical curves, solutions of (6.93) must be shown to exist. They can be obtained explicitly for $\varepsilon = 0$ (Blume et al. 1971) when (6.92) can be expressed as

$$x = \frac{\mathfrak{z}}{\mathfrak{z} + \sqrt{1 - m^2}}\,, \qquad \mathfrak{z} = \exp(\bar{\mu}/T)\,. \tag{6.94}$$

Substitution into (6.71) yields $\mathcal{H}$ as a function of m given by

$$\mathcal{H} = \frac{1}{2} T \ln\left(\frac{1+m}{1-m}\right) - \frac{zJ\mathfrak{z}m}{\mathfrak{z} + \sqrt{1 - m^2}}\,. \tag{6.95}$$

From (6.93) and (6.95)

$$\bar{\mu} = \frac{1}{2}T\ln\left(\frac{4T - zJ}{4T}\right),$$
$$\mathcal{H} = \frac{1}{2}T\ln\left[\frac{zJ - 2T + \sqrt{zJ(zJ-3T)}}{zJ - 2T - \sqrt{zJ(zJ-3T)}}\right] - \sqrt{zJ(zJ-3T)} \tag{6.96}$$

for the critical curve in the $\mathcal{H} > 0$ half-space. Changing $\mathcal{H}$ to $-\mathcal{H}$ gives the mirror image critical curve in the $\mathcal{H} < 0$ half-space. At $T = \frac{1}{2}zJ$, (6.96) yields $\mathcal{H} = 0$ and $\bar{\mu} = -T\ln(2)$ while (6.95) and (6.94) then give $m = 0$ and $x = \frac{1}{3}$. This means that (6.96) and its mirror image terminate on the $\mathcal{H} = 0$ plane at the tricritical point $\mathsf{P_t}$, as may be seen from putting $\varepsilon = 0$ in (6.85) and (6.92). Also, from (6.96), $\mathcal{H} \to \infty$ as $T \to \frac{1}{4}zJ + 0$ and it can be shown from (6.96) and (6.95) that $\sqrt{1-m^2} \simeq 3$ so that by (6.94), $x \to \frac{1}{2}$. For the $\varepsilon = 0$ case this agrees with the argument above about the large $|\mathcal{H}|$ limit. For $\varepsilon \neq 0$, (6.92) and (6.71) can be used to express m and $\mathcal{H}$ as functions of x. Then equations for the critical curves can be derived from relations similar to (6.93), but with x replacing m. However, these do not yield an explicit solution like (6.96) and must be solved numerically.

It can be seen that in a three-dimensional space of thermodynamic fields, like the $(\bar{\mu}, \mathcal{H}, T)$ space considered here, a tricritical point terminates three distinct critical curves as well as a line of triple points. This in fact explains the description 'tricritical', introduction by Griffiths (1970). If thermodynamic variables are adjusted to maintain three-phase equilibrium as the temperature increases, that is the system is kept on the line PE in Fig. 6.11, then the tricritical point occurs when the three coexisting phases coalesce.

The extension into $(\bar{\mu}, \mathcal{H}, T)$ space of the phase pattern of Fig. 6.8(a), with a critical end-point Q, is sketched in Fig. 6.12. The ordered region on the plane of symmetry, $\mathcal{H} = 0$, again becomes an Ω_+–Ω_- first-order transition surface bounded above by a critical curve. The segment QE becomes a line of Ω_+–Ω_-–Δ_1 triple points while QC is the intersection of a Δ_1–Δ_2 first-order transition surface with the plane $\mathcal{H} = 0$. A critical end-point is thus the termination of a line of triple points and a single critical curve. If thermodynamic variables are adjusted to keep three phases in equilibrium as the temperature is raised, then a critical end-point occurs when two phases become identical but remain distinct from the third phase. In the present case, Ω_+ and Ω_- coalesce into Δ_2 but remain distinct from Δ_1.

The three-dimensional topology corresponding to the intermediate $\mathcal{H} = 0$ phase pattern of Fig. 6.8(b) is rather complicated. D becomes a *quadruple point* where the four phases Δ_1, Δ_2, Ω_+ and Ω_- coexist. It is the meeting point of the Δ_1–Ω_+–Ω_- and Δ_2–$\Omega/_+$–Ω_- triple point lines DE and $\mathsf{DP_t}$ respectively in the $\mathcal{H} = 0$ plane with Δ_1–Δ_2–Ω_+ and Δ_1–Δ_2–Ω_- triple point lines which are mirror images of each other in this plane. For $3J < \varepsilon < 3.8019J$ the Δ_1–Δ_2–$\Omega_\pm$ lines terminate at critical end-points where the

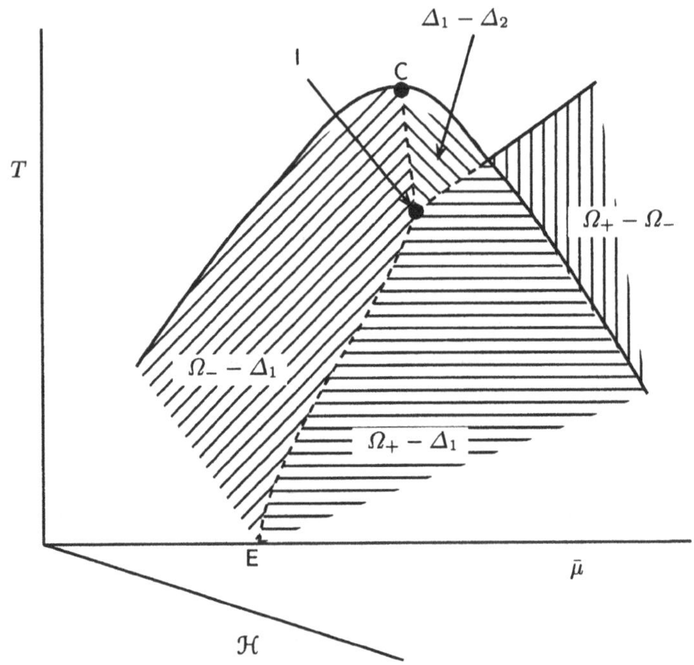

Fig. 6.12. As for Fig. 6.11, corresponding to Fig. 6.8(a); Q is the critical end-point. The broken curve QE is a line of triple points in the plane of symmetry $\mathcal{H} = 0$.

Δ_2 and $\Omega_\pm$ phases coalesce. When $\varepsilon = 3J$ (Mukamel and Blume 1974) there are three tricritical points, Δ_2–Ω_+–Ω_-, Δ_1–Δ_2–Ω_+ and Δ_1–Δ_2–Ω_-. For $2.7747 < \varepsilon < 3J$ there are again two critical end-points where the Δ_1 and Δ_2 phases now coalesce. The value $\varepsilon = 3J$ corresponds to the completely symmetrical 3-state Potts model (see Sect. 6.10).

The field space for the metamagnet can be extended into three dimensions by postulating a *staggered field* $\mathcal{H}_s$, conjugate to the order parameter ψ, which acts in opposite senses at the sites of the two sublattices. For $\lambda > 3/5$ the 'tricritical' phase diagram in $(\mathcal{H}, T, \mathcal{H}_s)$ space is similar to Fig. 6.11, the Ω_+ and Ω_- phases in equilibrium at the symmetry plane $\mathcal{H}_s = 0$ corresponding to the states specified in (4.10). For $\lambda < \frac{3}{5}$, four ordered phases Ω_+, Ω_-, Ω'_+ and Ω'_- must be envisaged, and when the diagram of Fig. 4.7 is extended into three dimensions, QC becomes a line of quadruple points where the four ordered phases are in equilibrium. The Ω_+–Ω'_- and Ω_-–Ω'_- critical lines terminate at C which is called a *bicritical end-point* (Kincaid and Cohn 1975). For general accounts of tricritical and other multicritical points see Aharony (1983), Lawrie and Sarbach (1984) and Pynn and Skjeltorp (1984).

6.10 The Unsymmetrical and Completely Symmetrical Models

In this section we discuss two contrasting cases. In the first the symmetry about the plane $\mathcal{H} = 0$ is removed whereas in the second a higher symmetry is introduced.

Consider the binary lattice gas model IV of Sect. 6.6.4, where the field variable $\mathcal{H}$ is replaced by $\mu_C - \mu_D$. This is equivalent to model I of Sect. 6.6.1, with the relation $\mu_C - \mu_D = 0$ defining the symmetry plane, provided that $\varepsilon_{CC} = \varepsilon_{DD}$. However, for real mixtures, symmetry conditions are satisfied only in special cases. When $\varepsilon_{CC} \neq \varepsilon_{DD}$ a term (see Example 6.4)

$$\frac{1}{4}(\varepsilon_{CC} - \varepsilon_{DD}) \sum_{\{i,j\}}^{(\text{n.n.})} (\tau_i\tau_j^2 + \tau_j\tau_i^2) \tag{6.97}$$

must be added to the Hamiltonian given by (6.49) and (6.67). Thus the Hamiltonian is no longer invariant under a simultaneous change of sign of $\mu_C - \mu_D$ and all τ_i; (that is, an interchange of the values of μ_C and μ_D and replacement of each C by a D and each D by a C.) We look at the effect on the three-dimensional phase pattern of Fig. 6.11 where, for model IV of Sect. 6.6.4, Δ labels a vapour phase and Ω_+ and Ω_- correspond to liquid C–D mixtures. When $\varepsilon_{CC} \neq \varepsilon_{DD}$ the Ω_+–Ω_- first-order transition surface no longer lies in the plane $\mu_C = \mu_D$ ($\mathcal{H} = 0$), which ceases to be a plane of symmetry. Again, the Δ–Ω_+ and Δ–Ω_- critical curves cease to be mirror images in the $\mu_C = \mu_D$ plane and no longer meet the Ω_+–Ω_- transition surface at the same point P_t. Suppose that the Δ–Ω_+ critical curve has slipped downwards relative to the T-axis and meets the Ω_+–Ω_- transition surface at A, where $T = T_a$. For $T > T_a$, the Ω_+ and Δ phases are identical and there remains only a Δ–Ω_- first-order transition surface, as shown in Fig. 6.13. The broken triple point line and the Δ–Ω_+ critical curve terminate at A, which is thus a critical end-point. In the absence of intrinsic symmetry, such as exists in magnetic systems, or an accidental symmetry, for instance the relation $\varepsilon_{CC} = \varepsilon_{DD}$, a line of triple points will normally terminate at a critical end-point. Suppose that an additional field is introduced by, say, adding another component to the system. By appropriately adjusting the new field we may be able to make A move up to point A′ on the Δ–Ω_- critical curve, and A′ will then become a tricritical point. A tricritical point thus appears at a special position on a line of critical end-points in a four-dimensional field space. It is helpful to recall the Gibbs phase rule (1.93) for fluid mixtures. For three coexisting phases, $\mathcal{D}(\kappa, 3) = \kappa - 1$, where κ is the number of components, implying a $(\kappa-1)$-dimensional domain of triple points in field space. We then expect $(\kappa - 2)$- and $(\kappa - 3)$-dimensional domains of critical end-points and tricritical points respectively. Thus, for lines of critical end-points and isolated tricritical points, $\kappa = 3$, which is a ternary solution. Experimentally, it is hard to find an isolated tricritical point for a fluid system

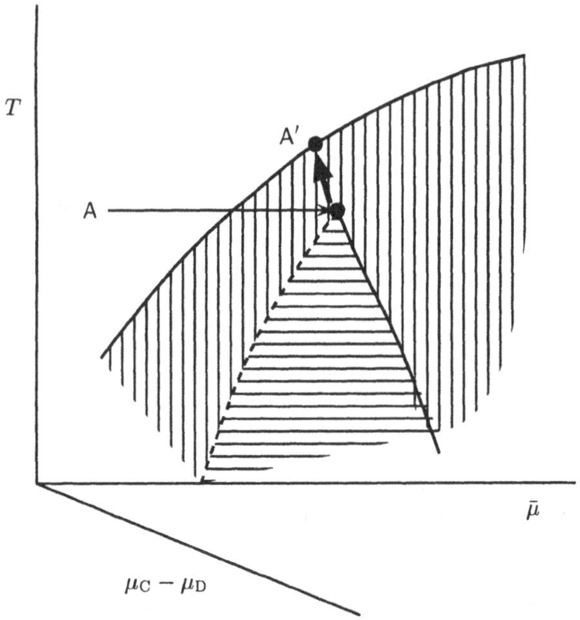

Fig. 6.13. Three-dimensional phase diagram for unsymmetrical binary lattice gas. The broken line (of triple points) does not lie in the plane $\mu_C - \mu_D = 0$.

so at least a quaternary ($\kappa = 4$) mixture is required, giving a line of tricritical points. One of these can then be located at, say, an arbitrary pressure. (For multicritical points in fluids, see Knobler and Scott (1984), or Rowlinson and Swinton (1982).)

The Potts model (Potts 1952, Wu 1982) was described in Example 2.2 (see also Sect. 10.14 and Volume 2, Sect. 3.5). The ferromagnetic 3-state Potts model can be related to the present theory by considering a completely symmetrical form of model III of Sect. 6.6.3 in which (6.60) is replaced by

$$\left.\begin{aligned} &\varepsilon_{BB} = \varepsilon_{CC} = \varepsilon_{DD} = u_0 - J\,, \\ &\varepsilon_{BC} = \varepsilon_{BD} = \varepsilon_{CD} = u_0 + J\,, \end{aligned}\right\} \quad J > 0\,. \tag{6.98}$$

The parameter u_0 gives a constant additive term Nu_0 in the energy and has no effect on the lattice distribution. Substitution of (6.98) into (6.62) gives

$$\varepsilon = 3J\,, \quad \mathcal{H} = \frac{1}{2}(\mu_C - \mu_D)\,, \qquad \mu = \frac{1}{2}(\mu_C + \mu_D) - \mu_B - 2zJ\,. \tag{6.99}$$

for the completely symmetrical ternary model. In the Potts model, B, C and D are interconvertible species of the same component so that $\mu_B = \mu_C = \mu_D$ and (6.99) becomes

$$\varepsilon = 3J, \qquad \mathcal{H} = 0, \qquad \mu = -2zJ. \tag{6.100}$$

From (6.98) the difference between the energies of like and unlike pairs is $-2J$, so that $J = \frac{1}{2}R$, where R is the parameter defined in Example 2.2. The completely symmetrical ternary model is sometimes regarded as a Potts model with fields.

To exhibit the properties of the completely symmetrical model use Cartesian coordinates

$$\left(\tfrac{\sqrt{3}}{2}[\mu_{\mathrm{C}} - \mu_{\mathrm{D}}], \tfrac{1}{2}[\mu_{\mathrm{C}} + \mu_{\mathrm{D}} - 2\mu_{\mathrm{B}}], T\right)$$

in field space. Hence, by (6.99), $\mu_{\mathrm{C}} - \mu_{\mathrm{D}} = 0$, $\mu = -2zJ$ on the T-axis which thus represents the Potts model with $\mu_{\mathrm{B}} = \mu_{\mathrm{C}} = \mu_{\mathrm{D}}$. The planes $\mu_{\mathrm{B}} - \mu_{\mathrm{D}} = 0$ and $\mu_{\mathrm{B}} - \mu_{\mathrm{C}} = 0$ meet the plane $\mu_{\mathrm{C}} - \mu_{\mathrm{D}} = 0$ on the T-axis at angles $\pm 2\pi/3$. The three-dimensional phase diagram has three-fold symmetry with respect to rotation about the T-axis,[7] a result which applies exactly as well as in the mean-field approximation. Since there is a first-order transition surface on the $\mu_{\mathrm{C}} - \mu_{\mathrm{D}} = 0$ plane there will be similar first-order transition surfaces on the $\mu_{\mathrm{B}} - \mu_{\mathrm{D}} = 0$ and $\mu_{\mathrm{B}} - \mu_{\mathrm{C}} = 0$ planes. The three transition surfaces meet in a line of triple points occupying the lower part of the T-axis.

Since $\varepsilon/J = 3$ lies in the intermediate range for the mean-field approximation, the phase diagram in the $\mu_{\mathrm{C}} - \mu_{\mathrm{D}} = 0$ plane is topologically similar to Fig. 6.8(b), the three-dimensional extension of which is discussed in Sect. 6.9. The line of triple points DE now lies on the T-axis and is joined at the quadruple point D by another line of triple points in the $\mu_{\mathrm{C}} - \mu_{\mathrm{D}} = 0$ plane which terminates at a tricritical point P_{t}. By the three-fold symmetry condition the phase patterns in the $\mu_{\mathrm{B}} - \mu_{\mathrm{D}} = 0$ and the $\mu_{\mathrm{B}} - \mu_{\mathrm{C}} = 0$ planes are identical with that in the $\mu_{\mathrm{C}} - \mu_{\mathrm{D}} = 0$ plane. Hence the two remaining lines of triple points which emanate from the quadruple point D lie respectively in the $\mu_{\mathrm{B}} - \mu_{\mathrm{D}} = 0$ and $\mu_{\mathrm{B}} - \mu_{\mathrm{C}} = 0$ planes, and each line terminates at a tricritical point. Hence when $\varepsilon/J = 3$ there are three tricritical points in the dilute Ising and equivalent models. At any value of T the Potts model is represented by a point on the T-axis. At high temperatures the state is disordered with $x_{\mathrm{B}} = x_{\mathrm{C}} = x_{\mathrm{D}} = \frac{1}{3}$, but when T is decreased, a transition will occur at the quadruple point D on the T-axis. The transition is first-order since the representative points must cross a first-order transition surface to one of the three equivalent states whose domains meet at the segment DE on the T-axis.

The mean-field approximation phase diagram is probably qualitatively correct for three-dimensional lattices (Wu 1982). However, it has been shown by Baxter (1973b) and Baxter et al. (1978) that the transition in the ferromagnetic ν-state Potts model on two-dimensional lattices is of higher order for $\nu \leq 4$. This implies that, in an exact solution of the dilute Ising model, the value $\varepsilon/J = 3$ is in a ε/J range corresponding to phase diagrams topolog-

[7] There is also symmetry with respect to reflection in the $\mu_{\mathrm{C}} - \mu_{\mathrm{D}} = 0$ and $\mu_{\mathrm{B}} - \mu_{\mathrm{C}} = 0$ planes. Hence the overall symmetry of the system is that of the permutation group $\mathcal{S}_3$.

ically similar to Figs. 6.8(c) and 6.11. From symmetry considerations the line of triple points $\mathsf{P_t E}$ occupies the lower part of the T-axis, and the three critical curves meeting at the tricritical point $\mathsf{P_t}$ lie respectively in the $\mu_C - \mu_D = 0$, $\mu_B - \mu_D = 0$ and $\mu_B - \mu_C = 0$ planes and are of identical form. For the Potts model, a higher-order transition occurs on the T-axis at $\mathsf{P_t}$. There is other evidence supporting this general picture (Straley and Fisher 1973, Wu 1982).

6.11 Alternative Forms for the Dilute Ising Model

6.11.1 Model A: Equilibrium Bond Dilution

In the *site-dilute* Ising model considered above, the fraction $1 - x$ of lattice sites not occupied by spins are 'inert' magnetically. In a *bond-dilute* model, on the other hand, all N sites are occupied by spins, but a fraction $1 - y$ of the nearest neighbour pairs are 'inert', with no magnetic interaction between the two spins. We can say that a *bond* exists between the spins of each interacting nearest neighbour pair and that for the non-interacting pairs the bond is *broken*. It is assumed that the distribution of unbroken and broken bonds among the $\frac{1}{2}zN$ lattice edges re-adjusts at each temperature to minimize the free energy. As will be shown in Sect. 9.2 the properties of the model can then be deduced from those of the pure Ising model on the same lattice. It turns out that a critical concentration y_c exists such that $T^*(y) \to 0$ as $y \to y_c + 0$, where $T^*(y)$ is the critical temperature for a fraction y of unbroken bonds. For $y < y_c$, no magnetic transition occurs.

6.11.2 Model B: Random Site Dilution

Suppose that a lattice mixture of xN A atoms with magnetic moment and $(1 - x)N$ magnetically inert B atoms is rapidly cooled (quenched) from a temperature T_1. It is assumed that the orientations of the spins on the A atoms can re-adjust rapidly enough to maintain thermal equilibrium, but that, due to the relative slowness of A–B diffusion, the positions of the A and B atoms on the lattice remain fixed. The A–B lattice configuration of any particular assembly is thus frozen. However, in the theory, it is necessary to consider a distribution of A–B configurations which is independent of the actual value of T but identical to the equilibrium distribution at T_1. It is usually assumed that T_1 is large enough for the equilibrium distribution at T_1 to be random, giving equal weight to all A–B configurations. Bell (1958) distinguished between this model and the equilibrium site-dilute model introduced in Sect. 6.6. The mean-field method, which implicitly assumes that the environment of each A atom is the same, is unsuitable for dealing with the random model. The first-order pair approximation described in Chap. 7 makes some allowance for short-range order effects, and with the pair numbers

N_{AA}, N_{BB} and N_{AB} fixed at their random values gives a critical concentration $x_c = 1/(z-1)$ (Bell 1958, Sato et al. 1959).

The accurate value of x_c can be derived from site percolation theory (Essam 1972, 1980). A cluster of A atoms is defined as a set such that there is at least one connected path of AA edges between any two atoms of the set. Long-range ordering of the spin orientations depends on the existence of infinite clusters. Thus, denoting the probability in the limit $N \to \infty$ of an A atom belonging to an infinite cluster by $p(x)$, the critical concentration x_c is the value for which $p(x) = 0$ when $x \leq x_c$ and $p(x) > 0$ for $x > x_c$. Hence x_c depends on the properties of random distributions on the lattice and not on the details of the AA interaction and is, for instance, the same for the Heisenberg model as for the Ising model. Accurate values for regular lattices differ from the approximate value $x_c = 1/(z-1)$; for example, on a square lattice, the accurate value is $x_c = \frac{1}{2}$ whereas $1/(z-1) = \frac{1}{3}$. However, percolation theory yields $x_c = 1/(z-1)$ for the loosely connected *Bethe lattice* described in Sect. 7.7, (Fisher and Essam 1961). For more recent references on random systems see Thompson (1985).

6.11.3 Model C: Random Bond Dilution

The random bond-dilute model bears the same relation to model A as model B does to the equilibrium site-dilute model. It is assumed that the $\frac{1}{2}zNy$ unbroken bonds and the $\frac{1}{2}zN(1-y)$ broken bonds are distributed randomly over the $\frac{1}{2}zN$ edges connecting nearest neighbour pairs of sites. In bond percolation theory (Essam 1972, 1980) a cluster of sites is a set such that there is at least one connected path of unbroken bonds between any two atoms of the set. Denoting the probability in the limit $N \to \infty$ of a site belonging to an infinite cluster by $p_b(y)$ the critical concentration y_c is the value for which $p_b(y) = 0$ when $y \leq y_c$ and $p_b(y) > 0$ when $y > y_c$. For the random cases it can be shown that $x_c \geq y_c$ for any lattice. For some $d = 2$ lattices, x_c and/or y_c can be deduced exactly by transformation methods, and for the remaining regular lattices (including $d = 3$ lattices), y_c can be obtained with reasonable accuracy by series methods. A *striped randomness* bond dilution model is reviewed by McCoy (1972). Let the Ising model interaction energy parameters for horizontal edges of the square lattice be J_1 and for vertical edges be J_2. The parameter J_1 has the same value over the whole lattice, whereas, with the rows of vertical edges labelled by the index ℓ, $J_2(\ell)$ is a random variable with a probability distribution $p(J_2)$. Critical properties differ considerably from those of the pure Ising model. For a review of models A, B and C as well as the equilibrium site-dilute model of Sect. 6.6 see Stinchcombe (1983).

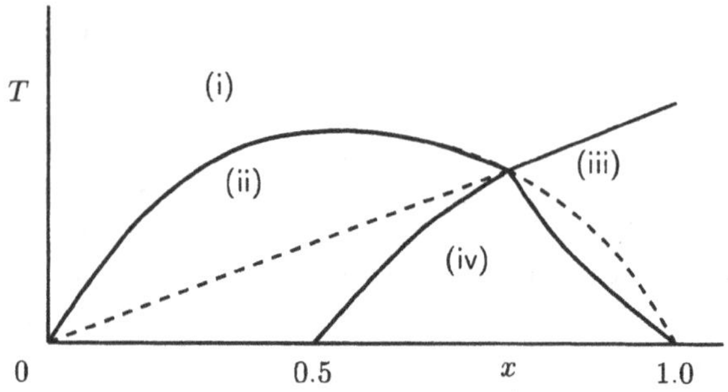

Fig. 6.14. Phase diagram, in mole fraction-temperature plane for $\varepsilon < -2J$. (Reprinted from Bell and Lavis (1965), by permission of the publisher Taylor and Francis Ltd.)

6.11.4 Model D: Equilibrium Site Dilution $\varepsilon < 0$

For negative ε and large enough magnitudes of $|\varepsilon|/J$ and $1-x$, the formation of AB pairs prevents a transition to a magnetized state and there is a critical concentration x_c. Using a first-order pair method, critical concentrations are found for the Heisenberg model (Bell and Fairbairn 1961a) and for the Ising model with $\varepsilon < -J$ (Bell 1975) even when A–B superlattice formation is not considered (see Example 7.5).

When A–B sublattice ordering can occur there are states of four types: (i) disordered; (ii) sublattices ordered but unmagnetized; (iii) magnetized without sublattice order; (iv) magnetized with sublattice order. Houska (1963) used the mean-field method with certain restrictive conditions to investigate the properties of Fe–Al alloys. Mean-field theories for A atoms carrying spins of any magnitude s and spin-spin and A–B interactions of arbitrary range were developed by Bell and Lavis (1965) and Lavis and Bell (1967), the former being concerned with 1:1 A–B sublattice structures and the latter with more elaborate ones. A first-order approximation was used by Lavis and Fairbairn (1966) with interactions confined to nearest-neighbour pairs. Superexchange effects were considered in the mean-field approximation by Bell and Lavis (1968). Curves of (i)–(ii), (i)–(iii), (ii)–(iv) and (iii)–(iv) transitions were derived giving a variety of phase diagrams in the (x, T) plane whose form depends on the relative magnitude of spin-spin and A–B interactions inside and between the sublattices. The simplest case is a loose-packed lattice with interactions confined to nearest-neighbour pairs. For $\varepsilon < -J$ and $x < \frac{1}{2}$ there are no AA nearest-neighbour pairs at $T = 0$ and the system is unmagnetized, giving a critical concentration $x_c = \frac{1}{2}$. A phase diagram for $\varepsilon < -2J$ is shown in Fig. 6.14, where the broken lines are undisturbed (i)–(ii) and (i)–(iii) transition curves.

Examples

6.1 For the one-dimensional binary mixture with all N sites occupied, show that, apart from a constant factor, the configurational canonical partition function is given by

$$Z = \sum_{\{X\}} g(M_{\mathrm{A}}, M_{\mathrm{B}}, X)\mathfrak{X}^{-X},$$

where $\mathfrak{X} = \exp(\varepsilon/T)$, ε is given by (6.8), $X = \frac{1}{2}N_{\mathrm{AB}}$ and g is defined by (2.126). By the maximum-term method (see Sect. 2.6), show that the equilibrium value of X is given by

$$\frac{X_{\max}}{N} = \frac{2x_{\mathrm{A}}x_{\mathrm{B}}}{1+\xi},$$

$$\xi = \sqrt{(x_{\mathrm{A}} - x_{\mathrm{B}})^2 + 4\mathfrak{X}x_{\mathrm{A}}x_{\mathrm{B}}}\,.$$

Show that $X_{\max}/N \to x_{\mathrm{A}}x_{\mathrm{B}}$, as $T \to \infty$. Given that $x_{\mathrm{A}} \geq x_{\mathrm{B}}$, show that $X_{\max}/N \to 0$, as $T \to 0$, if the inequality (6.17) applies and that $X_{\max}/N \to x_{\mathrm{B}}$, as $T \to 0$, if the inequality (6.21) applies. Show that the configurational heat capacity per site at constant composition is given by

$$c = 4k_{\mathrm{B}}\frac{(\ln \mathfrak{X})^2\,\mathfrak{X}x_{\mathrm{A}}^2x_{\mathrm{B}}^2}{\xi(1+\xi)^2}\,.$$

For $x_{\mathrm{A}} = x_{\mathrm{B}} = \frac{1}{2}$, show that this reduces to

$$c = \frac{1}{4}k_{\mathrm{B}}\frac{(\ln \mathfrak{X})^2}{(\mathfrak{X}^{1/4} + \mathfrak{X}^{-1/4})^2}\,.$$

Show that this expression has the same value whether ε is positive or negative and, putting $4J = |\varepsilon|$, show that it is equivalent to the expression (2.83) for the zero-field heat capacity of the one-dimensional Ising model.

6.2 From equation (6.32) where θ satisfies the equilibrium relation (6.34), show that

$$\left(\frac{\partial \Delta a}{\partial x_{\mathrm{A}}}\right)_{\theta} = \frac{T \ln f(\theta)}{1+s}\,, \qquad f(\theta) = \frac{(1+s\theta)(1-\theta)^s}{(1-ys\theta)(1+y\theta)^s}\,.$$

For $s = 3$, $y = \frac{1}{3}$, show that $f(\theta) < 1$, where $0 < \theta < 1$.

6.3 If the Ising model configurational energy for the plain square lattice is modified to

$$-J\sum_{\{i,j\}}^{(\text{n.n.})} \sigma_i\sigma_j - J_4 \sum_{\{i,j,k,\ell\}}^{(\text{squares})} \sigma_i\sigma_j\sigma_k\sigma_\ell\,,$$

where $J > 0$, $J_4 > 0$ and the second summation is over all squares of four sites such that (i, j), (j, k), (k, ℓ) and (ℓ, i) are nearest-neighbour pairs, then the Helmholtz free energy per site in the mean-field approximation is

$$a = T\left[\frac{1+m}{2}\ln\left(\frac{1+m}{2}\right) + \frac{1-m}{2}\ln\left(\frac{1-m}{2}\right)\right]$$
$$- 2Jm^2 - J_4 m^4 .$$

Show that a Landau expansion of a gives

$$a + T\ln(2) = \sum_{n=1}^{\infty} B_{2n}(T) m^{2n} ,$$

where

$$B_{2n}(T) = \begin{cases} \frac{1}{2}T - 2J , & n = 1 , \\ \frac{1}{12}T - J_4 , & n = 2 , \\ \frac{T}{2n(2n-1)} , & n \geq 3 . \end{cases}$$

Hence show that, if $J_4 > \frac{1}{3}J$, there is no second-order transition at the temperature $T = 4J$ where $B_2(T) = 0$, but instead a first-order transition at a temperature T_0 from $m = 0$ to $m = \pm m_0$. Defining $x_0 = \frac{1}{4}T_0/J$ and $\lambda = (3J_4/J) - 1$, show by truncating the series after the $n = 3$ term that, approximately,

$$x_0(x_0 - 1) = \frac{5}{48}(\lambda + 1 - x_0)^2 , \qquad m_0^2 = \frac{5(\lambda + 1 - x_0)}{4x_0} .$$

6.4 For the models described in Sect. 6.6, label sites for which $\tau_i = 1$, $\tau_i = -1$ and $\tau_i = 0$ by the indices 1, 2 and 3 respectively. Show that

$$N_1 = \frac{1}{2}\sum_{\{i\}}(\tau_i + 1)\tau_i ,$$
$$N_2 = \frac{1}{2}\sum_{\{i\}}(\tau_i - 1)\tau_i ,$$
$$N_3 = \sum_{\{i\}}(1 - \tau_i^2) ,$$

where the summations are over all sites. Similarly show that the pair numbers N_{ij} are given by

$$N_{11} = \frac{1}{4} \sum_{\{i,j\}}^{(\text{n.n.})} (\tau_i + 1)(\tau_j + 1)\tau_i\tau_j \,,$$

$$N_{22} = \frac{1}{4} \sum_{\{i,j\}}^{(\text{n.n.})} (\tau_i - 1)(\tau_j - 1)\tau_i\tau_j \,,$$

$$N_{12} = \frac{1}{2} \sum_{\{i,j\}}^{(\text{n.n.})} \tau_i\tau_j(\tau_i\tau_j - 1) \,.$$

Show that, if component A comprises species 1 and 2 ($\tau_i = \pm 1$), then

$$N_{\text{A}} = \sum_{\{i\}} \tau_i^2 \,, \qquad N_{\text{AA}} = \sum_{\{i,j\}}^{(\text{n.n.})} \tau_i^2\tau_j^2 \,.$$

The Hamiltonian of model II, Sect. 6.6.2 can be written in the form

$$\widehat{H} = -N_{\text{AA}}\varepsilon - (N_{11} + N_{22} - N_{12})J - \mu N_{\text{A}} - (N_1 - N_2)\mathcal{H} \,.$$

Show that this is equivalent to (6.49) .
The Hamiltonian for the unsymmetrical form of model IV, Sect. 6.6.4 is

$$\widehat{H} = \varepsilon_{\text{CC}}N_{\text{CC}} + \varepsilon_{\text{DD}}N_{\text{DD}} + \varepsilon_{\text{CD}}N_{\text{CD}} - \mu_{\text{C}}N_{\text{C}} - \mu_{\text{D}}N_{\text{D}} \,.$$

Identifying C and D respectively with $\tau_i = 1$ and $\tau_i = -1$ and the vacant sites with $\tau_i = 0$, show that

$$\begin{aligned}\hat{H} = {} & \frac{1}{4}(\varepsilon_{\text{CC}} + \varepsilon_{\text{DD}} + 2\varepsilon_{\text{CD}}) \sum_{\{i,j\}}^{(\text{n.n.})} \tau_i^2\tau_j^2 \\ & + \frac{1}{4}(\varepsilon_{\text{CC}} + \varepsilon_{\text{DD}} - 2\varepsilon_{\text{CD}}) \sum_{\{i,j\}}^{(\text{n.n.})} \tau_i\tau_j \\ & + \frac{1}{4}(\varepsilon_{\text{CC}} - \varepsilon_{\text{DD}}) \sum_{\{i,j\}}^{(\text{n.n.})} (\tau_i^2\tau_j + \tau_i\tau_j^2) \\ & - \frac{1}{2}(\mu_{\text{C}} + \mu_{\text{D}}) \sum_{\{i\}} \tau_i^2 - \frac{1}{2}(\mu_{\text{C}} - \mu_{\text{D}}) \sum_{\{i\}} \tau_i \,.\end{aligned}$$

Show that if the symmetry condition $\varepsilon_{\text{CC}} = \varepsilon_{\text{DD}}$ is satisfied then this becomes equivalent to (6.49), using the substitutions given in (6.66).

6.5 By substituting from (6.72) into (6.69) show that

$$f = T\ln(1 - x) + \frac{1}{2}z\varepsilon x^2 + \frac{1}{2}zJx^2m^2 \,.$$

Suppose that, at the disordered phase critical temperature $T = \frac{1}{4}z\varepsilon$ and $\mathcal{H} = 0$, the phase in which $x = \frac{1}{2}$, $m = 0$ is in equilibrium with a phase in which $x = x_1 > \frac{1}{2}$, $m = m_1 > 0$. Using the above expression for f and equation (6.92) for $\bar{\mu}$ show that the equilibrium conditions are

$$2\ln[2(1-x_1)] + 4x_1^2 - 1 + 4(J/\varepsilon)x_1^2 m_1^2 = 0,$$

$$2\ln\left\{\frac{x_1}{1-x_1}\right\} - 4(2x_1 - 1) + \ln(1 - m_1^2) = 0\,,$$

to be solved in conjunction with

$$\ln\left[\frac{1+m_1}{1-m_1}\right] - 8(J/\varepsilon)x_1 m_1 = 0\,.$$

Show that these relations yield $x_1 = 0.84059$, $m_1 = 0.67192$, $\varepsilon/J = 2.7747$.

6.6 Denoting the fractions of sites on which $\tau_i = 1$, $\tau_i = -1$ and $\tau_i = 0$ by x_1, x_2 and x_3 respectively, show from (6.98) that for the ferromagnetic Potts model ($\mu_{\text{B}} = \mu_{\text{C}} = \mu_{\text{D}}$) the free energy is given in the mean-field approximation by

$$f = \frac{1}{2}zu_0 - \frac{1}{2}zJ(x_1^2 + x_2^2 + x_3^2 - 2x_1x_2 - 2x_2x_3 - 2x_3x_1)$$
$$+ T(x_1 \ln x_1 + x_2 \ln x_2 + x_3 \ln x_3)\,.$$

Introducing an order variable s by putting $x_1 = x_2 = \frac{1}{3}(1-s)$, $x_3 = \frac{1}{3}(1+2s)$, show that

$$f = \frac{1}{6}z(3u_0 + J) - T\ln(3) - \frac{2}{3}zJs^2$$
$$+\frac{1}{3}T[2(1-s)\ln(1-s) + (1+2s)\ln(1+2s)]\,.$$

By expanding in a Landau series in powers of s, show that there is a first-order transition at some temperature greater than $\frac{2}{3}zJ$.

7. Cluster Variation Methods

7.1 Introduction

Cluster variation methods are closed-form approximations which, unlike the mean-field method, make some allowance for short-range order effects.[1] In *first-order* approximations, energies and weights for the microsystems on a *cluster* or group of sites are treated exactly, but the correlations due to sites shared between groups are handled by plausible assumptions. Equivalent versions were invented independently by various authors, notably Guggenheim (1935), (the *quasi-chemical method*), Bethe (1935), (the *Bethe method*) and Kasteleyn and van Kranendonk (1956a, 1956b, 1956c), (the *constant-coupling method*). We base our approach on a first-order method due to Guggenheim and McGlashan (1951) which allows any suitable group of sites to be used and enables quite complicated forms of ordering and intermolecular attraction to be incorporated in a natural way. For the particular problem of sublattice ordering, a similar method had been used by Yang (1945) and Li (1949). Kikuchi (1951) carried the process of approximation to a higher level by minimizing the free energy with respect to the occupation probabilities of subgroups of sites, including nearest-neighbour pairs, as well as those of the main group and this procedure was placed on a systematic basis by Hijmans and de Boer (1955, 1956) (see also Domb 1960, Burley 1972, Ziman 1979, Young 1981, Schlijper 1983 and Morita 1984). The cluster counting procedures used by Hijmans and de Boer (1955) have now been used in the finite-lattice series expansion method of de Neef and Enting (1977) (see Volume 2, Sect. 7.7).

Because they incorporate short-range ordering, cluster variation methods improve on the mean-field (zeroth-order) method as regards the accuracy of phase diagrams and critical parameters. Sometimes they can treat effects which the mean-field method cannot and which are not yet accessible to exact analysis. However, the behaviour they predict near critical points is still 'classical', giving the same critical exponents as in mean-field theory. In fact, this is bound to be true for any method in which the free energy densities

[1] Some authors apply the term 'mean field' to all closed-form approximations. We reserve it for the randomized entropy method introduced in Chap. 3.

depend analytically on thermodynamic variables and order parameters. We illustrate this point for the Ising model at the end of Sect. 7.3.

An interesting twist to the story is that, though first-order theory is approximate for $d = 2$ or $d = 3$ regular lattices, like those shown in Appendix A.1, it is exact for the interior sites of more loosely connected arrays. This is well known for the first-order method based on a nearest-neighbour pair of sites (Baxter 1982a, Chap. 4), and in Sect. 7.7 we make an extension to a general group of sites like that used in Sect. 7.2.

7.2 A First-Order Method Using a General Site Group

The lattice is divided into N_g similar basic clusters of p sites in such a way that each nearest-neighbour pair belongs to only one group. Since the total number of nearest-neighbour pairs is $\frac{1}{2}zN$, where N is the number of sites,

$$N_g = \frac{zN}{2q}, \tag{7.1}$$

where q is the number of nearest-neighbour pairs in a group.

7.2.1 Equivalent Sites

All the p sites in a basic group are taken to be equivalent so that each site belongs to $pN_g/N = (zp)/(2q)$ different groups. Where the group is a single nearest-neighbour pair then $q = 1$ and $p = 2$ so that, as would be expected, $N_g = \frac{1}{2}zN$ and each site belongs to z pairs. Figure 7.1 shows a division of the square lattice into the groups, shown with faces shaded, each being a square so that $q = p = 4$. Here, with $z = 4$, $N_g = \frac{1}{2}N$ and each site belongs to two groups.

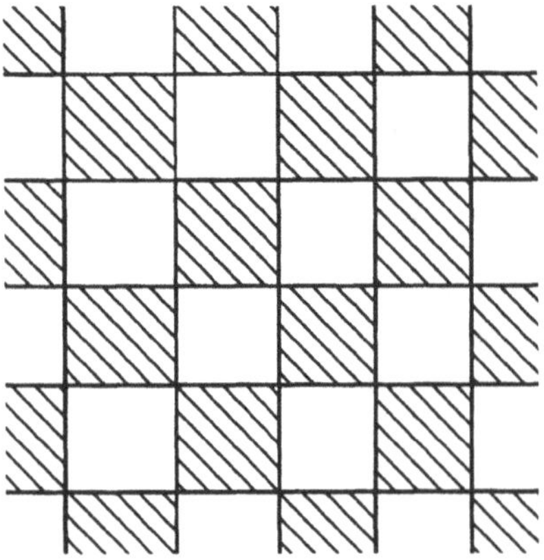

Fig. 7.1. Basic groups of four sites on a square lattice.

We suppose that there are κ species present, an atom or molecule of a species occupying just one site. The species may be independent components or dipoles in a given orientation or, as in the case of the lattice gas of Chap. 5, holes. Since there are κ choices for each site, the total number of possible configurations on the basic group is p^{κ}. The configurations are divided into types labelled by the index ℓ, each type being characterized by the numbers $\mathrm{p}_{j\ell}$ of each species j on sites of the group, where

$$\sum_{j=1}^{\kappa} \mathrm{p}_{j\ell} = \mathrm{p}\,, \tag{7.2}$$

and by a configurational energy ε_ℓ. The number of configurations of type ℓ is denoted by ω_ℓ and it is assumed that they are equally likely. Suppose that the fraction of the N_{g} groups occupied by any configuration of type ℓ is ψ_ℓ. Then the fraction of sites occupied by species j is

$$x_j = \mathrm{p}^{-1} \sum_{\{\ell\}} \mathrm{p}_{j\ell}\omega_\ell\psi_\ell\,. \tag{7.3}$$

The normalization condition

$$\sum_{j=1}^{\kappa} x_j = \sum_{\{\ell\}} \omega_\ell\psi_\ell = 1 \tag{7.4}$$

follows from (7.2) and (7.3).

The central feature of the method is an approximation for the number $\Omega(\psi_\ell)$ of configurations on the lattice of N sites. If the N_{g} groups had no sites in common then the number of configurations for given ψ_ℓ would be

$$\Omega_0(\psi_\ell) = \frac{N_{\mathrm{g}}!}{\prod_{\{\ell\}} [(N_{\mathrm{g}}\psi_\ell)!]^{\omega_\ell}}\,. \tag{7.5}$$

To allow approximately for the fact that each site is shared by $(\mathrm{zp})/(2\mathrm{q})$ groups, we put, in the absence of sublattice ordering,

$$\Omega(\psi_\ell) = \Omega_0(\psi_\ell)\left[\prod_{j=1}^{\kappa}(Nx_j)!/N!\right]^{\left(\frac{\mathrm{zp}}{2\mathrm{q}}-1\right)}, \tag{7.6}$$

Nx_j being the number of sites occupied by species j. From Stirling's formula and (7.1) it follows that

$$N^{-1}\ln\Omega(\psi_\ell) = -\frac{\mathrm{z}}{2\mathrm{q}}\sum_{\{\ell\}}\omega_\ell\psi_\ell\ln\psi_\ell + \left(\frac{\mathrm{zp}}{2\mathrm{q}} - 1\right)\sum_{j=1}^{\kappa} x_j \ln x_j\,. \tag{7.7}$$

For the random distribution, (which will occur in the limit $T = \infty$),

$$\psi_\ell = \prod_{j=1}^{\kappa} x_j^{\mathrm{p}_{j\ell}}\,. \tag{7.8}$$

From (7.3) and (7.7),

$$N^{-1}\ln\Omega(\psi_\ell) = -\sum_{j=1}^{\kappa} x_j \ln x_j\,, \tag{7.9}$$

which is correct for a random distribution. It is possible to proceed in the reverse direction and deduce the correcting factor in (7.6) by requiring that (7.9) be satisfied (Guggenheim 1952).

We now construct a non-equilibrium free energy for a given set of ψ_ℓ, using a distribution in which the configurational Hamiltonian is

$$\widehat{H} = N_{\mathrm{g}}\sum_{\{\ell\}} \omega_\ell \varepsilon_\ell \psi_\ell - N\sum_{j=1}^{\kappa} \xi_j x_j\,. \tag{7.10}$$

Here the first term is the configurational energy while in the second term the ξ_i are either fields, linearly dependent on fields or zero. From (7.1), the free energy per site is

$$f = -N^{-1}T\ln\Omega(\psi_\ell) + \frac{\mathrm{z}}{2\mathrm{q}}\sum_{\ell} \omega_{\{\ell\}}\varepsilon_\ell\psi_\ell - \sum_{j=1}^{\kappa}\xi_j x_j\,. \tag{7.11}$$

The free energy must be minimized at constant temperature; so using (7.7) and the normalization conditions (7.4),

$$\frac{2\mathrm{q}}{\mathrm{z}}\delta f = \sum_{\{\ell\}} \omega_\ell (T\ln\psi_\ell + \varepsilon_\ell)\delta\psi_\ell - \mathrm{p}\sum_{j=1}^{\kappa}\left[T\left(1-\frac{2\mathrm{q}}{\mathrm{pz}}\right)\ln x_j + \frac{2\mathrm{q}}{\mathrm{pz}}\xi_j\right]\delta x_j\,. \tag{7.12}$$

The procedure now depends on the distribution, that is, the choice of independent thermodynamic variables and order parameters. The final results will, of course, independent of this choice. For the distribution in which the only independent extensive variable is the number of sites N, which is identified with the reduced volume V (7.3) is used to express δx_j in terms of a linear sum of the ψ_ℓ. From (7.12),

$$\frac{2\mathrm{q}}{\mathrm{z}}\delta f = \sum_{\{\ell\}} \omega_\ell\left\{T\ln\psi_\ell + \varepsilon_\ell - \sum_{j=1}^{\kappa}\mathrm{p}_{j\ell}\left[T\left(1-\frac{2\mathrm{q}}{\mathrm{pz}}\right)\ln x_j + \frac{2\mathrm{q}}{\mathrm{pz}}\xi_j\right]\right\}\delta\psi_\ell\,. \tag{7.13}$$

The ψ_ℓ are connected by the normalization relation (7.4) and, selecting a reference configuration denoted by index r, the fraction ψ_r can be regarded as a dependent variable such that

$$\omega_r\delta\psi_r = -\sum_{\{\ell\neq r\}}\omega_\ell\delta\psi_\ell\,. \tag{7.14}$$

Substituting this relation into (7.13) and equating the coefficients of $\delta\psi_\ell$ to zero for $\ell\neq r$ gives the equilibrium conditions

$$\frac{\psi_\ell}{\psi_r} = \exp\left(\frac{\varepsilon_r - \varepsilon_\ell}{T}\right) \prod_{j=1}^{\kappa} \left[x_j^{1-2q/(zp)} \exp\left(\frac{2q\xi_i}{zpT}\right)\right]^{p_{j\ell}-p_{jr}} . \tag{7.15}$$

The composition of the assembly can be expressed in terms of the $\kappa - 1$ ratios x_j/x_κ $(j \neq \kappa)$. Since, from (7.2),

$$p_{\kappa\ell} - p_{\kappa r} = -\sum_{j=1}^{\kappa-1}(p_{j\ell} - p_{jr}) \tag{7.16}$$

the equilibrium conditions (7.15) can be rewritten in the form

$$\frac{\psi_\ell}{\psi_r} = \exp\left(\frac{\varepsilon_r - \varepsilon_\ell}{T}\right) \prod_{j=1}^{\kappa-1} \left\{\left(\frac{x_j}{x_\kappa}\right)^{\left(1-\frac{2q}{zp}\right)} \exp\left[\frac{2q}{zpT}(\xi_j - \xi_\kappa)\right]\right\}^{p_{j\ell}-p_{jr}} . \tag{7.17}$$

By substituting (7.15) back into (7.11) it follows that, at equilibrium,

$$f = \frac{z}{2q}\left\{T\ln\psi_r + \varepsilon_r - \sum_{j=1}^{\kappa} p_{jr}\left[\left(1 - \frac{2q}{zp}\right)T\ln x_j + \frac{2q\xi_j}{zp}\right]\right\} . \tag{7.18}$$

Since the choice of r is arbitrary, (7.18) is true for any r and (7.15)–(7.17) for any pair ℓ, r.

7.2.2 Sublattice Ordering

The first-order method can easily be adapted to cases where there is sublattice ordering. It will be supposed that there are equal sublattices labelled a and b and that the basic group has $\frac{1}{2}p$ sites in each sublattice. The sublattice mole fractions are given by

$$\begin{aligned} x_j^{(a)} &= 2p^{-1}\sum_{\{\ell\}} p_{j\ell}^{(a)}\omega_\ell\psi_\ell ,\\ x_j^{(b)} &= 2p^{-1}\sum_{\{\ell\}} p_{j\ell}^{(b)}\omega_\ell\psi_\ell , \end{aligned} \tag{7.19}$$

where $p_{j\ell}^{(a)}$ and $p_{j\ell}^{(b)}$ are respectively the number of members of species j on the a and b sites of a group in configuration ℓ. For the configuration number, (7.6) is replaced by

$$\Omega(\psi_\ell) = \Omega_0(\psi_\ell)\left[\frac{\left(\frac{1}{2}N!\right)}{\prod_{j=1}^{\kappa}\left(\frac{1}{2}Nx_j^{(a)}\right)!}\frac{\left(\frac{1}{2}N!\right)}{\prod_{j=1}^{\kappa}\left(\frac{1}{2}Nx_j^{(b)}\right)!}\right]^{-\left(\frac{zp}{2q}-1\right)} \tag{7.20}$$

After writing the corresponding expression for f we have

$$\frac{2\mathrm{q}}{z}\delta f = \sum_{\{\ell\}} \omega_\ell (T \ln \psi_\ell + \varepsilon_\ell)\delta\psi_\ell$$
$$-\mathrm{p}\sum_{j=1}^{\kappa}\left[T\left(1-\frac{2\mathrm{q}}{\mathrm{p}z}\right)\ln(x_j^{(a)})\delta x_j^{(a)} + \ln(x_j^{(b)})\delta x_j^{(b)} + \frac{2\mathrm{q}}{\mathrm{p}z}\xi_j \delta x_j\right], \tag{7.21}$$

which reduces to (7.12) for $x_j^{(a)} = x_j^{(b)}$ $(j = 1, \ldots, \kappa)$.

7.3 The Pair Approximation and the Ising Model

In the simplest type of first-order approximation (also called the *Bethe pair method*) the group is an nearest-neighbour pair, giving $\mathrm{p} = 2$, $\mathrm{q} = 1$. This is now applied to the Ising model ferromagnet taking up and down spins as species 1 and 2 respectively so that in the constant-field Hamiltonian

$$\xi_1 = \mathcal{H}, \qquad \xi_2 = -\mathcal{H}, \tag{7.22}$$

using the notation of (7.10). The four possible configurations on the pair are 11, 12, 21, and 22, but, in the absence of sublattice ordering, 12 and 21 are equivalent. With an obvious notation the configuration fractions ψ_ℓ are ψ_{11}, ψ_{22} and ψ_{12} with $\omega_{11} = \omega_{22} = 1$, $\omega_{12} = 2$, $\varepsilon_{11} = \varepsilon_{22} = -J$ and $\varepsilon_{12} = J$. Equations (7.3) reduce to

$$\begin{aligned} x_1 &= \frac{1+m}{2} = \psi_{11} + \psi_{12}, \\ x_2 &= \frac{1-m}{2} = \psi_{12} + \psi_{22}, \end{aligned} \tag{7.23}$$

where m is the relative magnetization. A parameter ϑ is defined by

$$\frac{\psi_{11}}{\psi_{22}} = \vartheta^2 = \left[\left(\frac{x_1}{x_2}\right)^{(z-1)/z} \exp\left(\frac{2\mathcal{H}}{zT}\right)\right]^2 \tag{7.24}$$

where the second relation follows from (7.22) and the equilibrium condition (7.17), with ψ_{11} and ψ_{22} substituted for ψ_ℓ and ψ_r respectively. Using (7.17),

$$\frac{\psi_{12}}{\psi_{22}} = \mathfrak{X}\vartheta, \tag{7.25}$$

where

$$\mathfrak{X} = \exp(-2J/T). \tag{7.26}$$

From (7.23)–(7.26)

$$m = \frac{\psi_{11} - \psi_{22}}{\psi_{11} + 2\psi_{12} + \psi_{22}} = \frac{\vartheta^2 - 1}{\vartheta^2 + 2\mathfrak{X}\vartheta + 1} = \frac{1 - \vartheta^{-2}}{1 + 2\mathfrak{X}\vartheta^{-1} + \vartheta^{-2}}. \tag{7.27}$$

For $\vartheta > 1$, the number of up spins is greater than the number of down spins, so that $m > 0$ and vice versa for $\vartheta < 1$. From (7.27), m is a monotonically increasing function of ϑ at given T and

$$m(\vartheta^{-1}) = -m(\vartheta) , \qquad m(1) = 0 . \tag{7.28}$$

The configurational energy u per site is

$$u = -\frac{1}{2}zJ\frac{\psi_{11} - 2\psi_{12} + \psi_{22}}{\psi_{11} + 2\psi_{12} + \psi_{22}} = -\frac{1}{2}zJ\frac{\vartheta^2 - 2\mathfrak{X}\vartheta + 1}{\vartheta^2 + 2\mathfrak{X}\vartheta + 1} . \tag{7.29}$$

Substituting first ψ_{11} and then ψ_{22} for ψ_r in (7.18) and halving the sum of the expressions obtained, the free energy density

$$\begin{aligned} f &= \frac{1}{4}[zT\ln(\psi_{11}\psi_{22}) - 2(z-1)T\ln(x_1x_2) - 2zJ] \\ &= \frac{1}{2}T(z-2)\ln(\vartheta + 2\mathfrak{X} + \vartheta^{-1}) \\ &\quad - \frac{1}{2}T(z-1)\ln[1 + \mathfrak{X}^2 + \mathfrak{X}(\vartheta + \vartheta^{-1})] - \frac{1}{2}zJ . \end{aligned} \tag{7.30}$$

Note that (7.30) is valid only for a value of ϑ satisfying the equilibrium relations (7.24) and (7.25). It cannot be treated as a non-equilibrium free energy for other values of ϑ.

To determine the equilibrium ϑ at given values of T and $\mathcal{H}$, the second equality of (7.24) is rewritten as

$$\exp(2\mathcal{H}/T) = \vartheta^z\left(\frac{x_2}{x_1}\right)^{z-1} = \phi(\vartheta) , \tag{7.31}$$

where, from (7.23) and (7.25),

$$\phi(\vartheta) = \vartheta\left(\frac{1 + \mathfrak{X}\vartheta}{\vartheta + \mathfrak{X}}\right)^{z-1} = [\phi(\vartheta^{-1})]^{-1}. \tag{7.32}$$

From (7.32), reversing the sign of $\mathcal{H}$ in (7.31) changes any solution ϑ to ϑ^{-1}. Hence, by (7.28), the symmetry relation (2.71) is satisfied. The symmetry relation (2.68) is also satisfied because the last expression for f in (7.30) is invariant under replacement of ϑ by ϑ^{-1}, and the same is true for u, as given by (7.29).

7.3.1 Zero Field

When $\mathcal{H} = 0$, (7.31) reduces to

$$\phi(\vartheta) = 1 . \tag{7.33}$$

For any value of T, (7.33) is satisfied by $\vartheta = 1$, giving $m = 0$. By (7.32) any other solutions exist in reciprocal pairs. By considering $d\phi/d\vartheta$ it can be shown (Example 7.2) that, for $\mathfrak{X} > (z-2)/z$, no such solutions exist, but, for $\mathfrak{X} < (z-2)/z$, there is a reciprocal pair $(\vartheta_0, \vartheta_0^{-1})$, corresponding, by (7.28), to $m = \pm m_s$ ($m_s > 0$). This indicates that the critical temperature is given by

$$\mathfrak{X}_c = \frac{z-2}{z}, \qquad T_c = \frac{2J}{\ln[z/(z-2)]}. \tag{7.34}$$

However, to confirm that a transition occurs, it must be shown that

$$f(\vartheta_0) - f(1) < 0, \tag{7.35}$$

where it follows from (7.30) that $f(\vartheta_0^{-1}) = f(\vartheta_0)$. Consider the case $z = 3$ where (7.33) is a cubic equation and it can be deduced that

$$\vartheta_0 + \vartheta_0^{-1} = \mathfrak{X}^{-2}(1 - 2\mathfrak{X} - \mathfrak{X}^2). \tag{7.36}$$

Substitution in (7.30) gives

$$f(\vartheta_0) - f(1) = -\tfrac{1}{2}T \ln\left[\frac{2(1-\mathfrak{X})^3}{(1-2\mathfrak{X})(1+\mathfrak{X})^2}\right]. \tag{7.37}$$

By manipulating the inequality $(1 - 3\mathfrak{X})^2 > 0$ it can be shown that, for $\mathfrak{X} < \mathfrak{X}_0 = \frac{1}{3}$ the argument of the logarithm is greater than unity so that (7.35) is satisfied. A similar proof is possible for $z = 4$ (Example 7.4), and for higher values of z the inequality (7.35) can be demonstrated numerically. An alternative approach is to consider the variation with respect to m of a Helmholtz free energy density $a(T, m)$.

7.3.2 The Critical Region

Pair approximation results depend entirely on the coordination number z and are thus, for instance, the same for the $d = 2$ triangular lattice as for the $d = 3$ simple cubic lattice. Critical parameters are compared with accurate values in Chap. 8, Table 8.1, and it can be seen that the pair approximation still under estimates the effects of short-range correlations for $T > T_c$, although results are better for $d = 3$ than for $d = 2$.

For the critical region it is convenient to use the variable $L = \mathcal{H}/T$ defined in (2.146). From (7.28), (7.31) and (7.32) m and L are odd functions of

$$\Delta = \ln \vartheta, \tag{7.38}$$

in terms of which they can be expanded in odd powers. From (7.27) and (7.31)

$$\frac{\mathcal{H}}{T} = L(\Delta) = \frac{z\mathfrak{X} - (z-2)}{2(1+\mathfrak{X})}\Delta + \frac{(z-1)\mathfrak{X}(1-\mathfrak{X})}{6(1+\mathfrak{X})^3}\Delta^3 + \mathrm{O}(\Delta^5), \tag{7.39}$$

$$m = \frac{\Delta}{1+\mathfrak{X}} + \mathrm{O}(\Delta^3). \tag{7.40}$$

For $T > T_c$, these relations yield

$$\left(\frac{\partial m}{\partial \mathcal{H}}\right)_{\mathcal{H}=0} = \lim_{\Delta \to 0} \frac{m}{\mathcal{H}} = \frac{2}{zT(\mathfrak{X} - \mathfrak{X}_c)} . \tag{7.41}$$

It can now be shown that the present theory yields the same classical values obtained in Sect. 3.6 for the critical exponents α, β, γ and δ, defined in Sect. 3.4. In the present case the field $\xi = \mathcal{H}$ and the conjugate density is m. Since

$$\mathfrak{X} - \mathfrak{X}_c \simeq \left(\frac{d\mathfrak{X}}{dT}\right)_c (T - T_c) \qquad \text{as } T - T_c \to 0 , \tag{7.42}$$

(7.41) yields $\gamma = 1$. At $T = T_c$, where $\mathfrak{X} = \mathfrak{X}_c \simeq (z-2)/z$, (7.39) and (7.40) give

$$\frac{\mathcal{H}}{T} = \frac{(z-1)(z-2)}{3z^2} m^3 + O(m^5) , \tag{7.43}$$

so that $\delta = 3$. For $\mathcal{H} = 0$ the non-zero solutions of (7.39) give

$$m \simeq \frac{\Delta}{1 + \mathfrak{X}_c} \simeq \left[\frac{3z^3(\mathfrak{X}_c - \mathfrak{X})}{2(z-1)(z-2)}\right]^{1/2} , \tag{7.44}$$

so that $\beta = \frac{1}{2}$. It can be shown (Example 7.4) that the zero-field heat capacity has a finite discontinuity at $T = T_c$, giving $\alpha = 0$.

7.3.3 The Linear Lattice

The pair method is exact for the linear ($d = 1$) lattice. For the Ising model this can be demonstrated by putting $z = 2$, $p = 2$, $q = 1$ and $\kappa = 2$ in (7.1) and (7.6) to yield the configuration number

$$\Omega(\psi_\ell) = \frac{(Nx_1)!(Nx_2)!}{(N\psi_{11})!(N\psi_{22})![(N\psi_{12})!]^2} . \tag{7.45}$$

With a different notation this is identical to $g(N_1, N_2, X)$, defined by (2.126), where $X = \frac{1}{2}zN\psi_{12} = N\psi_{12}$. However, as shown in Sect. 2.6, $g(N_1, N_2, X)$ can be treated as an exact configuration number in the thermodynamic limit. It is easy to extend the equivalence to a model with an arbitrary number κ of species.

7.4 Phase Transitions in Amphipathic Monolayers

Amphipathic molecules have polar head groups with an affinity for water and hydrocarbon chains which resist immersion in water. Such molecules form stable or permanent metastable monolayers at the air–water surface. One important way of investigating their properties is to plot isotherms of surface pressure against surface area per molecule. Two-dimensional vapour–liquid first-order transitions occur at very low surface pressures, and at higher surface pressure in the appropriate temperature range a further transition is observed. The states on the lower pressure (higher area) and higher pressure (lower area) sides of this transition are usually termed *liquid-expanded* and *liquid-condensed* respectively. As the temperature is lowered, the separate vapour–liquid-expanded and liquid-expanded–liquid-condensed transitions coalesce into a single first-order transition. Increase in carbon chain length has a similar effect. These phase phenomena occur both for single chain compounds such as fatty acids with simple head groups and for the phospho-lipids, which have complex head groups attached to double chains. On a large number of experimental pressure–area isotherms the liquid-expanded–liquid-condensed transition is shown as second-order with the area per molecule (and hence surface density) continuous but with a discontinuity in the slope. The rate of change of pressure against area is smallest in magnitude on the liquid-expanded side of the transition. Similar liquid-expanded–liquid-condensed transitions are observed in phospho-lipid monolayers at the oil–water interface, though in this case the low surface pressure vapour–liquid-expanded transitions are absent. As the temperature decreases, the slope on the liquid-expanded side of the transition decreases for both air–water and oil–water monolayers. Pallas and Pethica (1985) assert that for air–water monolayers the observed slope should be zero, giving a horizontal part of the isotherm, which represents a first-order liquid-expanded–liquid-condensed transition. The fact that this has been missed in so many experiments is attributed to lack of purity in the chemicals used and poor humidity control. However, these authors accept the slope discontinuity second-order transition as genuine for oil–water monolayers.

The transition from the liquid-condensed to the liquid-expanded state is thought to be due to disordering of the chain conformations, a process which can take two forms. One of these is *isomerism* or chain 'bending' which is the basis of a theory of the transition due to Nagle (1975, 1986), discussed in Volume 2, Chap. 9. The other is cooperative rotation of the carbon chains, which are planar in the lowest energy isomer of each molecule, about axes perpendicular to the interface. Kirkwood (1943) was the first author to base a theory on this 'rotational' or 'orientational' effect. Here we present the simplest type of orientational model (Bell et al. 1978) which is a lattice gas with vacant sites and molecules with two orientational states. This is equivalent to the zero-field case of the Ising model lattice gas, that is model II of Sect. 6.6.2. The detailed treatment will differ from that of Chap. 6 in that the emphasis is on pressure as a function of density, and the first-order pair approximation is used rather than the mean-field method.

As in Sect. 7.3 the two orientational species will be labelled '1' and '2', and the new hole species accordingly has the label '3'. Hence, if ρ denotes the number density,

$$x_1 + x_2 = \rho\,, \qquad x_3 = 1 - \rho\,. \tag{7.46}$$

We shall use the grand distribution so that, since $\mathcal{H} = 0$,

$$\xi_1 = \xi_2 = \mu\,, \qquad \xi_3 = 0\,, \qquad f = -P\,, \tag{7.47}$$

dropping the index g used in Sect. 6.6, which is no longer necessary. From Sect. 6.6.2 the energies of the various types of pair are

$$\begin{aligned} &\varepsilon_{11} = \varepsilon_{22} = -\varepsilon - J\,, \\ &\varepsilon_{12} = -\varepsilon + J\,, \\ &\varepsilon_{13} = \varepsilon_{23} = \varepsilon_{33} = 0\,. \end{aligned} \tag{7.48}$$

We now introduce parameters w and y by

$$\frac{\psi_{11}}{\psi_{22}} = w^{2(z-1)}, \qquad \frac{\psi_{33}}{\psi_{22}} = y^2 \tag{7.49}$$

and, from (7.17), (7.46)–(7.52),

$$\frac{x_1}{x_2} = w^z\,, \qquad x_2 = \frac{\rho}{1+w^z}\,, \tag{7.50}$$

$$y = \mathfrak{z}\left(\frac{x_3}{x_2}\right)^{(z-1)/z} \exp\left(-\frac{\mu}{zT}\right), \tag{7.51}$$

where

$$\mathfrak{z} = \exp\left(\frac{\varepsilon + J}{2T}\right). \tag{7.52}$$

It can be seen, from (7.50), that $w = 1$ in the disordered state with $x_1 = x_2$. With $w > 1$ or $w < 1$ we have ordered states with $x_1 > x_2$ or $x_1 < x_2$ respectively, corresponding to $m > 0$ or $m < 0$ in Sect. 7.3. The use of w^{z-1} in place of the parameter ϑ of Sect. 7.3 avoids fractional powers. From (7.17) and (7.49)

$$\frac{\psi_{12}}{\psi_{22}} = \mathfrak{X}\, w^{z-1}\,, \qquad \frac{\psi_{13}}{\psi_{22}} = \mathfrak{z}\, w^{z-1} y\,, \qquad \frac{\psi_{23}}{\psi_{22}} = \mathfrak{z}\, y\,, \tag{7.53}$$

where $\mathfrak{X}$ is defined in (7.26). We now express the intensive variables ρ, μ and P in terms of w and y. From (7.3),

$$\begin{aligned} &x_1 = \psi_{11} + \psi_{12} + \psi_{13}\,, \\ &x_2 = \psi_{22} + \psi_{12} + \psi_{23}\,, \\ &x_3 = \psi_{33} + \psi_{13} + \psi_{23}\,, \end{aligned} \tag{7.54}$$

and, from (7.46), (7.49)–(7.54),

$$\frac{1-\rho}{\rho} = \frac{y(y + \mathfrak{z}\, w^{z-1} + \mathfrak{z})}{(1+w^z)(1 + \mathfrak{z}\, w^{z-1} + \mathfrak{z}\, y)}\,. \tag{7.55}$$

From (7.46) and(7.51)

$$\mu/T = (z-1)\ln(x_3/x_2) - z\ln(y/3)$$

$$= (z-1)\ln[(1+w^z)(1-\rho)/\rho] - z\ln(y/3) \tag{7.56}$$

and, from (7.18), with $\psi_r = \psi_{33}$, (7.46) (7.47) and (7.54),

$$P/T = (z-1)\ln x_3 - \frac{1}{2}z\ln\psi_{33}$$

$$= \frac{1}{2}[(z-2)\ln x_3 + z\ln(x_3/\psi_{33})]$$

$$= \frac{1}{2}\{(z-2)\ln(1-\rho) + z\ln[(y+3w^{z-1}+3)/y]\}\,. \tag{7.57}$$

Since $n = 3$ in the sense of Sect. 1.8 and $\mathcal{H} = 0$, we have, after T is specified, only one remaining independent intensive variable, which can be ρ, μ or P. It follows that for the ordered region, where w is a variable parameter, a relation between y and w is needed. This is given, from (7.51) and (7.54), by

$$w = \frac{w^{z-1} + \mathfrak{X} + 3y}{1 + \mathfrak{X}w^{z-1} + 3y}\,, \tag{7.58}$$

or

$$y = \frac{w^{z-1} - w - (w^z - 1)\mathfrak{X}}{3(w-1)}\,. \tag{7.59}$$

In the disordered state, where $w = 1$, (7.59) is indeterminate but we can obtain y as a function of T and ρ by solving (7.55) as a quadratic equation in y and retaining the positive root to give

$$y = \rho^{-1}\{(1-2\rho)3 + [(1-2\rho)^2 3^2 + 2\rho(1-\rho)(1+\mathfrak{X})]^{1/2}\}\,. \tag{7.60}$$

From (7.55), $y = 0$ implies $\rho = 1$ and the theory becomes identical with that of Sect. 7.3 for the zero-field Ising model. This is confirmed by putting $y = 0$ in (7.58) which then becomes equivalent to (7.33) when ϑ is substituted for w^{z-1}. Now consider the (ρ, T) plane. It can be verified from (7.55) and (7.58) that, if there is a solution $w \neq 1$, then w^{-1} is also a solution for the same value of ρ. On the $\rho = 1$ axis there is such a reciprocal pair of solutions for $T < T_c$, where T_c is the Ising model critical temperature given by (7.34). Taking $w > 1$ for definiteness and fixing $T < T_c$, it is found that w decreases with ρ until a value $\rho = \rho^*$ is reached where $w = 1$. This is the critical value of ρ at the given T; that is the density at which the order-disorder second-order transition occurs. To obtain ρ^* as a function of T, first take the limit of the right-hand side of (7.59) as $w \to 1$ giving

$$y^* = (z - 2 - z\mathfrak{X})/3\,. \tag{7.61}$$

Substitution of $w = 1$ and $y = y^*$ into (7.55) yields

$$\frac{\rho^*}{1-\rho^*} = \frac{2(z-1)(3y^*+2)}{zy^*(y^*+23)}\,. \tag{7.62}$$

Since, by (7.52) and (7.61), $\mathfrak{z}$ and y^* are functions of T, (7.62) is the equation of the critical curve in the (ρ, T) plane. From (7.57) the corresponding surface pressure is

$$P^* = \frac{1}{2}T\{(z-2)\ln(1-\rho^*) + z\ln[1 + 2\mathfrak{z}/y^*]\}. \tag{7.63}$$

Equations (7.62) and (7.63) give the order-disorder transition point on any $T < T_c$ isotherm in the (ρ, P) plane. There is a discontinuous decrease in $(\partial P/\partial \rho)_T$ as ρ increases through ρ^*.

We now consider vapour–liquid first-order transitions, which involve equilibrium between two disordered phases. In the disordered state, $w = 1$ so that

$$\psi_{11} = \psi_{22}, \qquad \psi_{12} = \psi_{22}, \qquad \psi_{13} = \psi_{23} = \mathfrak{z}\, y\psi_{22}. \tag{7.64}$$

By substituting in (7.11) it can be shown that, apart from an extra term $\ln(2)$ in the entropy per molecule, the model is equivalent to a simple lattice gas with an effective value of ε/T given by

$$\exp[(-\varepsilon/T)_{\text{eff}}] = \frac{\exp(-\varepsilon/T)}{\cosh(J/T)}. \tag{7.65}$$

It follows that, if the density of one conjugate phase is ρ_1, then that of the other is $1 - \rho_1$ and that the chemical potential for conjugate phases at temperature T is given by

$$\mu/T = -\ln(2) - \frac{1}{2}z(\varepsilon/T)_{\text{eff}}. \tag{7.66}$$

The critical point for disordered phase separation is thus defined by

$$\rho = \frac{1}{2}, \qquad \left[\frac{\exp(-\varepsilon/T)}{\cosh(J/T)}\right]^{1/2} = \frac{z-2}{z}. \tag{7.67}$$

The tricritical point occurs on the critical line at the temperature $T = T_t$ where $(\partial P/\partial \rho)_T = 0$ at $\rho = \rho^* + 0$. For $T < T_t$, $(\partial P/\partial \rho)_T < 0$ in a range of ρ above ρ^* implying, by (6.57), that $(\partial \mu/\partial \rho)_T < 0$, giving an instability loop like that shown in Fig. 6.9, with x replaced by ρ, and leading to separation between ordered and disordered phases. With some manipulation it can be deduced from (7.55), (7.57), (7.59), (7.61) and (7.62) that the relation between temperature and the parameter y^* at the tricritical point is given by

$$3(y^*)^2 - 2[2(z-2) - (2z-1)\mathfrak{z}^2]y^* + 8\mathfrak{z} = 0. \tag{7.68}$$

From (7.61), this has a solution $T = T_t > 0$ for any $\varepsilon/J > -1$. For the highest range of ε/J the tricritical point falls into the instability region initiated by the vapour–liquid critical point and does not appear on the equilibrium phase diagram which displays a critical end-point and is similar to that shown in Fig. 6.8(a). When ε/J is reduced, the phase diagram first becomes similar to that of Fig. 6.8(b), with a triple point and a tricritical point, and then to that of Fig. 6.8(c) with a tricritical point only. The latter disappears at $\varepsilon/J = -1$.

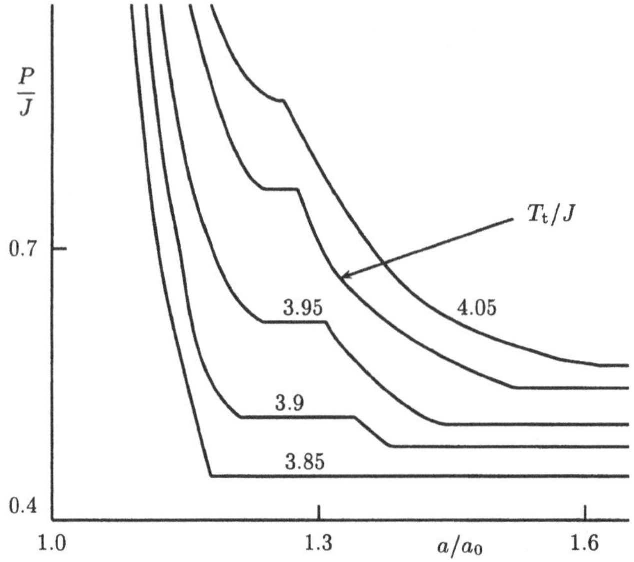

Fig. 7.2. Surface pressure-area isotherms for $z = 6$, $\varepsilon/J = 3.2$. Each isotherm is labelled with the value of T/J.

For $\varepsilon/J < -1$ there is no first-order transition line and the critical curve meets the ρ-axis at $\rho = (z-1)/(2z-3)$, (Example 7.5). The parameter ranges giving (a)- or (b)-type behaviour, where there are two successive transitions on certain isotherms, are relevant for air–water interface monolayers just as the (c) range is for ^{3}He–^{4}He mixtures, (see Sect. 6.8). The attraction between the carbon chains of the monolayer molecules is likely to be much diminished when the chains are in a hydrocarbon oil medium rather than air. In fact, the parameter ratios relevant to monolayers at the oil–water interface lie in the range $\varepsilon/J < -1$, where all transitions are second order.[2]

For $z = 6$ it is found that the changeover from the (a)-type phase pattern to (b) occurs at $\varepsilon/J = 3.6129$, and from (b) to (c) at 2.8192. These values can be compared to the mean-field values of 3.8019 and 2.7747 respectively obtained in Sect. 6.8. Figure 7.2 shows pressure–area isotherms calculated for $z = 6$ and $\varepsilon/J = 3.2$, for which the tricritical temperature is $T_t = 4.0134J$. Equations (7.59) and (7.60) are used for the ordered and disordered states respectively. The area per monolayer molecule and the minimum area per monolayer molecule are denoted respectively by a and a_0 so that $a/a_0 = \rho^{-1}$ and the surface pressure field $P = \widetilde{P}a_0$, where $\widetilde{P}$ is the surface pressure. The isotherms for $T > T_t$ and $T = T_t$ display a first-order vapour–liquid-extended transition and then, at a lower value of a, a second-order liquid-extended–liquid-condensed transition. The next two isotherms display two first-order

[2] Sublattice segregation of molecules and holes is possible for $\varepsilon/J < -1$ but is unlikely for monolayers at a fluid interface.

transitions whereas for the lowest one, the vapour–liquid-extended and liquid-extended–liquid-condensed transitions have merged into a single first-order transition. It can be seen that comparatively small alterations in T/J or ε/J can change a first-order liquid-extended–liquid-condensed transition to a second-order one or vice versa.

By using (7.62) and (7.63), values of $\mathfrak{X}$ and $\mathfrak{Z}$, and hence of J and ε, can be deduced from the observed values of $\widetilde{P}$ and a at a second-order transition point on any isotherm, provided that a value can be assumed for a_0. The transitions on the 10, 15 and 20°C isotherms for a phospho-lipid (di C_16 lecithin) monolayer at an oil (n-heptane)–water interface were used in this way by Bell et al. (1978) who found negative ε and positive J varying appreciably with temperature, ε/J values being about -3 or -4. The temperature variation is not surprising since monolayer molecules, especially the phospho-liquids, have a large number of internal degrees of freedom so that $-\varepsilon$ and $-J$ are properly free energy differences containing entropy terms rather than simple energy parameters. Since both $-\varepsilon$ and $-J$ increase with temperature it follows that at the oil–water interface there is entropy loss involved in bringing two molecules to their minimum separation and then in adjusting their mutual orientation to that of minimum free energy. Baret and Firpo (1983) used a mean-field method on a monolayer model similar to that of this section, which they termed a 'spin-1 Ising model'. They considered the effect of incorporating a term equivalent to (6.97), which removes the symmetry between states 1 and 2. Another 'spin-1 Ising model' for monolayers was treated by Banville et al. (1986). Albrecht et al. (1978) considered the possibility of a tricritical point in a monolayer model.[3] The model of Nagle (1986) gives rise to an unusual kind of tricritical point. Tricritical points were emphasized by Firpo et al. (1981) whose model, like a number of other monolayer lattice gas models, involved molecular states in which more than one lattice site is occupied. A simple monomer-dimer model of this type is treated in Volume 2, Chap. 8.

7.5 A Lattice Gas Model for Fluid Water

The feature of the water molecule whose cooperative effect leads to the anomalous properties of liquid and supercritical water is the hydrogen bond (see Appendix A.3). A number of statistical mechanical models, attempting to incorporate this feature have been developed. The simplest of these is the two-dimensional triangular lattice model of Bell and Lavis (1970) (see Example 7.6). In this section we consider a somewhat more realistic three-dimensional model, based on the body-centred cubic lattice and due to Bell (1972). This is a lattice gas model of M molecules on $N(\geq M)$ sites with each molecule

[3] For comments, see Bell et al. (1981), who reviewed theories of amphipathic monolayers up to that date.

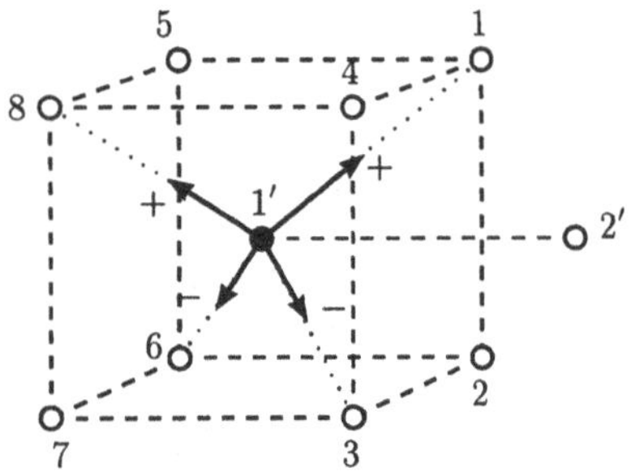

Fig. 7.3. Sites on the body-centred cubic with a water molecule placed on site 1′. (Reprinted from Bell (1972), by permission of the publisher IOP Publishing Ltd.)

having four bonding arms pointing to the vertices of a tetrahedron. These arms are directed towards four of the eight nearest-neighbour sites and two arms have positive polarity and two negative, implying twelve distinguishable orientations (Fig. 7.3). Two nearest-neighbour molecules are connected by a hydrogen bond when a positive arm of one and a negative arm of the other lie along the segment connecting their sites. This occurs in 18 of the 144 possible orientational states of the two molecules. An open ice Ic structure (see Appendix A.3), with one diamond sublattice of the body-centred cubic vacant and the other occupied by a bonded network, is thus possible. In the present fluid state theory this arrangement occurs only as short-range order, but it is of crucial importance since it is the breakdown of the local open structure by increasing temperature which produces anomalous effects. The difference between the locally similar ice Ic (diamond) and ice I (wurzite) structures is not likely to be significant in the fluid state. Locally ordered regions with the close-packed ice VII structure will exist at high pressures together with a range of other types of short-range order including the 'cage' structure in which unbonded molecules lie in the interstices of an open bonded network. However, the types of local order in real water are considerably more complex. The reduction of the ten or so observed ice structures (Eisenberg and Kauzmann 1969, Fletcher 1970) to two indicates the approximations made in neglecting hydrogen bond distortion from the tetrahedral angles and the possibility of ordered hydrogen positions.

An interaction energy $-\varepsilon$ ($\varepsilon > 0$) is assumed for each unbonded nearest-neighbour pair of molecules which changes to $-\varepsilon - w$ ($w > 0$) when a hydrogen bond is formed. To ensure that the ice VII structure is stable only at high pressures, an energy penalty against close packing is needed. This is imposed by associating an energy $\frac{1}{3}u$ ($u > 0$) with the occupation by three molecules of a triad of sites where two second neighbours share a nearest neighbour (for example, the triads $121'$ or $1'2'1$ in Fig. 7.3). There are no occupied triads in the open structure, but a total of $12N$ when all sites of the body-centred cubic lattice are occupied. As pointed out in connection with the one-dimensional water-like model of Sect. 5.5, density maxima occur on isobars for pressures at which the open structure is stable when $T = 0$. By considering ground-state enthalpies it can be shown that, if

$$\varepsilon < 2u < \varepsilon + w\,, \tag{7.69}$$

then, at $T = 0$, the open (ice Ic) structure is stable for $0 < P < P_0$ and the close-packed (ice VII) structure for $P_0 < P$ where P is the pressure field and

$$P_0 = 4u - 2\varepsilon\,. \tag{7.70}$$

A number density ρ, equal to 1 for close packing, is defined by

$$\rho = M/N. \tag{7.71}$$

A mean-field approximation would smear out the short-range order phenomena in which we are interested so we use a first-order approximation based on a non-planar group of four sites, like $121'2'$ in Fig. 7.3, two belonging to each of the diamond sublattices of the body-centred cubic lattice. There are four nearest-neighbour site pairs in the group, $11'$, $12'$, $21'$ and $22'$, so that $\mathrm{p} = \mathrm{q} = 4$. Since $\mathrm{z} = 8$, $\frac{1}{2}\mathrm{z}/\mathrm{q} = 1$ and $\frac{1}{2}\mathrm{zp}/\mathrm{q} = 4$. For fluid states it can be assumed that the twelve orientational molecular species are equally likely, a mole fraction $\frac{1}{12}\rho$ of each being present together with a fraction $1 - \rho$ of the hole species. Hence the orientational species do not have to be considered separately provided that appropriate statistical weights are assigned. Table 7.1 shows, for each group configuration ℓ, the configuration number or weight ω_ℓ, the number of molecules p_ℓ and the energy ε_ℓ. The line segments in the configuration diagrams represent hydrogen bonds. As an example of the reasoning involved in deriving the weights, we shall evaluate ω_4 and ω_5. Now, $\omega_4 + \omega_5 = 4 \times 144 = 576$ since there are four ways of placing a nearest-neighbour pair of molecules on the group of sites and any two molecules have 144 distinguishable orientational states. Since hydrogen bonding occurs in one eighth of the orientational states of an nearest-neighbour pair, $\omega_4 = 576/8 = 72$ and hence $\omega_5 = 504$. There are four site triads in the group (i.e. $121'$, $122'$, $1'2'1$, $1'2'2$ in the group $121'2'$ in Fig. 7.3) so that the ratio of nearest-neighbour site pairs to triads is 1:1. Since in the lattice as a whole this ratio is 1:3 we adjust by using $3 \times \frac{1}{3}u = u$ as the triad energy in our approximation. This ensures that the correct high-pressure ground-state energy is attained at $T = 0$.

Using the grand distribution we have, in the notation of Sect. 7.2, $\xi_i = \mu$ for molecules and $\xi_i = 0$ for holes while the equilibrium $f = -P$. A parameter ϑ is defined by

$$\vartheta^2 = \frac{\psi_8}{\psi_4} = \exp\left(\frac{2\varepsilon_4 - 2\varepsilon_8 + \mu}{2T}\right)\left[\frac{\rho}{12(1-\rho)}\right]^{3/2}, \tag{7.72}$$

where the last expression follows from (7.17) with $\ell = 8$ and $r = 4$. Reference to Table 7.1 shows that ϑ^2 expresses the ratio of ice VII-type to ice Ic-type configurations on the basic group of four sites. Applying (7.17) again and substituting from (7.72) we have, for any configuration ℓ,

Table 7.1. Reproduced from Bell (1972), by permission of the publisher IOP Publishing Ltd.

Configuration index ℓ	Configuration	ω_ℓ	ε_ℓ	p_ℓ
1		1	0	0
2		48	0	1
3		288	0	2
4		72	$-w-\varepsilon$	2
5		504	$-\varepsilon$	2
6		1728	$-w-2\varepsilon+u$	3
7		5184	$-2\varepsilon+u$	3
8		648	$-2w-4\varepsilon+4u$	4
9		9072	$-w-4\varepsilon+4u$	4
10		11016	$-4\varepsilon+4u$	4

$$\psi_\ell = \psi_4 \vartheta^{\mathrm{p}_\ell - 2} \exp\left[\frac{(4-\mathrm{p}_\ell)\varepsilon_4 + (\mathrm{p}_\ell - 2)\varepsilon_8 - 2\varepsilon_\ell}{2T}\right] . \tag{7.73}$$

For this system, equations (7.3) become, using (7.73),

$$\begin{aligned} \rho &= \frac{1}{4}\sum_{\{\ell\}} \mathrm{p}_\ell \omega_\ell \psi_\ell = \psi_4 \mathfrak{X}^{-4} \mathfrak{Z}^{-1} \vartheta^{-2} X(\vartheta) , \\ 1-\rho &= \frac{1}{4}\sum_{\{\ell\}} (4-\mathrm{p}_\ell)\omega_\ell \psi_\ell = \psi_4 \mathfrak{X}^{-4} \mathfrak{Z}^{-1} \vartheta^{-2} Y(\vartheta) , \end{aligned} \tag{7.74}$$

where

$$X(\vartheta) = 12\mathfrak{X}^3\,\vartheta[\mathfrak{Y}\mathfrak{Z}^3 + 3\mathfrak{X}\mathfrak{Z}(1+7\mathfrak{Y}^2+4\mathfrak{X}^2\mathfrak{Y}^2)\vartheta + 108\mathfrak{X}^2\mathfrak{Y}(1+3\mathfrak{Y}^2)\vartheta^2 + 54\mathfrak{X}\mathfrak{Z}(1+14\mathfrak{Y}^2+17\mathfrak{Y}^4)\vartheta^3\}\,, \tag{7.75}$$

$$Y(\vartheta) = 3^5 + 36\mathfrak{X}^3\mathfrak{Y}\mathfrak{Z}^3\,\vartheta + 36\mathfrak{X}^4\mathfrak{Z}(1+7\mathfrak{Y}^2+4\mathfrak{X}^2\mathfrak{Y}^2)\vartheta^2 + 432\mathfrak{X}^5\mathfrak{Y}(1+3\mathfrak{Y}^2)\vartheta^3 \tag{7.76}$$

and

$$\mathfrak{X} = \exp\left(-\frac{\varepsilon}{2T}\right), \qquad \mathfrak{Y} = \exp\left(-\frac{w}{2T}\right), \qquad \mathfrak{Z} = \exp\left(-\frac{u}{T}\right). \tag{7.77}$$

Dividing the first relation of (7.74) by the second yields

$$\frac{\rho}{1-\rho} = \frac{X(\vartheta)}{Y(\vartheta)}\,. \tag{7.78}$$

It can be shown that, at given T, the ratio $X(\vartheta)/Y(\vartheta)$ is a monotonically increasing function of ϑ. From (7.72) and (7.74) the configurational chemical potential μ is given by

$$\exp\left(\frac{\mu}{T}\right) = \vartheta^4\mathfrak{X}^{12}\mathfrak{Y}^4\mathfrak{Z}^{-8}\left[\frac{12Y(\vartheta)}{X(\vartheta)}\right]^3. \tag{7.79}$$

Putting $f = -P$ and $r = 4$ in equation (7.18) for the equilibrium free energy density yields

$$-P = T\ln\psi_4 + \varepsilon_4 - \frac{3}{2}T\left[\ln\left(\frac{\rho}{12}\right) + \ln(1-\rho)\right] - \frac{1}{2}\mu\,. \tag{7.80}$$

After a little manipulation, including adding the two relations of (7.74) to obtain an expression for ψ_4, we obtain

$$\exp\left(\frac{P}{T}\right) = \frac{[X(\vartheta)]^3}{3^5\,[X(\vartheta)+Y(\vartheta)]^2}\,. \tag{7.81}$$

Equations (7.78) and (7.81) constitute an implicit equation of state giving ρ in terms of P and T. Best results are found when u lies in the middle of the range given by (7.69). That is, when $2u = \frac{1}{2}w + \varepsilon$, which gives, from (7.70), $P_0 = w$. There is liquid–vapour phase separation, and density maxima occur on (ρ, T) isobars in the liquid state and in the supercritical state for $P < P_0$, as shown in Fig. 7.4. The temperature of maximum density varies very little with pressure. A compressibility minimum also occurs on isobars in the liquid state, and the mean number of nearest neighbours of a given molecule is between 4 and 5 at low pressures (Bell 1972). It can be seen that reasonable qualitative agreement with experiment is obtained in spite of the simplications inherent in the model. The same model was used by Lavis and Christou (1977) to study dielectric properties and Wilson and Bell (1978) for an investigation of aqueous solutions. Bell and Salt (1976) used the mean-field approximation on a similar model to discuss transitions to ordered (ice-like) states. Weres and Rice (1972) used a model based on the body-centred cubic lattice, making rather different assumptions and were mainly concerned with local properties. These authors had reservations about Bell's triad term u

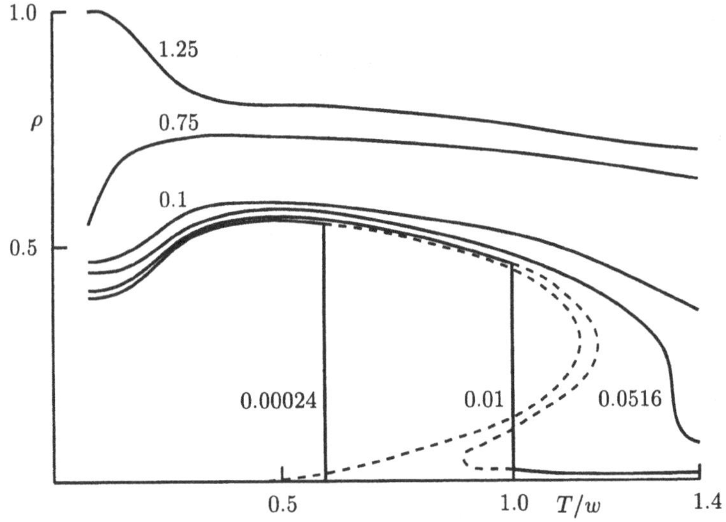

Fig. 7.4. Density–temperature isobars for $\varepsilon/w = 2$ and $u/w = 5/4$. Each isobar is labelled with the value of P/w. (Reprinted from Bell (1972), by permission of the publisher IOP Publishing Ltd.)

which they considered made excessively large contributions to the energies of less dense structures. Other authors who used the body-centred cubic lattice with a different approach were O'Reilly (1973) and Fleming and Gibbs (1974) (see also the review by Stillinger (1975)). A square lattice model with polarized bonds was discussed by Lavis and Christou (1979). Whitehouse et al. (1984) examined the Bell (1972) model treated above using Monte Carlo numerical methods and concluding that 'the cooperativity of the model is seriously underestimated by the first-order approximation, although the general behaviour of the model is correctly represented'. These authors consider that it would be more realistic to have $u = 0$, $\varepsilon < 0$, pointing out that the inequality (7.69) could still be satisfied and a positive P_0 obtained. Shinmi and Huckaby (1986) have used a similar model with non-polar bonding arms and repulsive bond energy to represent carbon tetrachloride.

Both the two-dimensional bonded lattice model of Bell and Lavis (1970) and the three-dimensional non-polar version of the Bell (1972) model, introduced by Shinmi and Huckaby (1986), can be expressed in Hamiltonian form as modified versions of the dilute Ising model introduced in Sect. 6.6.1. The modifications consist in the introduction of certain terms involving more that two sites. Based on this approach Southern and Lavis (1980) and Lavis and Southern (1984) used finite-size cluster real-space renormalization group methods to investigate the two models.[4]

---

[4] For an account of real-space renormalization group methods see Volume 2, Chap. 6.

7.6 1:1 Ordering on the Face-Centred Cubic Lattice

We consider an A–B mixture on the face-centred cubic lattice with nearest-neighbour pair energies obeying the inequality (6.21). The canonical distribution is used and it is assumed that $M_A = M_B$. The field $\mathcal{H}$ in the equivalent antiferromagnet is thus zero. The face-centred cubic lattice is close-packed but can be divided into equal, but not interpenetrating, sublattices $\mathfrak{a}$ and $\mathfrak{b}$, which are alternate planes of sites perpendicular to one of the cubic axes. The sublattices are each composed of two of the four equivalent simple cubic sublattices of the face-centred cubic lattice and hence, in the notation of Sect. 4.5, $z_{\mathfrak{aa}} = z_{\mathfrak{bb}} = 4$, $z_{\mathfrak{ab}} = z_{\mathfrak{ba}} = 8$. The configurational energy of an ordered mixture with the A atoms concentrated on $\mathfrak{a}$ sites and the B atoms on $\mathfrak{b}$ sites will thus be lower than that of a random mixture. This layer type of sublattice ordering for an equimolar mixture will be studied in this section.

In Sect. 6.5 a mean-field theory for 1:3 sublattice ordering on the face-centred cubic was developed and it was found that transitions are first-order, which is in accordance with experiment on the Au–Cu system. However, as we pointed out in Sect. 6.5, mean-field results are unsatisfactory in other ways and in particular they predict a second-order transition for an equimolar mixture. The first-order pair method is even more unsatisfactory, and the smallest group which adequately represents the symmetry of the sublattice structure is a tetrahedron of two $\mathfrak{a}$ sites and two $\mathfrak{b}$ sites. This is shown in Fig. 7.5 where the full lines are the six face-centred cubic lattice edges connecting nearest-neighbour pairs in the group. It can be seen that there are one $\mathfrak{aa}$, one $\mathfrak{bb}$ and four $\mathfrak{ab}$ pairs, giving the same pair ratios as in the whole lattice.

In the notation of Sect. 7.2, $z = 12$, $p = 4$ and $q = 6$. The various configurations on the tetrahedral site group are shown in Table 7.2. Using (6.9) for the configurational energy the canonical Hamiltonian is

$$\widehat{H} = E = \text{constant} - \frac{1}{2} N_{AB}|\varepsilon| \,, \qquad |\varepsilon| = \varepsilon_{AA} + \varepsilon_{BB} - 2\varepsilon_{AB} > 0 \,. \tag{7.82}$$

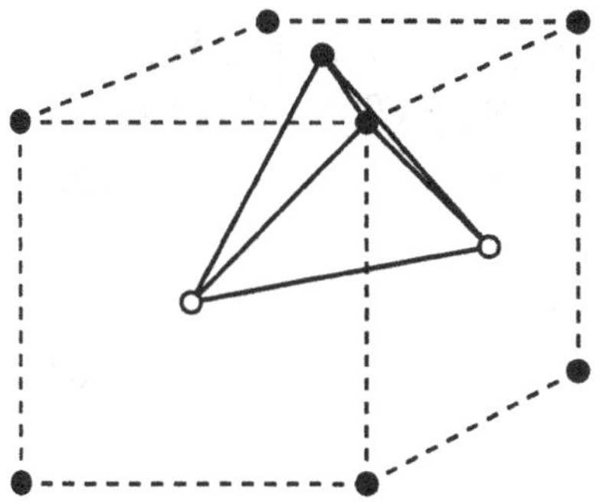

Fig. 7.5. The basic tetrahedron group of sites on the face-centred cubic lattice. Sites on the $\mathfrak{a}$ sublattice are denoted by • and those on the $\mathfrak{b}$ sublattice by ∘.

Table 7.2. Tetrahedron site group configurations on the face-centred cubic lattice.

Configuration index ℓ	Configuration $\mathfrak{aa}\vert\mathfrak{bb}$	ω_ℓ	$2\varepsilon_\ell/\vert\varepsilon\vert$
1	AA—BB	1	−4
2	AA—AB	2	−3
3	AB—BB	2	−3
4	AA—AA	1	0
5	BB—BB	1	0
6	AB—AB	4	−4
7	AB—AA	2	−3
8	BB—AB	2	−3
9	BB—AA	1	−4

The constant can be disregarded and the energy ratios in Table 7.2 are simply the numbers of unlike pairs multiplied by −1. Since $x_{\mathrm{A}} = x_{\mathrm{B}}$ the sublattice mole fractions can be written as

$$\begin{aligned} x_{\mathrm{A}\mathfrak{a}} = x_{\mathrm{B}\mathfrak{b}} &= \frac{1+s}{2}\,, \\ x_{\mathrm{A}\mathfrak{b}} = x_{\mathrm{B}\mathfrak{a}} &= \frac{1-s}{2}\,, \end{aligned} \tag{7.83}$$

where s is an order parameter. Again, by symmetry in the equimolar mixture,

$$\psi_2 = \psi_3\,, \qquad \psi_4 = \psi_5\,, \qquad \psi_7 = \psi_8\,. \tag{7.84}$$

Equations (7.83) and (7.84) are connected since, by using (7.19), the relations $x_{\mathrm{A}\mathfrak{a}} = x_{\mathrm{B}\mathfrak{b}}$ and $x_{\mathrm{A}\mathfrak{b}} = x_{\mathrm{B}\mathfrak{a}}$ can be deduced from (7.84). It also follows that

$$\begin{aligned} \frac{1+s}{2} &= \psi_1 + 3\psi_2 + \psi_4 + 2\psi_6 + \psi_7\,, \\ \frac{1-s}{2} &= \psi_2 + \psi_4 + 2\psi_6 + 3\psi_7 + \psi_9\,. \end{aligned} \tag{7.85}$$

From (7.21), (7.83) and (7.84) the variation in the Helmholtz free energy per site can be written as

$$\begin{aligned} \delta a = {} & (T\ln\psi_1 + \varepsilon_1)\delta\psi_1 + 4(T\ln\psi_2 + \varepsilon_2)\delta\psi_2 + 2(T\ln\psi_4 + \varepsilon_4)\delta\psi_4 \\ & + 4(T\ln\psi_6 + \varepsilon_6)\delta\psi_6 + 4(T\ln\psi_7 + \varepsilon_7)\delta\psi_7 + (T\ln\psi_9 + \varepsilon_9)\delta\psi_9 \\ & - \frac{3}{2}T\ln\left(\frac{1+s}{1-s}\right)\delta s\,. \end{aligned} \tag{7.86}$$

It is useful to define parameters ϑ and $\mathfrak{X}$ by

$$\vartheta^4 = \frac{\psi_1}{\psi_9}\,, \qquad \mathfrak{X} = \exp\left(\frac{|\varepsilon|}{2T}\right)\,. \tag{7.87}$$

We regard ψ_1 and ψ_9 as dependent variables given in terms of the independent variables ψ_2, ψ_4, ψ_6, ψ_7 and s by (7.85). Then, equating $\partial a/\partial\psi_\ell$ to zero for $\ell = 2, 4, 6, 7$

$$\psi_2 = \psi_9 \mathfrak{X}\vartheta^3 , \qquad \psi_4 = \psi_9 \mathfrak{X}^4 \vartheta^2 ,$$
$$\psi_6 = \psi_9 \vartheta^2 , \qquad \psi_7 = \psi_9 \mathfrak{X}\vartheta . \tag{7.88}$$

Substitution in (7.85) yields the relation

$$\frac{1+s}{1-s} = \frac{\vartheta(\vartheta^3 + 3\mathfrak{X}\vartheta^2 + \mathfrak{X}^4\vartheta + 2\vartheta + \mathfrak{X})}{\mathfrak{X}\vartheta^3 + \mathfrak{X}^4\vartheta^2 + 2\vartheta^2 + 3\mathfrak{X}\vartheta + 1} . \tag{7.89}$$

It follows that s increases monotonically with ϑ. Provided that (7.88) is satisfied, (7.86) and (7.85) yield

$$\left(\frac{\partial a}{\partial s}\right)_T = \frac{1}{2}T\left[\ln\left(\frac{\psi_1}{\psi_9}\right) - 3\ln\left(\frac{1+s}{1-s}\right)\right] = \frac{1}{2}T\ln\phi(\vartheta) , \tag{7.90}$$

where, from (7.89),

$$\phi(\vartheta) = \vartheta\left(\frac{\mathfrak{X}\vartheta^3 + \mathfrak{X}^4\vartheta^2 + 2\vartheta^2 + 3\mathfrak{X}\vartheta + 1}{\vartheta^3 + 3\mathfrak{X}\vartheta^2 + \mathfrak{X}^4\vartheta + 2\vartheta + \mathfrak{X}}\right)^3 . \tag{7.91}$$

It can be seen that $\phi(\vartheta)$ as given by (7.91) obeys the equilibrium condition

$$\phi(\vartheta) = 1 , \tag{7.92}$$

which is satisfied by $\vartheta = 1$ for all T, with any other solutions in reciprocal pairs $(\vartheta, \vartheta^{-1})$.

The solution $\vartheta = 1$ gives $s = 0$ by (7.89), whereas ϑ and ϑ^{-1} with $\vartheta > 1$ give s and $-s$ ($s > 0$) respectively. We now have to determine whether the latter type of solution exists and, if so, whether it is stable and how the transition occurs. Although, numerical solutions of (7.92) are ultimately necessary, useful information can be obtained analytically in the neighbourhood of $s = 0$. Equation (7.89) can be rewritten in the form

$$s = \frac{\sinh(2\Delta) + 2\mathfrak{X}\sinh(\Delta)}{\cosh(2\Delta) + 4\mathfrak{X}\cosh(\Delta) + \mathfrak{X}^4 + 2} ,$$
$$\Delta = \ln\vartheta = \frac{1}{4}\ln\left(\frac{\psi_1}{\psi_9}\right) . \tag{7.93}$$

Hence Δ is an odd function of s and can be expanded in a series of odd powers of s, as can $\partial a/\partial s$. Obtaining the first two terms in the expansion of Δ, substituting in (7.90) and integrating with respect to s yields

$$a(s) - a(0) = B_2(T)s^2 + B_4(T)s^4 + \mathrm{O}(s^6) , \tag{7.94}$$

where

$$\begin{aligned} B_2(T) &= \frac{1}{2}T\mathfrak{X}(\mathfrak{X}^2 - \mathfrak{X} + 1)\,, \\ B_4(T) &= \frac{1}{96}T[\mathfrak{X}(\mathfrak{X}+1)^2(\mathfrak{X}^2 - 2\mathfrak{X} + 3)^2(5\mathfrak{X} - 2\mathfrak{X}^2 - \mathfrak{X}^3) - 24]\,. \end{aligned} \tag{7.95}$$

A Landau expansion has been derived for the first-order approximation rather than, as is more usual, for the mean-field approximation. Since $B_2(T) > 0$ and $a(s)$ thus has a minimum at $s = 0$ for all $T > 0$ it follows that there can be no second-order transition at which the equilibrium value of s changes continuously from zero to non-zero values. However, $B_4(T)$ changes from positive to negative as T decreases, suggesting the possibility that $\partial a/\partial s$ may become negative for a range of $s > 0$. If this happens, a maximum and a minimum will appear in the curve of a against s between $s = 0$ and $s = 1$, since $\partial a/\partial s \to \infty$ as $s \to 1$. At the temperature when the value of a at this minimum becomes equal to $a(0)$, a first-order transition will take place. Since $a(-s) = a(s)$, corresponding phenomena will occur for $s < 0$ and the transition can be to either one of a pair of values $(-s, s)$. Numerical solution of (7.92) and calculation of the curve of a against s confirms the existence of a first-order transition which occurs at $T = 0.365|\varepsilon|$ and takes s from zero to ± 0.983 (Li 1949). The first-order transition is thus of the symmetrical type as opposed to the unsymmetrical first-order transition to a unique value of s which was found in Sect. 6.5 for 3:1 ordering on the face-centred cubic.

This treatment is confined to the $x_A = x_B = \frac{1}{2}$ mixture, but Li (1949) used a similar first-order method, based on a tetrahedron of sites, to investigate the whole composition range. He found that all transitions are first-order and are associated with changes in composition (as in Sect. 6.5) except at the compositions $x_A = \frac{1}{4}, \frac{1}{2}$ and $\frac{3}{4}$. There are three distinct sets of phase separation curves with gaps between them and azeotropic-type maxima (like point P in Fig. 1.7) at $x_A = \frac{1}{4}, \frac{1}{2}$ and $\frac{3}{4}$ respectively. Kikuchi (1974) treated the whole composition range using a version of the Kikuchi method with the site tetrahedron as basic group. The chief difference between his phase diagram and Li's is the disappearance of the gaps. Danielian (1961) discovered (using the language of the antiferromagnet) that the completely ordered layer sublattice structure is not a unique ground state for the equimolar mixture (see also Phani et al. 1980). There are Ω_0 states of minimum energy where $\ln(\Omega_0/2) \sim N^{1/3}$ so that $N^{-1}\ln(\Omega_0) \to 0$ as $N \to \infty$. It is difficult to predict the effect on an exact solution since the situation differs from that in the exactly solved triangular lattice antiferromagnet (see Sect. 8.12). In the latter, the ground state is also infinitely degenerate but Ω_0 is so large that $N^{-1}\ln(\Omega_0)$ tends to a non-zero limit as $N \to \infty$ and there is a zero-point entropy. It is interesting to note (Binder 1987) that Monte Carlo methods give a phase diagram like Li's but that with the introduction of second-neighbour interactions it becomes more like Kikuchi's.

In the Cu–Au alloy there are phase separation curve maxima at the Cu_3Au, CuAu and $CuAu_3$ compositions, but the $CuAu_3$ transition temperature is much lower than that for Cu_3Au, indicating that the model's A–B symmetry is unrealistic. For the equimolar alloy, the layer sublattice structure, known as CuAuI, involves an atomic distribution in the layer planes quite different from that in planes perpendicular to the layers. This causes a change in lattice symmetry type from cubic to tetragonal, a-a or b-b nearest-neighbour distances becoming larger than a-b distances. This change may stabilize the layer structure relative to other low-energy configurations. For an interval above the temperature of transition to CuAuI, an orthorhombic structure, known as CuAuII, is stable. This corresponds to one of the alternative minimum energy states and resembles CuAuI, except that at every fifth unit cell in one lattice direction parallel to the layers, the planes occupied by Cu and Au are interchanged, creating a 'stepped' layer structure, (Irani 1972, Parsonage and Staveley 1978). The lattice retains its cubic symmetry when the 3:1 or 1:3 sublattice structures are stable, near the Cu_3Au or $CuAu_3$ compositions.

7.7 Homogeneous Cacti

Rushbrooke and Scoins (1955) adapted the Mayer cluster expansion approach to lattice systems and found that with all closed circuits of nearest-neighbour pairs excluded from consideration their results were equivalent to those of the first-order pair approximation. This implies (Domb 1960) that the pair method is accurate for *Bethe lattices.* It is convenient at this point to introduce some graph theory terminology,[5] identifying sites with *vertices* and lines connecting the members of a nearest-neighbour pair with *edges*, the *valency* of a vertex being the number of edges incident there. A *Cayley tree* is a graph with no closed circuits of edges, and the Bethe lattice is a Cayley tree homogeneous in the sense that all except the outer vertices have the same valency z. The case $z = 3$ is illustrated in Fig. 7.6 where the outer vertices are indicated by circles. In a regular lattice of N sites with a 'reasonable' shape the number N_b of outer or boundary sites is proportional to $N^{1/2}$ for $d = 2$ and $N^{2/3}$ for $d = 3$ so that $N_b/N \to 0$ in the thermodynamic limit $N \to \infty$. This is not so for the Bethe lattice. Suppose that there are m edges between a site of the central pair and an outer site in a graph of the type shown in Fig. 7.6 (where $m = 2$). Then

$$\begin{aligned} N_b &= 2^{m+1}, \\ N &= 2 + 2^2 + \cdots + 2^{m+1} = 2(2^{m+1} - 1) \end{aligned} \tag{7.96}$$

and hence $N_b/N \to \frac{1}{2}$ as $N \to \infty$. For general z, this limiting ratio is $(z-2)/(z-1)$ so that the limiting value of N_b/N is zero only for the $z = 2$ Bethe lattice which is just the open linear lattice of Sect. 2.4.1. For any Bethe

[5] For detailed accounts of graph theory terminology see Temperley (1981) and Volume 2, Appendix A.7.

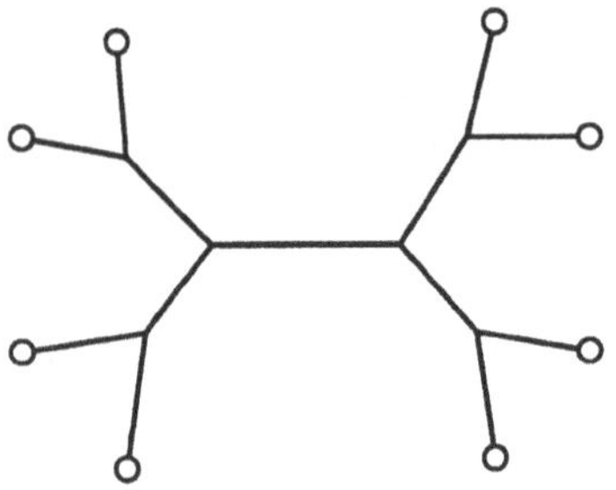

Fig. 7.6. Cayley tree with $z = 3$.

lattice the total number of edges is $N - 1$ so that in the thermodynamic limit the mean coordination number or valency is 2. This results from the high proportion of sites which are outer vertices with valency unity.

For $z > 2$, a Bethe lattice system is thus likely to have very different properties from those of a regular lattice system. This is confirmed by Eggarter's (1974) derivation of the partition function for the zero-field Ising model. Suppose we first sum over the spin variables σ_i for the outer sites (i.e. those indicated by circles in Fig. 7.6), each of which is connected by a single edge to a site in the next layer. Since (2.77) applies, a factor $2\cosh(J/T)$ is obtained for each outer site irrespective of the spin variable values for the next layer. Summing over the latter again produces a factor $2\cosh(J/T)$ for each site, and so on till the relation (2.79) is reproduced for the Bethe lattice with any value of z. This implies a continuous heat capacity and no spontaneous magnetization for $T > 0$ which is at variance with the pair method results obtained in Sect. 7.3 and appears to contradict the inference mentioned at the beginning of the present section. However, in a treatment of the nearest-neighbour exclusion model on a Bethe lattice, Runnels (1967) pointed out that it is results for interior sites deep within the lattice which can be obtained from the first-order pair method. These are quite different from results obtained by averaging over all sites of the lattice. Using Eggarter's partition function for the zero-field Ising model is equivalent to averaging over the entire Bethe lattice. Baxter (1982a) showed that for this model, pair method relations are applicable to sites deep within the lattice in the limit $N \to \infty$.

We now generalize by showing that the first-order theory of Sect. 7.2 is exact for interior sites of a lattice equivalent to a graph which may be termed a *homogeneous Husimi tree* or *homogeneous cactus*. This consists entirely of figures identical to the site group of Sect. 7.2, a central group being surrounded by m layers of other groups connected so that all vertices except the outer ones of the last layer have the same valency z. When the group is simply a pair of vertices joined by an edge, the Bethe lattice of Fig. 7.6 is regained. Figure 7.7 shows a cactus composed of squares of four sites with $z = 4$ and $m = 2$. (The squares are shaded, and the decrease in their size is due to the exigencies of drawing on a plane.) Temperley (1965) considered a nearest-neighbour exclusion model on a cactus of triangles using the

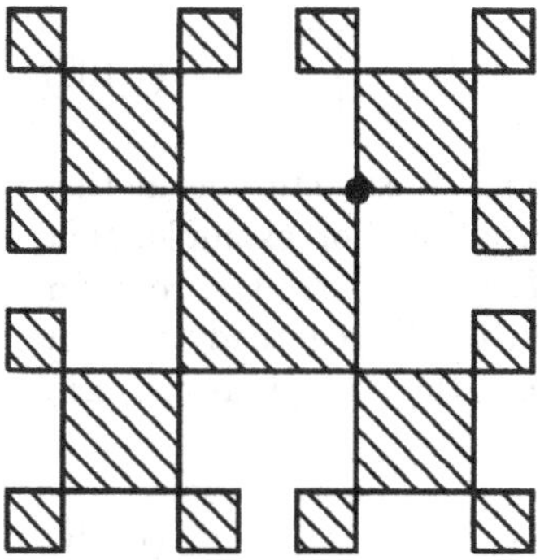

Fig. 7.7. Cactus of squares of four sites with $\mathrm{z} = 4$.

Rushbrooke-Scoins technique. For the general group, using the notation of Sect. 7.2, $(\mathrm{zp})/(2\mathrm{q})$ groups are attached at each inner vertex. For a polygon group this number reduces to $\frac{1}{2}\mathrm{z}$, since $\mathrm{p} = \mathrm{q}$, and two edges from each group are incident at the vertex. In Fig. 7.7, with $\mathrm{z} = 4$, each inner vertex belongs to just two groups.

We construct a partition function for the κ-species model of Sect. 7.2, using the $\eta = n - 1$ distribution. The partition function is based on the central group. Each site of this group belongs to $(\mathrm{zp})/(2\mathrm{q}) - 1$ other groups and may be regarded as a 'root' for $(\mathrm{zp})/(2\mathrm{q}) - 1$ sub-cacti of m layers, each joined to the central group at the root. Let $g_j(\mathrm{m})$ be the partition function for a subcactus of m layers with the root occupied by a member of species j, the summation being taken over all possible occupations of the subcactus sites apart from the root. Then, for the entire cactus, the partition function is

$$Z = \sum_{\{\ell\}} \omega_\ell \exp(-\varepsilon_\ell/T) \prod_{j=1}^{\kappa} \left\{ [g_j(\mathrm{m})]^{\frac{\mathrm{zp}}{2\mathrm{q}}-1} \exp(\xi_i/T) \right\}^{\mathrm{p}_{j\ell}}, \tag{7.97}$$

where the summation is over all configurations on the central group of sites. The ratio of the probabilities of configurations of types ℓ and r respectively is thus given by

$$\frac{\psi_\ell}{\psi_r} = \frac{\exp(-\varepsilon_\ell/T) \prod_{j=1}^{\kappa} \left\{ [g_j(\mathrm{m})]^{\frac{\mathrm{zp}}{2\mathrm{q}}-1} \exp(\xi_i/T) \right\}^{\mathrm{p}_{j\ell}}}{\exp(-\varepsilon_r/T) \prod_{j=1}^{\kappa} \left\{ [g_j(\mathrm{m})]^{\frac{\mathrm{zp}}{2\mathrm{q}}-1} \exp(\xi_i/T) \right\}^{\mathrm{p}_{jr}}}. \tag{7.98}$$

Now define

$$\phi_j(\mathrm{m}) = g_j(\mathrm{m})/g_\kappa(\mathrm{m}), \qquad j = 1, \ldots, \kappa - 1. \tag{7.99}$$

Then, from (7.98) and (7.99),

$$\frac{\psi_\ell}{\psi_r} = \exp\left(\frac{\varepsilon_r - \varepsilon_\ell}{T}\right) \prod_{j=1}^{\kappa-1} \left\{\exp\left(\frac{\xi_j - \xi_\kappa}{T}\right) [\phi_j(m)]^{\frac{zp}{2q}-1}\right\}^{p_{j\ell}-p_{jr}} . \tag{7.100}$$

The partition function for the entire cactus can be constructed in another way, based on one particular site of the central group (for example, that indicated by a black circle in Fig. 7.7). As before, this site is a root for $(zp)/(2q) - 1$ subcacti of m layers. However, it can also be regarded as a root for a subcactus of $m + 1$ layers, the first layer being the central group itself. Hence

$$Z = \sum_{j=1}^{\kappa} \exp(\xi_j/T) [g_j(m)]^{\frac{zp}{2q}-1} g_j(m+1) . \tag{7.101}$$

It follows that

$$\frac{x_j}{x_\kappa} = \exp\left(\frac{\xi_j - \xi_\kappa}{T}\right) [\phi_j(m)]^{\frac{zp}{2q}-1} \phi_j(m+1), \quad j = 1, \ldots, \kappa - 1 . \tag{7.102}$$

We now suppose that $\phi_j(m) \to \phi_j^*$, where ϕ_j^* is finite, when $m \to \infty$. Equations (7.100) and (7.102) become respectively

$$\frac{\psi_\ell}{\psi_r} = \exp\left(\frac{\varepsilon_r - \varepsilon_\ell}{T}\right) \prod_{j=1}^{\kappa-1} \left\{\exp\left(\frac{\xi_j - \xi_\kappa}{T}\right) [\phi_j^*]^{\frac{zp}{2q}-1}\right\}^{p_{j\ell}-p_{jr}} , \tag{7.103}$$

$$\frac{x_j}{x_\kappa} = \exp\left(\frac{\xi_j - \xi_\kappa}{T}\right) [\phi_j^*)]^{\frac{zp}{2q}} , \quad j = 1, \ldots, \kappa - 1 . \tag{7.104}$$

Substitution of (7.104) into (7.103) then yields (7.17) again so that the latter is an exact result for the central group of a homogeneous cactus. By adapting this method it can be shown that (7.17) applies to any group in layer n, provided that n is kept fixed as $m \to \infty$. This condition, in fact, defines an *interior* group and we have thus shown that first-order results are exact for the interior of a homogeneous cactus.

Equations (7.103) and (7.104) are similar to the basic relations of the Bethe approach, though the latter was developed as an approximation for regular lattices. In the Bethe method a factor $\mathfrak{Z}_j^z$ is postulated in the probability of occupation of a site by a member of species j to allow for the effect of the lattice environment. For a site of a group it is assumed that the corresponding factor is $\mathfrak{Z}_j^{z-2q/p}$, $z - 2q/p$ being the number of edges incident at the site from outside the group. Equations (7.103) and (7.104) result if we substitute ϕ_j^* for $(\mathfrak{Z}_j/\mathfrak{Z}_\kappa)^{2q/p}$ in the Bethe relations.

Examples

7.1 Apply the pair approximation to the Ising model ferromagnet on the $d = 1$ lattice by putting $z = 2$ in Sect. 7.3. Show that, for all T and $\mathcal{H}$, (7.31) gives only one physically meaningful value for ϑ which is

$$\vartheta = \frac{1}{2\mathfrak{X}}\left\{\mathfrak{Y}^2 - 1 + [(\mathfrak{Y}^2-1)^2 + 4\mathfrak{X}^2\mathfrak{Y}^2]^{1/2}\right\},$$

where $\mathfrak{Y} = \exp(\mathcal{H}/T)$. Hence show that the pair approximation yields the accurate relation (2.99) between m and $\mathcal{H}$.

7.2 With $\phi(\vartheta)$ defined by (7.32), show that

$$\frac{\mathrm{d}\phi}{\mathrm{d}\vartheta} = \frac{(1+\mathfrak{X}\vartheta)^{\mathrm{z}-2}}{(\mathfrak{X}+\vartheta)^{\mathrm{z}}}[\mathfrak{X}(\vartheta-1)^2 + (\mathfrak{X}+1)(\mathrm{z}\mathfrak{X}-\mathrm{z}+2)\vartheta].$$

Hence show that when $\mathfrak{X} > (\mathrm{z}-2)/\mathrm{z}$ the only solution of the equation $\phi(\vartheta) = 1$ for positive ϑ is $\vartheta = 1$ but that a pair of other solutions appears when $\vartheta < (\mathrm{z}-2)/\mathrm{z}$.

7.3 For the Ising model ferromagnet in the pair approximation, show that the internal energy per site

$$u = -\frac{\mathrm{z}J(1-\mathfrak{X})}{2(1+\mathfrak{X})},$$

when $\mathcal{H} = 0$ and $\mathfrak{X} > (\mathrm{z}-2)/\mathrm{z}$. Hence show that the corresponding zero-field configurational heat capacity per site is given by

$$c_0 = k_{\mathrm{B}}\frac{\mathrm{d}u}{\mathrm{d}T} = \frac{2k_{\mathrm{B}}\mathrm{z}\mathfrak{X}}{(1+\mathfrak{X})^2}\left(\frac{J}{T}\right)^2 = \tfrac{1}{2}k_{\mathrm{B}}\mathrm{z}\left[\frac{J/T}{\cosh(J/T)}\right]^2.$$

Show that c_0 is a monotonically decreasing function of T when $\mathfrak{X} > (\mathrm{z}-2)/\mathrm{z}$ and $\mathrm{z} \geq 3$ and that

$$c_0 \to k_{\mathrm{B}}\frac{\mathrm{z}^2(\mathrm{z}-2)}{8(\mathrm{z}-1)^2}\left[\ln\left(\frac{\mathrm{z}}{\mathrm{z}-2}\right)\right]^2$$

as

$$\mathfrak{X} \to \frac{\mathrm{z}-2}{\mathrm{z}} + 0.$$

7.4 Apply the pair approximation to the Ising model ferromagnet on the square lattice by putting $\mathrm{z} = 4$ in Sect. 7.3. Show that (7.33) has solutions $\vartheta = \pm 1$ for all temperatures, but that for $\mathfrak{X} < \mathfrak{X}_{\mathrm{c}} = \frac{1}{2}$ there is a pair of positive real solutions $(\vartheta_0, \vartheta_0^{-1})$ for which

$$\vartheta_0 + \vartheta_0^{-1} = (1 - 3\mathfrak{X}^2)/\mathfrak{X}^3.$$

Hence show, using (7.29), that for $\mathfrak{X} < \frac{1}{2}$,

$$\frac{u}{J} = -\frac{8}{1-\mathfrak{X}^2} + \frac{4}{1-2\mathfrak{X}^2} + 2,$$

and

$$c_0 = 32k_{\mathrm{B}}\mathfrak{X}^2\left[\frac{1}{(1-2\mathfrak{X}^2)^2} - \frac{1}{(1-\mathfrak{X}^2)^2}\right]\left(\frac{J}{T}\right)^2.$$

By using these results with those of Example 3, show that u is continuous at $\mathfrak{X} = \mathfrak{X}_c = \frac{1}{2}$ and find the magnitude of the discontinuity in c_0.

7.5 When $\varepsilon > -J$, show from (7.61) and (7.62) that $\rho^* \to 0$ as $T \to 0$ on the critical curve. When $\varepsilon \leq -J$, show that $\rho^* \to \rho_c > 0$ as $T \to 0$ on the critical curve, where $\rho_c = 2(z-1)/(z^2-2)$ for $\varepsilon = -J$ and $\rho_c = (z-1)/(2z-3)$ for $\varepsilon < -J$. It is to be assumed in all cases that $J > 0$.

7.6 In the two-dimensional water-analogue lattice gas model of Bell and Lavis (1970), mentioned in Sect. 7.5, each molecule has three identical bonding arms in the triangular lattice plane at angles of $2\pi/3$ to each other. These bonding arms point to three of the six nearest-neighbour sites so that the molecule has two distinguishable orientations. When the bonding arms of a pair of nearest-neighbour molecules lie on the same line, a bond is formed and the pair interaction energy is $-(\varepsilon+w)$ where $\varepsilon > 0$ and $w > 0$. For an unbonded nearest-neighbour pair the interaction energy is $-\varepsilon$, so that w is the energy associated with a bond.
The reader can verify, with the aid of a sketch, that it is possible to form a completely bonded network on a honeycomb sublattice of the triangular lattice. The open ground state consists of such a network with the remaining one third of the lattice sites vacant. There are a large number of close-packed ground states with all lattice sites occupied, but in all of them, exactly one of the three nearest-neighbour pairs on every triangle of sites is bonded. Show that at $T = 0$ the open ground state is stable for $P < w - 3\varepsilon$, where the two-dimensional pressure field $P = \widetilde{P}a_0$, a_0 being the area per site. Apply a first-order method, with a triangle of sites as basic group, to this model, adopting a similar approach to that of Sect. 7.5, where the $d = 3$ model is considered. Let ϑ be the ratio of the probability of a triangle configuration with all three sites occupied and one pair of molecules bonded to that of a configuration with one site vacant and the other two occupied by a bonded pair. Also, let

$$\mathfrak{X} = \exp(-\varepsilon/T)\,, \quad \mathfrak{Y} = \exp(-w/T)\,.$$

Show that the number density is given by

$$\frac{\rho}{1-\rho} = \frac{X(\vartheta)}{Y(\vartheta)}\,,$$

where

$$X(\vartheta) = 2\vartheta[(3+\mathfrak{Y})\vartheta^2 + (1+3\mathfrak{Y})\vartheta + \mathfrak{Y}\mathfrak{X}^{-1}]\,,$$

$$Y(\vartheta) = (1+3\mathfrak{Y})\vartheta^2 + 4\mathfrak{Y}\mathfrak{X}^{-1}\vartheta + \mathfrak{Y}\mathfrak{X}^{-3}\,,$$

and that the chemical potential and two-dimensional pressure respectively are given by

$$\mu = -6\varepsilon - T\left\{2\ln\left[\frac{\rho}{2(1-\rho)}\right] - 3\ln(\vartheta)\right\},$$

$$P = T\ln[Y(\vartheta)(1-\rho)] + w - 3\varepsilon.$$

8. Exact Results for Two-Dimensional Ising Models

8.1 Introduction

We already have exact results for the one-dimensional Ising model (Chap. 2 and Sect. 4.1) and exact transcriptions between models, for instance from the zero-field Ising ferromagnet to the zero-field antiferromagnet (Sect. 4.2) or from the Ising model to the simple lattice fluid or two-component mixture (Chaps. 5 and 6). In Sects. 8.2–8.6, *dual* and *star-triangle* transformations are derived, connecting the partition function for the zero-field Ising model on a given lattice to that of the zero-field Ising model on a related lattice at a different temperature.[1] Some associated relations connecting nearest-neighbour correlations are obtained, a particularly important one being derived in Sect. 8.6. We then derive exact results for critical conditions and thermodynamic functions for the two-dimensional zero-field Ising model. There are a large number of approaches to this, all mathematically intricate.[2] The original method of Onsager (1944), Kaufman (1949) and Kaufman and Onsager (1949) depended on transfer matrices, of which the 2×2 matrix appearing in the theory of the one-dimensional Ising model in Sect. 2.4 is a simple example. Other workers have employed the Pfaffian (see Volume 2, Chap. 8, Green and Hurst 1964, McCoy and Wu 1973) or combinatorial methods (Kac and Ward 1952, Vdovichenko 1965a, 1965b). We shall use a method due to Baxter and Enting (1978), which is closely related to the preceding transformation theory and has the advantage of giving results simultaneously for the square, triangular and honeycomb lattices. The exact expressions for configurational energy and heat capacity depend on elliptic integrals and these protean quantities appear in the literature in a number of different forms. We have tried to use expressions valid for the whole temperature range and closely related to the formulae given for the partition function or free energy.

For most of this chapter we assume, for definiteness, a ferromagnetic interaction. However in the last section, the antiferromagnet is discussed, especially the antiferromagnet on the triangular lattice. This lattice is close-

[1] Such transformations are discussed by Syozi (1972) and Baxter (1982a).

[2] There is an excellent short survey of this work by Temperley (1972). For other applications of transformation methods see Maillard (1985).

packed so the behaviour of the zero-field antiferromagnet cannot be deduced from that of the ferromagnet and it displays some curious properties.

Exact solutions exist for various two-dimensional models apart from the Ising model and are considered in Chaps. 9 and 10 and Volume 2, Chaps. 5 and 8. However, exact methods have their limitations. There are no exact solutions for three-dimensional lattices and, not even for the two-dimensional Ising model in non-zero field.

8.2 The Low-Temperature Form and the Dual Lattice

Consider the zero-field Ising model ferromagnet whose Hamiltonian is given by (2.66) with $\mathcal{H} = 0$. This can also be written as

$$\widehat{H} = E = -N_{\mathrm{E}}J + J \sum_{\{i,j\}}^{(\mathrm{n.n.})} (1 - \sigma_i\sigma_j)\,, \tag{8.1}$$

where N_{E} is the number of nearest-neighbour site pairs, which may be replaced by $\frac{1}{2}zN$ in the thermodynamic limit. Each unlike nearest-neighbour pair of spins makes a contribution $2J$ to the sum in (8.1) whereas a like nearest-neighbour pair gives a zero contribution. Substitution into (2.67) gives the configurational partition function in the form

$$Z = \exp(K N_{\mathrm{E}}) \sum_{\Lambda} \exp(-2K N_{\mathrm{U}})\,, \tag{8.2}$$

where, as in (2.146)[3] $K = J/T$, the sum is over all spin configurations and N_{U} denotes the number of unlike nearest-neighbour pairs. Each term involves a power of the variable $\exp(-2K)$, which is zero at $T = 0$ and unity at $T = \infty$. Since this variable is small at low temperatures, (8.2) is a *low-temperature form*. If terms in Z or $\ln Z$ are arranged in increasing powers of $\exp(-2K)$ then we have a low-temperature series (see Volume 2, Chap. 7).

The terms of the sum in (8.2) can be given a geometrical interpretation by using the idea of a *dual lattice*. First note that the graph theory terminology introduced for the Bethe lattice in Sect. 7.7 can be used for the regular two-dimensional lattices shown in Appendix A.1. Each site is a *vertex*, the members of each nearest-neighbour pair are connected by an *edge* and the interior of each elementary polygon of edges (for example, a square of four edges for the square lattice) is a *face*. The resulting graph is termed the *full lattice graph*. The dual lattice corresponds to the graph theoretical dual (Temperley 1981) of the full lattice graph. It is obtained by placing a point

[3] Parameters formed from ratios of energies or fields to temperature are often called *couplings*. These will be used extensively in this chapter and Chap. 9. An alternative formulation of thermodynamics to the field-density representation of Sect. 1.8 is a coupling-density representation (see, for example, Volume 2, Sect. 1.4.).

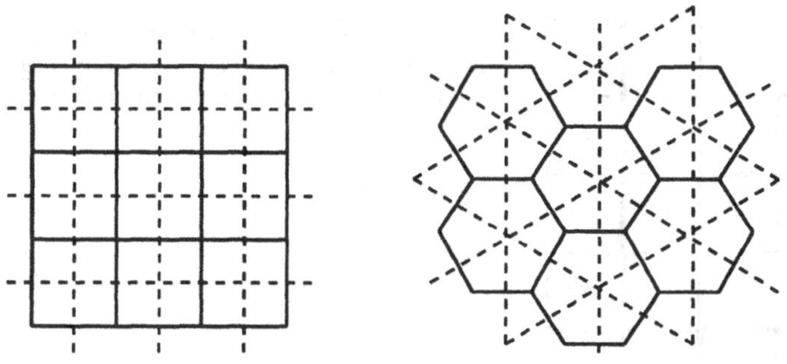

(a) (b)

Fig. 8.1. Dual lattices: **(a)** square (self-dual); **(b)** honeycomb-triangular.

representing a dual lattice site in each face of the original lattice. Points in adjacent faces of the original lattice form a nearest-neighbour pair of sites in the dual lattice and are connected by a dual lattice edge, shown by a broken segment in Fig. 8.1. It can be seen from Fig. 8.1(a) that the square lattice is *self-dual*, since its dual is also a square lattice. The dual of the honeycomb lattice is the triangular lattice, as can be seen from Fig. 8.1(b), and since the duality relation is a reciprocal one, the dual of the triangular lattice is the honeycomb lattice. Denote the numbers of sites and edges respectively in the original lattice by N and N_{E}, and the number of sites in the dual lattice by N^*. From the method of construction, N_{E} is also the number of edges in the dual lattice. Since N^* is the number of faces of the original lattice, Euler's relation for planar graphs (Temperley 1981) yields[4]

$$N + N^* = N_{\mathrm{E}}. \tag{8.3}$$

Take a distribution of positive and negative spins on the original lattice, each site of which lies in a face of the dual lattice with nearest-neighbour pairs occupying adjacent faces of the dual lattice. When the spins of a nearest-neighbour pair are unlike, replace the broken line segment representing the dual lattice edge separating them (see Fig. 8.2) by a full line segment. Thus each spin distribution on the sites of the original lattice is represented by a set of lines on the edges of the dual lattice. (Figure 8.2 gives an illustration for the self-dual square lattice.) The lines and the vertices they connect form a *configuration graph* which is a subgraph of the full dual lattice graph. The form of the configuration graphs can be deduced by noting that when the sites on the boundary of any face of the original lattice are traversed in order, there is either no change of spin sign or an even number of changes. Hence the number of lines incident at the dual lattice vertex inside the face is either zero or even. Apart from sequences terminating at the lattice boundary, the lines of any configuration graph thus form closed polygons (see Fig. 8.2). Reversal

[4] For finite plane graphs, the right-hand side of (8.3) is $N_{\mathrm{E}} + 1$ or $N_{\mathrm{E}} + 2$, but in the thermodynamic limit of very large N and N^* we can use N_{E}.

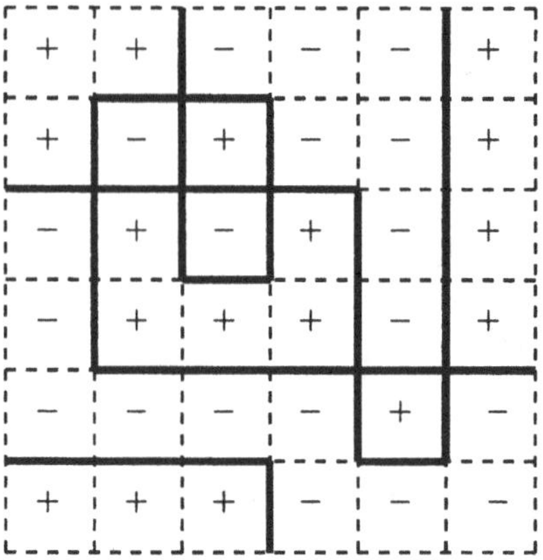

Fig. 8.2. A spin configuration on the square lattice and the corresponding configuration graph on the dual lattice.

of all spins on the original lattice sites does not change the distribution of like and unlike nearest-neighbour pairs, and hence leaves the polygon graph on the dual lattice unaltered. Thus each polygon graph corresponds to a pair of spin distributions. Putting $N_{\mathrm{U}} = \nu$, where ν is the number of lines in a polygon graph, (8.2) can be expressed in the form

$$Z = 2\exp(KN_{\mathrm{E}}) \sum^{(\mathrm{p.g.})} \exp(-2K\nu)\,, \tag{8.4}$$

where the summation is over all polygon graphs.

8.3 The High-Temperature Form and the Dual Transformation

Since the product $\sigma_i\sigma_j$ can take only the two values ±1,

$$\exp(K\sigma_i\sigma_j) = \cosh(K) + \sigma_i\sigma_j \sinh(K) = \cosh(K)(1 + \sigma_i\sigma_j\tau)\,, \tag{8.5}$$

where

$$\tau = \tanh(K)\,. \tag{8.6}$$

Thus, when $\mathcal{H} = 0$, the configurational partition function (2.67) for the Ising model can be expressed as

$$Z = [\cosh(K)]^{N_{\mathrm{E}}} \sum_{\Lambda} \prod_{\{i,j\}}^{(\mathrm{n.n.})} (1 + \sigma_i\sigma_j\tau)\,. \tag{8.7}$$

Each of the products in (8.7) contains N_{E} factors and can be expanded binomially into $2^{N_{\mathrm{E}}}$ terms. However, many of these terms vanish when the sum over spin values is performed. For instance, a linear term $\sigma_i\sigma_j\tau$ disappears when either σ_i or σ_j is summed over the values ±1. The condition for a term to have a non-zero value after the summation over spin values is easily understood geometrically. Suppose we draw the lattice with broken segments for edges. Now, after binomial expansion, a typical term in (8.7) can be written as

$$(\sigma_{i_1}\sigma_{j_1})(\sigma_{i_2}\sigma_{j_2})\cdots(\sigma_{i_n}\sigma_{j_n})\tau^n \tag{8.8}$$

and can be represented by replacing the broken segments corresponding to the nearest-neighbour pairs $i_1j_1, i_2j_2, \ldots, i_nj_n$ by full line segments. If the nearest-neighbour pairs have no sites in common then the term vanishes after summation. The simplest type of non-vanishing term on the square lattice is

$$(\sigma_i\sigma_j)(\sigma_j\sigma_\ell)(\sigma_\ell\sigma_m)(\sigma_m\sigma_i)\tau^4 = \sigma_i^2\sigma_j^2\sigma_\ell^2\sigma_m^2\tau^4 \,, \tag{8.9}$$

where the lines form an elementary square. Summation over σ_i, σ_j, σ_ℓ and σ_m yields a factor 16, and summation over the spin variables not present in this term yields a further factor 2^{N-4} so that finally $2^N\tau^4$ is obtained as the contribution from a single square. It is now not difficult to derive the general rule for non-vanishing terms. Each spin variable σ_i must be present as an even power which means that either no lines or an even number of lines meet at each lattice site. Thus only polygon graphs as defined in Sect. 8.2 contribute to (8.7), which can be expressed as

$$Z = [\cosh(K)]^{N_E} 2^N \sum^{(\text{p.g.})} \tau^\nu, \tag{8.10}$$

where ν denotes the total number of lines composing a given graph. In this context, polygon graphs are referred to as *zero-field graphs* (Volume 2, Appendix A.7). The variable $\tau = \tanh(K) = \tanh(J/T)$ is unity at $T = 0$ and zero at $T = \infty$. Since τ is thus small at high temperatures, (8.10) is a *high-temperature form*. If the terms in Z or $\ln Z$ are arranged in ascending powers of τ we have a *high-temperature series* (see Volume 2, Chap. 7).

In (8.4) the low-temperature form was expressed in terms of a sum over certain sets of lines or graphs on the dual lattice. Except at the boundaries, the rules for permissible sets of lines are exactly the same as for the high-temperature form (8.10) for an Ising ferromagnet on the dual lattice. Thus for large N and N^*, that is in the thermodynamic limit, the sum in (8.4) is equal to the sum in (8.10) for the dual lattice Ising model, provided that

$$\exp(-2K) = \tanh(K^*) \,, \tag{8.11}$$

where K^* denotes the value of K for the dual lattice. This relation, which is meaningful only for positive K and K^*, implies that

$$\exp(-KN_E)Z(N, K) = 2^{-N^*}[\cosh(K^*)]^{-N_E}Z^*(N^*, K^*) \,, \tag{8.12}$$

where we have introduced the dependence of Z on N and K explicitly, and Z^* is the configurational partition function for the dual lattice. A factor $\frac{1}{2}$, which has a negligible effect in the thermodynamic limit, is omitted from the left-hand side of (8.12). Both (8.11) and (8.12) can be expressed in symmetrical form, as would be expected from the symmetrical nature of the lattice-dual relation. From (8.3) and (8.11),

$$\sinh(2K)\sinh(2K^*) = 1\,, \tag{8.13}$$

$$2^{N^*}[\cosh(K^*)]^{N_E}\exp(-KN_E) = [2\sinh(2K^*)]^{N^*/2}[2\sinh(2K)]^{-N/2} \tag{8.14}$$

and (8.12) becomes

$$[2\sinh(2K)]^{-N/2}Z(N,K) = [\sinh(2K^*)]^{-N^*/2}Z^*(N^*,K^*)\,. \tag{8.15}$$

From (2.20) and (2.66)

$$-\frac{U}{N_U J} = -\frac{2u}{zJ} = \langle\sigma_i\sigma_j\rangle_{\text{n.n.}} = \frac{1}{N_E}\frac{\partial \ln Z}{\partial K}\,, \tag{8.16}$$

where u is the configurational energy per site. Combining this result with (8.15) (see Example 8.1) gives

$$\frac{\langle\sigma_i\sigma_j\rangle_{\text{n.n.}}}{\coth(2K)} + \frac{\langle\sigma_i\sigma_j\rangle^*_{\text{n.n.}}}{\coth(2K^*)} = 1\,, \tag{8.17}$$

which is a relation between nearest-neighbour correlations. For the self-dual square lattice, (8.15) becomes

$$[\sinh(2K)]^{-N/2}Z(N,K) = [\sinh(2K^*)]^{-N/2}Z(N,K^*)\,. \tag{8.18}$$

If $K = K^* = K_c$ then, from (8.13), K_c is given by

$$\sinh^2(2K_c) = 1\,. \tag{8.19}$$

The transformation (8.13) connects any value of K in the high-temperature interval $[0, K_c]$ with one in the low-temperature interval $[K_c, \infty]$, and vice versa. A critical point arises from a singularity in Z, and thus from (8.18) a critical point in $[0, K_c]$ implies a critical point in $[K_c, \infty]$. Thus, if there is a unique critical point for the square lattice, it must correspond to

$$K = K_c\,, \qquad T_c = J/K_c = 2J/\ln(1+\sqrt{2})\,. \tag{8.20}$$

Intuitively a single critical temperature seems likely since for the Ising ferromagnet there is only one simple type of long-range order and this corresponds to the $T = 0$ ground state. (For a proof, see Sect. 8.9.) Substituting $K = K^* = K_c$ in (8.17), the critical point values of the nearest-neighbour correlation and the configurational energy per site for the square lattice are given by

$$u_c = -\tfrac{1}{2}zJ[\langle\sigma_i\sigma_j\rangle_{\text{n.n.}}]_c = -J\coth(2K_c)\,, \tag{8.21}$$

where $z = 4$. Numerical values for T_c and u_c are given in Table 8.1.

Table 8.1. Critical parameters for the Ising model ferromagnet of dimension d and coordination number z =. Results for $d = 3$ are those obtained by series methods and are taken from Domb (1974). The first-order pair values can be derived from formulae given in Sect. 7.2.

Lattice	d	z	$T_c/(zJ)$	$-u_c/(\frac{1}{2}zJ)$	$[\ln(2) - s_c]/\ln(2)$
Honeycomb	2	3	0.5062	0.7698	0.6181
Plane square	2	4	0.5673	0.7071	0.5579
Triangular	2	6	0.6068	0.6667	0.5235
Diamond	3	4	0.6761	0.4370	0.2640
Simple cubic	3	6	0.7518	0.3308	0.1951
Body-centred cubic	3	8	0.7942	0.2732	0.1603
Face-centred cubic	3	12	0.8163	0.2475	0.1485
Mean-field	–	–	1.0	0.0	0.0
First-order pair	–	3	0.6068	0.5000	0.2831
First-order pair	–	4	0.7213	0.3333	0.1634
First-order pair	–	6	0.8221	0.2000	0.0871
First-order pair	–	8	0.8690	0.1429	0.0591
First-order pair	–	12	0.9141	0.0909	0.0358

8.4 The Star-Triangle Transformation

The dual transformation maps the triangular lattice into the honeycomb and vice versa, and hence does not yield a high-temperature–low-temperature relation like (8.13). However such relations can be obtained by using the *star-triangle* transformation together with the dual transformation. The star-triangle transformation is obtained as follows. First note from Appendix A.1 that the honeycomb lattice is loose-packed and divides into two triangular sublattices, whose sites are represented by white and black circles respectively in Fig. 8.3. Consider a particular white site and denote the spin variable there by δ. The spin on the white site interacts only with the spins on the three neighbouring black sites, where the spin variables are denoted by α, β and γ. These interactions give a factor

$$\exp[G\delta(\alpha + \beta + \gamma)]\,, \tag{8.22}$$

where

$$G = (J/T)_{\text{honeycomb}} \tag{8.23}$$

in each term of Z. In zero field this is the only factor involving δ, and summing over $\delta = \pm 1$ gives

$$\sum_{\delta=\pm 1} \exp[G\,\delta(\alpha + \beta + \gamma)] = 2\cosh[G(\alpha + \beta + \gamma)]\,. \tag{8.24}$$

The star-triangle transformation depends on defining parameters R and K such that, for $\alpha = \pm 1,\ \beta = \pm 1,\ \gamma = \pm 1$,

$$2\cosh[G(\alpha + \beta + \gamma)] = \mathsf{R}\exp[K(\alpha\beta + \beta\gamma + \gamma\alpha)]\,. \tag{8.25}$$

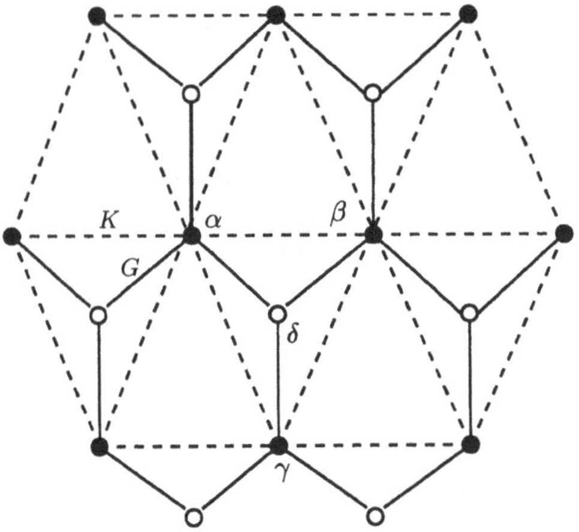

Fig. 8.3. The honeycomb lattice, sublattices and the star-triangle transformation.

This is possible because (8.25) corresponds to only two independent relations. Since both sides are invariant under a simultaneous change of sign of α, β and γ and any permutation of α, β and γ, these two relations are obtained by putting $\alpha = \beta = \gamma = 1$ and $\alpha = \beta = -\gamma = 1$ respectively giving

$$\begin{aligned} 2\cosh(3G) &= \mathrm{R}\exp(3K)\,, \\ 2\cosh(G) &= \mathrm{R}\exp(-K)\,. \end{aligned} \tag{8.26}$$

Eliminating R and K successively,

$$\begin{aligned} \exp(4K) &= 2\cosh(2G) - 1\,, \\ \mathrm{R}^4 &= 16\cosh(3G)\cosh^3(G) = \exp(4K)[\exp(4K)+3]^2\,. \end{aligned} \tag{8.27}$$

The right-hand side of (8.25) corresponds to an Ising model with interaction parameter K between nearest-neighbour pairs on the triangular sublattice of black sites, such neighbouring pairs being represented by the broken segments in Fig. 8.3. It follows that if, in the configurational partition function for a honeycomb lattice of $2N$ sites, we sum over all spin variables corresponding to white sublattice sites and take the thermodynamic limit so that boundary effects can be disregarded then we obtain the star-triangle transformation relation

$$Z^{(\mathrm{H})}(2N, G) = \mathrm{R}^N Z^{(\mathrm{T})}(N, K)\,. \tag{8.28}$$

Here 'H' and 'T' refer to honeycomb and triangular lattices respectively and the parameters G, K and R are connected by (8.27).

Now suppose that the dual transformation is performed on the triangular lattice Ising model, mapping it into a honeycomb lattice Ising model. From (8.15), using K' and G in place of K and K^* respectively,

$$[2\sinh(2K')]^{-N/2} Z^{(\mathrm{T})}(N, K') = [2\sinh(2G)]^{-N} Z^{(\mathrm{H})}(2N, G)\,, \tag{8.29}$$

since in this case $N^* = 2N$. By (8.13), K' and G satisfy the relation

$$\sinh(2K')\sinh(2G) = 1\,. \tag{8.30}$$

If the star-triangle transformation is now performed on the honeycomb lattice, the triangular lattice is recovered and, combining (8.28), (8.29) and (8.30),

$$[\sinh(2K')]^{-N/2}Z^{(\mathrm{T})}(N, K') = 2^{-N/2}[\sinh(2K')]^{N}\mathrm{R}^{N}Z^{(\mathrm{T})}(N, K)\,. \tag{8.31}$$

Eliminating G between (8.30) and the first relation of (8.27) gives

$$[\exp(4K) - 1][\exp(4K') - 1] = 4\,. \tag{8.32}$$

Equation (8.31) can now be expressed in symmetrical form. Using (8.32) and the second relation of (8.27),

$$[\sinh(2K')]^4\mathrm{R}^4 = 4/[\sinh(2K)]^2 \tag{8.33}$$

and, substituting in (8.31),

$$[\sinh(2K)]^{-N/2}Z^{(\mathrm{T})}(N, K) = [\sinh(2K')]^{-N/2}Z^{(\mathrm{T})}(N, K')\,. \tag{8.34}$$

This is identical in form to the square lattice dual transformation relation (8.18), but the two K values are now connected by (8.32) instead of (8.13). Like (8.13), equation (8.32) relates values of K in a high-temperature interval $[0, K_\mathrm{c}]$ with those in a low-temperature interval $[K_\mathrm{c}, \infty]$ where K_c is now given by

$$[\exp(4K_\mathrm{c}) - 1]^2 = 4\,. \tag{8.35}$$

By similar reasoning to that used in the square case, if there is a unique critical point it must be at $K = K_\mathrm{c}$ where the two intervals meet and the critical temperature is then

$$T_\mathrm{c} = J/K_\mathrm{c} = 4J/\ln(3)\,. \tag{8.36}$$

By a star-triangle transformation followed by a dual transformation it can be shown (Example 8.2) that, for the honeycomb lattice,

$$[\sinh(2G')]^{-N/2}Z^{(\mathrm{H})}(N, G') = [\sinh(2G)]^{-N/2}Z^{(\mathrm{H})}(N, G)\,, \tag{8.37}$$

where

$$[\cosh(2G) - 1][\cosh(2G') - 1] = 1\,. \tag{8.38}$$

The meeting point of the high- and low-temperature intervals connected by (8.38) is at $G = G_\mathrm{c}$, where

$$[\cosh(2G_\mathrm{c}) - 1]^2 = 1 \tag{8.39}$$

and, if there is a unique critical point, it occurs at temperature

$$T_\mathrm{c} = J/G_\mathrm{c} = 2J/\ln(2 + \sqrt{3})\,. \tag{8.40}$$

It can be verified (Example 8.3) that (8.21) for the square lattice critical point energy applies also for both the triangular and honeycomb lattices. Numerical values for T_c and u_c are given in Table 8.1.

8.5 The Star-Triangle Transformation with Unequal Interactions

In the next three sections, following Baxter and Enting (1978), we develop a functional equation for the nearest-neighbour correlation in two-dimensional lattices. A necessary preliminary is to consider the star-triangle transformation when the interactions are different in the three honeycomb lattice directions so that the parameter G of Sect. 8.4 is replaced by G_1, G_2 and G_3 as shown in Fig. 8.4. Equation (8.25) is now generalized to

$$2\cosh[G_1\alpha + G_2\beta + G_3\gamma] = \mathsf{R}\exp[K_1\beta\gamma + K_2\gamma\alpha + K_3\alpha\beta]\,. \tag{8.41}$$

As parameter K has to be replaced by K_1, K_2 and K_3, which correspond to the three triangular lattice directions, there are four unknowns on the right-hand side of (8.41). However, unlike (8.25), (8.41) is not invariant under permutations of α, β and γ and is equivalent to four independent relations which may be obtained by putting $\alpha = \beta = \gamma = 1$, $-\alpha = \beta = \gamma = 1$, $\alpha = -\beta = \gamma = 1$ and $\alpha = \beta = -\gamma = 1$ respectively. It is useful to define

$$\begin{aligned} \mathsf{x} &= \cosh[G_1 + G_2 + G_3]\,, \\ \mathsf{x}_i &= \cosh[G_1 + G_2 + G_3 - 2G_i]\,, \qquad i = 1,2,3, \\ \mathsf{X} &= \sqrt{\mathsf{x}\mathsf{x}_1\mathsf{x}_2\mathsf{x}_3}\,, \end{aligned} \tag{8.42}$$

and then (8.41) gives

$$\begin{aligned} \mathsf{R}\exp(K_1 + K_2 + K_3) &= 2\mathsf{x}\,, \\ \mathsf{R}\exp(K_1 - K_2 - K_3) &= 2\mathsf{x}_1\,, \\ \mathsf{R}\exp(-K_1 + K_2 - K_3) &= 2\mathsf{x}_2\,, \\ \mathsf{R}\exp(-K_1 - K_2 + K_3) &= 2\mathsf{x}_3\,. \end{aligned} \tag{8.43}$$

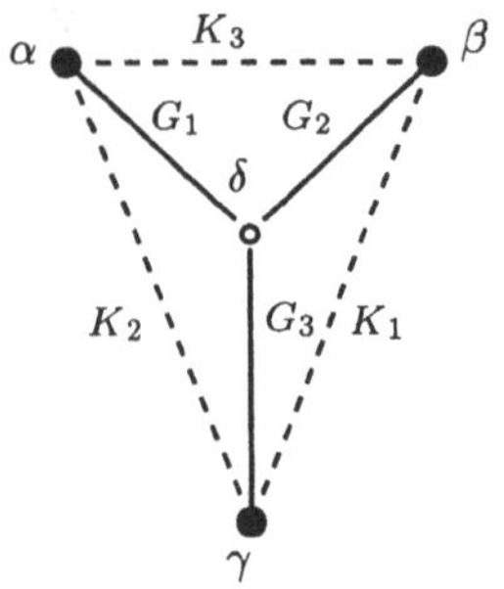

Fig. 8.4. The star-triangle transformation with unequal interactions.

Taking (i, j, ℓ) to be a permutation of $(1, 2, 3)$, the parameters on the right-hand side of (8.41) can now be expressed in terms of G_1, G_2 and G_3 by

$$\exp(4K_i) = \frac{\mathsf{x}\mathsf{x}_i}{\mathsf{x}_j\mathsf{x}_\ell}, \qquad i = 1, 2, 3, \qquad \mathsf{R} = 2\sqrt{\mathsf{X}}. \tag{8.44}$$

The aim is now to express (8.44) in an alternative form and to obtain inverse relations giving the honeycomb parameters G_i in terms of the triangular parameters K_i. Let

$$\mathsf{c}_i = \cosh(2G_i), \qquad \mathsf{s}_i = \sinh(2G_i), \qquad i = 1, 2, 3. \tag{8.45}$$

Standard hyperbolic function relations then give

$$\begin{aligned} \mathsf{x}\mathsf{x}_i &= \tfrac{1}{2}[\cosh(2G_i) + \cosh(2G_j + 2G_\ell)] = \tfrac{1}{2}(\mathsf{c}_i + \mathsf{c}_j\mathsf{c}_\ell + \mathsf{s}_j\mathsf{s}_\ell), \\ \mathsf{x}_j\mathsf{x}_\ell &= \tfrac{1}{2}[\cosh(2G_i) + \cosh(2G_j - 2G_\ell)] = \tfrac{1}{2}(\mathsf{c}_i + \mathsf{c}_j\mathsf{c}_\ell - \mathsf{s}_j\mathsf{s}_\ell). \end{aligned} \tag{8.46}$$

By putting $i = 1$, $j = 2$, $\ell = 3$ in (8.46) and substituting in the expression for X given by (8.42),

$$\mathsf{X} = \frac{1}{2}\sqrt{2\mathsf{c}_1\mathsf{c}_2\mathsf{c}_3 + \mathsf{c}_1^2 + \mathsf{c}_2^2 + \mathsf{c}_3^2 - 1}. \tag{8.47}$$

From (8.43)

$$\begin{aligned} \mathsf{R}^2 \exp(2K_i) &= 4\mathsf{x}\mathsf{x}_i, \\ \mathsf{R}^2 \exp(-2K_i) &= 4\mathsf{x}_j\mathsf{x}_\ell \end{aligned} \tag{8.48}$$

and then, using (8.46), it is easy to prove that

$$\sinh(2K_i) = \frac{\mathsf{s}_j\mathsf{s}_\ell}{2\mathsf{X}}, \qquad \cosh(2K_i) = \frac{\mathsf{c}_i + \mathsf{c}_j\mathsf{c}_\ell}{2\mathsf{X}}, \qquad i = 1, 2, 3. \tag{8.49}$$

It can now be shown (Example 8.5) that

$$\cosh(2G_i) = \cosh(2K_j)\cosh(2K_\ell) + \sinh(2K_j)\sinh(2K_\ell)\coth(2K_i), \qquad i = 1, 2, 3. \tag{8.50}$$

This is the inverse relation giving G_i in terms of K_1, K_2 and K_3. From the first relation of (8.49),

$$\sinh(2K_i)\sinh(2G_i) = \frac{\mathsf{s}_i\mathsf{s}_j\mathsf{s}_\ell}{2\mathsf{X}} = \frac{\mathsf{s}_1\mathsf{s}_2\mathsf{s}_3}{2\mathsf{X}}, \tag{8.51}$$

since (i, j, ℓ) is a permutation of $(1, 2, 3)$. The last expression is clearly independent of i and hence (8.51) can be written

$$\sinh(2K_i)\sinh(2G_i) = k^{-1}\,, \qquad i = 1, 2, 3, \tag{8.52}$$

which is a symmetrical form of the relation between the honeycomb and triangle parameters. The last expression of (8.51) gives k^{-1} in terms of the honeycomb parameters G_i. Before obtaining k in terms of the triangle parameters we define

$$\begin{aligned} \bar{c}_i &= \cosh(2K_i)\,, \\ & \qquad\qquad i = 1, 2, 3. \\ \bar{s}_i &= \sinh(2K_i)\,, \end{aligned} \tag{8.53}$$

Then, putting $i = 1$ in (8.52) and (8.50),

$$k^2 = \frac{1}{(\bar{c}_2\bar{c}_3\bar{s}_1 + \bar{s}_2\bar{s}_3\bar{c}_1 - \bar{s}_1)(\bar{c}_2\bar{c}_3\bar{s}_1 + \bar{s}_2\bar{s}_3\bar{c}_1 + \bar{s}_1)}\,. \tag{8.54}$$

Using standard hyperbolic function relations, k can then be expressed in symmetrical form as

$$k = \frac{(1-\tau_1^2)(1-\tau_2^2)(1-\tau_3^2)}{4[(\tau_1+\tau_2\tau_3)(\tau_2+\tau_3\tau_1)(\tau_3+\tau_1\tau_2)(1+\tau_1\tau_2\tau_3)]^{1/2}}\,, \tag{8.55}$$

where

$$\tau_i = \tanh(K_i)\,. \tag{8.56}$$

8.6 A Linear Relation for Correlations

Let $p(\alpha, \beta, \gamma)$ be the probability that the spin variables on the triangle of sites shown in Fig. 8.3 or Fig. 8.4 have given values α, β and γ respectively. Then

$$p(\alpha, \beta, \gamma) = \frac{Z(\alpha, \beta, \gamma)}{Z}\,, \tag{8.57}$$

where $Z(\alpha, \beta, \gamma)$ denotes the part of the partition function Z corresponding to the given α, β and γ. It is important, in this section and the next one, to note that $p(\alpha, \beta, \gamma)$ (and indeed the distribution probability for any group of black sites) is not affected by a star-triangle transformation since the latter involves summation over white site spin variables only. If $p(\alpha, \beta, \gamma, \delta)$ similarly denotes a probability distribution for the set of four sites

$$p(\alpha, \beta, \gamma, \delta) = p(\delta|\alpha, \beta, \gamma)p(\alpha, \beta, \gamma) \tag{8.58}$$

where $p(\delta|\alpha, \beta, \gamma)$ is the conditional probability of δ taking a given value for prescribed α, β and γ. Since spin δ interacts with spins α, β and γ only and since $p(1|\alpha, \beta, \gamma) + p(-1|\alpha, \beta, \gamma) = 1$,

$$\begin{aligned} p(\delta|\alpha, \beta, \gamma) &= \frac{\exp[\delta(G_1\alpha + G_2\beta + G_3\gamma)]}{2\cosh(G_1\alpha + G_2\beta + G_3\gamma)} \\ &= \frac{1}{2}[1 + \delta\tanh(G_1\alpha + G_2\beta + G_3\gamma)]\,. \end{aligned} \tag{8.59}$$

Let

$$\tanh(G_1\alpha + G_2\beta + G_3\gamma) = w_1\alpha + w_2\beta + w_3\gamma - w\alpha\beta\gamma\,, \tag{8.60}$$

where w_1, w_2, w_3 and w are parameters determined below. Equation (8.60) is invariant under a simultaneous change of sign of α, β and γ and hence, like (8.41) above, is equivalent to four independent relations. In terms of the variables

$$\begin{aligned} y &= \sinh(G_1 + G_2 + G_3)\,, \\ y_i &= \sinh(G_1 + G_2 + G_3 - 2G_i)\,, \qquad i = 1,2,3, \end{aligned} \tag{8.61}$$

these are

$$\begin{aligned} y &= x(w_1 + w_2 + w_3 - w)\,, \\ y_1 &= x_1(-w_1 + w_2 + w_3 + w)\,, \\ y_2 &= x_2(w_1 - w_2 + w_3 + w)\,, \\ y_3 &= x_3(w_1 + w_2 - w_3 + w)\,, \end{aligned} \tag{8.62}$$

where x and x_i are defined by (8.42). The coefficients w and w_i must now be evaluated. From (8.62),

$$\begin{aligned} 4w &= -\frac{y}{x} + \frac{y_1}{x_1} + \frac{y_2}{x_2} + \frac{y_3}{x_3} \\ &= \frac{-(yx_1 - y_1x)x_2x_3 + (y_2x_3 + y_3x_2)xx_1}{X^2}\,. \end{aligned} \tag{8.63}$$

Using (8.46) with standard hyperbolic function relations,

$$w = \frac{s_1(xx_1 - x_2x_3)}{4X^2} = \frac{s_1s_2s_3}{4X^2} = \frac{2X}{s_1s_2s_3}\left(\frac{s_2s_3}{2X}\right)\left(\frac{s_3s_1}{2X}\right)\left(\frac{s_1s_2}{2X}\right)\,. \tag{8.64}$$

Similarly,

$$w_1 = \frac{1}{4}\left(\frac{y}{x} - \frac{y_1}{x_1} + \frac{y_2}{x_2} + \frac{y_3}{x_3}\right) = \frac{s_1(c_1 + c_2c_3)}{4X^2} = \frac{w(c_1 + c_2c_3)}{s_2s_3}\,. \tag{8.65}$$

Then the last expression of (8.51) gives k^{-1}, using (8.49) and, generalizing from w_1 to w_i, we have

$$\begin{aligned} w &= k\sinh(2K_1)\sinh(2K_2)\sinh(2K_3)\,, \\ w_i &= w\coth(2K_i), \qquad i = 1,2,4\,. \end{aligned} \tag{8.66}$$

Combining (8.58), (8.59) and (8.60),

$$p(\alpha,\beta,\gamma,\delta) = \frac{1}{2}[1 + \delta(w_1\alpha + w_2\beta + w_3\gamma - w\alpha\beta\gamma)]p(\alpha,\beta,\gamma)\,. \tag{8.67}$$

Now multiply by $\gamma\delta$ and sum over $\alpha,\beta,\gamma,\delta = \pm1$. Since $\delta^2 = \gamma^2 = 1$,

$$\sum_{\{\alpha,\beta,\gamma,\delta\}} \gamma\delta p(\alpha,\beta,\gamma,\delta) = \frac{1}{2} \sum_{\{\alpha,\beta,\gamma,\delta\}} (\gamma\delta + w_1\alpha\gamma + w_2\beta\gamma + w_3 - w\alpha\beta) p(\alpha,\gamma)\,, \tag{8.68}$$

which gives the linear relation between correlations

$$\langle\gamma\delta\rangle = w_1\langle\alpha\gamma\rangle + w_2\langle\beta\gamma\rangle + w_3 - w\langle\alpha\beta\rangle\,. \tag{8.69}$$

The expectation $\langle\gamma\delta\rangle$ is an nearest-neighbour correlation for the honeycomb Ising model with parameters G_1, G_2 and G_3. Each expectation on the right-hand side can be regarded either as a second-neighbour correlation for the honeycomb lattice or as a nearest-neighbour correlation for a triangular lattice Ising model with parameters K_1, K_2, K_3 related to G_1, G_2, G_3 by the equations in Sect. 8.5.[5]

8.7 Baxter and Enting's Transformation and the Functional Equation

From Fig. 8.3 it can be seen that instead of performing the star-triangle transformation by summing over the spin variables associated with the white sites the summation could equally well have been taken over those associated with the black sites. In other words, up-pointing stars could have been used instead of down-pointing stars. In the reverse triangle-star transformation there is similarly a choice between placing new spins in down-pointing or up-pointing triangles. These alternatives are exploited very ingeniously in a transformation illustrated in Fig. 8.5. The starting point is the honeycomb lattice, illustrated by the full lines in Fig. 8.5(a), which is regarded as wound round a cylinder with the right-hand side joined to the left. The central row of vertical edges containing sites i, j, m and n is denoted by R. A star-triangle transformation is performed using up-pointing stars above R and down-pointing stars below R and producing the edges represented by dotted segments in Fig. 8.5(a). The new lattice consists of the unchanged vertical edges of R, the dotted edges and the original honeycomb edges at the upper and lower boundaries. It embodies a row of rectangular faces containing R, triangular faces above and below R and, at the boundaries, kite-shaped quadrilaterals formed by the original honeycomb boundary edges and the top and bottom rows of (dotted) triangle edges. The next step is a triangle-star transformation in which a new spin is put into all down-pointing triangles above R and all up-pointing triangles below R. The result is the lattice of Fig. 8.5(b). Both the above transformations leave untouched the original

[5] The derivation of (8.69) given here is that of Baxter and Enting (1978), but it can also be obtained by putting $s_g = s_3$ in equation (81) of Fisher (1959a). The expressions for w and w_i, though not the derivation above, are due to Baxter and Enting.

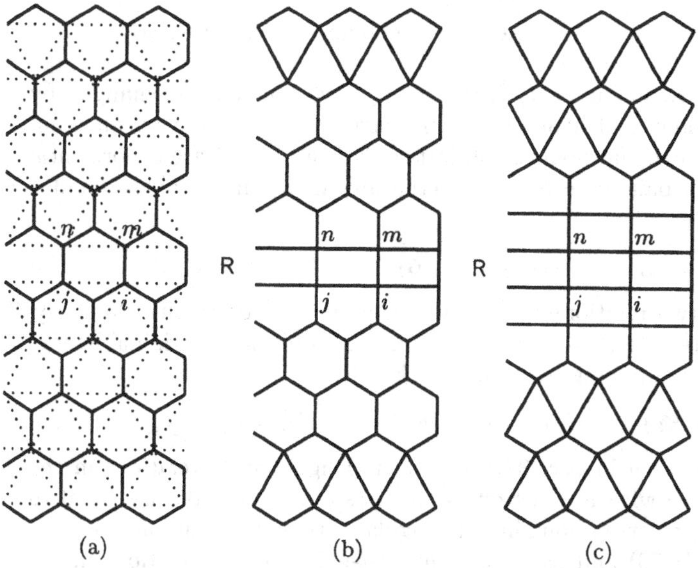

Fig. 8.5. The effect of repeated star-triangle and triangle-star transformations on the honeycomb lattice. (Reprinted from Baxter and Enting (1978), by permission of the publisher IOP Publishing Ltd.)

sites of R (including i, j, m and n) and preserve symmetry about R. This will also be true of all subsequent transformations. Star-triangle and triangle-star transformations of the kind described above are now performed alternately, each pair of transformations producing a new row of rectangular faces above and below R and a new row of upward or downward pointing kites at each boundary. This process is continued until the entire lattice is occupied by a rectangular region and two regions of kite-shaped faces, apart from two rows of pentagons where the regions meet. This situation is shown in Fig. 8.5(c).

Suppose that the interaction parameters for the original hexagonal lattice are G_1, G_2 and G_3, as shown in Fig. 8.4, and that they are all positive. The interaction parameter for vertical edges in the rectangular parts of the final lattice is then G_3 and, for horizontal edges, K_3, where K_3 is given by (8.44) or (8.49) with $i = 3$. The rectangular region is thus equivalent to a square lattice with unequal interactions in the two mutually perpendicular lattice directions. In the thermodynamic limit, the numbers of faces in each row and column of the original honeycomb lattice tend to infinity and thus so do the numbers of faces in each row and column of the rectangular region. The sites i, j, m and n lie deep in the interior of the latter and hence the spin correlations between them must be equal to those of the square lattice. Let $g(K, G)$ denotes the horizontal nearest-neighbour correlation in a square lattice with interaction parameter K in the horizontal direction and G in the vertical direction then $\langle \sigma_i \sigma_j \rangle = g(K_3, G_3)$. By considering a 90° rotation of the square lattice, the vertical nearest-neighbour correlation is $g(G, K)$ and thus $\langle\sigma_i\sigma_m\rangle = g(G_3, K_3)$.

The sites of R, including i, j, m and n, have remained unchanged by the transformations and hence the spin distribution probabilities on them are those of the original honeycomb lattice, where i and j are a horizontal second-neighbour pair corresponding to the sites with spin variables α and β in Fig. 8.4. Thus

$$\langle\alpha\beta\rangle = g(K_3, G_3)\,, \qquad \langle\beta\gamma\rangle = g(K_1, G_1)\,, \qquad \langle\gamma\alpha\rangle = g(K_2, G_2)\,, \tag{8.70}$$

where the last two relations follow by symmetry considerations. Similarly, since i and m are a vertical pairs of neighbouring sites in the original honeycomb lattice corresponding to δ and γ in Fig. 8.4,

$$\langle\gamma\delta\rangle = g(G_3, K_3)\,, \qquad \langle\alpha\delta\rangle = g(G_1, K_1)\,, \qquad \langle\beta\delta\rangle = g(G_2, K_2)\,. \tag{8.71}$$

However, as was seen in Sec. 8.5, the second-neighbour correlations in the honeycomb lattice with interaction parameters G_1, G_2 and G_3 are equal to nearest-neighbour correlations in a triangular lattice with interactions K_1, K_2 and K_3. So (8.70) and (8.71) give nearest-neighbour correlations in the triangular and honeycomb lattices respectively in terms of those of a square lattice with unequal interactions in the two lattice directions. The parameters G_i and K_i are connected by the relations given in Sect. 8.5 and, from (8.49), our assumption that the G_i are all positive implies that the K_i are also all positive.

Substitution of (8.70) and (8.71) into the linear correlation relation (8.69) yields the functional equation

$$g(G_3, K_3) = w_1 g(K_2, G_2) + w_2 g(K_1, G_1) + w_3 - w g(K_3, G_3)\,. \tag{8.72}$$

Baxter and Enting (1978) showed that (8.72) effectively determines the form of $g(K, G)$ and thus yields the nearest-neighbour correlations and hence the configurational energies for the zero-field Ising model on the square, triangular and honeycomb lattices. In the next section an outline is given of the derivation of $g(K, G)$ from equation (8.72) following the general lines of the proof given by Baxter and Enting.

8.8 The Solution of the Functional Equation

In terms of the square lattice nearest-neighbour correlation $g(K, G)$ we define

$$f(K|k) = \frac{g(K, G)}{\coth(2K)}\,, \tag{8.73}$$

where K is termed the *amplitude* of $f(K|k)$ and

$$k = \frac{1}{\sinh(2K)\sinh(2G)} \tag{8.74}$$

is its *modulus*. For fixed k, $K = \infty$ implies $G = 0$ and vice versa. Now since $g(K, G)$ is $\langle\sigma_i\sigma_j\rangle$ for a pair of neighbouring sites with interaction parameter K, $g(\infty, 0) = 1$ and $g(0, \infty) = 0$. It follows that, where k is kept fixed,

$$f(\infty|k) = 1, \qquad f(0|k) = 0. \tag{8.75}$$

8.8.1 A Preliminary Result for $f(K|k)$

Now substitute (8.73) in the functional equation (8.72), noting that, by (8.52), the modulus k is the same for all terms. After using the second relation of (8.66)

$$\begin{aligned}\coth(2G_3)f(G_3|k) = -w\{&f(K_3|k)[\coth(2K_1)\coth(2K_2) + \coth(2K_3)]\\ &- \coth(2K_1)\coth(2K_2)[f(K_1|k) + f(K_2|k)\\ &+ f(K_3|k)] - \coth(2K_3)\}.\end{aligned} \tag{8.76}$$

From (8.50) and (8.52) with $i = 3$ and the first relation of (8.66),

$$\coth(2G_3) = w[\coth(2K_1)\coth(2K_2) + \coth(2K_3)]. \tag{8.77}$$

We now use (8.77) to substitute for $w\coth(2K_3)$ on the right-hand side of (8.76). After using (8.66) for w, (8.76) can then be rearranged to give

$$\begin{aligned}b(K_3|k)\mathrm{sech}(2K_1)\mathrm{sech}(2K_2)\mathrm{sech}(2K_3) = k[&f(K_1|k) + f(K_2|k)\\ &+ f(K_3|k) - 1],\end{aligned} \tag{8.78}$$

where

$$b(K_3|k) = \coth(2K_3)\coth(2G_3)[f(K_3|k) + f(G_3|k) - 1]. \tag{8.79}$$

That G_3 can be omitted from the arguments of $b(K_3|k)$ follows from (8.52) which gives G_3 as a function of K_3 and k. However, by permuting the interactions in Fig. 8.4 we could derive (8.78) with K_2 in place of K_3. Since all other factors in (8.78) are symmetric in K_1, K_2 and K_3 it can be deduced that

$$b(K_2|k) = b(K_3|k). \tag{8.80}$$

From (8.55) and (8.56) it is possible for K_2 to vary with k and K_3 fixed. In fact with $K_3 = 0$ and $\tau_3 = 0$, K_2 can be made to vary over its whole range from 0 to ∞ by varying τ_1 from 1 to 0. Hence $b(K|k)$ must be independent of the value of its argument K and can be written as $b(k)$.

The $f(K|k)$ functions in (8.78) have four distinct arguments K_1, K_2, K_3 and G_3. We aim to derive a relation for f with one argument only.

We now wish to obtain an expression for $f'(K|k)$, the partial derivative of $f(K|k)$ with respect to K at constant k. Taking K_3 and k as constant, differentiation of (8.78) with respect to K_1 gives

$$\frac{\partial K_2}{\partial K_1} = -\frac{kf'(K_1|k) + 2b(k)\tanh(2K_1)\mathrm{sech}(2K_1)\mathrm{sech}(2K_2)\mathrm{sech}(2K_3)}{kf'(K_2|k) + 2b(k)\tanh(2K_2)\mathrm{sech}(2K_1)\mathrm{sech}(2K_2)\mathrm{sech}(2K_3)}. \tag{8.81}$$

With $\mathcal{K}_3$ and k constant, $\mathcal{G}_3$ is also constant and differentiation of the expression for $\cosh(2\mathcal{G}_3)$ given by (8.50), together with substitution from the corresponding expressions for $\cosh(2\mathcal{G}_1)$ and $(\cosh 2\mathcal{G}_2)$, yields

$$\frac{\partial \mathcal{K}_2}{\partial \mathcal{K}_1} = -\frac{\sinh(2\mathcal{K}_2)\cosh(2\mathcal{G}_2)}{\sinh(2\mathcal{K}_1)\cosh(2\mathcal{G}_1)} . \tag{8.82}$$

Eliminating $\partial \mathcal{K}_2/\partial \mathcal{K}_1$ between (8.81) and (8.82) and using (8.50) gives

$$a(\mathcal{K}_1|k) = a(\mathcal{K}_2|k) , \tag{8.83}$$

where

$$a(\mathcal{K}|k) = \frac{1}{2}f'(\mathcal{K}|k)\coth(2\mathcal{G}) + b(k)\tanh^2(2\mathcal{K}). \tag{8.84}$$

Equation (8.83) is true for all values of $\mathcal{K}_1$ and $\mathcal{K}_2$, and reasoning similar to that used for $b(k)$ shows that $a(\mathcal{K}|k)$ is likewise independent of the modulus $\mathcal{K}$ and may be written as $a(k)$. Then (8.84) gives

$$f'(\mathcal{K}|k) = \frac{2[a(k) - b(k)\tanh^2(2\mathcal{K})]}{\coth(2\mathcal{G})} = \frac{2[a(k) - b(k)\tanh^2(2\mathcal{K})]}{[1 + k^2\sinh^2(2\mathcal{K})]^{1/2}} . \tag{8.85}$$

Integration of (8.85) yields

$$f(\mathcal{K}|k) = a(k)A(\mathcal{K}|k) - b(k)B(\mathcal{K}|k) , \tag{8.86}$$

where, since $f(0|k) = 0$,

$$\begin{aligned} A(\mathcal{K}|k) &= \int_0^{2\mathcal{K}} \frac{\mathrm{d}x}{\sqrt{1 + k^2\sinh^2(x)}} , \\ B(\mathcal{K}|k) &= \int_0^{2\mathcal{K}} \frac{\tanh^2 x \mathrm{d}x}{\sqrt{1 + k^2\sinh^2(x)}} . \end{aligned} \tag{8.87}$$

Since $f(\infty|k) = 1$, (8.86) gives

$$1 = a(k)A(\infty|k) - b(k)B(\infty|k) . \tag{8.88}$$

From (8.73), (8.86) and (8.88),

$$g(K, L) = \coth(2\mathcal{K})\left\{\frac{A(\mathcal{K}|k)}{A(\infty|k)} - b(k)\left[B(\mathcal{K}|k) - B(\infty|k)\frac{A(\mathcal{K}|k)}{A(\infty|k)}\right]\right\} . \tag{8.89}$$

We thus have the required formula for the nearest-neighbour correlation once we have formulae for $A(\infty|k)$, $B(\infty|k)$ and $b(k)$.

8.8.2 Expressions for $A(\infty|k)$ and $B(\infty|k)$

We now derive formulae for $A(\infty|k)$ in terms of $A(\mathrm{K}|k)$ and $A(\mathrm{G}|k)$ and $B(\infty|k)$ in terms of $B(\mathrm{K}|k)$ and $B(\mathrm{G}|k)$, which depend solely on the relation between k, K and G given in (8.74).

The first step is to put $y = \sinh x$ in (8.87) which yields

$$A(\mathrm{K}|k) = \int_0^{\sinh(2\mathrm{K})} \frac{\mathrm{d}y}{\sqrt{(1+y^2)(1+k^2y^2)}}\,,$$
$$B(\mathrm{K}|k) = \int_0^{\sinh(2\mathrm{K})} \frac{y^2\mathrm{d}y}{\sqrt{(1+y^2)^3(1+k^2y^2)}}\,. \tag{8.90}$$

For $A(\mathrm{K}|k)$, a second transformation $y = 1/(kz)$ then gives

$$A(\mathrm{K}|k) = \int_{1/(k\sinh(2\mathrm{K}))}^{\infty} \frac{\mathrm{d}z}{\sqrt{(1+z^2)(1+k^2z^2)}}$$
$$= \int_{\sinh(2\mathrm{G})}^{\infty} \frac{\mathrm{d}y}{\sqrt{(1+y^2)(1+k^2y^2)}}\,, \tag{8.91}$$

where we have replaced the dummy variable z by y in the last expression. This is equivalent to

$$A(\infty|k) = A(\mathrm{K}|k) + A(\mathrm{G}|k)\,. \tag{8.92}$$

For $B(\mathrm{K}|k)$, the transformation $y = 1/(kz)$ yields

$$B(\mathrm{K}|k) = \int_{\sinh(2\mathrm{G})}^{\infty} \frac{\mathrm{d}y}{\sqrt{(1+y^2)(1+k^2y^2)^3}}\,. \tag{8.93}$$

Now, replacing K by G in the second relation of (8.90) and integrating by parts,

$$B(\mathrm{G}|k) = -\left[\frac{y}{\sqrt{(1+k^2y^2)(1+y^2)}}\right]_0^{\sinh(2\mathrm{G})}$$
$$+ \int_0^{\sinh(2\mathrm{G})} \frac{\mathrm{d}y}{\sqrt{(1+y^2)(1+k^2y^2)^3}}\,. \tag{8.94}$$

Since $B(\infty|k)$ is the limit of $B(\mathrm{G}|k)$ as $\mathrm{G} \to \infty$ for fixed k, (8.94) yields

$$B(\infty|k) = \int_0^{\infty} \frac{\mathrm{d}y}{\sqrt{(1+y^2)(1+k^2y^2)^3}}\,. \tag{8.95}$$

Adding (8.93) to (8.94) and substituting for k^2 in the integrated part from (8.73)

$$B(\infty|k) = B(\mathrm{K}|k) + B(\mathrm{G}|k) + \tanh(2\mathrm{K})\tanh(2\mathrm{G}). \tag{8.96}$$

Expressions for $A(\mathrm{K}|k)$ and $B(\mathrm{K}|k)$ in terms of elliptic integrals are given in Appendix A.2.

8.8.3 An Expression for $b(k)$

The square lattice Ising model with horizontal and vertical interaction parameters K and G respectively has a zero-field Hamiltonian given by

$$\widehat{H}/T = -K \sum_{\{i,j\}}^{\text{(n.n.h.)}} \sigma_i \sigma_j - G \sum_{\{i,j\}}^{\text{(n.n.v.)}} \sigma_i \sigma_j \,, \tag{8.97}$$

where the first and second summations are over horizontal and vertical nearest-neighbour pairs respectively. Since $g(K,G)$ and $g(G,K)$ denote horizontal and vertical correlations respectively,

$$g(K,G) = \frac{1}{N}\frac{\partial \ln Z(K,G)}{\partial K}\,, \qquad g(G,K) = \frac{1}{N}\frac{\partial \ln Z(K,G)}{\partial G}\,, \tag{8.98}$$

where $Z(K,G)$ denotes the configuration partition function for a square lattice with unequal interactions. Hence

$$\frac{\partial g(K,G)}{\partial G} = \frac{1}{N}\frac{\partial^2 \ln Z(K,G)}{\partial K \partial G} = \frac{\partial g(G,K)}{\partial K}\,. \tag{8.99}$$

From (8.73),

$$\begin{aligned} \frac{\partial g(K,G)}{\partial G} &= \coth(2K)\frac{\partial f(K|k)}{\partial k}\frac{\partial k}{\partial G} \\ &= -2k\coth(2K)\coth(2G)\frac{\partial f(K|k)}{\partial k}\,. \end{aligned} \tag{8.100}$$

Derivation of a similar expression for $\partial g(G,K)|\partial K$ and substitution in (8.99) gives

$$\frac{\partial f(K|k)}{\partial k} = \frac{\partial f(G|k)}{\partial k}\,. \tag{8.101}$$

From (8.79), (without the index 3),

$$f(G|k) = \frac{b(k)}{\coth(2K)\coth(2G)} + 1 - f(K|k)\,. \tag{8.102}$$

Differentiating both sides of this relation with respect to k at constant G and using (8.101) and (8.84) gives after some rearrangement,

$$k\frac{\partial f(K|k)}{\partial k} = \frac{1}{2}\left[a(k) - b(k) + K\frac{\mathrm{d}b(k)}{\mathrm{d}k}\right] C(K|k)\,, \tag{8.103}$$

where $C(K|k)$ is defined by (A.26). Also it follows from (8.86) that

$$\begin{aligned} \frac{\partial f(K|k)}{\partial k} &= \frac{\mathrm{d}a(k)}{\mathrm{d}k}A(K|k) - \frac{\mathrm{d}b(k)}{\mathrm{d}k}B(K|k) \\ &\quad + a(k)\frac{\partial A(K|k)}{\partial k} - b(k)\frac{\partial B(K|k)}{\partial k}\,. \end{aligned} \tag{8.104}$$

With some manipulation it can be shown from (8.87) that

$$k\frac{\partial A(K|k)}{\partial k} = B(K|k) - A(K|k) + C(K|k),$$

$$k\frac{\partial[(k')^2 B(K|k)]}{\partial k} = (k')^2 B(K|k) - A(K|k) + C(K|k),$$

(8.105)

where $k' = \sqrt{1-k^2}$ (see (A.6)). Combining (8.103), (8.104) and (8.105) we finally obtain

$$\left[B(K|k) + \frac{1}{2}C(K|k)\right]\left[-k\frac{db(k)}{dk} + a(k) - (1+k^2)(k')^{-2}b(k)\right]$$
$$+ A(K|k)\left[k\frac{da(k)}{dk} - a(k) + (k')^{-2}b(k)\right] = 0.$$

(8.106)

$A(K|k)$ and $B(K|k) + \frac{1}{2}C(K|k)$ are linearly independent and since (8.106) is true for all K it follows that the coefficients of $A(K|k)$ and $B(K|k)+\frac{1}{2}C(K|k)$ are zero. This yields a pair of first-order differential equations for $a(k)$ and $b(k)$, or equivalently for $a(k)$ and $c(k) = (k')^{-2}b(k)$.

$$k\frac{da(k)}{dk} = a(k) - c(k),$$

$$k(k')^2\frac{dc(k)}{dk} = a(k) - (k')^2 c(k).$$

(8.107)

For $k < 1$ these equations can be solved in terms of complete elliptic integrals. Eliminating $a(k)$,

$$\frac{d}{dk}\left[k(k')^2\frac{dc(k)}{dk}\right] - kc(k) = 0.$$

(8.108)

From (A.10), the general solution of (8.108) gives

$$b(k) = (k')^2 c(k) = (k')^2[\zeta\mathcal{K}(k) + \eta\mathcal{K}(k')],$$

(8.109)

where ζ and η are constants. By definition, $|g(K,G)| \leq 1$ for all K and G, and hence all K and k. However $\mathcal{K}(k') \to \infty$, as $k \to 0$, $(k' \to 1)$. It follows from (8.73), (8.86) and (8.87) that, if $\eta \neq 0$, $|g(K,G)|$ will become unbounded as $k \to 0$ for fixed K. Thus it must be the case that $\eta = 0$. Then, from the first relation of (8.107), (8.109) and (A.8),

$$a(k) = \zeta\mathcal{E}(k), \qquad b(k) = \zeta(k')^2\mathcal{K}(k).$$

(8.110)

The constant ζ can be evaluated by putting $K = \infty$ in (A.24) and substituting it and (8.110) into (8.88) to give

$$\zeta[\mathcal{E}(k)\mathcal{K}(k') + \mathcal{E}(k')\mathcal{K}(k) - \mathcal{K}(k)\mathcal{K}(k')] = 1.$$

(8.111)

From the identity (A.11), $\zeta = 2/\pi$ giving

$$b(k) = 2(k')^2\mathcal{K}(k)/\pi.$$

(8.112)

Since the modulus of an elliptic integral cannot be greater than 1, this procedure must be modified for $k > 1$. It can be shown that then

$$b(m) = -2(m')^2\mathcal{K}(m)/(m\pi), \quad \text{where} \quad m = k^{-1}. \tag{8.113}$$

An expression valid for all k can be obtained by using the modulus

$$k_1 = \frac{2\sqrt{k}}{1+k} = \frac{2\sqrt{m}}{1+m}, \tag{8.114}$$

where $k_1 < 1$ for all $k \neq 1$ and $k_1 = 1$ for $k = 1$. Using Landen's transformation in the form (A.13), (8.112) and (8.113) both take the form

$$b(k) = 2(1-k)\mathcal{K}(k_1)/\pi. \tag{8.115}$$

8.9 Critical Behaviour

For equal interactions in the various lattice directions the configurational energy per site is

$$u = -\frac{1}{2}zJ\langle\sigma_i\sigma_j\rangle_{\text{n.n.}} = -\frac{1}{2}zJg(K,G), \tag{8.116}$$

where $g(K,G)$ is given by (8.89); $K = G$ for the square lattice, and K and G are connected by (8.27) for the triangular lattice. For the honeycomb lattice, $g(K,G)$ is replaced by $g(G,K)$. Since $A(K|k)$ and $B(K|k)$ are analytic except at $K = \infty$, a critical point can arise only from a singularity in $b(k)$. However, $b(k)$ itself is analytic unless $k_1 = 1$. From (A.5), near $k_1 = 1$,

$$\mathcal{K}(k_1) \simeq \frac{1}{2}\ln\left[\frac{16}{(k_1')^2}\right] = \frac{1}{2}\ln\left(\frac{16}{1-k_1^2}\right) \simeq \frac{1}{2}\ln\left(\frac{8}{1-k_1}\right). \tag{8.117}$$

Hence, near $k_1 = 1$, (8.115) gives

$$b(k) \simeq \frac{1-k}{\pi}\ln\left(\frac{8}{1-k_1}\right). \tag{8.118}$$

Now k increases steadily from 0 at $T = 0$ to ∞ at $T = \infty$, passing through the value 1 at a temperature denoted by T_c, while k_1 attains its maximum value of 1 at T_c. Hence $1-k$ is proportional to $T_c - T$ whereas $1 - k_1$ is proportional to $(T_c - T)^2$ near T_c. Hence, from (8.89), (8.116) and (8.118),

$$u = u_c + D(T - T_c)\ln|T - T_c| \tag{8.119}$$

near T_c, where u_c is the value of u at $T = T_c$ and D is a constant appropriate to the particular two-dimensional lattice. So u is continuous at $T = T_c$ but the zero-field configurational heat capacity c_0, which is proportional to du/dT, diverges to infinity as $-\ln|T - T_c|$, both above and below the critical point. Since $x^z \ln x \to 0$ as $x \to 0$ for any $z > 0$ it follows that the critical indices $\alpha = \alpha' = 0$. Since $\alpha = \alpha' = 0$ also for the mean-field theory, where c_0 has a finite discontinuity, it is obvious that these critical index values can correspond to quite different types of behaviour.

The analysis presented above confirms the existence of a unique critical point for a given lattice. It is easy to verify that the singularity condition $k = 1$

gives the same values of T_c as those deduced from transformation theory, on the assumption of a unique critical point for each lattice, in Sect. 8.3 and 8.4. It is convenient to adopt a notation used by Domb (1960) in which the high- and low-temperature parameter values connected by a transformation are written as (K, K_{inv}). Thus for the square lattice, $K_{\text{inv}} = K^*$, given by (8.13), for the triangular lattice, $K_{\text{inv}} = K'$, given by (8.32) and for the honeycomb lattice, $G_{\text{inv}} = G'$, given by (8.38). For the square lattice with $K = G$, from (8.74) $k = 1/\sinh^2(2K)$ and hence, by (8.13),

$$k = \frac{\sinh(2K_{\text{inv}})}{\sinh(2K)} . \tag{8.120}$$

For the triangular and honeycomb lattices k is given by (8.74) where K and G are connected by the star-triangle transformation (8.27). However, for the triangular lattice $\sinh(2G) = 1/\sinh(2K')$ and hence (8.120) applies. Again for the honeycomb lattice, $\sinh(2K) = 1/\sinh(2G')$ and (8.120) thus applies if J/T for this lattice is also denoted by K. The critical point condition used in Sects. 8.3 and 8.4 was $K = K_{\text{inv}}$, and it follows immediately from (8.120) that this is equivalent to $k = 1$. Values of $T_c/(zJ)$ are given in Table 8.1 together with those of other critical parameters, discussed below. Corresponding quantities for $d = 3$ lattices, obtained by series methods and given in Domb (1974), are included for comparison, as are results from standard approximations.

The expression for the energy u_c at the critical point given by (8.21), applies for the triangular and honeycomb lattices as well as for the square lattice (Example 8.3) and it will be shown in Sects. 8.10 and 8.11 that (8.21) follows from (8.89) and (8.116). At $T = \infty$, where the spin distribution is random, $u = 0$ whereas, in the ground state at $T = 0$, $u = -\frac{1}{2}zJ$. The ratio $-u_c/(\frac{1}{2}zJ)$, given in Table 8.1, is thus the proportion of the total energy change from $T = \infty$ to $T = 0$ which occurs when $T > T_c$ and is therefore due entirely to short-range ordering effects. In the mean-field approximation, u_c has the random value zero, as can be seen from (3.14).

It is useful to define a partition function λ per spin (or per site) by

$$Z = \lambda^N . \tag{8.121}$$

Since the constant magnetic field distribution is being used with $\mathcal{H} = 0$ it follows from (2.59) that, with $\mathcal{H} = 0$,

$$f = u - Ts = -T\ln(\lambda) , \tag{8.122}$$

where f and s are respectively the configurational free energy and entropy per site. Thus

$$s = u/T + \ln(\lambda). \tag{8.123}$$

At $T = \infty$, where the spin distribution is random, $\lambda = 2$ and $u = 0$ so that $s = \ln(2)$. At $T = 0$, where the spins are completely ordered, $s = 0$. Hence if s_c denotes the value of s at the critical point $T = T_c$ the ratio $(\ln(2) - s_c)/\ln(2)$, values of which are presented in Table 8.1, gives the proportion of the total entropy change from $T = \infty$ to $T = 0$ which occurs when $T > T_c$. Exact expressions for s_c are obtained in Sects. 8.10 and 8.11.

The direct method of deriving the spontaneous magnetization is to use (3.11), as was done in mean-field theory. This requires an expression for the equilibrium magnetization as a function of T and $\mathcal{H}$. However such an expression is not available since the exact partition function is known only for $\mathcal{H} = 0$. One way out is to calculate the dependence of the zero-field correlations $\langle\sigma_i\sigma_j\rangle$ on the separation between sites i and j and to use (3.2). We shall not attempt to present the long and complicated algebraic derivation (McCoy and Wu 1973) but simply give the compact and elegant result

$$m_s = (1-k^2)^{1/8} = [1-\sinh^2(2K_{\text{inv}})\sinh^2(2K)]^{1/8}, \tag{8.124}$$

which applies for the square, triangular and honeycomb lattices. Since $1-k$ is proportional to $T_c - T$ for $T_c - T \ll T_c$ this immediately yields $\beta = \frac{1}{8}$, where the critical exponent β is defined in (3.55).

To derive the critical exponent γ, defined by (3.58), we need the limit of $\partial m/\partial\mathcal{H}$ as $\mathcal{H} \to 0$, just above the critical point. Since the partition function is available only for $\mathcal{H} = 0$, a direct derivation of $\partial m/\partial\mathcal{H}$ is impossible. However, it is possible to use (2.115) which was shown in the one-dimensional case to yield the same expression for $(\partial m/\partial\mathcal{H})_{\mathcal{H}=0}$ as the direct method. For $T \geq T_c$ the second member of (2.115) is true for any dimension and with (2.116) gives

$$T\frac{\partial m}{\partial\mathcal{H}} = 1 + \sum_{\{k\neq j\}} \langle\sigma_j\sigma_k\rangle\,, \tag{8.125}$$

where j denotes a typical site and the summation is over all other sites. It has been assumed that in the thermodynamic limit all sites can be regarded as equivalent. Suppose that the distance between sites j and ℓ, in lattice constant units, is r ($r = 1$ for a nearest-neighbour pair) and that the angle made by the vector from j to ℓ with a fixed direction is θ. Then it is deduced from transfer matrix considerations that, for large r and $T - T_c \ll T_c$,

$$\langle\sigma_j\sigma_\ell\rangle \simeq w(\theta)r^{-1/4}\exp[-D(\theta)(T-T_c)r]\,. \tag{8.126}$$

This implies a large correlation range just above T_c and the first term on the right-hand side of (8.125) is negligible in comparison to the sum. Replacing the latter by an integral

$$T\frac{\partial m}{\partial\mathcal{H}} = \int_0^\infty \mathrm{d}r \int_0^{2\pi} \mathrm{d}\theta\, w(\theta)r^{3/4}\exp[-D(\theta)(T-T_c)r]\,. \tag{8.127}$$

Replacing the variable r by $x = D(T-T_c)r$ and integrating first with respect to x and then with respect to θ,

$$\frac{\partial m}{\partial\mathcal{H}} \sim (T-T_c)^{-7/4} \tag{8.128}$$

and thus $\gamma = \frac{7}{4}$.[6] There remains the difficulty that in a direct derivation of $\partial m/\partial\mathcal{H}$ the thermodynamic limit would in principle be approached before letting $\mathcal{H} \to 0$, whereas in the method discussed, $\mathcal{H}$ is set equal to zero before the thermodynamic limit is taken. However Abraham (1973) has shown that this inversion of limit operations makes no difference. The values $\alpha = 0$, $\beta = \frac{1}{8}$ and $\gamma = \frac{7}{4}$ satisfy the Essam–Fisher scaling law (3.87). If (3.77) is assumed then $\delta = 15$, which is the generally accepted value.

8.10 Thermodynamic Functions for the Square Lattice

Thermodynamic functions for the two-dimensional Ising model are discussed at length by Domb (1960), Green and Hurst (1964), McCoy and Wu (1973) and Baxter (1982a). The fundamental result used here is the expression (8.89) for the nearest-neighbour correlation. For the square lattice with equal interactions in the two lattice directions $K = G$ and (8.92) and (8.96) become

$$A(\infty|k) = 2A(K|k)\,,$$
$$B(\infty|k) = 2B(K|k) + \tanh^2(2K) \tag{8.129}$$

and (8.89) then becomes

$$g(K,K) = \frac{1}{2}\coth(2K)[1 + b(k)\tanh^2(2K)]\,. \tag{8.130}$$

Substituting $k = 1/\sinh^2(2K)$ in the expressions (8.115) for $b(k)$ and (8.114) for k_1, and $z = 4$ in (8.116)

$$u = -J\coth(2K)\left\{1 + \frac{2}{\pi}\mathcal{K}(k_1)[2\tanh^2(2K) - 1]\right\}\,, \tag{8.131}$$

$$k_1 = \frac{2\sinh(2K)}{1 + \sinh^2(2K)} = 2\sinh(2K)\mathrm{sech}^2(2K)\,, \tag{8.132}$$

applicable for the entire temperature range. From (8.118), $b(k) = 0$ at $T = T_c$ and equation (8.21), for the critical value of u, is thus verified for the square lattice. It is of interest to look at the two extremes of the temperature range $T = 0$ and $T = \infty$, for both of which $k_1 = 0$ and hence from (A.4), $2\mathcal{K}(k_1) = \pi$. As $T \to 0$, $K \to \infty$ and hence $\tanh(2K) \to 1$ so that, by (8.131), $u \to -2J$, confirming that the zero-temperature state is completely ordered. The $T \to \infty$ case is more troublesome since $K \to 0$ and $\coth(2K) \simeq (2K)^{-1} \to \infty$. However, using (A.4) with $k_1 \simeq 4K$, the curly bracketed part of (8.131) is, to order K^2,

$$1 + (8K^2 - 1)(1 + 4K^2) \simeq 4K^2\,. \tag{8.133}$$

[6] This derivation is essentially that of Fisher (1959b).

Thus $u/J \simeq -2\mathcal{K} \to 0$, as $T \to \infty$, as would be expected in a random distribution. The configurational heat capacity can easily be deduced from (8.131) (see Example 8.7). Using (2.26), and noting that in zero field the enthalpy per site $h = u$,

$$u = -\frac{J}{N}\frac{\partial \ln(Z)}{\partial \mathcal{K}} = -J\frac{\partial \ln(\lambda)}{\partial \mathcal{K}}, \tag{8.134}$$

where λ is defined by (8.121). The form of u and the condition $\lambda = 2$ at $T = \infty$, ($\mathcal{K} = 0$) completely determine λ since, from (8.134),

$$J \ln(\lambda/2) = -\int_0^{\mathcal{K}} u \mathrm{d}\mathcal{K}. \tag{8.135}$$

It can be verified (Example 8.6) that when substituted in (8.134) the expression

$$\ln(\lambda/2) = \ln\{\cosh(2\mathcal{K})\} + \frac{1}{\pi}\int_0^{\pi/2} \ln\left\{\tfrac{1}{2}[1 + (1 - k_1^2 \sin^2\psi)^{1/2}]\right\} \mathrm{d}\psi \tag{8.136}$$

yields the form of u given by (8.131). Also, putting $\mathcal{K} = 0$ in (8.136) gives $\lambda = 2$. Hence (8.136) is the required integral of (8.131). An alternative formula for the partition function is

$$\begin{aligned}\ln(\lambda/2) &= \frac{1}{2\pi^2}\int_0^{\pi} \mathrm{d}\theta \int_0^{\pi} \mathrm{d}\phi \, \ln\{\cosh^2(2\mathcal{K}) - \sinh(2\mathcal{K})[\cos\theta + \cos\phi]\} \\ &= \ln[\cosh(2\mathcal{K})] + \frac{1}{2\pi^2}\int_0^{\pi} \mathrm{d}\theta \int_0^{\pi} \mathrm{d}\phi \ln\left\{1 - \tfrac{1}{2}k_1[\cos\theta + \cos\phi]\right\}.\end{aligned} \tag{8.137}$$

For $k_1 < 1$ the integrand is finite throughout the domain of integration, but for $k_1 = 1$ it tends to $-\infty$ when $\theta \to 0$, $\phi \to 0$. Since $k_1 < 1$ for $T \neq T_c$ but $k_1 = 1$ at $T = T_c$ the divergence in the integrand is associated with the critical point. It can be shown that (8.137) is equivalent to (8.136) by first changing the variables of integration to

$$\omega_1 = \tfrac{1}{2}(\theta + \phi), \qquad \omega_2 = \tfrac{1}{2}(\theta - \phi). \tag{8.138}$$

The domain of integration transforms to a square in the (ω_1, ω_2) plane bounded by the lines $\omega_1 + \omega_2 = \pi$, $\omega_1 + \omega_2 = 0$, $\omega_1 - \omega_2 = \pi$ and $\omega_1 - \omega_2 = 0$. By using the symmetry of the expression

$$\cos\theta + \cos\phi = 2\cos\omega_1 \cos\omega_2, \tag{8.139}$$

it can be shown that an alternative domain of integration is the rectangle $0 < \omega_1 < \pi$, $0 < \omega_2 < \frac{1}{2}\pi$. Since $\mathrm{d}\theta\mathrm{d}\phi = 2\mathrm{d}\omega_1\mathrm{d}\omega_2$

$$\ln(\lambda/2) = \ln[\cosh(2\mathcal{K})] + \frac{1}{\pi^2}\int_0^{\pi/2} \mathrm{d}\omega_2 \int_0^{\pi} \mathrm{d}\omega_1 \ln[1 - k_1 \cos\omega_1 \cos\omega_2]. \tag{8.140}$$

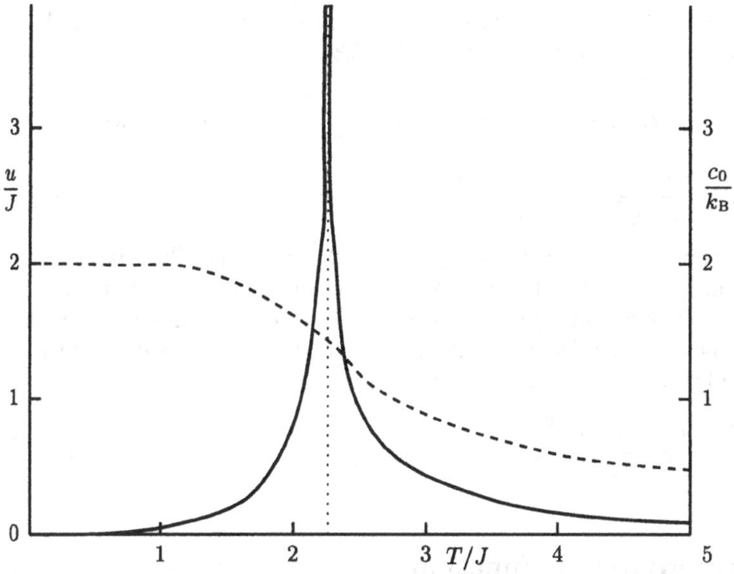

Fig. 8.6. Configurational internal energy (*broken line*) and heat capacity (*continuous line*) for zero field plotted against temperature for the square lattice ferromagnet.

Integration over ω_1 can be performed using the standard result

$$\int_0^\pi \ln(y - x\cos\omega)\mathrm{d}\omega = \pi \ln\left[\frac{y + (y^2 - x^2)^{1/2}}{2}\right], \quad x < y, \tag{8.141}$$

and a trivial transformation from ω_2 to $\psi = \frac{1}{2}\pi - \omega_2$ then recovers (8.136). A different single integral expression is obtained by applying (8.141) directly to (8.137), without the preliminary transformation from (θ, ϕ) to $(\omega_1, \omega_)$. Integration with respect to ϕ yields

$$\begin{aligned}\ln(\lambda/2) &= \ln(\cosh 2K) + \frac{1}{2}\ln\left(\frac{1}{4}k_1\right) \\ &\quad + \frac{1}{2\pi}\int_0^\pi \mathrm{d}\theta \ln\{2k_1^{-1} - \cos\theta + [(2k_1^{-1} - \cos\theta)^2 - 1]^{1/2}\} \\ &= \frac{1}{2}\ln\left[\frac{1}{2}\sinh(2K)\right] + \frac{1}{2\pi}\int_0^\pi \mathrm{d}\theta\, \mathrm{arccosh}(2k_1^{-1} - \cos\theta)\,. \end{aligned} \tag{8.142}$$

This is the form obtained directly by several authors from the largest eigenvalue of the transfer matrix (Domb 1960, Green and Hurst 1964) although in the version of the transfer matrix method given by Baxter (1982a) the result is an expression like (8.136) but with ψ replaced by $\frac{1}{2}\pi - \theta$.

The configurational entropy can be obtained by substituting (8.131) and (8.136) in (8.123). As $T \to 0$ both terms become infinite. However, it may be shown from (8.136) that $\ln\lambda - 2K \to 0$ as $T \to 0$ $(K \to \infty)$ and from (8.131), using (A.4), that $T^{-1}u + 2K \to 0$ as $T \to 0$. Hence, as expected, $s \to 0$. At the critical point, using (8.123), (8.21) and (8.136),

$$\begin{aligned} s_c &= -K_c \coth(2K_c) + \frac{1}{2}\ln[2\cosh^2(2K_c)] + \frac{1}{\pi}\int_0^{\pi/2} d\psi \ln(1+\cos\psi) \\ &= \ln(2) - \frac{\ln(1+\sqrt{2})}{\sqrt{2}} + \frac{1}{\pi}\int_0^{\pi/2} d\psi \ln(1+\cos\psi)\,, \end{aligned} \tag{8.143}$$

the critical parameter values being given by (8.19) and (8.21). The critical entropy ratio is given in Table 8.1. The configurational energy and heat capacity for the square lattice ferromagnet are plotted against temperature in Fig. 8.6. Using (8.13) the general relation (8.124) for the relative magnetization of the square lattice ferromagnet takes the form

$$m_s = \left[\frac{\sinh^4(2K) - 1}{\sinh^4(2K)}\right]^{1/8}. \tag{8.144}$$

8.11 Thermodynamic Functions for the Triangular and Honeycomb Lattices

By (8.70), (8.71) and (8.116) we require $g(K,G)$ for the triangular lattice and $g(G,K)$ for the honeycomb lattice, G and K being connected by the star-triangle relation (8.27) in both cases. It is thus necessary to use for the coefficients $A(K|k)$ and $B(K|k)$ not only the equations (8.92)and (8.96), which depend on the relation (8.74), but also equations (A.39) and (A.41) which arise from the star-triangle relation between K and G. Combining these

$$A(\infty|k) = \begin{cases} 3A(K|k)\,, \\ \frac{3}{2}A(G|k)\,, \end{cases} \tag{8.145}$$

$$B(\infty|k) = \begin{cases} 3B(K|k) + \tanh(2K)\tanh(2G)[1+\tanh(2K)]\,, \\ \frac{3}{2}B(G|k) + \tanh(2K)\tanh(2G)\left[1 - \frac{1}{2}\tanh(2K)\right]. \end{cases} \tag{8.146}$$

Substitution into (8.89) then yields

$$g(K,G) = \frac{1}{3}\coth(2K)\left\{1 + b(k)\tanh(2K)\tanh(2G)\left[1+\tanh(2K)\right]\right\}, \tag{8.147}$$

$$g(G,K) = \frac{2}{3}\coth(2G)\left\{1 + b(k)\tanh(2K)\tanh(2G)\left[1 - \frac{1}{2}\tanh(2K)\right]\right\}. \tag{8.148}$$

From (8.114), (8.115), (8.116) and (8.27) it follows that, for the triangular lattice with equal interactions in the three lattice directions, the configurational energy per site is given by

$$u = -J\coth(2K)\left\{1 + \frac{4\mathcal{K}(k_1)\mathfrak{X}(\mathfrak{X}-3)}{\pi\left[4\mathfrak{X}^{1/2} + (\mathfrak{X}-1)^{3/2}(\mathfrak{X}+3)^{1/2}\right]}\right\}, \tag{8.149}$$

where

$$\mathfrak{X} = \exp(4K),$$

$$k_1 = \frac{4(\mathfrak{X}-1)^{3/4}(\mathfrak{X}^2+3\mathfrak{X})^{1/4}}{4\mathfrak{X}^{1/2} + (\mathfrak{X}-1)^{3/2}(\mathfrak{X}+3)^{1/2}} \tag{8.150}$$

and for the honeycomb lattice, with equal interactions in the three lattice directions,

$$u = -J\coth(2G)\left\{1 + \frac{\mathcal{K}(k_1)(\mathrm{c}-2)(\mathrm{c}+1)}{\pi[\mathrm{s}(\mathrm{c}-1) + (2\mathrm{c}-1)^{1/2}]}\right\}, \tag{8.151}$$

where $\mathrm{c} = \cosh(2G)$, $\mathrm{s} = \sinh(2G)$ and

$$k_1 = \frac{2\mathrm{s}^{1/2}(\mathrm{c}-1)^{1/2}(2\mathrm{c}-1)^{1/4}}{\mathrm{s}(\mathrm{c}-1) + (2\mathrm{c}-1)^{1/2}}. \tag{8.152}$$

These expressions are valid over the entire temperature range. In both cases, k_1 has its maximum value of 1 at $T = T_\mathrm{c}$, giving rise to the critical point singularity, while $k_1 = 0$ at $T = 0$ and $T = \infty$. As for the square lattice, it can be shown that (8.149) and (8.151) give $u = 0$ at $T = \infty$ and $u = -\frac{1}{2}zJ$ at $T = 0$. It is easy to verify that (8.21) applies at $T = T_\mathrm{c}$ in both cases.

The expressions for the free energy are connected by (8.135) with the forms derived above for u. Since we are no longer concerned with the star-triangle transformation (though see Example 8.8) the symbol K is used to denote J/T for both the triangular and honeycomb lattices. For the triangular lattice

$$\ln(\lambda/2) = \frac{1}{8\pi^2}\int_0^{2\pi} d\theta \int_0^{2\pi} d\phi\, \ln\{\bar{\mathrm{c}}^3 + \bar{\mathrm{s}}^3 - \bar{\mathrm{s}}[\cos\theta + \cos\phi + \cos(\theta+\phi)]\}, \tag{8.153}$$

where $\bar{\mathrm{c}} = \cosh(2K)$, $\bar{\mathrm{s}} = \sinh(2K)$. For the honeycomb lattice,

$$\ln(\lambda/2) = \frac{1}{16\pi^2}\int_0^{2\pi} d\theta \int_0^{2\pi} d\phi\, \ln\left\{\bar{\mathrm{c}}^3 + 1 - \bar{\mathrm{s}}^2[\cos\theta + \cos\phi + \cos(\theta+\phi)]\right\} - \frac{1}{4}\ln(2) \tag{8.154}$$

It follows immediately that $\lambda = 2$ for $T = \infty$ ($K = 0$). We shall now sketch a proof that (8.134) is satisfied by the expressions given for λ and u. A first step is to put (8.153) and (8.154) into a common form by writing

$$\ln(\lambda/2) = \frac{1}{2}\ln \bar{\mathrm{s}} + \frac{1}{4}\ln\chi + \frac{\chi}{8\pi^2}\int_0^{2\pi} \mathrm{d}\theta \int_0^{2\pi} \mathrm{d}\phi \, \ln[\eta - \cos\theta - \cos\phi - \cos(\theta+\phi)] \, , \tag{8.155}$$

where

$$\chi = \begin{cases} 1 \text{ for the triangular lattice,} \\ \frac{1}{2} \text{ for the honeycomb lattice,} \end{cases} \tag{8.156}$$

$$\eta = \begin{cases} \dfrac{\bar{\mathrm{c}}^3 + \bar{\mathrm{s}}^3}{\bar{\mathrm{s}}} = \dfrac{\mathfrak{X}^2 + 3}{2(\mathfrak{X} - 1)} \text{ for the triangular lattice,} \\ \dfrac{\bar{\mathrm{c}}^3 + 1}{\bar{\mathrm{s}}^2} = \dfrac{\bar{\mathrm{c}}^2 - \bar{\mathrm{c}} + 1}{\bar{\mathrm{c}} - 1} \text{ for the honeycomb lattice.} \end{cases} \tag{8.157}$$

In both cases, $\eta = \infty$ at $T = 0$ and $T = \infty$, and η has a minimum value of 3 at $T = T_c$. For non-zero $T \neq T_c$ the integrand in (8.155) is therefore finite over the whole domain of integration. The singularity at $T = T_c$ arises because of the infinity in the integrand when $\theta = \phi = 0$ or $\theta = \phi = 2\pi$.

We shall now express $\ln(\lambda/2)$ in terms of an integral with respect to a single variable. First write

$$\cos\theta + \cos(\theta + \phi) = 2\cos\left(\theta + \frac{1}{2}\phi\right)\cos\left(\frac{1}{2}\phi\right) . \tag{8.158}$$

However, $\cos\left(\theta + \frac{1}{2}\phi\right)$ can be replaced by $\cos\theta$ in (8.155) because the integration with respect to θ is taken over a whole period 2π. Then use of (8.141) enables an integration over θ to be performed, giving

$$\ln(\lambda/2) = \frac{1}{2}\ln(\bar{\mathrm{s}}) + \frac{1}{4}\ln(\chi) + \frac{\chi}{2\pi}\int_0^{\pi} \mathrm{d}\theta \, \ln\left[\frac{1}{2}\{\eta - \cos\phi + [(\eta - \cos\phi)^2 - 4\cos^2(\phi/2)]^{1/2}\}\right] . \tag{8.159}$$

Then, with the substitution $x = \cosh\phi$, (8.134) gives

$$\frac{u}{J} = -\coth(2K) - \frac{\chi}{2\pi}\frac{\mathrm{d}\eta}{\mathrm{d}K}\int_{-1}^{1} \frac{\mathrm{d}x}{\sqrt{(1 - x^2)[(x - \eta)^2 - 2x - 2]}} \, , \tag{8.160}$$

which is easily transformed to (8.149) or (8.151) using (A.28). From (8.123), (8.21) and (8.159) the critical point entropy is given by

$$s_c = -K_c\coth(2K_c) + \frac{1}{4}\ln[16\chi\sinh^2(2K_c)] + \frac{\chi}{2\pi}\int_0^{\pi} \mathrm{d}\phi \, \ln\left[\frac{1}{2}\{3 - \cos\phi + [(3 - \cos\phi)^2 - 2\cos\phi - 2]^{1/2}\}\right] . \tag{8.161}$$

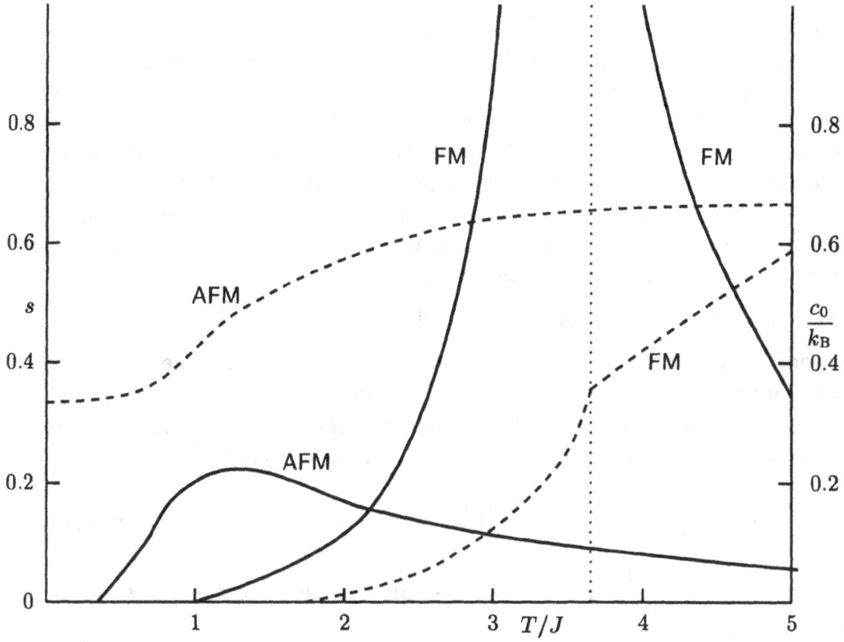

Fig. 8.7. Configurational entropy (*broken lines*) and heat capacity (*continuous lines*) for zero field plotted against temperature for the triangular lattice ferromagnet (FM) and antiferromagnet (AFM).

The entropy ratios presented in Table 8.1 are derived from this. In both cases, $T^{-1}u+\frac{1}{2}zK$ and $\ln\lambda-\frac{1}{2}zK$ tend to zero with T so that $s\to 0$. Plots of entropy and heat capacity against temperature for the triangular ferromagnet and antiferromagnet are shown in Fig. 8.7. For the triangular and honeycomb lattices respectively, (8.124), (8.32) and (8.38) yield

$$m_s=\left[\frac{(\mathfrak{X}+1)^3(\mathfrak{X}-3)}{(\mathfrak{X}-1)^3(\mathfrak{X}+3)}\right]^{1/8}, \tag{8.162}$$

$$m_s=\left[\frac{\bar{c}^3(\bar{c}-2)}{(\bar{c}+1)(\bar{c}-1)^3}\right]^{1/8}. \tag{8.163}$$

8.12 The Antiferromagnet

For the antiferromagnet the nearest-neighbour pair interaction $-J\sigma_i\sigma_j$ of the ferromagnet is changed to $J\sigma_i\sigma_j$ $(J>0)$, which corresponds to the transformation $K\to -K$ $(K>0)$. In Sect. 4.2 it was shown from the form of the Hamiltonian that, for loose-packed lattices, the zero-field configurational partition function is the same for the antiferromagnet as for the ferromagnet. It follows that thermodynamic quantities such as the zero-field internal energy and heat capacity and the critical temperature are also the same. The square and honeycomb lattices are loose-packed, and we now verify that the expressions for the corresponding partition functions are invariant to the transformation $K \to -K$. This follows for the honeycomb lattice since, in (8.154), the quantities $\bar{\mathrm{c}}^3 + 1$ and $\bar{\mathrm{s}}^2$ are unaltered by the transformation. In the first expression for the square lattice partition function, given in (8.137), the change from K to $-K$ alters the integrand to

$$\ln\{\cosh^2(2K) + \sinh(2K)[\cos\theta + \cos\phi]\}\,.$$

However a change of variables to $\theta' = \pi - \theta$, $\phi' = \pi - \phi$ restores the integral to its original form, and the required invariance property is thus demonstrated.

For the triangular lattice, which is close-packed, the situation is quite different. Equation (8.153) for the partition function becomes

$$\begin{aligned}\ln(\lambda/2) &= \frac{1}{8\pi^2}\int_0^{2\pi} \mathrm{d}\theta \int_0^{2\pi} \mathrm{d}\phi\, \ln\{\bar{\mathrm{c}}^3 - \bar{\mathrm{s}}^3 + \bar{\mathrm{s}}[\cos\theta + \cos\phi + \cos(\theta+\phi)]\} \\ &= \tfrac{1}{2}\ln\bar{\mathrm{s}} + \frac{1}{8\pi^2}\int_0^{2\pi} \mathrm{d}\theta \int_0^{2\pi} \mathrm{d}\phi\, \ln[\eta' + \cos\theta + \cos\phi + \cos(\theta+\phi)]\,,\end{aligned} \tag{8.164}$$

where

$$\eta' = \frac{\bar{\mathrm{c}}^3 - \bar{\mathrm{s}}^3}{\bar{\mathrm{s}}} = \frac{\mathfrak{X}^{-2} + 3}{2(1 - \mathfrak{X}^{-1})}. \tag{8.165}$$

At $K = 0$ ($T = \infty$), $\eta' = \infty$, and η' decreases steadily as K increases, attaining its least value of $\frac{3}{2}$ at $K = \infty$ ($T = 0$). Now the minimum value of $\cos\theta + \cos\phi + \cos(\theta+\phi)$ is $-\frac{3}{2}$ for $\theta = \phi = \frac{2}{3}\pi$. Thus a singularity appears in the integrand in (8.164) only at $T = 0$, indicating that there is no non-zero critical temperature. A transformation similar to that producing equation (8.159) gives

$$\begin{aligned}\ln(\lambda/2) = \tfrac{1}{2}\ln(\bar{\mathrm{s}}) + \frac{1}{2\pi}\int_0^{\pi} \mathrm{d}\phi\, \ln\big\{\tfrac{1}{2}\big[\eta' + \cos\phi \\ + \sqrt{(\eta + \cos\phi)^2 - 4\cos^2(\phi/2)}\big]\big\}\,.\end{aligned} \tag{8.166}$$

Then (8.134) and the substitution $x = \cos\phi$ yield

$$\frac{u}{J} = -\coth(2K) - \frac{1}{2\pi}\frac{\mathrm{d}\eta'}{\mathrm{d}K}\int_{-1}^{1} \frac{\mathrm{d}x}{\sqrt{(1 - x^2)[(x + \eta')^2 - 2x - 2]}}\,. \tag{8.167}$$

The quadratic $(x + \eta')^2 - 2x - 2$ is positive for all x when $\eta' > \frac{3}{2}$. It may be written as $(x + \alpha)(x + \beta)$ where $\alpha = \eta' - 1 + \mathrm{i}\sqrt{2\eta' - 3}$ and β is its complex conjugate. Using (A.30),

$$u = -J\coth(2K)\left[1 - \frac{4\mathfrak{X}(3 - \mathfrak{X})^{1/2}}{\pi(1 + \mathfrak{X})^{3/2}}\mathcal{K}(k)\right], \tag{8.168}$$

where now $\mathfrak{X} = \exp(-4\kappa)$ and

$$k = \frac{(1-\mathfrak{X})^{3/2}(3+\mathfrak{X})^{1/2}}{(1+\mathfrak{X})^{3/2}(3-\mathfrak{X})^{1/2}} \,. \tag{8.169}$$

The modulus $k = 0$ at $\kappa = 0$ and increases monotonically with κ to approach the value 1 as $\kappa \to \infty$. There is, therefore, no singularity and no critical point at a non-zero temperature. The heat capacity is a continuous function of T with a maximum and approaches zero as $T \to 0$ or $T \to \infty$. For $\mathfrak{X} \ll 1$, $k' \sim 4/\sqrt{3\mathfrak{X}}$ and using (A.5) it can be seen that $u \to -J$ as $\kappa \to \infty$. It is not difficult to prove the slightly stronger result that

$$\lim_{T\to 0} \frac{u+J}{T} = 0 \,. \tag{8.170}$$

So far the behaviour of the triangular antiferromagnet, with a maximum rather than a singularity in the heat capacity, looks rather like that of a linear Ising model. However, there is an important difference which is crucial to physical understanding of the behaviour of the model. From (8.166), since $\eta' \to \frac{3}{2}$ as $T \to 0$,

$$\lim_{T\to 0}\left(\ln\lambda - \frac{J}{T}\right) = \frac{1}{2\pi}\int_0^{\pi} \mathrm{d}\phi \ln\left\{\frac{3}{2} + \cos\phi + \left[\left(\frac{3}{2}+\cos\phi\right)^2 - 4\cos^2(\phi/2)\right]^{1/2}\right\}. \tag{8.171}$$

From (8.170) and(8.171) it follows that the entropy at $T = 0$ has a non-zero value[7] given by

$$\begin{aligned} s &= \lim_{T\to 0}\left(\ln\lambda + \frac{u}{T}\right) \\ &= \frac{1}{2\pi}\int_0^{\pi} \mathrm{d}\phi \ln\left\{\frac{3}{2} + \cos\phi + \left[\left(\frac{3}{2}+\cos\phi\right)^2 - 2\cos\phi - 2\right]^{1/2}\right\} \\ &= 0.32306 \,. \end{aligned} \tag{8.172}$$

Plots of the zero-field antiferromagnetic entropy and heat capacity against temperature are shown in Fig. 8.7. For the explanation of this zero-point entropy we must look at the configurations giving the zero-temperature energy $-J$ per site. A minimum energy state for the triangular antiferromagnet is one in which each triangle of neighbouring sites is occupied by two + spins and one − spin or vice versa. Out of the total of $3N$ neighbouring site pairs, $2N$ will then be occupied by like spins and N by unlike spins, giving $U = -2NJ + NJ = -NJ$. Hence $u = -J$ is the least possible energy per site. It might be thought that this minimum energy is associated with a highly ordered configuration like that shown in Fig. 8.8(a). However there are many other minimum energy configurations, as can be seen from Fig. 8.8(b) where

[7] We agree with Domb (1960) rather than Wannier (1950) about the magnitude of this quantity.

Fig. 8.8. Minimum energy configurations for the triangular lattice antiferromagnet.

a triangle lattice is divided into a honeycomb sublattice and the sites at the centres of the hexagons, which are shown as open circles. If the honeycomb sublattice is occupied alternately by + and − spins as shown in Fig. 8.8(b), then any distribution of spins on the circle sites gives $u = -J$. There are $2^{N/3}$ minimum energy configurations of this type which thus give a contribution of $\frac{1}{3}\ln(2)$ to s. Other minimum energy configurations make up the entropy value to that given in (8.172).

The ground states are *frustrated* in that, although the interaction between the spins of each nearest-neighbour pair is antiferromagnetic in character, one third of the pairs are forced to adopt a parallel or ferromagnetic configuration. It is natural to assume that the resulting infinite degeneracy of the minimum energy state causes the absence of a transition point for $T > 0$. However, from other examples it appears that the connection between frustration, or infinite degeneracy of the ground state, and the absence of a transition to a long-range ordered state is not a simple one (Mouritsen 1984).

Expressions for $\ln(\lambda/2)$ for the Ising model on the honeycomb, triangular and square lattices with unequal interactions are derived by the Pfaffian method in Volume 2, Chap. 8.

Examples

8.1 From (8.15), show that

$$N - \frac{1}{\coth(2K)}\frac{\partial \ln Z}{\partial K} = -N^* + \frac{1}{\coth(2K^*)}\frac{\partial \ln Z^*}{\partial K^*},$$

where K and K^* are connected by (8.13). Hence prove (8.17).

8.2 By a star-triangle transformation from a honeycomb lattice of N sites with interaction parameter G to a triangular lattice of $\frac{1}{2}N$ sites and then a dual transformation to a honeycomb lattice of N sites with interaction parameter G', prove (8.37) and (8.38).

8.3 Given that $Z(N, K)$ is the partition function for the triangular lattice, show, from (8.34) and (8.32), that

$$-N\coth(2K) + \frac{\partial \ln Z}{\partial K} = \left[N\coth(2K') - \frac{\partial \ln Z'}{\partial K'}\right]\frac{1-\exp(-4K')}{1-\exp(-4K)},$$

where Z' denotes $Z(N, K')$. Hence, using (8.16) and noting that $N_{\mathrm{E}} = \frac{1}{2}zN = 3N$, show that

$$\frac{[1-\exp(-4K)]\langle\sigma_i\sigma_j\rangle_{\mathrm{n.n.}} + [1-\exp(-4K')]\langle\sigma_i\sigma_j\rangle'_{\mathrm{n.n.}}}{2+\exp(-4K)+\exp(-4K')} = \frac{1}{3}\,.$$

By a similar procedure, show that for the honeycomb lattice, where $N_{\mathrm{E}} = \frac{1}{2}zN = \frac{3}{2}N$,

$$\frac{\sinh(2G)\langle\sigma_i\sigma_j\rangle_{\mathrm{n.n.}}}{\cosh(2G)+1} + \frac{\sinh(2G')\langle\sigma_i\sigma_j\rangle'_{\mathrm{n.n.}}}{\cosh(2G)+1} =$$
$$\frac{2}{3}\left[\frac{\cosh(2G)}{\cosh(2G)+1} + \frac{\cosh(2G)}{\cosh(2G')+1}\right].$$

Hence prove that, for both the triangular and honeycomb lattices, the critical value of the configurational energy is given by

$$u_{\mathrm{c}} = -J\coth(2K_{\mathrm{c}})\,,$$

assuming a unique critical point and replacing K_{c} by G_{c} in the honeycomb case.

8.4 Show that the relations (8.44) between triangle and honeycomb parameters for the unequal interactions star-triangle transformation reduce to (8.27) if $K_i = K$, $G_i = G$ $(i = 1, 2, 3)$.

8.5 Show using (8.47) and (8.49) that, if K_i and G_i $(i = 1, 2, 3)$ are respectively the triangle and honeycomb parameters connected by the star-triangle transformation, then

$$\cosh(2G_i) = \cosh(2K_j)\cosh(2K_\ell) + \sinh(2K_j)\sinh(2K_\ell)\coth(2K_i)\,,$$

where (i, j, ℓ) is a permutation of $(1, 2, 3)$.

8.6 With k_1 for the square lattice defined by (8.132), show that

$$\frac{1}{k_1}\frac{\mathrm{d}k_1}{\mathrm{d}K} = \frac{2\{1-2\tanh^2(2K)\}}{\tanh(2K)}\,.$$

Obtain the formula (8.131) for u by substituting the expression for λ given by (8.136) into the relation

$$\frac{\partial\ln(\lambda)}{\partial K} = -\frac{u}{J}.$$

8.7 Show that the complementary modulus k_1' to the k_1 defined by (8.132) satisfies the relation

$${k_1'}^2 = [2\tanh^2(2K) - 1]^2.$$

Show that the zero-field configurational heat capacity c_0 per site is given by

$$c_0 = -k_{\mathrm{B}} K^2 \frac{\partial(u/J)}{\partial K} .$$

Hence, using (8.131) and (A.8), deduce that, for the square lattice,

$$c_0 = \frac{4k_{\mathrm{B}}}{\pi} [K \coth(2K)]^2 \{\mathcal{K}(k_1) - \mathcal{E}(k_1) + [\tanh^2(2K) - 1] \left[\frac{1}{2}\pi + \mathcal{K}(k_1)\{2\tanh^2(2K) - 1\}\right]\} .$$

The first term in the square brackets gives the singular part of c_0.

8.8 Show that the star-triangle relation (8.28) between the partition functions of the triangular and honeycomb lattices may be written

$$\ln \lambda^{(\mathrm{H})}(G) = \frac{1}{2} \ln \mathrm{R} + \frac{1}{2} \ln \lambda^{(\mathrm{T})}(K).$$

Use this relation with the star-triangle relation (8.27) between lattice parameters to deduce equation (8.154) for $\lambda^{(\mathrm{H})}$ from equation (8.153) for $\lambda^{(\mathrm{T})}$.

9. Applications of Transform Methods

9.1 The Decoration Transformation

In this chapter we describe how a number of models, mathematically equivalent to the Ising model, can be derived by transformation methods and, in particular, by the *decoration transformation*. Accurate results will be derived for cooperative phenomena, for example transitions on critical curves and the associated exponents, ferrimagnetic first-order transitions in the $\mathcal{H} \neq 0$ region and water-like behaviour (when $d > 1$) which have until now been treated by approximation methods. New phenomena, such as *lower critical points* and *critical double-points*, will also arise.

Decoration is based on a regular lattice of N sites with coordination number z, which is called the *primary lattice*. An additional *decorating or secondary site*, is then placed between each nearest-neighbour pair of primary sites (see Fig. 9.1). Hence there is one secondary site for each of the $\frac{1}{2}zN$ edges of the primary lattice. An assembly on a decorated lattice is equivalent to a standard Ising model on the primary lattice when:

(i) There are two occupation states for any primary site, represented by $\sigma_i = \pm 1$ for the primary site labelled by i.

(ii) A microsystem on a secondary site can interact only with the microsystems on the two adjacent primary sites.[1]

(iii) A microsystem on a primary site can interact only with the microsystems on the adjacent secondary sites and the nearest-neighbour primary sites. The two types of interaction are termed primary–secondary and primary–primary respectively. Using the discrete variable σ_{ij} to label the occupation states of the secondary site between the nearest-neighbour primary sites i and j, the Hamiltonian for the decorated lattice model can then be written

[1] The single secondary site on each edge of the primary lattice can be replaced by a group of secondary sites, (Fisher 1959a, Syozi 1972). This condition must then be augmented with the phrase 'or with the microsystems on the other secondary sites of the same group'. However in the decorated models treated in this chapter there is only one secondary site per edge.

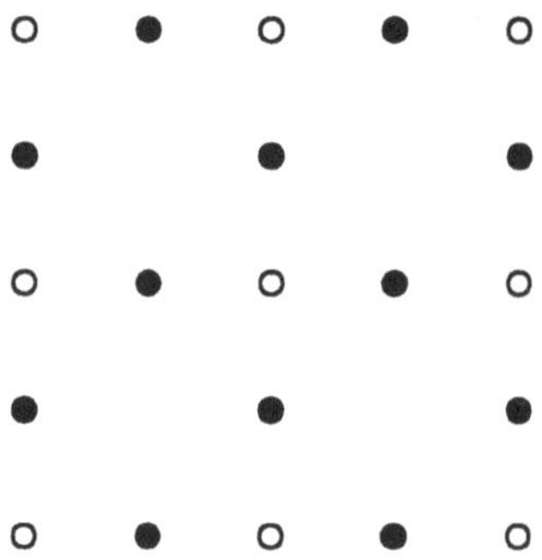

Fig. 9.1. Decorated square lattice: white sites are primary and black sites are secondary.

$$\widehat{H} = \sum_{\{i,j\}}^{(\text{n.n.p.})} \hat{h}(\sigma_i, \sigma_j; \sigma_{ij}) + \sum_{\{i\}}^{(\text{p.})} \xi(\sigma_i) \,, \tag{9.1}$$

where the first summation is over primary nearest-neighbour pairs, the second is over primary sites and the $\xi(\sigma_i)$ are field related quantities. The primary–primary interaction is not present in all decorated models.

Since each σ_{ij} occurs in only one term of $\widehat{H}$ a partial summation over the σ_{ij} can be performed to yield, for the partition function Z of the decorated model,

$$\begin{aligned} Z &= \sum_{\{\sigma_i, \sigma_{ij}\}} \exp(-\widehat{H}/T) \\ &= \sum_{\{\sigma_i\}} \exp\Big[-\sum_{\{i\}}^{(\text{p.})} \xi(\sigma_i)/T\Big] \prod_{\{i,j\}}^{(\text{n.n.p.})} \phi(\sigma_i, \sigma_j) \,, \end{aligned} \tag{9.2}$$

where

$$\phi(\sigma_i, \sigma_j) = \sum_{\{\sigma_{ij}\}} \exp[-\hat{h}(\sigma_i, \sigma_j; \sigma_{ij})/T] \,. \tag{9.3}$$

With condition (i), there are three forms of $\phi(\sigma_i, \sigma_j)$, $\phi(+1,+1)$, $\phi(-1,-1)$ and $\phi(+1,-1)$. From (9.2),

$$\begin{aligned} Z = \sum_{\{\sigma_i\}} & [\phi(+1,+1)]^{N_{++}} [\phi(-1,-1)]^{N_{--}} [\phi(+1,-1)]^{N_{+-}} \\ & \times \exp\{-[N_+ \xi(+1) + N_- \xi(-1)]/T\} \,, \end{aligned} \tag{9.4}$$

where

$$N_\pm = \frac{1}{2} \sum_{\{i\}}^{(\text{p.})} (1 \pm \sigma_i) = \frac{1}{2} \Big(N \pm \sum_{\{i\}}^{(\text{p.})} \sigma_i \Big) \tag{9.5}$$

are the numbers of primary sites for which $\sigma_i = \pm 1$ and

$$N_{++} = \frac{1}{4} \sum_{\{i,j\}}^{(\text{n.n.p.})} (1+\sigma_i)(1+\sigma_j) = \frac{1}{8}\Big[Nz + 2z\sum_{\{i\}}^{(\text{p.})}\sigma_i + 2\sum_{\{i,j\}}^{(\text{n.n.p.})}\sigma_i\sigma_j\Big],$$

$$N_{--} = \frac{1}{4} \sum_{\{i,j\}}^{(\text{n.n.p.})} (1-\sigma_i)(1-\sigma_j) = \frac{1}{8}\Big[Nz - 2z\sum_{\{i\}}^{(\text{p.})}\sigma_i + 2\sum_{\{i,j\}}^{(\text{n.n.p.})}\sigma_i\sigma_j\Big],$$

$$N_{+-} = \frac{1}{2} \sum_{\{i,j\}}^{(\text{n.n.p.})} (1-\sigma_i\sigma_j) = \frac{1}{4}\Big[Nz - 2\sum_{\{i\}}^{(\text{n.n.p.})}\sigma_i\sigma_j\Big], \tag{9.6}$$

are the respective numbers of $++$, $--$ and $+-$ nearest-neighbour pairs on the primary lattice. From (2.67),

$$Z = \mathsf{R}^{N/4} Z_{\text{p}}^{(\text{I.M})}(K_{\text{e}}, L_{\text{e}}), \tag{9.7}$$

where

$$\mathsf{R} = [\phi(+1,+1)\phi(-1,-1)]^{z/2}[\phi(+1,-1)]^{z}\exp\{-2[\xi(+1)+\xi(-1)]/T\} \tag{9.8}$$

and $Z_{\text{p}}^{(\text{I.M.})}$ is the partition function for an Ising model on the primary lattice with the quantities J/T and H/T, which appear in (2.67), replaced by the effective couplings K_{e} and L_{e} respectively, where

$$\begin{aligned} \exp(4K_{\text{e}}) &= \frac{\phi(+1,+1)\phi(-1,-1)}{[\phi(+1,-1)]^2}, \\ \exp(4L_{\text{e}}) &= \left[\frac{\phi(+1,+1)}{\phi(-1,-1)}\right]^{z}\exp\{2[\xi(-1)-\xi(+1)]/T\}. \end{aligned} \tag{9.9}$$

The general form of these relations is due to Fisher (1959a). The equivalent primary lattice Ising model with couplings K_{e} and L_{e} is a ferromagnet or antiferromagnet according as K_{e} is positive or negative. For $K_{\text{e}} > 0$, any critical point or curve in the space of decorated model thermodynamic variables maps into the point $K_{\text{e}} = K_{\text{c}}$, $L_{\text{e}} = 0$ in the $(K_{\text{e}}, L_{\text{e}})$ space of the Ising model while any first-order transition line or surface maps into the segment $L_{\text{e}} = 0$ $K_{\text{e}} > K_{\text{c}}$. Two useful relations follow immediately from (2.67). The first one is

$$g_{\text{p}}(K_{\text{e}}, L_{\text{e}}) = \langle\sigma_i\sigma_j\rangle_{\text{p}} = \frac{2}{zN}\frac{\partial \ln Z_{\text{p}}^{(\text{I.M.})}}{\partial K_{\text{e}}} = -\frac{2}{zN}u^{(\text{I.M.})}(K_{\text{e}}, L_{\text{e}}), \tag{9.10}$$

where i and j are a nearest-neighbour pair on the primary lattice and $u^{(\text{I.M.})}(K_{\text{e}}, L_{\text{e}})$ is the configurational energy per site in the equivalent Ising model. Similarly we have

$$\langle\sigma_i\rangle_{\text{p}} = N^{-1}\frac{\partial \ln Z_{\text{p}}^{\text{I.M.})}}{\partial L_{\text{e}}} = m^{(\text{I.M.})}(K_{\text{e}}, L_{\text{e}}), \tag{9.11}$$

where $m^{(\mathrm{I.M.})}(K_{\mathrm{e}}, L_{\mathrm{e}})$ is the relative magnetization for the equivalent Ising model.

9.2 Dilute Decorated Models

9.2.1 A Superexchange Model

We consider a mixture model with all primary sites occupied by Ising spins and with a fraction y of the secondary sites occupied by solute atoms, the remaining secondary sites being vacant. The solute atoms have no magnetic moment but they have the *superexchange* property that a solute atom between the nearest-neighbour pair of primary sites i and j induces a primary–primary interaction $-J_{\mathrm{p}}\sigma_i\sigma_j$ ($J_{\mathrm{p}} > 0$). When the secondary site between i and j is unoccupied, the spins on i and j do not interact.

Denoting the solute chemical potential by μ, we use the distribution corresponding to the choice of μ, $\mathcal{H}$ and T as independent intensive variables. Let $\sigma_{ij} = 1$ when the secondary site is occupied by a solute atom, and $\sigma_{ij} = 0$ when it is unoccupied. Then

$$\hat{h}(\sigma_i, \sigma_j; \sigma_{ij}) = -\sigma_{ij}(\mu + J_{\mathrm{p}}\sigma_i\sigma_j) \tag{9.12}$$

and (9.3) gives

$$\begin{aligned} \phi(+1,+1) = \phi(-1,-1) &= \lambda \exp(K_{\mathrm{p}}) + 1\,, \\ \phi(+1,-1) &= \lambda \exp(-K_{\mathrm{p}}) + 1\,, \end{aligned} \tag{9.13}$$

where

$$K_{\mathrm{p}} = J_{\mathrm{p}}/T\,, \qquad \lambda = \exp(\mu/T)\,. \tag{9.14}$$

Since the primary sites are occupied by Ising spins,

$$\xi(\pm 1) = \mp \mathcal{H}, \tag{9.15}$$

where $\mathcal{H} = m_0 \widetilde{\mathcal{H}}$, m_0 being the primary site dipole moment (see Sect. 1.7). From (9.9), K_{e} and L_{e} are given by

$$\exp(2K_{\mathrm{e}}) = \frac{\phi(+1,+1)}{\phi(+1,-1)} = \frac{\lambda \exp(K_{\mathrm{p}}) + 1}{\lambda \exp(-K_{\mathrm{p}}) + 1}\,, \qquad L_{\mathrm{e}} = \mathcal{H}/T\,. \tag{9.16}$$

Since $\lambda > 0$, it follows from (9.16) that $K_{\mathrm{e}} < K_{\mathrm{p}}$. From (9.7), (9.8), (9.13) and (9.15) the partition function for the decorated model is

$$Z = [\phi(+1,+1)\phi(+1,-1)]^{zN/4}\, Z_{\mathrm{p}}^{(\mathrm{I.M.})}(K_{\mathrm{e}}, L_{\mathrm{e}})\,. \tag{9.17}$$

Since, in each term of Z there is a factor λ for each occupied secondary site, the equilibrium value of y is given by

$$y = \left(\tfrac{1}{2}zN\right)^{-1} \lambda \left(\frac{\partial \ln(Z)}{\partial \lambda}\right)_{T,\mathcal{H}} \tag{9.18}$$

and from (9.10), (9.16) and (9.17), (noting that $\partial L_\mathrm{e}/\partial\lambda = 0$),

$$\begin{aligned} y &= \frac{1}{2}\lambda \frac{\partial \ln[\phi(+1,+1)\phi(+1,-1)]}{\partial\lambda} + \lambda\left(\frac{1}{2}zN\right)^{-1}\frac{\partial K_\mathrm{e}}{\partial\lambda}\frac{\partial \ln Z_\mathrm{p}^{(\mathrm{I.M.})}}{\partial K_\mathrm{e}} \\ &= \frac{1}{2}\lambda \frac{\partial \ln[\phi(+1,+1)\phi(+1,-1)]}{\partial\lambda} \\ &\quad + \frac{1}{2}\lambda g_\mathrm{p}(K_\mathrm{e}, L_\mathrm{e})\frac{\partial \ln[\phi(+1,+1)/\phi(+1,-1)]}{\partial\lambda} . \end{aligned} \tag{9.19}$$

This can be re-expressed as

$$y = \frac{1}{2}\lambda\langle 1 + \sigma_i\sigma_j\rangle_\mathrm{p}\frac{\partial \ln[\phi(+1,+1)]}{\partial\lambda} + \frac{1}{2}\lambda\langle 1 - \sigma_i\sigma_j\rangle_\mathrm{p}\frac{\partial \ln[\phi(+1,-1)]}{\partial\lambda}, \tag{9.20}$$

where i and j are a nearest-neighbour pair on the primary lattice. The terms in the final expression have a physical interpretation: $\frac{1}{2}\langle 1 - \sigma_i\sigma_j\rangle_\mathrm{p}$ is the fraction of primary lattice edges for which $\sigma_i \neq \sigma_j$ while $\lambda\partial \ln\phi(+1,-1)/\partial\lambda$ is the fraction of secondary sites on such edges which are occupied by solute atoms. The first term corresponds similarly to the case $\sigma_i = \sigma_j$, noting that $\phi(+1,+1) = \phi(-1,-1)$.

9.2.2 The Equilibrium Bond Dilute Ising Model

Since the only property of the solute atom in Sect. 9.2.1 is to induce an interaction between the spins on the adjacent primary sites, the model is equivalent to the equilibrium bond dilution model of Sect. 6.11.1 (Rapaport 1972). Occupied and unoccupied secondary sites correspond respectively to unbroken and broken bonds between nearest-neighbour pairs on the primary lattice, and y is the fraction of unbroken bonds. The solute activity λ can be expressed in terms of K_e and K_p from (9.16), and substitution in (9.19) yields

$$y = \mathsf{C}(K_\mathrm{e}, K_\mathrm{p}) + g_\mathrm{p}(K_\mathrm{e}, L_\mathrm{e})\mathsf{S}(K_\mathrm{e}, K_\mathrm{p}), \tag{9.21}$$

where

$$\begin{aligned} \mathsf{C}(K_\mathrm{e}, K_\mathrm{p}) &= \frac{\sinh(K_\mathrm{e})\cosh(K_\mathrm{p} - K_\mathrm{e})}{\sinh(K_\mathrm{p})}, \\ \mathsf{S}(K_\mathrm{e}, K_\mathrm{p}) &= \frac{\sinh(K_\mathrm{e})\sinh(K_\mathrm{p} - K_\mathrm{e})}{\sinh(K_\mathrm{p})}. \end{aligned} \tag{9.22}$$

Equation (9.21) is an implicit relation giving K_e in terms of y, T and $\mathcal{H}$. In the equivalent Ising model there is a second-order transition when K_e passes through the critical value K_c at $L_\mathrm{e} = 0$. With $\mathcal{H} = 0$ and $K_\mathrm{e} = K_\mathrm{c}$ in (9.21)

$$\begin{aligned} y = \sinh(K_\mathrm{c})\,\{&\coth(K_\mathrm{p})[\cosh(K_\mathrm{c}) - g_\mathrm{p}(K_\mathrm{c},0)\sinh(K_\mathrm{c})] \\ &+ g_\mathrm{p}(K_\mathrm{c},0)\cosh(K_\mathrm{c}) - \sinh(K_\mathrm{c})\} \end{aligned} \tag{9.23}$$

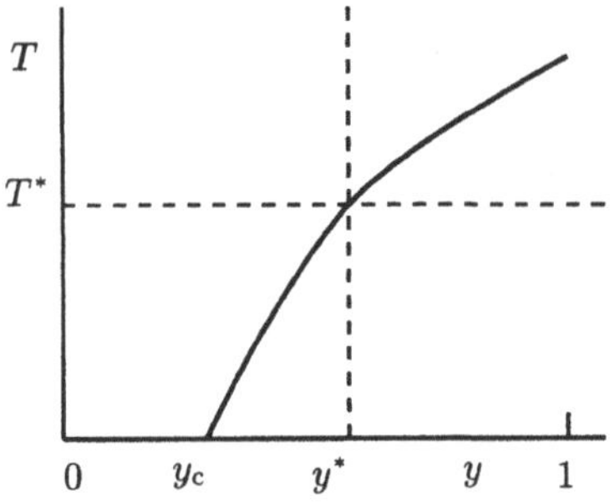

Fig. 9.2. The critical curve for equilibrium bond dilute model.

is the curve of second-order transitions (critical curve) in the (y, T) plane, which is sketched in Fig. 9.2. The bond dilute model is equivalent to the standard Ising model when $y = 1$, and it can be verified from (9.23) that $y = 1$ corresponds to $K_p = K_c$ $(T = J_p/K_c)$. As T decreases from J_p/K_c, y decreases steadily and $dy/dT = \infty$ when the curve meets the $T = 0$ axis at $y = y_c$, given by

$$y_c = \frac{1 + g_p(K_c, 0)}{1 + \coth(K_c)} . \tag{9.24}$$

For $y < y_c$ there is no transition at any value of T. The internal energy per primary site is

$$\begin{aligned} u &= -\frac{1}{2} z J_p \langle \sigma_{ij} \sigma_i \sigma_j \rangle = -\frac{1}{2} z J_p \left(\frac{1}{2} z N\right)^{-1} \left(\frac{\partial \ln(Z)}{\partial K_p}\right)_{\lambda, \mathcal{H}} \\ &= \frac{1}{2} z J_p \{ S(K_e, K_p) + g_p(K_e, L_e) C(K_e, K_p) \} . \end{aligned} \tag{9.25}$$

9.3 Heat Capacity and Exponent Renormalization

In the bond dilute Ising model, the response functions behave in a rather surprising way at the critical curve shown in Fig. 9.2. To analysis this behaviour we begin by defining T^* as the zero-field critical temperature given by (9.23) for $y = y^*$ where $y_c < y^* < 1$. Suppose that the critical curve in the (y, T) plane is approached along the line $y = y^*$ (Fig. 9.2) from the disordered region where $T > T^*$. The heat capacity per primary site is

$$c(T, y^*) = k_B \left(\frac{\partial u}{\partial T}\right)_y = -\frac{k_B K_p^2}{J_p} \left(\frac{\partial u}{\partial K_p}\right)_y . \tag{9.26}$$

Now define

$$g'_{\mathrm{p}}(K_{\mathrm{e}}) = \frac{\partial g_{\mathrm{p}}(K_{\mathrm{e}},0)}{\partial K_{\mathrm{e}}},$$

$$\mathrm{C_e}(K_{\mathrm{e}},K_{\mathrm{p}}) = \frac{\partial \mathrm{C}(K_{\mathrm{e}},K_{\mathrm{p}})}{\partial K_{\mathrm{e}}}, \tag{9.27}$$

$$\mathrm{C_p}(K_{\mathrm{e}},K_{\mathrm{p}}) = \frac{\partial \mathrm{C}(K_{\mathrm{e}},K_{\mathrm{p}})}{\partial K_{\mathrm{p}}}$$

with similar definitions of $\mathrm{S_e}(K_{\mathrm{e}},K_{\mathrm{p}})$ and $\mathrm{S_p}(K_{\mathrm{e}},K_{\mathrm{p}})$. From (9.25)

$$\left(\frac{\partial u}{\partial K_{\mathrm{p}}}\right)_y = -\frac{1}{2}zJ_{\mathrm{p}}\left[\mathrm{S_p} + g_{\mathrm{p}}\mathrm{C_p} + \left(\frac{\partial K_{\mathrm{e}}}{\partial K_{\mathrm{p}}}\right)_y (\mathrm{S_e} + g_{\mathrm{p}}\mathrm{C_e} + g'_{\mathrm{p}}\mathrm{C})\right]. \tag{9.28}$$

With $y = y^* =$ constant and $\mathcal{H} = 0$, differentiation of (9.21) with respect to K_{p} gives

$$0 = \mathrm{C_p} + g_{\mathrm{p}}\mathrm{S_p} + \left(\frac{\partial K_{\mathrm{e}}}{\partial K_{\mathrm{p}}}\right)_y (\mathrm{C_e} + g_{\mathrm{p}}\mathrm{S_e} + g'_{\mathrm{p}}\mathrm{S}). \tag{9.29}$$

Eliminating $\partial K_{\mathrm{e}}/\partial K_{\mathrm{p}}$ between (9.28) and (9.29) and using (9.26),

$$\begin{aligned} c(T,y^*) &= \frac{1}{2}zk_{\mathrm{B}}K_{\mathrm{p}}^2\left[\mathrm{S_p} + g_{\mathrm{p}}\mathrm{C_p} - \frac{(\mathrm{C_p} + g_{\mathrm{p}}\mathrm{S_p})(\mathrm{S_e} + g_{\mathrm{p}}\mathrm{C_e} + g'_{\mathrm{e}}\mathrm{C})}{\mathrm{C_e} + g_{\mathrm{p}}\mathrm{S_e} + g'_{\mathrm{e}}\mathrm{S}}\right] \\ &= \frac{1}{2}zk_{\mathrm{B}}K_{\mathrm{p}}^2\left[\mathrm{S_p} + g_{\mathrm{p}}\mathrm{C_p} - \frac{\mathrm{C}(\mathrm{C_p} + g_{\mathrm{p}}\mathrm{S_p})}{\mathrm{S}}\right. \\ &\quad \left. - \frac{[\mathrm{C_p} + g_{\mathrm{p}}\mathrm{S_p}][\mathrm{SS_e} - \mathrm{CC_e} + g_{\mathrm{e}}(\mathrm{SC_e} - \mathrm{CS_e})]}{\mathrm{S}(\mathrm{C_e} + g_{\mathrm{p}}\mathrm{S_e} + g'_{\mathrm{p}}\mathrm{S})}\right]. \end{aligned} \tag{9.30}$$

From (9.10) the heat capacity per site in the equivalent Ising model is

$$c^{(\mathrm{I.M.})}(K_{\mathrm{e}}) = \frac{1}{2}zk_{\mathrm{B}}K_{\mathrm{e}}^2 g'_{\mathrm{p}}(K_{\mathrm{e}}). \tag{9.31}$$

Since it is known that, in two and three dimensions, $c^{(\mathrm{I.M.})}(K_{\mathrm{e}})$ diverges to infinity as $K_{\mathrm{e}} \to K_{\mathrm{c}}$, $g'_{\mathrm{p}}(K_{\mathrm{e}}) \to \infty$ as $\Delta K_{\mathrm{e}} = K_{\mathrm{c}} - K_{\mathrm{e}} \to 0$. As all other quantities on the right-hand side of (9.30) are finite at $K_{\mathrm{e}} = K_{\mathrm{c}}$ it follows that $c(T^*,y^*)$, the bond dilute model heat capacity at the transition point for $y = y^* < 1$, is itself finite.

9.3.1 Three-Dimensional Lattices

In three-dimensional lattices $g'_{\mathrm{p}}(K_{\mathrm{e}}) \sim (\Delta K_{\mathrm{e}})^{-\alpha}$ and hence $g_{\mathrm{p}}(K_{\mathrm{e}},0) \sim (\Delta K_{\mathrm{e}})^{1-\alpha}$ as $\Delta K_{\mathrm{e}} \to +0$ with the Ising model critical exponent $\alpha \doteq \frac{1}{8}$ (Sect. 3.6). Defining $\Delta c(T,y^*)$ as $c(T,y^*) - c(T^*,y^*)$, so that Δc includes any singular part of c, it follows from (9.30) that since $\alpha < \frac{1}{2}$, the leading term in Δc is proportional to $1/g'_{\mathrm{p}}(K_{\mathrm{e}})$ and hence to $(\Delta K_{\mathrm{e}})^{\alpha}$ giving

$$\Delta c(T,y^*) \sim (\Delta K_{\mathrm{e}})^{\alpha}, \qquad \text{as } \Delta K_{\mathrm{e}} \to 0\ . \tag{9.32}$$

Let $\Delta K_\mathrm{p} = K_\mathrm{p}^* - K_\mathrm{p}$, where $K_\mathrm{p}^* = J_\mathrm{p}/T^*$. Then (9.29) yields

$$\frac{\partial \Delta K_\mathrm{e}}{\partial K_\mathrm{p}} = \frac{\partial K_\mathrm{e}}{\partial K_\mathrm{p}} \sim (\Delta K_\mathrm{e})^\alpha , \qquad \text{as } \Delta K_\mathrm{p} \to 0 . \tag{9.33}$$

Hence $\Delta K_\mathrm{p} \sim (\Delta K_\mathrm{e})^{1-\alpha}$ and

$$\Delta K_\mathrm{e} \sim (\Delta K_\mathrm{p})^{1/(1-\alpha)} , \qquad \text{as } \Delta K_\mathrm{p} \to 0 . \tag{9.34}$$

With (9.32) this gives

$$\Delta c(T, y^*) \sim (\Delta K_\mathrm{p})^{\alpha/(1-\alpha)} \sim (T - T^*)^{\alpha/(1-\alpha)} , \qquad \text{as } T \to T^* + 0 . \tag{9.35}$$

Since the only magnetic moments in the system belong to the primary site spins, the relative magnetization is given by

$$m(K_\mathrm{p}, \mathcal{H}) = \langle \sigma_i \rangle_\mathrm{p} = m^{(\mathrm{I.M.})}(K_\mathrm{e}, L_\mathrm{e}). \tag{9.36}$$

It follows, using the second relation of (9.16), that

$$\left(\frac{\partial m}{\partial \mathcal{H}}\right)_{\mathcal{H}=0} = \frac{1}{T}\left(\frac{\partial m^{(\mathrm{I.M.})}}{\partial L_\mathrm{e}}\right)_{L_\mathrm{e}=0} \sim (\Delta K_\mathrm{e})^{-\gamma} , \qquad \text{as } \Delta K_\mathrm{e} \to +0 , \tag{9.37}$$

where γ is the Ising model exponent. With (9.34) this gives

$$\left(\frac{\partial m}{\partial \mathcal{H}}\right)_{\mathcal{H}=0} = (\Delta K_\mathrm{p})^{-\gamma/(1-\alpha)} \sim (T - T^*)^{-\gamma/(1-\alpha)} , \qquad \text{as } T \to T^* + 0 . \tag{9.38}$$

Again, since

$$m^{(\mathrm{I.M.})}(K_\mathrm{c}, L_\mathrm{e}) \sim L_\mathrm{e}^{1/\delta} , \qquad \text{as } L_\mathrm{e} \to 0 , \tag{9.39}$$

where δ is the Ising Model exponent, we have, from (9.16) and (9.36),

$$m(T^*, \mathcal{H}) \sim L_\mathrm{e}^{1/\delta} \sim \mathcal{H}^{1/\delta} . \tag{9.40}$$

Labelling exponents in the $d = 3$ bond dilute model at fixed $y = y^* < 1$ by the index r, we have, from (9.35), (9.38) and (9.40),

$$\alpha_\mathrm{r} = -\frac{\alpha}{1-\alpha} , \qquad \gamma_\mathrm{r} = \frac{\gamma}{1-\alpha} , \qquad \delta_\mathrm{r} = \delta . \tag{9.41}$$

The exponent transformations of (9.41) and (9.45) below are termed *exponent renormalization.*

For $T < T^*$ ($K_\mathrm{p} > K_\mathrm{p}^*$), similar methods yield relations like (9.34), (9.35) and (9.38) with α', γ' and $T^* - T$ replacing α, γ and $T - T^*$ respectively. From (9.35),

$$\left(\frac{\partial c}{\partial T}\right)_y \sim (T - T^*)^{-(1-2\alpha)/(1-\alpha)} , \qquad \text{as } T \to T^* + 0 , \tag{9.42}$$

and there is a similar relation with α replaced by α' when $T \to T^* - 0$. Since $\alpha' = \alpha \doteq \frac{1}{8}$, the Ising model infinite singularity in the heat capacity at the critical point is replaced by a cusp with vertical tangents.

For $K_p > K_p^*$ the spontaneous magnetization of the bond dilute Ising model is, from (9.36) and (9.16),

$$m_s(K_p) = \lim_{\mathcal{H} \to 0} \pm(K_p, \mathcal{H}) = m_s^{(\mathrm{I.M.})}(K_e) . \tag{9.43}$$

Now $m_s^{(\mathrm{I.M.})}(K_e) \sim |\Delta K_e|^\beta$ as $\Delta K_e \to -0$, β being the Ising model exponent. Thus, from (9.43),

$$m_s \sim (T^* - T)^{\beta/(1-\alpha')}, \qquad \text{as } T \to T^* - 0 . \tag{9.44}$$

Hence the critical exponents, when $T < T^*$, are given by

$$\alpha'_r = -\frac{\alpha'}{1-\alpha'}, \qquad \gamma'_r = \frac{\gamma'}{1-\alpha'}, \qquad \beta_r = \frac{\beta}{1-\alpha}. \tag{9.45}$$

From (9.41) and (9.45) it is easily verified that, if the Ising exponents satisfy the Widom scaling law (3.77) and the Essam–Fisher scaling law (3.87) then

$$\begin{aligned} &\gamma'_r = \beta_r(\delta_r - 1), \\ &\alpha'_r + 2\beta_r + \gamma'_r = 2. \end{aligned} \tag{9.46}$$

These exponents also satisfy the scaling laws. Also $\alpha'_r = \alpha_r$ and $\gamma'_r = \gamma_r$ if $\alpha' = \alpha$ and $\gamma' = \gamma$.

The critical curve can also be approached along the line $T = T^*$, see Fig. 9.2). With $K_p = K_p^*$, (9.21) yields

$$\left(\frac{\partial K_e}{\partial y}\right)_{T=T^*} = \frac{1}{C_e + g_p S_e + g'_p S} \sim (\Delta K_e)^\alpha, \qquad \text{as } \Delta K_e \to 0 , \tag{9.47}$$

where $\Delta K_e = 0$ now corresponds to $\Delta y = y^* - y = 0$. Hence we have,

$$\Delta K_e \sim (\Delta y)^{1/(1-\alpha)}, \tag{9.48}$$

which is similar to (9.34).

From (9.32), $\Delta c \sim (\Delta y)^{\alpha/(1-\alpha)}$ as $\Delta y \to 0$ and we again find $\alpha_r = -\alpha/(1-\alpha)$. For $y > y^*$, the spontaneous magnetization $m_s \sim (\Delta K_e)^\beta \sim |\Delta y|^{\beta/(1-\alpha')}$ so that, again, $\beta_r = \beta/(1-\alpha')$. It can similarly be shown that the other constant-temperature exponents are equal to those at constant y.

For $y = 1$ the exponents must revert to the standard Ising values. In the present formalism, putting $y = 1$ in (9.21) or (9.23) yields $K_e = K_p$ so that (9.34) is replaced by $\Delta K_e = \Delta K_p$. There are thus finite discontinuities in the exponents, since the limits as $y \to 1 - 0$ are not equal to the values at $y = 1$. It is remarkable that for any non-zero value of $1 - y$, however small, there is a qualitative change in the properties of the model, the infinite discontinuity in the heat capacity being replaced by a cusp. However, the smaller the value of $1 - y$, the narrower becomes the range around the critical point where the exponent changes have a significant numerical effect(Fisher and Scesney 1970).[2]

The exponent renormalization results from taking y as an independent variable which can change at constant T or remain constant when T changes. It is not simply a consequence of the existence of broken bonds. If the latter are regarded as defects with an activation energy $\Delta\varepsilon$, then $\lambda = \exp(\Delta\varepsilon/T)$ and K_e, as given by (9.16), depends only on T. Thus $\Delta K_e \sim \Delta K_p$ and no renormalization occurs. The concentration y in zero field becomes a function of T. In fact, it is more realistic physically to take y as an independent variable in the mixture interpretation of this model given in Sect. 9.2.1 than in the broken bond interpretation in Sect. 9.2.2. In the mixture case, the mole fraction of solute is $zy/(2+z)$ and is a legitimate thermodynamic density.

9.3.2 Two-Dimensional Lattices

In two dimensions $\alpha = 0$ and there is no renormalization. However, as we deduced from (9.30), the transition-point heat capacity is finite for $y < 1$. From (8.119) and (9.10), $g'_p(K_e) \sim -\ln|\Delta K_e|$ as $\Delta K_e \to 0$ and thus

$$\left.\begin{aligned} \Delta c(T, y^*) &\sim -\frac{1}{\ln|\Delta T|}\,, \\ \frac{\mathrm{d}|\Delta c|}{\mathrm{d}T} &\sim \frac{1}{|\Delta T|(\ln|\Delta T|)^2}\,, \end{aligned}\right\} \quad \text{as } \Delta T = T - T^* \to 0\ . \tag{9.49}$$

There is thus a cusp with vertical tangents at the transition point, as in the $d = 3$ case. For the $d = 2$ case this represents a further variety of $\alpha = 0$ behaviour discussed in Sect. 3.4.

Renormalization relations like (9.41) and (9.45) were derived by Essam and Garelick (1967) for the *Syozi dilution model* (Syozi 1972, Essam 1972 and Example 9.1), which is slightly more elaborate than the one treated here. Widom (1967) derived the expression $\gamma/(1-\alpha)$ for the compressibility exponent in a decorated mixture model with the density of one component fixed. Fisher (1968) used thermodynamic arguments, in conjunction with examples from statistical mechanics, to establish general conditions for exponent renormalization. These were the existence of a 'hidden variable' subject to a constraint, for instance the presence of an equilibrium distribution of a fixed number of impurity atoms. 'Ideal' exponents are obtained when the largest possible number of fields are taken as independent variables, that is $\eta = n - 1$, and are renormalized when a field is replaced as an independent variable by a density. The replacement of μ by y in the model of Sect. 9.2 is an example. Version II of the dilute Ising model in Sect. 6.6.2 can be regarded as a compressible Ising model since the volume per spin $v = v_0/\rho$, where v_0 is the volume per site, and v can thus be adjusted by altering the pressure

[2] These authors used a different model, but the same general principles must apply here.

or chemical potential. Renormalization is to be expected if v is taken as an independent variable. The same applies to the compressible Ising model of Baker and Essam (1971). Renormalization will not occur in a closed-form approximation since there is then a finite discontinuity in the heat capacity, giving $\alpha = 0$ for $d = 3$.

9.4 Fisher's Decorated Antiferromagnetic Model

Suppose that on a decorated square lattice (Fig. 9.1) the microsystems on the N horizontal edge secondary sites have different properties from those on the N vertical edge secondary sites. Labelling quantities connected with the horizontal and vertical edges by the subscripts h and v respectively, we can define $\phi_{\mathrm{h}}(\pm 1, \pm 1)$ and $\phi_{\mathrm{v}}(\pm 1, \pm 1)$ by sums like (9.3) but restricted to the secondary microstates on the horizontal and vertical edges respectively. Then, by a derivation similar to that of (9.7),

$$Z = \mathsf{R}_{\mathrm{hv}}^{N/4} Z_{\mathrm{p}}^{(\mathrm{I.M})}(K_{\mathrm{eh}}, K_{\mathrm{ev}}, L_{\mathrm{e}}) , \tag{9.50}$$

where

$$\begin{aligned} \mathsf{R}_{\mathrm{hv}} = {} & \phi_{\mathrm{h}}(+1,+1)\phi_{\mathrm{v}}(+1,+1)\phi_{\mathrm{h}}(-1,-1)\phi_{\mathrm{v}}(-1,-1) \\ & \times [\phi_{\mathrm{h}}(+1,-1)\phi_{\mathrm{v}}(+1,-1)]^2 \exp\{-2[\xi(+1)+\xi(-1)]/T\} , \end{aligned} \tag{9.51}$$

where $Z_{\mathrm{p}}^{(\mathrm{I.M.})}$ refers to an anisotropic Ising model on the primary sites with interaction parameters K_{eh} and K_{ev} respectively in the horizontal and vertical directions (see (8.97) for the zero-field case). Here

$$\begin{aligned} &\exp(4K_{\mathrm{ex}}) = \frac{\phi_{\mathrm{x}}(+1,+1)\phi_{\mathrm{x}}(-1,-1)}{[\phi_{\mathrm{x}}(+1,-1)]^2} , \qquad \mathrm{x} = \mathrm{h},\mathrm{v} , \\ &\exp(2L_{\mathrm{e}}) = \left[\frac{\phi_{\mathrm{h}}(+1,+1)\phi_{\mathrm{v}}(+1,+1)}{\phi_{\mathrm{h}}(-1,-1)\phi_{\mathrm{v}}(-1,-1)}\right] \exp\{[\xi(-1)-\xi(+1)]/T\} . \end{aligned} \tag{9.52}$$

Suppose that both primary and secondary sites are occupied by Ising spins with primary–secondary interactions. The interaction constants are J_{h} and J_{v}, and the magnetic moments of the secondary spins $r_{\mathrm{h}} m_0$ and $r_{\mathrm{v}} m_0$ for horizontal and vertical edges respectively, m_0 being a standard magnetic moment. Then

$$\hat{h}_{\mathrm{x}}(\sigma_i, \sigma_j; \sigma_{ij}) = -J_{\mathrm{x}}\sigma_{ij}(\sigma_i + \sigma_j) - r_{\mathrm{x}}\sigma_{ij}\mathcal{H} , \qquad \mathrm{x} = \mathrm{h},\mathrm{v} , \tag{9.53}$$

$\sigma_{ij} = \pm 1$ being the spin component on the secondary site. Let

$$G_{\mathrm{x}} = J_{\mathrm{x}}/T, \qquad F_{\mathrm{x}} = r_{\mathrm{x}}\mathcal{H}/T , \qquad \mathrm{x} = \mathrm{h},\mathrm{v} , \tag{9.54}$$

and then, from (9.3),

$$\begin{aligned} &\phi_{\mathrm{x}}(\pm 1, \pm 1) = 2\cosh(2G_{\mathrm{x}} \pm F_{\mathrm{x}}) , \\ & \qquad\qquad\qquad\qquad\qquad \mathrm{x} = \mathrm{h},\mathrm{v} . \\ &\phi_{\mathrm{x}}(+1,-1) = 2\cosh(F_{\mathrm{x}}) , \end{aligned} \tag{9.55}$$

Generalizing (9.10),

$$\frac{\partial \ln Z_{\mathrm{p}}^{(\mathrm{I.M.})}}{\partial K_{\mathrm{ex}}} = \langle\sigma_i, \sigma_j\rangle_{\mathrm{p,x}} = g_{\mathrm{p,x}}(K_{\mathrm{eh}}, K_{\mathrm{ev}}, L_{\mathrm{e}}), \qquad \mathrm{x} = \mathrm{h}, \mathrm{v}, \tag{9.56}$$

where $\langle\sigma_i, \sigma_j\rangle_{\mathrm{p,x}}$ for x=h,v are the nearest-neighbour correlations in the equivalent Ising model for horizontal and vertical edges respectively. Using (9.50), the relative magnetizations for the two types of secondary sites are given by

$$\begin{aligned} m_{\mathrm{x}} &= \frac{1}{N}\frac{\partial \ln Z}{\partial F_{\mathrm{x}}} \\ &= \frac{1}{4}\frac{\partial}{\partial F_{\mathrm{x}}}[\ln \phi_{\mathrm{x}}(+1,+1) + \ln \phi_{\mathrm{x}}(-1,-1) + 2\ln \phi_{\mathrm{x}}(+1,-1)] \\ &\quad + \frac{\partial \ln Z_{\mathrm{p}}^{(\mathrm{I.M.})}}{\partial K_{\mathrm{ex}}}\frac{\partial K_{\mathrm{ex}}}{\partial F_{\mathrm{x}}} + \frac{\partial \ln Z_{\mathrm{p}}^{(\mathrm{I.M.})}}{\partial L_{\mathrm{e}}}\frac{\partial L_{\mathrm{e}}}{\partial F_{\mathrm{x}}}, \qquad \mathrm{x} = \mathrm{h}, \mathrm{v} \end{aligned} \tag{9.57}$$

and substituting from (9.11), (9.52), (9.55), (9.56) and into (9.57)

$$\begin{aligned} m_{\mathrm{x}} &= \frac{1}{4}[1 + g_{\mathrm{p,x}}(K_{\mathrm{eh}}, K_{\mathrm{ev}}, L_{\mathrm{e}})] \\ &\quad \times [\tanh(2G_{\mathrm{x}} + F_{\mathrm{x}}) - \tanh(2G_{\mathrm{x}} - F_{\mathrm{x}}) - 2\tanh(F_{\mathrm{x}})] + \tanh(G_{\mathrm{x}}) \\ &\quad + \frac{1}{2}m^{(\mathrm{I,M})}(K_{\mathrm{eh}}, K_{\mathrm{ev}}, L_{\mathrm{e}})[\tanh(2G_{\mathrm{x}} + F_{\mathrm{x}}) + \tanh(2G_{\mathrm{x}} - F_{\mathrm{x}})], \\ &\qquad \mathrm{x} = \mathrm{h}, \mathrm{v}. \end{aligned} \tag{9.58}$$

The basic assumptions of Fisher's (1960) model are:

(a) The magnetic moments of all primary site spins are zero and those of all secondary site spins are equal, so that $r_{\mathrm{h}} = r_{\mathrm{v}} = 1$. Hence

$$\xi(+1) = \xi(-1) = 0, \qquad F_{\mathrm{h}} = F_{\mathrm{v}} = F = \mathcal{H}/T. \tag{9.59}$$

(b) The horizontal Ising interactions are ferromagnetic in character whereas the vertical Ising interactions are antiferromagnetic, the interaction constants being of equal magnitude. Thus $J_{\mathrm{h}} = -J_{\mathrm{v}} > 0$ and

$$G_{\mathrm{h}} = -G_{\mathrm{v}} = G = J_{\mathrm{h}}/T. \tag{9.60}$$

Substitution from (9.59) and (9.60) into (9.55) yields

$$\begin{aligned} \phi_{\mathrm{h}}(\pm 1, \pm 1) &= \phi_{\mathrm{v}}(\mp 1, \mp 1) = 2\cosh(2G \pm F), \\ \phi_{\mathrm{h}}(+1, -1) &= \phi_{\mathrm{v}}(+1, -1) = 2\cosh(F). \end{aligned} \tag{9.61}$$

From (9.52) and (9.59),

$$\begin{aligned} K_{\mathrm{eh}} &= K_{\mathrm{ev}} = K_{\mathrm{e}} = \frac{1}{4}\ln\left\{\frac{\phi_{\mathrm{h}}(+1,+1)\phi_{\mathrm{h}}(-1,-1)}{[\phi_{\mathrm{h}}(+1,-1)]^2}\right\}, \\ L_{\mathrm{e}} &= 0. \end{aligned} \tag{9.62}$$

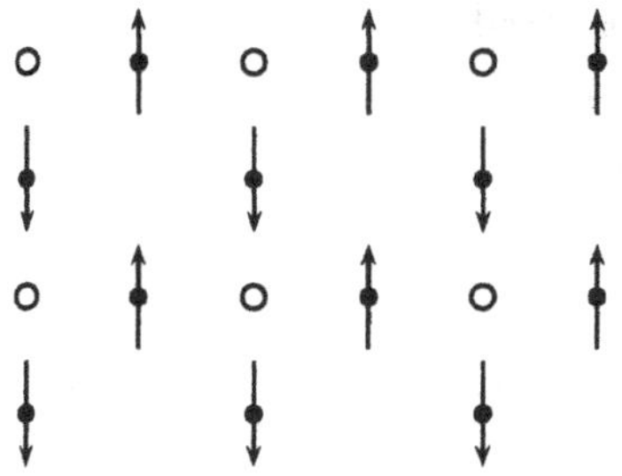

Fig. 9.3. The ground state of Fisher's antiferromagnetic model.

This shows the remarkable feature of this model, which is that the entire (G, F) plane is mapped into the $(K_e, 0)$ line of the Ising model ferromagnet.[3] The exact results for the decorated model *at all values of* T *and* $\mathcal{H}$ are obtainable from the known exact results (see Sect. 8.10) for the square lattice Ising ferromagnet *in zero field.*

In the $\mathcal{H} = 0$ ground state for the decorated model, the primary site spins, which have zero magnetic moment, are all parallel whereas the secondary site magnetic dipoles on the horizontal and vertical edges respectively are oppositely oriented (see Fig. 9.3). This is an antiferromagnetically ordered state with magnetization on two sublattices cancelling to give zero overall magnetization. The ground-state energy per primary site is $u = -4J_h$. By similar considerations to those of Sect. 4.2 it can be shown that there is a critical field $\mathcal{H}_c = 2J_h$ and that the antiferromagnetic ground state is stable at $T = 0$ for $-\mathcal{H}_c < \mathcal{H} < \mathcal{H}_c$. For $\mathcal{H} > \mathcal{H}_c$ or $\mathcal{H} < -\mathcal{H}_c$ the stable state at $T = 0$ is paramagnetic with all magnetic moments aligned with the field. The Fisher model is sometimes referred to as a *superexchange antiferromagnet* since the only function of the primary site spins is to couple the secondary site magnetic moments.

The behaviour of the Fisher model is similar to that indicated for an Ising antiferromagnet on a loose-packed lattice by series or mean-field methods. By substitution from (9.61) into the first relation of (9.62) the latter can be expressed in the form

$$\sinh^2(2G) = \cosh^2(F)[\exp(4K_e) - 1]. \tag{9.63}$$

This gives $K_e > 0$ for all G and F, so that, although the decorated model is an antiferromagnet, its equivalent Ising model is a ferromagnet. Since $L_e = 0$ for all T and $\mathcal{H}$, $m^{(\mathrm{I.M.})}$ in (9.58) must be replaced by $m_s^{(\mathrm{I.M.})}$, the spontaneous magnetization. Again, from (9.62), $g_{p,h} = g_{p,v} = g_p(K_e, 0)$ as defined by (9.10). Substituting (9.59) and (9.60) into (9.58) then yields

[3] It is shown below that $K_e > 0$ for all G and F.

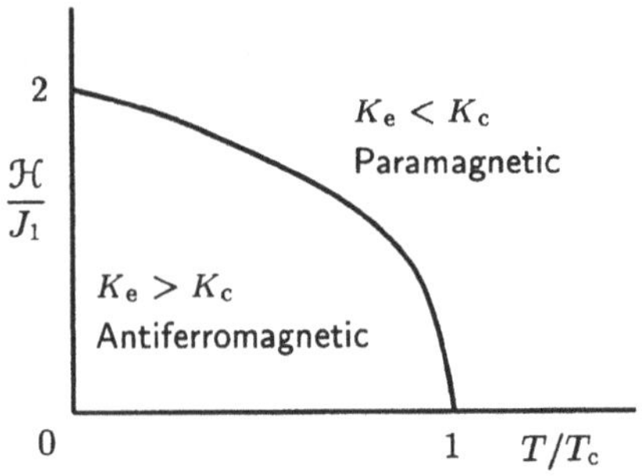

Fig. 9.4. The critical curve for Fisher's model.

$$\begin{aligned} m &= \frac{1}{2}(m_h + m_v) \\ &= \frac{1}{4}[1 + g_p(K_e, 0)][\tanh(2G + F) - \tanh(2G - F) - 2\tanh(F)] \\ &\quad + \tanh(F)\,, \end{aligned} \tag{9.64}$$

$$\begin{aligned} \psi &= \frac{1}{2}(m_h - m_v) \\ &= \frac{1}{2}m_s^{(\text{I.M.})}(K_e, 0)\{\tanh(2G + F) + \tanh(2G - F)\}\,, \end{aligned}$$

where m is the overall relative magnetization and ψ a sublattice order parameter defined as in (4.32). Since the primary lattice is square, $g_p = -\frac{1}{2}u^{(\text{I.M.})}/J_h$, where $u^{(\text{I.M.})}$ is given by (8.131) and (8.132) and $m_s^{(\text{I.M.})}$ is given by (8.144) with K_e replacing K in both cases. For $K_c > K_e$, $m_s^{(\text{I.M.})} = 0$.

From (9.50), (9.61) and (9.62) the configurational energy per primary site is given by

$$\begin{aligned} u = -\frac{J_h}{N}\frac{\partial \ln Z}{\partial G} &= -J_h[1 + g_p(K_e, 0)] \\ &\quad \times [\tanh(2G + F) + \tanh(2G - F)]\,. \end{aligned} \tag{9.65}$$

The condition $K_e = K_c$ yields, from (8.20),

$$\sinh^2(2G) = \cosh^2(F)[\exp(4K_c) - 1] = 2\cosh^2(F)(1 + \sqrt{2}) \tag{9.66}$$

between $G = J_h/T$ and $F = \mathcal{H}/T$. It can be shown that $\mathcal{H}$ is a monotonically decreasing function of T on the curve in the $(T, \mathcal{H})$ plane defined by (9.66); (sketched in Fig. 9.4). Inside the curve $K_e > K_c$ and hence by (9.64), $\psi \neq 0$ while outside the curve, $K_e < K_c$ and $\psi = 0$. It follows that the curve defined by (9.66) is the *critical curve* between the antiferromagnetic (sublattice ordered) and paramagnetic regions in the $(T, \mathcal{H})$ plane. From (9.66), $F/2G \to 1$ as $T \to 0$ and the critical curve meets the positive $\mathcal{H}$-axis at $\mathcal{H} = 2J_h = \mathcal{H}_c$, as would be expected from ground-state considerations. The critical curve meets the T-axis orthogonally at $T = T_c$, where T_c is the Néel temperature. Putting $F = 0$ in (9.66), gives

$$\begin{aligned}&\sinh(G_c) = \sinh(2J_h/T_c) = \sqrt{2}(1+\sqrt{2})^{1/2}\,,\\&T_c = 1.3084 J_h\,.\end{aligned} \tag{9.67}$$

On the T-axis it can be seen, from (9.64) with $F = 0$, that

$$m = 0,$$

$$m_h = -m_v = \begin{cases} 0\,, & T > T_c\,,\\ m_s^{(\mathrm{I.M})}(K_e, 0)\tanh(2G)\,, & T < T_c\,. \end{cases} \tag{9.68}$$

As $T \to 0$, $m_h = -m_v \to 1$.

We now consider the magnetic susceptibility. From (9.64),

$$\begin{aligned}16T\left(\frac{\partial m}{\partial \mathcal{H}}\right)_T &= 4[1+g_p(K_e,0)][\mathrm{sech}^2(2G+F)+4\mathrm{sech}^2(2G-F)]\\&\quad+8[1-g_p(K_e,0)]\mathrm{sech}^2(F)\\&\quad+g_p'(K_e)[\tanh(2G+F)-\tanh(2G-F)-2\tanh(F)]^2\,,\end{aligned} \tag{9.69}$$

$g_p'(K_e)$ being given by (9.27). With $F = 0$, the zero-field susceptibility

$$\chi_0 = C\left(\frac{\partial m}{\partial \mathcal{H}}\right)_T = \frac{C}{2T}\{\mathrm{sech}^2(2G)[1+g_p(K_e,0)]+1-g_p(K_e,0)\}\,, \tag{9.70}$$

where C is defined by (4.46). From (9.70), χ_0 is continuous for all T and tends to zero as $T \to 0$ or $T \to \infty$. From (8.119), $g_p'(K_e) \sim \ln|\Delta K_e|$ as $\Delta K_e = K_c - K_e \to 0$ so that the χ_0 against T curve has a vertical tangent at $T = T_c$. Its maximum is at a temperature $T > T_c$, as series methods indicate is the case for a loose-packed lattice Ising antiferromagnets.

Take a point $(T^*, \mathcal{H}^*)$, with $\mathcal{H}^* \neq 0$, on the critical curve (see Fig. 9.4). From (9.69), $\partial m/\partial\mathcal{H} \sim g_p'(K_e) \sim \ln|\Delta K_e|$ as $\Delta K_e \to 0$, provided that $\mathcal{H} \neq 0$. Hence as $T \to T^* \pm 0$ on the constant-field line $\mathcal{H} = \mathcal{H}^* \neq 0$, the magnetic susceptibility $\chi_T(T) \sim g_p'(K_e) \sim \ln|T - T^*|$. Thus χ_T has a logarithmic infinity at the critical temperature for all $\mathcal{H}$ such that $0 < |\mathcal{H}| < \mathcal{H}_c$ and its continuity on the $\mathcal{H} = 0$ axis is an exceptional case. It follows similarly that as $\mathcal{H} \to \mathcal{H}^* \pm 0$ on the constant-temperature line $T = T^*$ then $\partial m/\partial\mathcal{H} \sim \ln|\mathcal{H} - \mathcal{H}^*|$. Thus if m is plotted against $\mathcal{H}$ for $T = T^* < T_c$ the curve has a vertical tangent at the critical field value $\mathcal{H}^*$, instead of the finite slope discontinuity predicted for the loose-packed lattice Ising antiferromagnet by mean-field theory (Sect. 4.3). It can be deduced from (9.65) that the constant-field heat capacity $c(T, \mathcal{H}^*) \sim \ln|T - T^*|$ as $T \to T^* \pm 0$ for all $\mathcal{H}^*$ such that $|\mathcal{H}^*| < \mathcal{H}_c$, including $\mathcal{H}^* = 0$ (when $T^* = T_c$). Again from (9.64) the order parameter $\psi \sim (T^* - T)^{1/8}$ or $(\mathcal{H}^* - \mathcal{H})^{1/8}$ as the critical curve is approached along the lines $\mathcal{H} = \mathcal{H}^*$ or $T = T^*$ respectively.

Versions of the Fisher model are possible for $d = 3$. For instance, a body-centred cubic lattice of N sites can be divided into planes of sites passing through one cubic axis and making angles of $\pi/4$ with the other cubic axes. There are $2N$ edges within these planes and $2N$ edges connecting sites in adjacent planes. For the susceptibility and heat capacity the $\ln|T - T^*|$ or $\ln|\mathcal{H} - \mathcal{H}^*|$ dependence near the critical line is replaced by dependence on $|T-T^*|^{-\alpha}$ or $|\mathcal{H}-\mathcal{H}^*|^{-\alpha}$ where $\alpha \doteq \frac{1}{8}$. The order parameter $\psi \sim (T^* - T)^{\beta}$ or $(\mathcal{H}^* - \mathcal{H})^{\beta}$ where $\beta \doteq \frac{5}{16}$. If m rather than $\mathcal{H}$ is kept constant as T varies, renormalization of the exponents occurs (Fisher 1968). Accurate transition curves away from the $\mathcal{H} = 0$ axis are also obtainable for the ferrimagnetic model of the next section, although the entire $(T, \mathcal{H})$ plane does not map into the $L_e = 0$ axis, as it does in the Fisher model.

9.5 The Decorated Lattice Ferrimagnet

In this type of model (Syozi and Nakono 1955, Syozi 1972, Bell 1974b) the secondary and primary sites are occupied by Ising spins with magnetic moments m_0 and rm_0 respectively. There is an antiferromagnetic primary–secondary interaction. The model considered here can be related to the ferrimagnetic model treated in Sect. 4.5 by putting $N_a = N$, $N_b = \frac{1}{2}zN$ and, thus, $s = z/2$, where z is the coordination number of the primary lattice. For conformity with Sect. 4.5 the labels 'a' and 'b' will be used in this section for quantities connected with the primary and secondary sites respectively.[4] From condition (2) of Sect. 9.1 there is no interaction between the b spins but we assume a direct ferromagnetic interaction between the spins of nearest-neighbour pairs on the a lattice. With interaction constants J_{ab} and J_{aa}

$$\hat{h}(\sigma_i, \sigma_j; \sigma_{ij}) = J_{ab}\sigma_{ij}(\sigma_i + \sigma_j) - J_{aa}\sigma_i\sigma_j - \mathcal{H}\sigma_{ij}\,, \qquad J_{ab}, J_{aa} > 0\,. \tag{9.71}$$

Unless stated otherwise, it will be assumed that $z > 2r$. For small $|\mathcal{H}|$ the ground state will be ferrimagnetically ordered with the b spins aligned with the field and the a spins against it. For larger $|\mathcal{H}|$ all spins will be aligned with the field in the ground state. At $T = 0$ transitions between the ground states will occur at $\mathcal{H} = \pm\mathcal{H}_c$. The critical field $\mathcal{H}_c$ can be obtained by putting $s = \frac{1}{2}z$, $z_{ba} = 2$ in (4.63), giving

$$\mathcal{H}_c = zJ_{ab}/r\,. \tag{9.72}$$

Defining

[4] The model shown in Fig. 4.8 is an example of a decorated lattice ferrimagnet based on a triangular lattice of a sites with the b sites decorating the edges of the a lattice.

$$
\begin{aligned}
&G = J_{\mathfrak{ab}}/T\,,\\
&G' = J_{\mathfrak{aa}}/T = \tau G\,, \qquad \tau = J_{\mathfrak{aa}}/J_{\mathfrak{ab}}\,,\\
&F = \mathcal{H}/T\,,
\end{aligned}
\tag{9.73}
$$

the partition function is given by (9.7) with

$$
\begin{aligned}
&\phi(\pm 1, \pm 1) = 2\cosh(2G \mp F)\exp(G')\,,\\
&\phi(+1, -1) = 2\cosh(F)\exp(-G')
\end{aligned}
\tag{9.74}
$$

and, since the $\mathfrak{a}$ site magnetic moment is equal to rm_0,

$$
\xi(\pm 1) = \mp r\mathcal{H}\,. \tag{9.75}
$$

The relative magnetization $m_{\mathfrak{a}}$ is equal to that of the equivalent Ising model whereas $m_{\mathfrak{b}}$ is given by (9.58) with $G_{\mathrm{x}} = -G$, $F_{\mathrm{x}} = F$ and $g_{\mathrm{p,x}} = g_{\mathrm{p}}$, defined by (9.10). Hence

$$
\begin{aligned}
m_{\mathfrak{a}} &= m^{(\mathrm{I.M.})}(K_{\mathrm{e}}, L_{\mathrm{e}}),\\
m_{\mathfrak{b}} &= \tanh(F) + \frac{1}{4}[1 + g_{\mathrm{p}}(K_{\mathrm{e}}, L_{\mathrm{e}})]\\
&\quad \times\,[\tanh(2G + F) - \tanh(2G - F) - 2\tanh(F)]\\
&\quad - \frac{1}{2}m^{(\mathrm{I.M.})}(K_{\mathrm{e}}, L_{\mathrm{e}})[\tanh(2G + F) + \tanh(2G - F)]\,.
\end{aligned}
\tag{9.76}
$$

The overall relative magnetization is given by

$$
m = \frac{2rm_{\mathfrak{a}} + zm_{\mathfrak{b}}}{2r + z}\,. \tag{9.77}
$$

As $T \to 0$, $m \to (z - 2r)/(z + 2r)$ when $|\mathcal{H}| < \mathcal{H}_{\mathrm{c}}$ whereas $m \to 1$ for $|\mathcal{H}| > \mathcal{H}_{\mathrm{c}}$. Now consider the $\mathcal{H} = 0$ axis. From (9.74), (9.75) and (9.9), $\mathcal{H} = F = 0$ implies $L_{\mathrm{e}} = 0$ and then putting $K_{\mathrm{e}} = K_{\mathrm{c}}$,

$$
\cosh(2G_{\mathrm{c}})\exp(2\tau G_{\mathrm{c}}) = \exp(2K_{\mathrm{c}})\,, \tag{9.78}
$$

for the critical temperature $T_{\mathrm{c}} = J_{\mathfrak{ab}}/G_{\mathrm{c}}$. At $\mathcal{H} = 0$, from (9.76),

$$
\begin{aligned}
m_{\mathfrak{a}} &= m^{(\mathrm{I.M.})}(K_{\mathrm{e}}, 0) = \pm m_{\mathrm{s}}^{(\mathrm{I.M.})}(K_{\mathrm{e}})\,,\\
m_{\mathfrak{b}} &= -\tanh(2G)m^{(\mathrm{I.M.})}(K_{\mathrm{e}}, 0) = \mp\tanh(2G)m_{\mathrm{s}}^{(\mathrm{I.M.})}(K_{\mathrm{e}})\,,
\end{aligned}
\tag{9.79}
$$

where $m_{\mathrm{s}}^{(\mathrm{I.M.})}(K_{\mathrm{e}})$ (> 0) is the spontaneous magnetization of the Ising ferromagnet. Hence (9.77) yields, for the spontaneous magnetization of the ferrimagnet,

$$
m_{\mathrm{s}} = \pm m_{\mathrm{s}}^{(\mathrm{I.M.})}(K_{\mathrm{e}})\left(\frac{2r - z\tanh(2G)}{2r + z}\right)\,. \tag{9.80}
$$

To avoid ambiguity we specify the limit $\mathcal{H} \to +0$ so that the $\pm$ sign must be chosen to make $m_s > 0$. When $T > T_c$, $K_e < K_c$ so that $m_s^{(\text{I.M.})} = 0$ while, as T decreases from T_c to 0, $\tanh(2G)$ increases from $\tanh(2G_c)$ to 1. Since $r < \frac{1}{2}z$, $m_a = -m_s^{(\text{I.M.})} = -1$ at $T = 0$, in accordance with our discussion of the ground state. We can distinguish three cases:

(a) $r < \frac{1}{2}z\tanh(2G_c)$,

(b) $r = \frac{1}{2}z\tanh(2G_c)$,

(c) $\frac{1}{2}z > r > \frac{1}{2}z\tanh(2G_c)$.

The plots of m against T are qualitatively similar to the curves for $r = 2$,r_0 and 2.6 respectively of Fig. 4.9. For case (c) there is a compensation temperature $T_{\text{comp}} < T_c$, given by

$$r = \frac{1}{2}z\tanh(2J_{ab}/T_{\text{comp}}), \tag{9.81}$$

where $m_s = 0$ because the first factor on the right-hand side of (9.80) is zero. For $T_{\text{comp}} < T < T_c$

$$m_a = +m_s^{(\text{I.M.})}(K_e) > 0, \qquad m_b < 0, \tag{9.82}$$

but, for $0 < T < T_{\text{comp}}$,

$$m_a = -m_s^{(\text{I.M.})}(K_e) < 0, \qquad m_b > 0. \tag{9.83}$$

Both m_a and m_b change sign discontinuously at $T = T_{\text{comp}}$ although $|m_a|$ and $|m_b|$ are continuous.

Since $m_s^{(\text{I.M.})}(K_e) \sim |\Delta K_e|^\beta$ as $\Delta K_e = K_e - K_c \to 0$, β being the Ising exponent, (9.80) gives $m_s \sim (T_c - T)^\beta$ as $T \to T_c - 0$, except in case (b). For case (b), $r - \frac{1}{2}z\tanh(2G) \sim T_c - T$ as $T \to T_c - 0$ and hence $m_s \sim (T_c - T)^{\beta+1}$. Since $0 < \beta < 1$ for both $d = 2$ and $d = 3$ lattices it follows that the plot of m_s against T at $\mathcal{H} = 0$ has a vertical tangent at $T = T_c$ except for case (b) where the tangent is horizontal. The zero-field susceptibility can be deduced from (9.76), (Example 9.2).

We now consider the $(T, \mathcal{H})$ plane of the decorated ferrimagnet. In addition to the segment $[0, T_c]$ of the $\mathcal{H} = 0$ axis, where m undergoes a discontinuous change of sign, there are first-order transition lines on which $|\mathcal{H}| \neq 0$ and for which exact relations can be derived (Bell 1974b). This arises from the fact that the mapping from the $(T, \mathcal{H})$ plane of the decorated ferrimagnet into the (K_e^{-1}, L_e) plane of the Ising model is not one-to-one. Putting $L_e = 0$ in the second relation of (9.9) and substituting from (9.74) and (9.75) gives

$$\left[\frac{\cosh(2G - F)}{\cosh(2G + F)}\right] \exp(4rF/z) = 1. \tag{9.84}$$

As would be expected, $F = 0$ maps into $L_e = 0$. However, when $r < \frac{1}{2}z$ there is another branch in the $(T, \mathcal{H})$ plane which maps into $L_e = 0$ and which is given, after rearrangement of (9.84), by

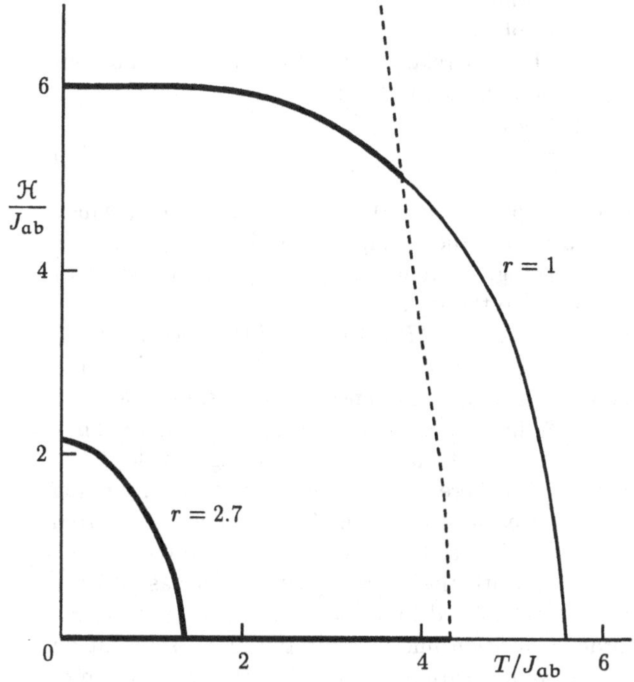

Fig. 9.5. The $L_e = 0$ locus and first-order transition (*heavy*) lines for the decorated ferrimagnet with $J_{aa} = J_{bb}$. (Reprinted from Bell (1974b), by permission of the publisher IOP Publishing Ltd.)

$$\exp(4G) = \frac{\sinh[(1 + 2r/z)F]}{\sinh[(1 - 2r/z)F]}. \tag{9.85}$$

This curve is clearly symmetrical about $\mathcal{H} = 0$ and meets the $T = 0$ axis where $\mathcal{H} = \pm zJ_{ab}/r = \pm\mathcal{H}_c$.

The branch (9.85) of the $L_e = 0$ locus cuts the T-axis orthogonally. Taking the limit of the right-hand side of (9.85) as $F \to 0$ gives, after a little rearrangement, the formula

$$r = \frac{1}{2}z\tanh(2G) \tag{9.86}$$

for the temperature at the point of intersection. The intersection of the branch (9.85) of the $L_e = 0$ locus with the T-axis occurs at a temperature $T > T_c$ for case (a) and at $T = T_{comp} < T_c$ for case (c). The two cases are illustrated by the curves for $r = 1$ and $r = 2.7$ respectively in Fig. 9.5. These are derived from the model of Fig. 4.8, where the b sites decorate the edges of a triangular lattice of a sites, with $J_{aa} = J_{ab}$, which gives $T_c = 4.4354J_{ab}$ and $\frac{1}{2}z\tanh(2G_c) = 1.2679$. In the $\mathcal{H} > 0$ half-plane, L_e is negative between the T-axis and the curve (9.85) and positive outside this curve, these signs being reversed in the $\mathcal{H} < 0$ half-plane.

By substituting (9.74) into the first relation of (9.9) and a little manipulation, the $K_e = K_c$ curve can be obtained in the form

$$\exp[4(K_c - G')] - 1 = \frac{\sinh^2(2G)}{\cosh^2(F)}. \tag{9.87}$$

Equation (9.78) is reproduced by putting $F = 0$. As $|F| \to \infty$, $G' \to K_c$ which implies that $T \to J_{aa}/K_c$ and defines the asymptotes of the $K_e = K_c$ curves as $|\mathcal{H}| \to \infty$. The broken line in Fig. 9.5 represents the $K_e = K_c$ curve, with $K_e > K_c$ to the left and $K_e < K_c$ to the right.

The first-order transition lines in the $(T, \mathcal{H})$ plane of the decorated ferrimagnet are the images of the segment $0 < K_e^{-1} < K_c^{-1}$, $L_e = 0$ from the (K_e^{-1}, L_e) plane of the Ising model. So these lines of discontinuity are the parts of the $L_e = 0$ locus which lie to the left of the $K_e = K_c$ curve. They are represented by heavy lines in Fig. 9.5 and include the segment $[0, T_c]$ of the T-axis in all cases. For case (a) there are two other first-order transition lines which are mirror images of each other in the T-axis. They start from the points $\mathcal{H} = \pm\mathcal{H}_c$ on the $\mathcal{H}$-axis and terminate at critical points at the intersections of the $L_e = 0$ locus with the $K_e = K_c$ curve. For case (c) the entire branch of the $L_e = 0$ locus defined by (9.85) lies in the $K_e > K_c$ region and is thus a first-order transition line crossing the $[0, T_c]$ segment on the T-axis at $T = T_{comp}$. The compensation point is thus the intersection of first-order transition lines as indicated by the mean-field theory of Sect. 4.5. The configurational energy is continuous over the segment $[0, T_c]$ of the T-axis but, together with the magnetization m, it is discontinuous at first-order transition points for which $\mathcal{H} \neq 0$.

The phase pattern is different when $J_{aa} = 0$. Although the $L_e = 0$ locus is unaffected there is a significant change in the $K_e = K_c$ curve. Putting $G' = 0$ in (9.87), it can be seen that this curve no longer goes to $F = \pm\infty$ for non-zero T but turns to meet the $\mathcal{H}$-axis at $\mathcal{H} = \pm 2J_{ab}$. This is illustrated by the broken curve in Fig. 9.6 which is drawn for a model similar to that of Fig. 9.5 but with $J_{aa} = 0$ so that $T_c = 1.7449 J_{ab}$ and $\frac{1}{2}z\tanh(2G_c) = 2.4495$. Since $2J_{ab} < \mathcal{H}_c$ when $r < \frac{1}{2}z$, the transition points on the $\mathcal{H}$-axis at $\mathcal{H} = \pm\mathcal{H}_c$ are now isolated since $K_e < K_c$ in the adjacent parts of the $(T, \mathcal{H})$ plane. For case (a) there are no transitions except on the T-axis since the branch of the $L_e = 0$ locus given by (9.85) lies entirely in the $K_e < K_c$ region of the $(T, \mathcal{H})$ plane. Case (c) with $r = 2.7$ is illustrated in Fig. 9.6. The first-order transition segment $[0, T_c]$ on the T-axis is crossed at the compensation point $T = T_{comp}$ by another first-order transition line on the branch of the $L_e = 0$ locus given by (9.85). This transition line terminates at critical points (like B in Fig. 9.6 and its mirror image) at the intersection of the $L_e = 0$ locus with the $K_e = K_c$ curves. These critical points are reached by *reducing* the temperature from T_{comp} and so are the first examples we encounter of *lower critical points*.

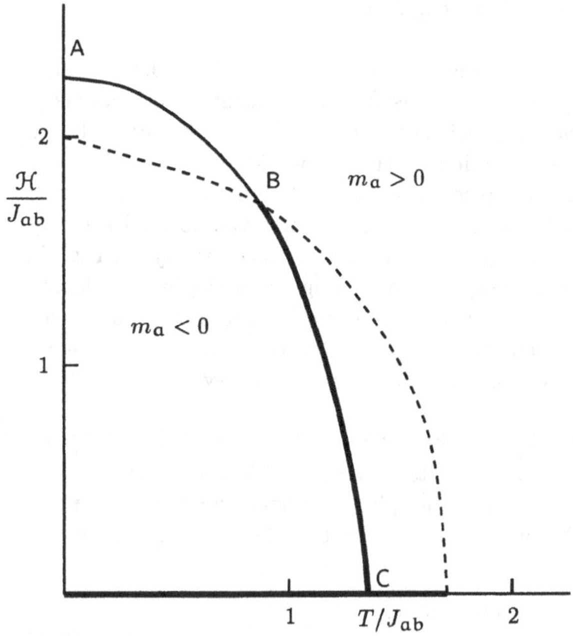

Fig. 9.6. As Fig. 9.5 for $J_{aa} = 0$, $r = 2.7$. (Reprinted from Bell (1974b), by permission of the publisher IOP Publishing Ltd.)

We now consider briefly the situation when $r \geq \frac{1}{2}z$. The first factor on the right-hand side of (9.80) is now positive and hence $m^{(\text{I.M.})} = +m_s^{(\text{I.M.})}$ for the entire range $0 < T < T_c$ with $\mathcal{H} = +0$. For $r = \frac{1}{2}z$, $m_s \to 0$ as $T \to 0$ and the plot of m_s against T is qualitatively similar to the $r = 3$ curve of Fig. 4.9(ii) whereas for $r > \frac{1}{2}z$ the plot is similar to the $r = 4$ curve of Fig. 4.9(ii). As $r \to \frac{1}{2}z - 0$ the branch of the $L_e = 0$ locus given by (9.85) shrinks to a segment on the $\mathcal{H}$-axis and it is absent for $r > \frac{1}{2}z$. The only transition line is the $(0, T_c)$ segment of the T-axis.

There is general qualitative agreement as regards the phase diagrams between the exact theory of the present section and the mean-field theory of Sect. 4.5, with $J_{bb} = 0$. However, critical parameter values differ considerably. For instance when $J_{aa} = 0$ the exact theory above gives $r > \frac{1}{2}z\tanh(2G_c) = 2.4495$ as a condition for the occurrence of a compensation point in the model of Fig. 4.8. The mean-field equation (4.73) gives $r > r_0 = \sqrt{s} = \sqrt{3}$.

9.6 The Kagomé Lattice Ising Model

As well as the decoration transformation discussed in Sect. 9.1 the star-triangle transformation of Sect. 8.4 is useful for relating other models to the Ising model on lattices for which accurate results are available. In the present section the partition function of the zero-field Ising model on the two-dimensional Kagomé lattice (see Appendix A.1) is related to the partition function of the zero-field Ising model on the honeycomb lattice by applying decoration and star-triangle transformations successively. In Fig. 4.8 the b sites, which decorate the edges of the triangular a site lattice, form a Kagomé lattice. Since the triangular and honeycomb lattices are dual to each other there is a one-to-one correspondence between their edges. As shown in Fig. 9.7, the secondary sites decorating the edges of a honeycomb lattice also form a Kagomé lattice.

Take a honeycomb lattice of N sites (the white sites of Fig. 9.7) and decorate each edge with a secondary site (one of the black sites of Fig. 9.7). Suppose all sites are occupied by Ising spins and that there is a primary–secondary interaction with interaction constant $J' > 0$. By (9.3), in zero magnetic field,

$$\begin{aligned} \phi(+1,+1) &= \phi(-1,-1) = 2\cosh(2\mathrm{G}')\,, \\ \phi(+1,-1) &= 2\,, \end{aligned} \tag{9.88}$$

where

$$\mathrm{G} = J'/T \tag{9.89}$$

and

$$\xi(+1) = \xi(-1) = 0\,. \tag{9.90}$$

Hence, by (9.9),

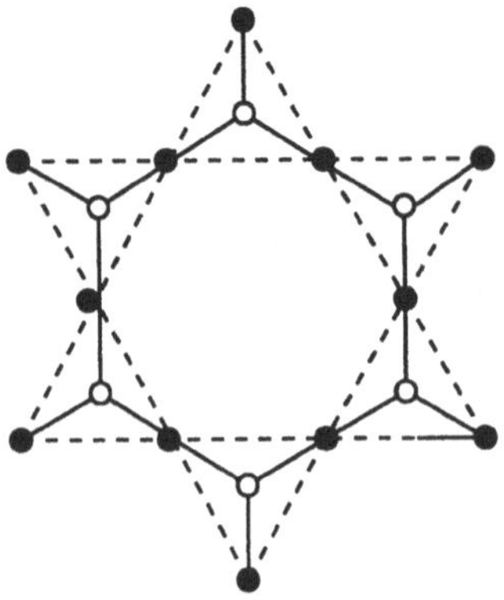

Fig. 9.7. The decorated honeycomb lattice.

$$\exp(2K_e) = \cosh(2G)\,, \qquad L_e = 0 \tag{9.91}$$

and, from (9.7), the partition function Z of the decorated model is given in terms of the partition function $Z^{(\mathrm{H})}$ of the zero-field honeycomb lattice Ising ferromagnet by

$$Z = [64\cosh^3(2G)]^{N/4} Z^{(\mathrm{H})}(N, K_e) \tag{9.92}$$

noting that, by (9.91), $K_e > 0$.

The next step is a star-triangle transformation which removes the white sites (those of the original honeycomb lattice) and replaces the interactions along the full-line segments of Fig. 9.7 by interactions along the broken segments. By (8.28) and (8.27),

$$Z = \mathsf{R}^N Z^{(\mathrm{KG})}(3N/2, K)\,, \tag{9.93}$$

where

$$\begin{aligned} \exp(4K) &= 2\cosh(2G) - 1\,, \\ \mathsf{R}^4 &= \exp(4K)[\exp(4K) + 3]^2\,. \end{aligned} \tag{9.94}$$

Since this implies $K > 0$, $Z^{(\mathrm{KG})}$ is the partition function for a zero-field Ising ferromagnet on the Kagomé lattice and $K = (J/T)_{\mathrm{KG}}$. Combining (9.92) and (9.93) and replacing N by $\frac{2}{3}N$ yields the relation

$$Z^{(\mathrm{KG})}(N, K) = \left\{\frac{8[\exp(4K) + 1]^3\exp(-4K)}{[\exp(4K) + 3]^2}\right\}^{N/6} Z^{(\mathrm{H})}(2N/3, K_e)\,, \tag{9.95}$$

where, from (9.91) and (9.94),

$$\exp(2K_e) = \frac{1}{2}[\exp(4K) + 1]\,. \tag{9.96}$$

From (8.39) the honeycomb Ising model critical temperature is given by $\cosh(2K_e) = 2$ or $\exp(2K_e) = 2 + \sqrt{3}$. Thus, from (9.96) the Kagomé Ising ferromagnet critical temperature is given by

$$\exp(4K_c) = 3 + 2\sqrt{3}\,, \qquad T_c = 2.1433J\,. \tag{9.97}$$

This is slightly lower than the value $T_c = 2.2692J$ for the square lattice, which also has coordination number $z = 4$.

From arguments similar to those used in Sects. 9.4 and 9.5, the zero-field relative magnetization on the decorating (black) sites is $\tanh(2G)m_s^{(\mathrm{H})}(K_e)$ where $m_s^{(\mathrm{H})}$ is the spontaneous magnetization for the honeycomb lattice Ising model. Since the black sites form the Kagomé lattice the spontaneous magnetization

$$\begin{aligned} m_s^{(\mathrm{KG})}(K) &= \tanh(2G)m_s^{(\mathrm{H})}(K_e) \\ &= \left[\frac{(1 + 3\mathfrak{X})(1 - \mathfrak{X})}{(1 + \mathfrak{X})^2}\right]^{1/2}\left[1 - 128\frac{\mathfrak{X}^3(1 + \mathfrak{X})^3(1 + 3\mathfrak{X}^2)}{(1 - \mathfrak{X}^6)(1 + 3\mathfrak{X})^2}\right]^{1/8}, \end{aligned} \tag{9.98}$$

where $\mathfrak{X} = \exp(-4K)$. The $\tanh(2G)$ factor was transformed using (9.91) and (9.96) while the expression (8.163) for $m_s^{(H)}$, with K_e substituted for G, was transformed using (9.96). The result (9.98) was obtained by Naya (1954).[5]

9.7 A Modified Star-Triangle Transformation and Three-Spin Correlations

In Sect. 8.4 the zero-field Ising model partition function for the honeycomb lattice was related to that for the triangular lattice by using a star-triangle transformation to remove half the honeycomb sites. In this section we generalize this procedure to the situation where the honeycomb Ising model is acted on by a field $\mathcal{H}$, it being assumed that all spins have the same magnetic moment m_0. In the limit $\mathcal{H} \to 0$ this yields a simple method of determining the correlation for the three spins on an elementary triangle of the triangular lattice which is used in the alternative ferrimagnetic model of the next section.

Putting $L = \mathcal{H}/T$, the expression $\cosh[G(\alpha + \beta + \gamma)]$ in (8.24) and (8.25) is modified to $\cosh[G(\alpha + \beta + \gamma) + L]$ which has four distinguishable values, corresponding to $\alpha = \beta = \gamma = \pm 1$ and $\alpha = \beta = -\gamma = \pm 1$. Accordingly four adjustable parameters are needed on the right-hand side of the relation corresponding to (8.25) which is modified to

$$\begin{aligned} 2\cosh[G(\alpha + \beta + \gamma) + L] = \mathsf{R}\exp[&K(\alpha\beta + \beta\gamma + \gamma\alpha) \\ &+ L_1(\alpha + \beta + \gamma) + K_3\alpha\beta\gamma] . \end{aligned} \tag{9.99}$$

The new parameters L_1 and K_3 correspond respectively to a field acting on the triangular lattice and a three-spin interaction energy. Defining

$$\theta(x, y) = \cosh(xG + yL) , \tag{9.100}$$

(9.99) yields the four relations

$$\begin{aligned} 2\theta(3, +1) &= \mathsf{R}\exp(3K + 3L_1 + K_3) , \\ 2\theta(3, -1) &= \mathsf{R}\exp(3K - 3L_1 - K_3) , \\ 2\theta(1, +1) &= \mathsf{R}\exp(-K + L_1 - K_3) , \\ 2\theta(1, -1) &= \mathsf{R}\exp(-K - L_1 + K_3) . \end{aligned} \tag{9.101}$$

Hence the four parameters R, K, L_1 and K_3 are given in terms of G and L by

[5] See Syozi 1972 for some related transformations.

$$\begin{aligned}
\mathrm{R}^8 &= 256\theta(3,+1)\theta(3,-1)[\theta(1,+1)\theta(1,-1)]^3\,,\\
\exp(8K) &= \frac{\theta(3,+1)\theta(3,-1)}{\theta(1,+1)\theta(1,-1)}\,,\\
\exp(8L_1) &= \frac{\theta(3,+1)\theta(1,+1)}{\theta(3,-1)\theta(1,-1)}\,,\\
\exp(8K_3) &= \frac{\theta(3,+1)[\theta(1,-1)]^3}{\theta(3,-1)[\theta(1,+1)]^3}\,.
\end{aligned} \tag{9.102}$$

When $L = 0$, (9.102) reduces to $L_1 = 0$, $K_3 = 0$ and the relations for R and K given by (8.27).

Taking the number of original honeycomb sites as $2N$, half of these (the white sites of Fig. 8.3) are removed by performing the star-triangle transformation over all the downward pointing triangles. From (9.99) this yields

$$\begin{aligned}
Z^{(\mathrm{H})}(2N,G,L) &= \mathrm{R}^N \sum_{\{\sigma_i\}}^{(\mathrm{T})} \exp[-\widehat{H}(\sigma_i)/T]\,,\\
\widehat{H}(\sigma_i)/T &= -K\sum_{\{i,j\}}^{(\mathrm{n.n.})}\sigma_i\sigma_j - (3L_1+L)\sum_{\{i\}}\sigma_j - K_3\sum_{\{i,j,\ell\}}\sigma_i\sigma_j\sigma_\ell\,.
\end{aligned} \tag{9.103}$$

The sum on the right-hand side of the first of these equations is the partition function for Ising spins on the triangular lattice of N sites (the black sites of Fig. 8.3) with a field and a three-spin interaction. The latter is represented by the term with coefficient K_3 in which the summation is over the N downward pointing triangles in the triangular lattice. In the field coefficient $3L_1 + L$, the term L is present because each black site spin has a magnetic moment m_0 and the $3L_1$ term because each black site belongs to three downward pointing triangles. For $L = 0$ the triangular lattice partition function reduces to that of a standard pair interaction Ising model in zero field. From (9.103),

$$\begin{aligned}
m^{(\mathrm{H})} &= \frac{1}{2N}\frac{\partial \ln Z^{(\mathrm{H})}}{\partial L}\\
&= \frac{1}{2}\left[\frac{\partial \ln \mathrm{R}}{\partial L} + \frac{3}{2}\langle\sigma_i\sigma_j\rangle\frac{\partial K}{\partial L} + m^{(\mathrm{T})}\left(1+3\frac{\partial L_1}{\partial L}\right) + \langle\sigma_i\sigma_j\sigma_\ell\rangle\frac{\partial K_3}{\partial L}\right]\,.
\end{aligned} \tag{9.104}$$

From (9.102), $\partial\mathrm{R}/\partial L = \partial K/\partial L = 0$ at $L = 0$ and, with $L = 0$ in (9.104),

$$\begin{aligned}
m_{\mathrm{s}}^{(\mathrm{H})}(G) = \frac{1}{2}\Big\{&m_{\mathrm{s}}^{(\mathrm{T})}(K) + \frac{3}{4}m_{\mathrm{s}}^{(\mathrm{T})}(K)[\tanh(3G)+\tanh(G)]\\
&+\frac{1}{4}\langle\sigma_i\sigma_j\sigma_\ell\rangle[\tanh(3G) - 3\tanh(G)]\Big\}\,,
\end{aligned} \tag{9.105}$$

where $m_{\mathrm{s}}^{(\mathrm{H})}$ and $m_{\mathrm{s}}^{(\mathrm{T})}$ are spontaneous magnetizations for the Ising ferromagnet on honeycomb and triangular lattices respectively. However, since the triangular lattice is formed from one sublattice of the honeycomb lattice by the star-triangle transformation it follows that $m_s^{(H)}(G) = m_s^{(T)}(K)$ where G and K are connected by (8.27).[6] The three-spin correlation in zero field is therefore given by

$$\begin{aligned}\langle\sigma_i\sigma_j\sigma_k\rangle &= m_s^{(T)}(K)\frac{4 - 3[\tanh(3G) + \tanh(G)]}{\tanh(3G) - 3\tanh(G)}, \\ &= m_s^{(T)}(K)\left\{3 + \frac{1+\mathfrak{X}}{1-\mathfrak{X}} - 2\left[\frac{1+3\mathfrak{X}}{(1-\mathfrak{X})^3}\right]^{1/2}\right\}, \qquad (9.106)\end{aligned}$$

where $\mathfrak{X} = \exp(-4K)$. Here $\langle\sigma_i\sigma_j\sigma_\ell\rangle$ is evaluated for a downward pointing triangle of sites, but for an isotropic triangular lattice the value for an upward pointing triangle of sites will be the same by symmetry. Using a different method, Baxter (1975) obtained the result (9.106) and the generalization to unequal interactions in the three triangular lattice directions. Enting (1977) developed an alternative approach and gave a brief outline of the variant of his method which is used here.

9.8 The Star-Triangle Ferrimagnet

An alternative ferrimagnet based on the triangular lattice is shown in Fig. 9.8.[7] The $\mathfrak{a}$ sites form a triangular sublattice of the original triangular lattice and the $\mathfrak{b}$ sites a honeycomb sublattice of the original triangular lattice so that $N_\mathfrak{b} = 2N_\mathfrak{a}$ and $s = 2$. Magnetic moments rm_0 and m_0 are associated respectively with the spins on the $\mathfrak{a}$ and $\mathfrak{b}$ sites. There are antiferromagnetic interactions between the spins of $\mathfrak{ab}$ nearest-neighbour pairs, that is along the edges indicated by full-line segments in Fig. 9.8, but no $\mathfrak{bb}$ nearest-neighbour interactions. Hence each $\mathfrak{b}$ spin interacts only with the $\mathfrak{a}$ spins at the vertices of the triangle with the $\mathfrak{b}$ site at its centre. A star-triangle transformation can thus be performed to transfer the interactions to the broken line segments in Fig. 9.8, which form the edges of the triangular lattice of $\mathfrak{a}$ sites.

The formalism of the previous section can be used, with some modifications. Since there is a $\mathfrak{b}$ site inside *every* triangle of $\mathfrak{a}$ sites, R, K and $3L_1$ in (9.103) must be replaced by R^2, $2K$ and $6L_1$ respectively. As the $\mathfrak{ab}$ interaction is antiferromagnetic, G must be replaced by $-G$ which, from (9.100) and (9.102) leaves R and K invariant but changes L_1 and K_3 to $-L_1$ and $-K_3$ respectively. Hence the partition function for the ferrimagnet is given by

[6] This result can be verified from (8.162) and (8.163).

[7] This is model (i) of Bell (1974a, 1974b). See also Lavis and Quinn (1983, 1987).

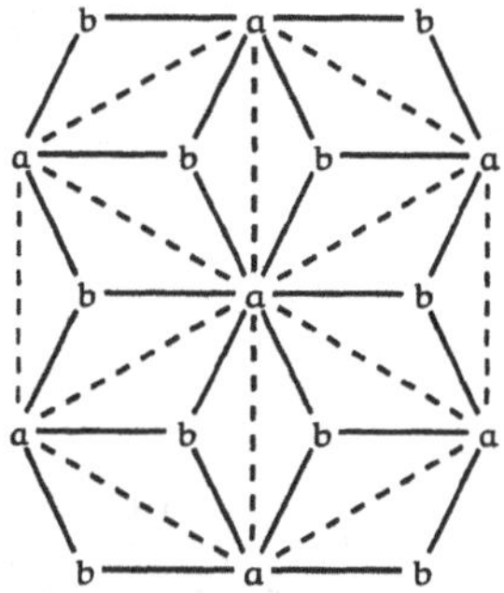

Fig. 9.8. The star-triangle ferrimagnet.

$$Z(G,L) = \mathsf{R}^{2N_a} \sum_{\{\sigma_i\}}^{(a)} \exp\{-\widehat{H}(\sigma_i)/T\}\,,$$
$$\widehat{H}(\sigma_i)/T = K' \sum_{\{i,j\}}^{(\text{n.n.})} \sigma_i\sigma_j + (L' - 6L_1)\sum_{\{i\}} \sigma_i - K_3 \sum_{\{i,j,\ell\}} \sigma_i\sigma_j\sigma_\ell\,, \tag{9.107}$$

where

$$K' = 2K\,, \qquad L' = rL\,, \tag{9.108}$$

R, K, L_1 and K_3 being defined by (9.102). The sum over $\{i,j,\ell\}$ is taken over all the $2N_a$ triangles of a sites.

Treating L and L', for the moment, as independent variables the zero-field relative magnetizations m_a and m_b, are given by

$$m_a = \frac{1}{N_a}\left(\frac{\partial \ln Z}{\partial L'}\right)_{L=L'=0} = m_s^{(T)}(K')\,,$$
$$m_b = \frac{1}{2N_a}\left(\frac{\partial \ln Z}{\partial L}\right)_{L=L'=0} = -3m_s^{(T)}(K')\left(\frac{\partial L_1}{\partial L}\right)_{L=0} - \langle\sigma_i\sigma_j\sigma_\ell\rangle\left(\frac{\partial K_3}{\partial L}\right)_{L=0}. \tag{9.109}$$

From (9.108) and either (9.102) or (8.27)

$$\exp(2K') = 2\cosh(2G) - 1\,. \tag{9.110}$$

The critical temperature for the ferrimagnet can be obtained by putting $K' = K_c$ where K_c is the critical parameter value for the triangular lattice. Thus

$$\cosh(2G_c) = \tfrac{1}{2}[\exp(2K_c) + 1] = \tfrac{1}{2}(\sqrt{3} + 1)\,,$$
$$T_c = J/G_c = 2.4055J\,. \tag{9.111}$$

From (9.109) the overall spontaneous magnetization m_s is given by

$$\begin{aligned}(r+2)m_{\mathrm{s}} &= rm_{\mathfrak{a}} + 2m_{\mathfrak{b}} \\ &= m_{\mathrm{s}}^{(\mathrm{T})}(K')\left\{r - \frac{3}{2}[\tanh(3G)+\tanh(G)]\right\} \\ &\quad - \frac{1}{2}\langle\sigma_i\sigma_j\sigma_\ell\rangle[\tanh(3G) - 3\tanh(G)]\,, \\ &= rm_{\mathrm{s}}^{(\mathrm{T})}(K') - \frac{2}{2\mathrm{c}-1}\left(\frac{\mathrm{c}-1}{\mathrm{c}+1}\right)^{1/2} \\ &\quad \times\left[3\mathrm{c}m_{\mathrm{s}}^{(\mathrm{T})}(K') - \langle\sigma_i\sigma_j\sigma_\ell\rangle(\mathrm{c}-1)\right]\,, \end{aligned} \tag{9.112}$$

where $\mathrm{c} = \cosh(2G)$. The last expression of (9.106) can be used for $\langle\sigma_i\sigma_j\sigma_\ell\rangle$, with K' substituted for K so that $\mathfrak{X} = \exp(-4K') = 1/(2\mathrm{c}-1)^2$. Finally

$$m_{\mathrm{s}} = \pm\frac{m_{\mathrm{s}}^{(\mathrm{T})}(K')[r - f(\mathrm{c})]}{r+2}\,, \tag{9.113}$$

where

$$f(\mathrm{c}) = \frac{(\mathrm{c}^2-1)^{1/2}}{\mathrm{c}(\mathrm{c}+1)}\left[3 + (2\mathrm{c}-1)\left(\frac{\mathrm{c}^3+1}{\mathrm{c}^3-\mathrm{c}}\right)^{1/2}\right]. \tag{9.114}$$

It has been assumed for simplicity that the triangular lattice spontaneous magnetization takes the positive value $m_{\mathrm{s}}^{(\mathrm{T})}$. However, putting $\mathcal{H} = +0$ implies $m \geq 0$, as in Sect. 9.5, and hence $+m_{\mathrm{s}}^{(\mathrm{T})}$ must be used in (9.113) when $r > f(\mathrm{c})$ and $-m_{\mathrm{s}}^{(\mathrm{T})}$ when $r < f(\mathrm{c})$. It can be verified that $f(\mathrm{c}) = 0$ at $T = \infty$ ($\mathrm{c} = 1$) and that $f(\mathrm{c})$ increases steadily as T decreases, assuming the value 2 at $T = 0$ ($\mathrm{c} = \infty$). At the critical temperature $T = T_{\mathrm{c}}$, where $\mathrm{c} = \frac{1}{2}(\sqrt{3}+1)$, $f(\mathrm{c}) = 4(\sqrt{3}/2)^{\frac{1}{2}}(2\sqrt{3}-3) = 1.72758$. Equation (9.113) was obtained by Lavis and Quinn (1983) using the results of Baxter (1975). For $2 > r > 1.72758$, $r = f(\mathrm{c})$ at some temperature T_{comp} in the range $(0, T_{\mathrm{c}})$ and, from (9.113), this defines a compensation point. The shapes of the curves of m_{s} against T are similar to those obtained in Sects. 4.5 and 9.5.

9.9 The Unsymmetrical Ising Model

When the Ising model represents a magnetic system it is natural to assume, as has been done until now, that the energies of ++ and −− pairs are equal and that the two spin states have equal statistical weight. However, in some other cases, it is useful to modify these conditions. It will be shown that the resulting unsymmetrical model is equivalent to a standard Ising model in a temperature dependent field.

Suppose that the $\sigma_i = 1$ and $\sigma_i = -1$ states represent different species of the same component and that ++ and −− nearest-neighbour pairs have unequal interaction energies. The zero-field Hamiltonian is then

$$\widehat{H} = \frac{1}{4} \sum_{\{i,j\}}^{(\text{n.n.})} \{J_{++}(1+\sigma_i)(1+\sigma_j) + J_{--}(1-\sigma_i)(1-\sigma_j)$$
$$+ J_{+-}[(1+\sigma_i)(1-\sigma_j) + (1-\sigma_i)(1+\sigma_j)]\}, \qquad (9.115)$$

which can be rearranged to give

$$\widehat{H} = \text{z}NR - J_\text{e} \sum_{\{i,j\}}^{(\text{n.n.})} \sigma_i\sigma_j + \Delta J \sum_{\{i\}} \sigma_i, \qquad (9.116)$$

where

$$R = \frac{1}{8}(2J_{+-} + J_{++} + J_{--}),$$
$$J_\text{e} = \frac{1}{4}(2J_{+-} - J_{++} - J_{--}), \qquad (9.117)$$
$$\Delta J = \frac{1}{4}\text{z}(J_{++} - J_{--}).$$

Suppose also that the states $\sigma_i = 1$ and $\sigma_i = -1$ are composed of substates, the degeneracies being ω_+ and ω_- respectively. Then the partition function is

$$Z = \sum_{\{\sigma_i\}} \omega_+^{N_+} \omega_-^{N_-} \exp(-\widehat{H}/T), \qquad (9.118)$$

where

$$N_\pm = \frac{1}{2} \sum_{\{i\}} (1 \pm \sigma_i). \qquad (9.119)$$

By comparison with (2.67)

$$Z = (\omega_+\omega_-)^{N/2} \exp[\text{z}NR] Z^{(\text{I.M.})}(T, \mathcal{H}_\text{e}), \qquad (9.120)$$

where

$$\mathcal{H}_\text{c} = \frac{1}{2} T \ln\left(\frac{\omega_+}{\omega_-}\right) - \Delta J. \qquad (9.121)$$

Here $Z^{(\text{I.M.})}$ is the partition function of a standard Ising model on the same lattice and at the same temperature as the unsymmetrical model but acted on by the field $\mathcal{H}_\text{e}$, given as a linear function of T by (9.121). The interaction constant J_e of the Ising model and the quantity ΔJ in (9.121) are given by (9.117).

Now assume $J_\text{e} > 0$ so that the equivalent Ising model is a ferromagnet. When ΔJ and $\ln(\omega_+/\omega_-)$ have the same sign, the line $\mathcal{H} = \mathcal{H}_\text{e}$ in the $(T, \mathcal{H})$ plane of the Ising model crosses the T-axis at $T = T_0$ where

$$T_0 = \frac{2\Delta J}{\ln(\omega_+/\omega_-)}. \qquad (9.122)$$

Let T_c be the zero-field Ising ferromagnet critical temperature for the given value of J_e. Then if $T_0 < T_c$ the point $T = T_0$ lies on the segment of discontinuity $(0, T_c)$ on the T-axis of the Ising model and the unsymmetrical model undergoes a first-order transition at $T = T_0$. Since the configurational energy contains the term $N\Delta J\langle\sigma_i\rangle$ there is a latent heat

$$\Delta U = 2N|\Delta J| m_s(T_0) = N|\ln(\omega_+/\omega_-)|T_0 m_s(T_0)\,, \tag{9.123}$$

where $m_s(T_0)$ is the spontaneous magnetization of the equivalent Ising model at $T = T_0$. When $T_0 > T_c$ or ΔJ and $\ln(\omega_+/\omega_-)$ have opposite signs or one of these quantities is non-zero but not the other then no transition occurs. Thus there is either a first-order transition or no transition at all, except in the physically improbable case $T_0 = T_c$. This theory is due to Bell and Fairbairn (1961b) who were concerned with the low temperature second-order transition in solid ortho-hydrogen. Denoting the axial-component angular momentum quantum number by M, under certain assumptions the $\mathsf{M} = \pm 1$ states can be identified with $\sigma_i = +1$ and the $\mathsf{M} = 0$ state with $\sigma_i = -1$, giving an unsymmetrical Ising model with $\omega_+/\omega_- = 2$. However, as we have seen, such a model does not display a second-order transition, and Bell and Fairbairn (1961b) also showed that, with the parameters given appropriate values, $T_0 > T_c$. They concluded that the transition cannot be a rotational one, and in later work (Bell and Fairbairn 1967) discussed sublattice ordering in the system. When the $\sigma_i = 1$ and $\sigma_i = -1$ states represent occupation by atoms of components A and B with degeneracy factors ω_A and ω_B the analysis showing equivalence to the Ising model is similar to that of Sect. 6.2 but the field $\mathcal{H}_e$ is now given by the relations

$$\begin{aligned} \mathcal{H}_e &= \frac{1}{2}\Delta\mu\,, \\ \Delta\mu &= \mu_A - \mu_B - \frac{1}{2}z(\varepsilon_{AA} - \varepsilon_{BB}) - T\ln(\omega_A/\omega_B)\,. \end{aligned} \tag{9.124}$$

For any $T < T_c$, where T_c is the critical temperature for the equivalent Ising ferromagnet, there is a first-order transition at the value of $\mu_A - \mu_B$ where $\mathcal{H}_e = 0$, the mole fractions in the conjugate phases being $x_A = \frac{1}{2}[1 \pm m_s(T)]$. The coexistence curve in the (x_A, T) plane is thus unaffected by the degeneracy factors, whose only effect is to add on a temperature dependent term to $\mu_A - \mu_B$. For $J = 0$ the dilute Ising model of Sect. 6.6.1 becomes a system of this type since, with the two spin orientations possible on sites occupied by A, $\omega_A/\omega_B = 2$. In this case the mean-field result that the coexistence curve is symmetrical about the line $x_A = \frac{1}{2}$ is exact. Griffiths (1967b) showed that when $J = 0$ the dilute Ising model is equivalent to a standard Ising model in a temperature dependent field.

9.10 A Competing Interaction Magnetic Model

Consider a decorated loose-packed lattice with the primary sites occupied by standard Ising spins with $s = \frac{1}{2}$ and the secondary sites by Ising spins of quantum number s_1, where s_1 can take any integer or half-integer value. As well as the primary–secondary interaction there is an antiferromagnetic primary–primary interaction. We assume that, in zero field,

$$\hat{h}(\sigma_i, \sigma_j; \sigma_{ij}) = -(J/q)\sigma_{ij}(\sigma_i + \sigma_j) + \alpha J \sigma_i \sigma_j\,, \tag{9.125}$$

where $J > 0$, $1 > \alpha > 0$ and $\sigma_{ij} = q, q-2, \ldots, -q$ with $q = 2s_1$ and $\sigma_i, \sigma_j = \pm 1$. The first term in $\hat{h}$ is scaled so that its largest magnitude is $2J$. Although the primary–secondary interaction term in (9.125) is taken to be ferromagnetic, an interaction of either sign tends to align the spins on the primary sites i and j. The primary–secondary interaction is thus in conflict with the antiferromagnetic primary–primary interaction which tends to make the spins on the primary sites i and j antiparallel.

From (9.125) the energy of a perfect ferromagnetic state with all primary site $\sigma_i = 1$ and all secondary site $\sigma_{ij} = q$ (or all $\sigma_i = -1$ and all $\sigma_{ij} = -q$) is $-\frac{1}{2}zN(2-\alpha)J$. It is also possible to have a state of perfect antiferromagnetic order on a loose-packed primary lattice in which all primary nearest-neighbour pairs have $\sigma_i + \sigma_j = 0$ and the energy is $-\frac{1}{2}zN\alpha J$. Since we assume $\alpha < 1$, the ground state attained at $T = 0$ is ferromagnetic. We thus expect ferromagnetic long-range order to be stable in a range $(0, T_c)$, where T_c is a critical temperature. However all the $q+1$ values of σ_{ij} are equally probable when the spins on primary sites i and j are antiparallel, contributing a term $k_B \ln(q+1)$ to the entropy. States with primary spins antiferromagnetically ordered have a high entropy. Ordered states with this unusual property can be destabilized by either an increase or a decrease in temperature but may be stable in a range (T_{c2}, T_{c1}), where T_{c1} and T_{c2} are respectively termed the *upper* and *lower critical temperatures*. For appropriate parameter values, this situation exists for the present model with the antiferromagnetic lower critical temperature $T_{c2} > T_{c1}$ where T_c is the critical temperature for ferromagnetic order.

Equations (9.3) and (9.125) give

$$\begin{aligned} \phi(\pm 1, \pm 1) &= \exp(-\alpha G)\sum_{\ell=0}^{q} \exp[(q-2\ell)2G/q]\,, \\ &= \exp(-\alpha G)\frac{\sinh[2(q+1)G/q]}{\sinh(2G/q)}\,, \\ \phi(+1,-1) &= (q+1)\exp(\alpha G)\,, \end{aligned} \tag{9.126}$$

where $G = J/T$. Since $\xi(+1) = \xi(-1) = 0$ in zero field, (9.9) yields

$$\exp(2K_e) = \exp(-2\alpha G)\frac{\sinh[2(q+1)G/q]}{(q+1)\sinh(2G/q)}, \tag{9.127}$$

$$L_e = 0$$

for the equivalent Ising model on the primary lattice. From (9.127),

$$\exp(2K_e) \simeq \frac{\exp[2(1-\alpha)G]}{q+1}, \qquad \text{as } G \to \infty . \tag{9.128}$$

Hence $K_e \to \infty$ as $T \to 0$, in accordance with the previous deduction of a ferromagnetic ground state. However, K_e does not increase monotonically with G since it follows from (9.127) that for small G (high T),

$$K_e = -\alpha G + \frac{q+2}{3q}G^2 + O(G^4). \tag{9.129}$$

For small G, $K_e < 0$ but, for large G, $K_e > 0$. To proceed further we need

$$\frac{dK_e}{dG} = -\alpha + \frac{q+1}{q}\coth[2(q+1)G/q] - \frac{1}{q}\coth(2G/q), \tag{9.130}$$

$$\frac{d^2K_e}{dG^2} = \frac{2}{[q\sinh(2G/q)]^2} - 2\left\{\frac{q+1}{q\sinh[2(q+1)G/q]}\right\}^2. \tag{9.131}$$

From the power-series expansion of the sinh function,

$$\sinh[2(q+1)G/q] > (q+1)\sinh[2G/q], \tag{9.132}$$

for all $G > 0$ so that $d^2K_e/dG^2 > 0$. It follows that K_e decreases from 0 as G increases from 0 to reach a unique minimum $K_e = K_{min} < 0$ at $G = G_{min}$ $(T = T_{min})$. Then K_e increases monotonically, passing through 0 at $G = G_0$ $(T = T_0)$ and through K_c (the critical value for the primary lattice Ising ferromagnet) at $G = G_c$ $(T = T_c < T_0)$ to approach infinity with G as $G \to \infty$.

From the above it follows that $K_e > K_c$ for the temperature interval $(0, T_c)$ giving long-range ferromagnetic order when $T < T_c$. If $K_{min} < -K_c$ then $K_e = -K_c$ at temperatures T_{c1} and T_{c2}, with $T_0 < T_{c2} < T_{min} < T_{c1}$. Thus $K_e < -K_c$ in the interval (T_{c2}, T_{c1}). Since we have specified a loose-packed primary lattice, there is long-range antiferromagnetic order when $T_{c2} < T < T_{c1}$. It follows that T_{c1} and T_{c2} are respectively upper and lower critical points. The value of K_{min}, which depends on q and α and which is given by the relation $dK_e/dG = 0$, is crucial. It can be shown from (9.130) that, with a fixed value of q, K_{min} has a greatest lower bound equal to $-\frac{1}{2}\ln(q+1)$ for the range $0 < \alpha < 1$. This bound is approached as $\alpha \to 1 - 0$. Hence antiferromagnetic critical points occur over some range of α if $\frac{1}{2}\ln(q+1) > K_c$, where K_c, of course, depends only on the primary lattice. With $q = 1$ this condition is satisfied for the $d = 3$ simple cubic and body-centred cubic lattices. With $q = 2$ it is also satisfied for the square and diamond lattices, and with $q = 3$ for the honeycomb lattice as well (see Table 8.1). As an example, with a square primary lattice, $q = 3$ and $\alpha = 0.8$ we have $T_{c1} = 1.0150J$, $T_{c2} = 0.4387J$, $T_0 = 0.1443J$ and $T_c = 0.0882J$. For the general case, since $dK_e/dG \neq 0$ when $K_e = \pm K_c$, $\Delta G \sim \Delta K_e$ and hence the heat capacity singularity at each critical point is of similar type to that at the critical point of the primary lattice Ising model. The $q = 1$ form of the present model was investigated by Nakono (1968) (see also Example 9.3). Syozi (1972) discusses related models.

The temperature $T = T_0$, where K_e changes sign, is of some interest. Since $K_e = 0$ at $T = T_0$ the primary lattice spins are completely disordered, as they are at $T = \infty$. Suppose that $\langle\sigma(\mathbf{0})\sigma(\boldsymbol{r})\rangle$ is the correlation between primary lattice spins whose relative position is represented by $\boldsymbol{r}$. As the temperature increases through T_0, K_e changes from positive to negative and $\langle\sigma(\mathbf{0})\sigma(\boldsymbol{r})\rangle$ changes from a positive quantity which decreases with $|\boldsymbol{r}|$ to one which alternates in sign, though its magnitude still decreases with $|\boldsymbol{r}|$. At $T = T_0$ itself $\langle\sigma(\mathbf{0})\sigma(\boldsymbol{r})\rangle = 0$ for all $\boldsymbol{r}$. For the $d = 1$ case of the present model, from (2.88),

$$\langle\sigma_i, \sigma_{i+\ell}\rangle = \begin{cases} (\tanh K_e)^\ell , & T < T_0 , \\ (-1)^\ell(\tanh|K_e|)^\ell , & T > T_0 . \end{cases} \tag{9.133}$$

The point $T = T_0$ is an example of a *disorder point*, other instances of which occur in the triangular Ising model with unequal interactions. Early work on disorder points is discussed by Stephenson (1970), who cites the $d = 1$, $q = 1$ case of the present model as one example (see also Georges et al. 1986). As in the present model, disorder points in other cases are due to competing interactions which promote different kinds of short-range order. Disorder points in the eight-vertex model are discussed briefly in Volume 2, Chap. 5.

9.11 Decorated Lattice Mixtures of Orientable Molecules

In Sects. 9.2–9.10 we were mainly concerned with magnetic systems. In the remainder of the chapter we shall consider non-magnetic mixtures and lattice gases.[8] The present section is concerned with mixtures which have closed-loop coexistence curves in the composition–temperature plane and thus both upper and lower critical solution temperatures. Phenomena of this type occur in real mixtures, for example glycerol-ethylbenzylamine and water-nicotine.

A decorated lattice is assumed with each primary and each secondary site occupied either by a molecule of component A or one of component B. A restricted grand distribution is used with the Hamiltonian given by (6.2). There is an interaction energy ε_{AA} for each primary–secondary pair of A molecules with ε_{AB} and ε_{BB} being similarly defined. The existence of a lower

[8] For a review of decorated models of such systems, see Wheeler (1977).

critical point is likely to be an entropic phenomenon, as we saw in Sect. 9.10. To achieve the necessary high entropy it is postulated that each A or B molecule on a secondary site has ω distinguishable orientation states. In two of these states an active *arm* is directed towards the molecule on one of the adjacent primary sites and terms τ_{AA}, τ_{AB} and τ_{BB} are then added to ε_{AA}, ε_{AB} and ε_{BB} respectively. In the other $\omega - 2$ states the secondary site molecule is said to be *disoriented.* The orientations of the molecules on the primary sites are taken as fixed so that these sites have only two distinguishable states, occupation by A or B respectively, which can be represented by $\sigma = +1$ and $\sigma = -1$. The system is thus equivalent to an Ising model on the primary lattice.

Symmetry between the A and B components can be obtained by putting[9]

$$\begin{aligned} &\varepsilon_{AA} = \varepsilon_{BB} = 0\,, \qquad &&\varepsilon_{AB} = \varepsilon, \\ &\tau_{AA} = \tau_{BB} = -\nu_1\varepsilon, \qquad &&\tau_{AB} = -\nu_2\varepsilon. \end{aligned} \tag{9.134}$$

Summing over the secondary site states

$$\begin{aligned} \phi(+1,+1) &= \lambda[\omega - 2 + 2\exp(\nu_1 Q)] \\ &\quad + \exp(-2Q)[\omega - 2 + 2\exp(\nu_2 Q)]\,, \\ \phi(-1,-1) &= \lambda\exp(-2Q)[\omega - 2 + 2\exp(\nu_2 Q)] \\ &\quad + \omega - 2 + 2\exp(\nu_1 Q)\,, \\ \phi(+1,-1) &= (\lambda + 1)\exp(-Q)[\omega - 2 + \exp(\nu_1 Q) + \exp(\nu_2 Q)]\,, \end{aligned} \tag{9.135}$$

where

$$Q = \varepsilon/T, \qquad \lambda = \exp[(\mu_A - \mu_B)/T]\,. \tag{9.136}$$

To evaluate the partition function we also need

$$\xi(+1) = -(\mu_A - \mu_B)\,, \qquad \xi(-1) = 0\,. \tag{9.137}$$

For phase equilibrium it is necessary to transform to the zero-field state of the equivalent Ising model by putting $\mu_A = \mu_B$ which gives $\lambda = 1$ and, from (9.135), $\phi(+1,+1) = \phi(-1,-1)$. Then substitution in (9.9) gives

$$\begin{aligned} \exp(2K_e) &= \frac{(\omega - 2)\cosh Q + \exp[(\nu_1 + 1)Q] + \exp[(\nu_2 - 1)Q]}{\omega - 2 + \exp(\nu_1 Q) + \exp(\nu_2 Q)}\,, \\ L_e &= 0\,. \end{aligned} \tag{9.138}$$

An expansion in powers of Q yields the relation

$$K_e = Q^2\frac{\omega - 2(\nu_2 - \nu_1)}{4\omega} + O(Q^3)\,, \tag{9.139}$$

[9] If $\varepsilon_{AA} = \varepsilon_{BB} \neq 0$ it is found that results depend only on the difference $\varepsilon_{AB} - \varepsilon_{AA}$, so that no generality is lost by taking $\varepsilon_{AA} = \varepsilon_{BB} = 0$.

which is useful at high temperatures. We now assume

$$\varepsilon > 0, \quad \nu_1 > 0, \quad \nu_2 > 0, \quad \nu_2 > \nu_1 + 1. \tag{9.140}$$

The lowest energy for a primary–secondary–primary triad of sites is $-(\nu_2 - 1)\varepsilon$, attained for occupational states AAB or ABB with the active arm of the secondary site molecule directed towards the *unlike* adjacent primary site molecule. Thus, on the primary lattice, AB nearest-neighbour pairs are favoured at low temperatures, corresponding to antiferromagnetic local order in the equivalent Ising model. For a loose-packed primary lattice the ground state is one of long-range sublattice order with As occupying one sublattice and Bs the other. The ground-state energy is $-\frac{1}{2}zN(\nu_2 - 1)\varepsilon$, where N is the number of primary lattice sites and z the primary lattice coordination number. However for a primary–secondary–primary triad of sites there are also occupation states AAA or BBB, with the molecule on the secondary site disoriented, which have zero energy but an entropy $k_B \ln(\omega - 2)$. Thus for high enough ω and T, AA or BB nearest-neighbour pairs on the primary lattices will be favoured and separation into A-rich and B-rich phases will occur, corresponding to ferromagnetic order in the equivalent Ising model. This is borne out by the behaviour of K_e. With the conditions (9.140), $K_e \to -\infty$ as $T \to 0$ ($Q \to \infty$). However, from (9.139), $K_e > 0$ at high temperatures if $\omega > 2(\nu_2 - \nu_1)$. The (T, K_e) plot then has the form shown in Fig. 9.9. For sufficiently large ω, the maximum value of $K_e > K_c$, the critical Ising parameter value for the primary lattice, as shown in Fig. 9.9. The values of T at which $K_e = K_c$ are denoted by T_{c1} and T_{c2} ($T_{c1} > T_{c2}$). By differentiating $\ln Z$ with respect to λ and then putting $\lambda = 1$, which implies $L_e = 0$, the mole fractions x_A in the conjugate phases are given by

$$x_A = \frac{1}{2} \pm \frac{m_s^{(\text{I.M.})}}{z+2} \times \left\{1 + \frac{z}{2} \frac{(\omega - 2)\sinh Q + \exp[Q(\nu_1 + 1)] - \exp[Q(\nu_2 - 1)]}{(\omega - 2)\cosh Q + \exp[Q(\nu_1 + 1)] + \exp[Q(\nu_2 - 1)]}\right\}, \tag{9.141}$$

where $m_s^{(\text{I.M.})}$ is the spontaneous magnetization of the primary lattice Ising model. For $T_{c1} > T > T_{c2}$, $m_s^{(\text{I.M.})} > 0$ and using the + and − signs in (9.141) gives the compositions of A-rich and B-rich conjugate phases. Outside this range of T, $m_s^{(\text{I.M.})} = 0$ and (9.141) gives $x_A = \frac{1}{2}$. Thus in the (x_A, T) plane there is a *closed-loop coexistence curve* symmetrical about the line $x_A = \frac{1}{2}$ with upper and lower critical temperatures for phase separation at $T = T_{c1}$ and $T = T_{c2}$ respectively. There is a *disorder point*, with $K_e = 0$, at some temperature $T_0 < T_{c2}$ (see Fig. 9.9). For a loose-packed primary lattice there is a transition to a sublattice ordered state when K_e passes through the value $-K_c$.

The $\nu_1 = 0$ case of the above theory was developed by Wheeler (1975, 1977) and Anderson and Wheeler (1978a) fitted theoretical results to the

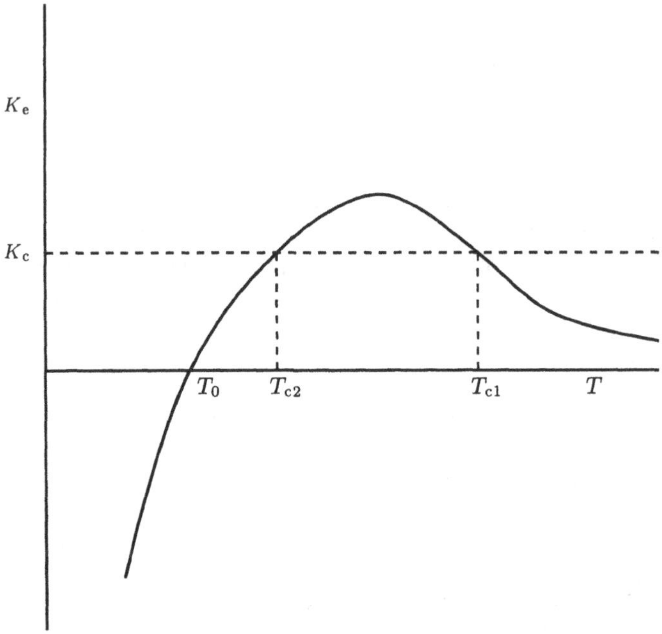

Fig. 9.9. Orientable molecule mixture with the inverse reduced temperature for the equivalent Ising model plotted against temperature. T_{c1} and T_{c2} are the upper and lower critical temperatures and T_0 is the disorder point.

experimental closed-loop curves for some organic mixtures, for example glycerol-ethylbenzylamine. They required the very large value $\omega = 5000$ for the number of secondary site orientation states. Although the model is highly schematic, this value of ω may reflect the large number of configurations available to fairly complex organic molecules. Bell and Fairbairn (1967) had earlier obtained a closed-loop coexistence curve from a regular lattice model with orientation dependent interaction energies. This is not equivalent to an Ising model, and a first-order approximation was used. Closed-loop coexistence curves are observed experimentally for some aqueous mixtures (in, for example, water-nicotine) but are highly unsymmetrical with respect to the two components. Anderson and Wheeler (1978b) fitted theoretical results obtained by taking $\varepsilon_{AA} \neq \varepsilon_{BB}$ and $\tau_{AA} \neq \tau_{BB}$.

Some interesting new phenomena appear when the symmetrical model is used without the last condition of (9.140) (Wheeler and Anderson 1980). For $\nu_1 + 1 > \nu_2$ but with the first three conditions of (9.140) still satisfied, the lowest energy for a primary–secondary–primary triad of sites is attained for occupations AAA or BBB with the active arm of the secondary site molecule directed to one of the adjacent primary site molecules. The ground state thus consists of pure A and pure B phases, the energy being $-\frac{1}{2}zN\nu_1\varepsilon$. First, let $\nu_1 + 1 = \nu_2$. Then, from (9.138), $K_e \to 0$ as $T \to 0$ and the (T, K_e)

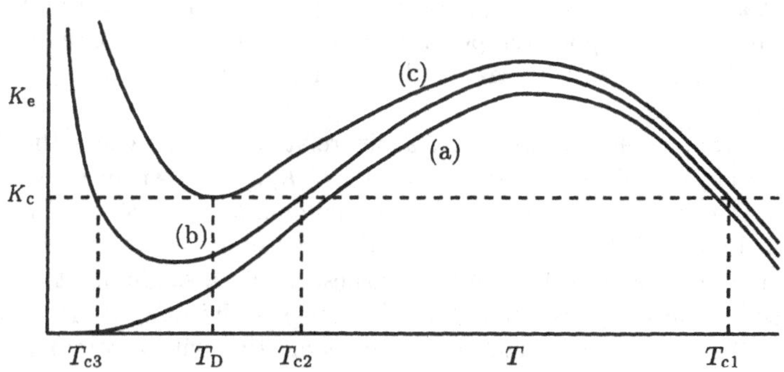

Fig. 9.10. As Fig. 9.9 but with curve (b) giving an additional upper critical temperature T_{c3} and curve (c) a critical double-point temperature T_D.

plot assumes the form sketched as curve (a) in Fig. 9.10. Then suppose that $\nu_1 + 1$ is slightly greater than ν_2. The (T, K_e) plot is sketched as curve (b) in Fig. 9.10 and remains close to curve (a) except for low values of T/ε where now $K_e \to \infty$. There is thus a low temperature interval $(0, T_{c3})$ where $K_e > K_c$, so that phase separation occurs. The coexistence diagram (broken curves) is sketched in Fig. 9.11. There is a closed-loop coexistence curve with upper critical temperature T_{c1} and lower critical temperature T_{c2}. Separated from the closed-loop curve by the temperature interval (T_{c3}, T_{c2}) is a low temperature coexistence curve with upper critical temperature T_{c3}.

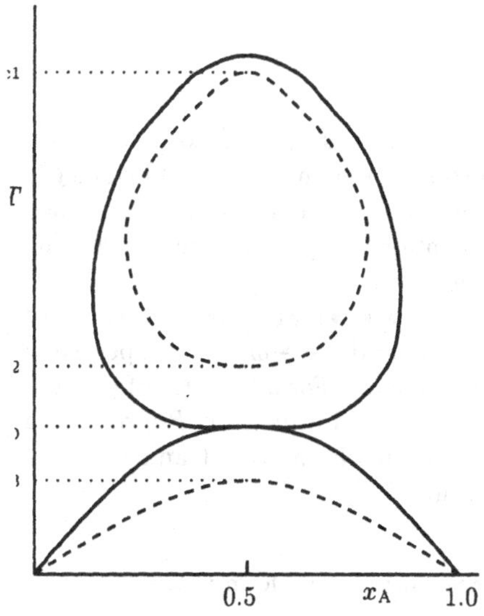

Fig. 9.11. Coexistence curves in the mole fraction–temperature plane; the full and broken curves correspond respectively to cases (c) and (b) of Fig. 9.10.

For both intervals (T_{c2}, T_{c1}) and $(0, T_{c3})$ there is separation into A-rich and B-rich phases, but for the upper temperature interval the phase separated state is entropically stabilized whereas for the lower temperature interval it is energetically stabilized.

Now suppose that $\nu_1 + 1$ is increased again relative to ν_2, raising the minimum of the (T, K_e) plot till it occurs at $K_e = K_c$, as shown in curve (c) of Fig. 9.10. Denote the temperature of the minimum at $K_e = K_c$ by T_D. The lower critical point of the upper branch of the coexistence curve and the upper critical point of the lower branch have coalesced into a single point at $T = T_D$. This is a *critical double-point*. In the full-line sketch in Fig. 9.11 the two branches of the coexistence curve are shown as meeting with a common tangent at the critical double-point. This is justified by the analysis below.

Let Δx_A denote the difference between the values of x_A in the conjugate phases. By (9.141), $\Delta x_A \sim m_s^{(\mathrm{I.M.})}$ as $\Delta K_e = K_e - K_c \to +0$ and $m_s^{(\mathrm{I.M.})} \sim (\Delta K_e)^{\beta}$ where β is the Ising model critical exponent. Now K_e has a minimum at $K_e = K_c$, $T = T_D$ so that $\Delta K_e \sim (T - T_D)^2$. This applies for both positive and negative $T - T_D$, and hence on both sides of the critical double-point we have $\Delta x_A \sim |T - T_D|^{\beta_D}$ where $\beta_D = 2\beta$. This result was derived for a somewhat different decorated solution model (see Sect. 9.13) by Bartis and Hall (1974). Since $\beta < \frac{1}{2}$ for both $d = 2$ and $d = 3$, $\beta_D < 1$ and the two coexistence branches meet with a common tangent parallel to the x_A-axis.[10] If $\nu_1 + 1$ is again increased relative to ν_2 the minimum K_e moves up to a value greater than K_c. The two branches of the coexistence curve are joined to form a rather bizarre 'neck' shape and there is now only one critical temperature.

9.12 The Decorated Lattice Gas

9.12.1 General Properties

We now treat liquid–vapour phase phenomena by using a decorated lattice gas model, which is more flexible in its properties than the simple lattice gas model of Sects. 5.3 and 5.4. In the one-component model of this section each primary site or secondary site is either vacant or occupied by a molecule. The grand distribution of Sect. 2.3.3 will be used.

Let the occupation states •–•–o, •–o–• and •–•–• for a primary–secondary–primary triad of sites have energies $-\frac{1}{2}\varepsilon$, $-w$ and $-\varepsilon - w + w_1$ respectively with all other occupation states having zero energy. For additivity of pair energies, $w_1 = 0$ but it is useful to include the non-additive case. Representing the occupied and vacant states for primary sites by $\sigma_i = +1$ and $\sigma_i = -1$ respectively and using the grand distribution

[10] If it were true that $\beta \geq \frac{1}{2}$, the coexistence curves would form a double cusp at their meeting point.

$$\begin{aligned}
&\phi(+1,+1) = \mathfrak{z}(1+\lambda\mathfrak{X}^2\mathfrak{z}_1)\,,\\
&\phi(-1,-1) = 1+\lambda\,,\\
&\phi(+1,-1) = 1+\lambda\mathfrak{X}\,,\\
&\xi(+1) = -\mu\,,\\
&\xi(-1) = 0\,,
\end{aligned} \tag{9.142}$$

where μ is the chemical potential and

$$\begin{aligned}
&\lambda = \exp(\mu/T)\,, &&\mathfrak{X} = \exp[\varepsilon/(2T)]\,,\\
&\mathfrak{z} = \exp(w/T)\,, &&\mathfrak{z}_1 = \exp(-w_1/T)\,.
\end{aligned} \tag{9.143}$$

Then, from (9.9),

$$\begin{aligned}
\exp(4K_\mathrm{e}) &= \frac{\mathfrak{z}(1+\lambda\mathfrak{X}^2\mathfrak{z}_1)(1+\lambda)}{(1+\lambda\mathfrak{X})^2}\,,\\
\exp(2L_\mathrm{e}) &= \lambda\left[\frac{\mathfrak{z}(1+\lambda\mathfrak{X}^2\mathfrak{z}_1)}{1+\lambda}\right]^{z/2}.
\end{aligned} \tag{9.144}$$

The density ρ (equal to 1 when all primary and secondary sites are occupied) is given by

$$\rho = \frac{\lambda}{(1+\frac{1}{2}z)N}\left(\frac{\partial \ln Z}{\partial\lambda}\right)_T, \tag{9.145}$$

where Z is the partition function for the decorated model, given by (9.7) and N the number of primary lattice sites. Using (9.142) and (9.144) in (9.145)

$$\begin{aligned}
\left(1+\tfrac{1}{2}z\right)\rho = \tfrac{1}{2} + \tfrac{1}{8}z\left[\frac{\lambda\mathfrak{X}^2\mathfrak{z}_1}{1+\lambda\mathfrak{X}^2\mathfrak{z}_1} + \frac{\lambda}{1+\lambda} + \frac{2\lambda\mathfrak{X}}{1+\lambda\mathfrak{X}}\right]\\
+\tfrac{1}{2}z g_\mathrm{p}(K_\mathrm{e},L_\mathrm{e})\lambda\left(\frac{\partial K_\mathrm{e}}{\partial\lambda}\right)_T + m^{(\mathrm{I.M.})}(K_\mathrm{e},L_\mathrm{e})\lambda\left(\frac{\partial L_\mathrm{e}}{\partial\lambda}\right)_T,
\end{aligned} \tag{9.146}$$

$g_\mathrm{p}(K_\mathrm{e},L_\mathrm{e})$ and $m^{(\mathrm{I.M.})}(K_\mathrm{e},L_\mathrm{e})$ being given by (9.10) and (9.11) respectively, and where

$$\begin{aligned}
\lambda\left(\frac{\partial K_\mathrm{e}}{\partial\lambda}\right)_T &= \frac{1}{4}\left[\frac{\lambda\mathfrak{X}^2\mathfrak{z}_1}{1+\lambda\mathfrak{X}^2\mathfrak{z}_1} + \frac{\lambda}{1+\lambda} - \frac{2\lambda\mathfrak{X}}{1+\lambda\mathfrak{X}}\right],\\
\lambda\left(\frac{\partial L_\mathrm{e}}{\partial\lambda}\right)_T &= \frac{1}{2} + \frac{1}{4}z\left[\frac{\lambda\mathfrak{X}^2\mathfrak{z}_1}{1+\lambda\mathfrak{X}^2\mathfrak{z}_1} - \frac{\lambda}{1+\lambda}\right].
\end{aligned} \tag{9.147}$$

The first-order transition lines in the (T,μ) plane are the parts of the $L_\mathrm{e}=0$ locus lying in the region where $K_\mathrm{e} > K_\mathrm{c}$, that is lying to one side of the

$K_e = K_c$ locus which is always a single connected curve. When, as will be assumed in the present section, the $L_e = 0$ locus is also a single connected curve, the part of it in the $K_e > K_c$ region forms a vapour–liquid transition line. The conjugate phase densities are given by taking $m^{(\text{I.M.})} = +m_s^{(\text{I.M.})}$ and $m^{(\text{I.M.})} = -m_s^{(\text{I.M.})}$, $m_s^{(\text{I.M.})}(K_e)$ being the spontaneous magnetization of the equivalent Ising model. The value of λ is that which satisfies the relation $L_e = 0$ at the given temperature T while $m_s^{(\text{I.M.})}(K_e)$ and $g_p(K_e, 0)$ are calculated for the zero-field Ising model at the corresponding value of K_e. The critical temperature T_c is defined by the intersection of the $L_e = 0$ and $K_e = K_c$ curves.

Let the densities of the conjugate liquid and vapour phases are denoted by ρ_ℓ and ρ_g respectively. Then, by (9.146), the *coexistence width*

$$\rho_\ell - \rho_g = m_s^{(\text{I.M.})}(K_e)\frac{2\lambda}{1 + \frac{1}{2}z}\left(\frac{\partial L_e}{\partial \lambda}\right)_T . \tag{9.148}$$

Let $\Delta T = T_c - T$ and $\Delta K_e = K_e - K_c$. Since $m_s^{(\text{I.M.})} \sim (\Delta K_e)^\beta$ as $K_e \to K_c + 0$ and since K_e and L_e are analytic functions of λ and T it follows that

$$\rho_\ell - \rho_g \sim (\Delta T)^\beta , \qquad \text{as } T \to T_c - 0 . \tag{9.149}$$

This is similar to the behaviour of the simple lattice gas described in Chap. 5, but there are important differences both globally and in the neighbourhood of the critical points. The hole–particle symmetry of the simple lattice gas is lost (except for a special case to be discussed in Sect. 9.12.3) and the coexistence curves in the (ρ, T) plane are no longer symmetrical about $\rho = \frac{1}{2}$. The *coexistence diameter*

$$\rho_d = \frac{1}{2}(\rho_\ell + \rho_g) \tag{9.150}$$

is, from (5.37), equal to $\frac{1}{2}$ for the simple lattice gas, but for the decorated model, from (9.146),

$$\left(1 + \frac{1}{2}z\right)\rho_d = \frac{1}{2} + \frac{1}{8}z\left[\frac{\lambda\mathfrak{X}^2\mathfrak{Z}_1}{1 + \lambda\mathfrak{X}^2\mathfrak{Z}_1} + \frac{\lambda}{1+\lambda} + \frac{2\lambda\mathfrak{X}}{1+\lambda\mathfrak{X}}\right] + \frac{1}{2}zg_p(K_e, 0)\lambda\left(\frac{\partial K_e}{\partial \lambda}\right)_T . \tag{9.151}$$

The empirical *law of the rectilinear diameter* (see Sect. 3.4) implies that $d\rho_d/dT$ remains finite as $T \to T_c - 0$. This law is satisfied by the simple lattice gas, since $d\rho_d/dT = 0$ for all $T < T_c$. However, for the decorated model, the right-hand side of (9.151) is not a linear function of T near the critical point. Unless $\partial K_e/\partial \lambda = 0$, the derivative $d\rho_d/dT$ contains a term proportional to $g_p'(K_e)$ defined by (9.27). Now

$$g_p'(K_e) \sim \begin{cases} (\Delta K_c)^{-\alpha'}, & \text{with } \alpha' \doteq \frac{1}{8} \text{ for } d = 3 , \\ -\ln|\Delta K_e| , & \text{for } d = 2 . \end{cases} \tag{9.152}$$

Hence

$$\frac{\mathrm{d}\rho_{\mathrm{d}}}{\mathrm{d}T} \sim \begin{cases} (\Delta T)^{-\alpha'}, & \text{with } \alpha' \doteq \frac{1}{8} \text{ for } d = 3 \ , \\ -\ln|\Delta T|, & \text{for } d = 2 \ . \end{cases} \tag{9.153}$$

In both cases, $\mathrm{d}\rho_{\mathrm{d}}/\mathrm{d}T \to \infty$ as $T \to T_{\mathrm{c}} - 0$. Mermin (1971a, 1971b) (see also Wheeler 1977) derived similar results for two types of lattice model, the first being the 'bar' model and the second the $w = w_1 = 0$ case of the decorated model discussed here. This indicates that a non-singular diameter is an artifact of the simple lattice fluid. As pointed out in Sect. 3.4, more general arguments for the existence of the singularity have been developed. However, experimentally the situation is still not completely resolved (Rowlinson and Swinton 1982).

9.12.2 A Water-Like Model

Bell and Sallouta (1975) developed a decorated model for water by taking the open, completely hydrogen-bonded structure as one in which all primary sites are occupied and all secondary sites empty. The interaction energy $-w$ ($w > 0$) of a primary–primary pair of molecules with the intervening secondary site empty is attributed to the formation of an hydrogen bond whereas the primary secondary pair energy $-\frac{1}{2}\varepsilon$ ($\varepsilon > 0$) is due to the non-bonding interaction. It is assumed that no hydrogen bond can form between an nearest-neighbour primary pair of molecules with the intervening secondary site occupied so that $w_1 = w$. The close-packed state is one in which all primary and all secondary sites are occupied. This model is less realistic structurally then the water model treated in Sect. 7.5 and less satisfactory for fluid phases in that the open structure is imposed *ab initio* rather than arising as short-range order due to bonding between appropriately placed and oriented molecules. However, it does have the advantage that, because of the equivalence to the Ising model, some accurate results can be obtained for comparison with those derived from standard approximations. The decorated model can be looked on as one of the class of *interstitial models* (see, for instance, Perram 1971) in which the open structure is disordered by displacement of molecules from 'framework' onto interstitial sites.

The results of Sects. 5.5 and 7.5 indicate that water-like properties are associated with the existence of a pressure range in which the open structure is stable at $T = 0$. In the present model the energies of perfect open and close-packed structures are respectively $-\frac{1}{2}\mathrm{z}Nw$ and $-\frac{1}{2}\mathrm{z}N\varepsilon$ or, in terms of the number of molecules M, $\frac{1}{2}\mathrm{z}Mw$ and $-\frac{1}{2}\mathrm{z}M\varepsilon/(1+\frac{1}{2}\mathrm{z})$. Denoting the volume (area for $d = 2$) per primary site by v_0, $V = \widetilde{V}/v_0$ and the reduced volume per molecule is $v = V/M$. For the open and close-packed states $v = 1$ and $v = 1/(1+\frac{1}{2}\mathrm{z})$ respectively. The pressure field $P = \widetilde{P}v_0$ is taken as an independent variable and from (1.84),

$$f = \mu = u - Ts + Pv = h - Ts\,, \tag{9.154}$$

where u and h are respectively the energy and enthalpy per molecule. For given P at $T = 0$ the equilibrium state is thus that of least $\mu = h = u + Pv$. Hence, in the open and close-packed states

$$\mu^{(\mathrm{O})} = P - \tfrac{1}{2}\mathrm{z}w\,, \qquad \mu^{(\mathrm{CP})} = \frac{P - \tfrac{1}{2}\mathrm{z}\varepsilon}{1 + \tfrac{1}{2}\mathrm{z}}\,. \tag{9.155}$$

An open–close-packed transition occurs when $\mu^{(\mathrm{O})} = \mu^{(\mathrm{CP})}$. From (9.155), the pressure and chemical potential at the transition are given by

$$P_0 = \left(1 + \tfrac{1}{2}\mathrm{z}\right) w - \varepsilon\,, \qquad \mu_0 = w - \varepsilon\,. \tag{9.156}$$

There is an interval $0 < P < P_0$ of stability for the open structure if $P_0 > 0$, which is equivalent to

$$\left(1 + \tfrac{1}{2}\mathrm{z}\right) w > \varepsilon\,. \tag{9.157}$$

If (9.157) applies there is a vapour–open transition at $P = 0$ when $T = 0$ and $\mu = \mu^{(\mathrm{O})} = -\frac{1}{2}\mathrm{z}w$. The inequality (9.157) is in fact equivalent to the condition $-\frac{1}{2}\mathrm{z}w < \mu_0$. It can be shown from the second relation of (9.144) with $\mathfrak{Z}_1 = 1/3$ that the $L_\mathrm{e} = 0$ locus in the (T, μ) plane then meets the $T = 0$ axis at $\mu = -\frac{1}{2}\mathrm{z}w$. The open–close-packed transition point $\mu = \mu_0$ on the $T = 0$ axis is an isolated singularity. It cannot be the terminus of any branch of the $L_\mathrm{e} = 0$ locus because at low temperatures the primary lattice is nearly full in both the open and close-packed states and so both correspond to $L_\mathrm{e} > 0$.

Coexistence curves in the (ρ, T) plane can be obtained from (9.146). Curve A in Fig. 9.12 is for the diamond lattice ($d = 3$, $\mathrm{z} = 4$, $1 + \frac{1}{2}\mathrm{z} = 3$) with $\varepsilon = 2.8w$, accurate series results being used for $m_\mathrm{s}^{(\mathrm{I.M.})}(K_\mathrm{e})$ and $g_\mathrm{p}(K_\mathrm{e}, 0)$. The terminal points $\rho = 0$ and $\rho = \frac{1}{3}$ correspond to the vapour and open structure phases respectively. The most striking feature, apart from the marked lack of symmetry about $\rho = \frac{1}{2}$, is the pronounced maximum on the liquid density branch. Curves B and C respectively, given for comparison, show the results of using first-order and mean-field formulae for $m_\mathrm{s}^{(\mathrm{I.M.})}(K_\mathrm{e})$ and $g_\mathrm{p}(K_\mathrm{e}, 0)$. Unfortunately the calculation of (ρ, T) isobaric curves requires Ising model results away from the $\mathcal{H} = 0$ axis. However, near the maximum on the liquid branch of the coexistence curve, the first-order plot is quite close to the accurate one. First-order calculations show maxima on the adjacent isobaric (ρ, T) curves. By considering perturbations from the perfect open structure, Bell and Sallouta (1975) showed that near $T = 0$ the density ρ increases with T on isobars for $0 < P < P_0$, provided that $\varepsilon > w$. This implies a density

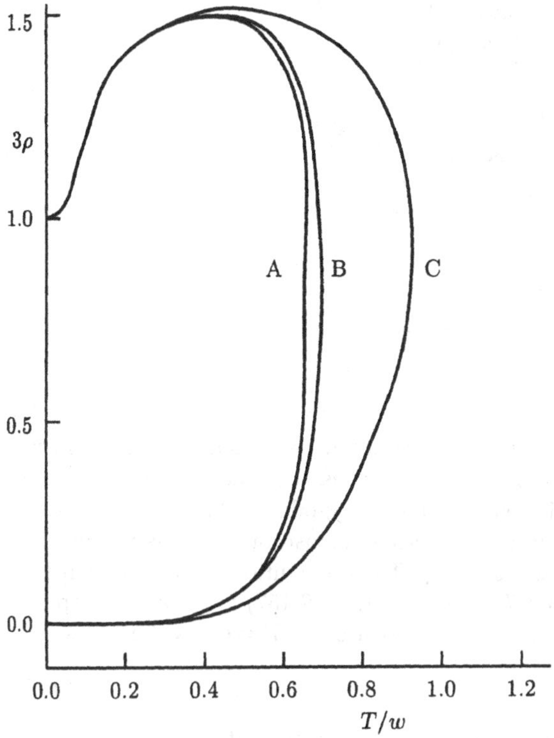

Fig. 9.12. Coexistence curves in the temperature–density plane for the water model. (Reprinted from Bell and Sallouta (1975), by permission of the publisher Taylor and Francis Ltd.)

maximum on any continuous isobar in this pressure range. Since P_0 is considerably greater than the critical pressure P_c, density anomalies occur well into the supercritical region, in accordance with experimental results. Compressibility minima are also found. There is thus good qualitative agreement between the results for this model and those for the model of Sect. 7.5, where first-order approximation methods had to be used exclusively.

9.12.3 Maxithermal, Critical Double and Cuspoidal Points

When the $K_e = K_c$ and $L_e = 0$ loci in the (T, μ) plane of the decorated model are both single curves and have a unique intersection, the phase diagram in this plane is topologically similar to that of a simple lattice gas, though as we have shown in Sect. 9.12.1 there are important differences in properties. However, for large negative values of ε the mapping from the (T, μ) plane of the decorated model to the (K_e^{-1}, L_e) plane of the Ising model is no longer one-to-one and the $L_e = 0$ locus has several distinct branches. The resulting phenomena are complicated and best approached by considering the transitions on the μ-axis. For the general case, (9.155) and (9.156) are still valid provided ε is replaced by $\varepsilon + w - w_1$. However we also have to consider an

Table 9.1. The values of P and μ at possible transitions between the vapour (V), open (O), close-packed (CP) and intermediate states (I).

Transition	P	μ
V–O	0	$-\frac{1}{2}zw$
V–CP	0	$-\frac{1}{2}z\frac{\varepsilon+w-w_1}{1+\frac{1}{2}zw}$
O–CP	$\frac{1}{2}zw-\varepsilon+w_1$	$w_1-\varepsilon$
O–I	$\frac{z^2w}{z-2}$	$\frac{zw}{z-2}$
I–CP	$\frac{1}{4}z^2(w_1-\varepsilon-w)$	$\frac{1}{2}z(w_1-\varepsilon-w)$

'intermediate' state in which all secondary sites are occupied and all primary sites are empty. The intermediate state energy is zero and $v = 2/z$ so that $h^{(\mathrm{I})} = \mu^{(\mathrm{I})} = 2P/z$. With this relation and the modified (9.155), Table 9.1, showing the values of P and μ at the possible transitions, is constructed. It will be assumed that $w > 0$ and $w \geq w_1$. These conditions can be shown to imply that $K_e > 0$ for all μ and T. The condition (9.157) for a range of open state stability and a vapour–open state transition at $T = 0$ now becomes

$$\frac{1}{2}zw > \varepsilon - w_1 , \tag{9.158}$$

If (9.158) is not satisfied then the close-packed state is stable at $T = 0$ for all $P > 0$. The vapour–close-packed transition point is the only one on the μ-axis and is the terminus of the $L_e = 0$ locus.

We turn from this rather uninteresting case to ask whether a range of stability is possible for the intermediate state at $T = 0$. Such a range exists if the value of μ (or P) at the intermediate–close-packed transition point is greater than the value of μ (or P) at the open–intermediate transition point. From Table 9.1 this is true if

$$w_1 - \varepsilon > \frac{zw}{z-2} . \tag{9.159}$$

Since $w > 0$, the condition (9.159) implies (9.158). So when (9.159) is satisfied there are three transition points on the μ-axis, vapour–open at $\mu = -\frac{1}{2}zw$, open–intermediate at $\mu = zw/(z-2)$ and intermediate–close-packed at $\mu = \frac{1}{2}z(w_1 - \varepsilon - w)$. These points are denoted by Q, R and S respectively in Fig. 9.13. Now, at low temperatures, the primary lattice is nearly empty in the vapour and intermediate states so that $L_e < 0$. The points Q, R and S all correspond to changes in the sign of L_e and it can be confirmed from (9.144) that, when (9.159) applies, Q, R and S are termini of the $L_e = 0$ locus. Thus the $L_e = 0$ locus has more than one branch.

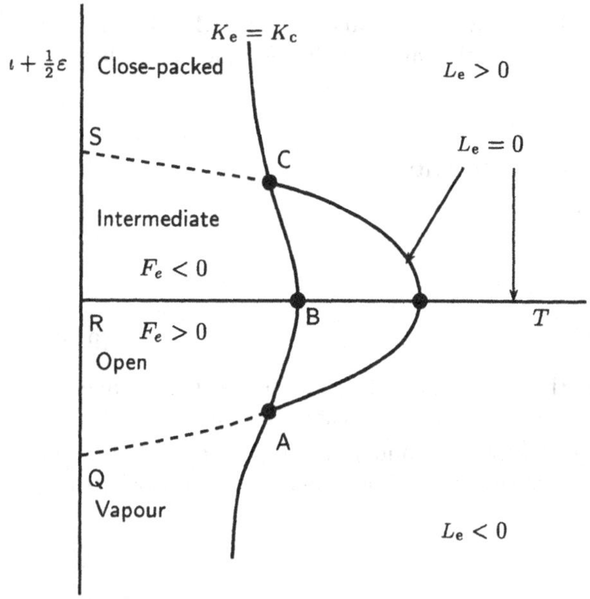

Fig. 9.13. The symmetric case of the decorated lattice gas showing the loci of $L_e = 0$ and $K_e = K_c$. The broken curves are the two first-order transition lines which do not lie on the T-axis.

We now consider the case $w_1 = 0$ when the pair energies are additive.[11] Defining

$$\mathfrak{Y} = \lambda \mathfrak{X} = \lambda \exp[\varepsilon/(2T)] = \exp[(2\mu + \varepsilon)/(2T)], \tag{9.160}$$

the first relation of (9.144) (with $\mathfrak{z}_1 = 1$) becomes

$$\exp(4K_e) = \frac{\mathfrak{z}\mathfrak{X}^{-1}(\mathfrak{Y} + \mathfrak{X})(\mathfrak{Y}^{-1} + \mathfrak{X})}{(\mathfrak{Y} + 1)(\mathfrak{Y}^{-1} + 1)}. \tag{9.161}$$

For given T, K_e is unaltered when $\mathfrak{Y}$ is replaced by $\mathfrak{Y}^{-1}$. The $K_e = K_c$ locus in the (T, μ) plane is given by the relation

$$\exp(4K_c - w/T) - 1 = \frac{\mathfrak{X} + \mathfrak{X}^{-1} - 2}{\mathfrak{Y} + \mathfrak{Y}^{-1} + 2}. \tag{9.162}$$

This curve

(i) is symmetrical about the line $\mu = -\frac{1}{2}\varepsilon$ ($\mathfrak{Y} = 1$),

(ii) has the line $T = w/(4K_c)$ as asymptote when $\mu \to \pm\infty$,

(iii) lies entirely to the right of this asymptote,

(iv) has its point of maximum T when $\mu = -\frac{1}{2}\varepsilon$.

---

[11] This has been treated extensively by Mulholland and Rehr (1974).

The general form is shown by the $K_e = K_c$ curve in Fig. 9.13. The $L_e = 0$ curve has no similar symmetry for arbitrary ε/w, but for the particular case

$$-\varepsilon = \frac{2zw}{z-2} \tag{9.163}$$

the second relation of (9.144) can be written as

$$\exp(2L_e) = \mathfrak{Y}\left[\frac{1+\mathfrak{Y}\mathfrak{X}}{\mathfrak{X}+\mathfrak{Y}}\right]^{z/2}, \tag{9.164}$$

giving

$$L_e(T, \mathfrak{Y}) = -L_e(T, \mathfrak{Y}^{-1}). \tag{9.165}$$

The locus $L_e = 0$, and thus the whole phase diagram in the (μ, T) plane, are symmetrical about the line $\mu = -\frac{1}{2}\varepsilon$. This is the case shown in Fig. 9.13. From the above, the Ising model $m^{(\text{I.M.})}(K_e, L_e)$ and $g_p(K_e, L_e)$ are respectively odd and even functions of $\mu + \frac{1}{2}\varepsilon$ at given T. It can then be shown from (9.146) that

$$\rho(T, \mathfrak{Y}) + \rho(T, \mathfrak{Y}^{-1}) = 1. \tag{9.166}$$

Thus there is particle–hole symmetry and the phase pattern in the (ρ, T) plane is symmetrical about $\rho = \frac{1}{2}$.

When ε is given by (9.163) the entire line $\mu = -\frac{1}{2}\varepsilon$ ($\mathfrak{Y} = 1$) is part of the $L_e = 0$ locus in the (T, μ) plane, ending at R. As (9.159) is satisfied, Q and S are also termini of the $L_e = 0$ locus. It can be shown that two other solutions of $L_e = 0$ separate from the solution $\mathfrak{Y} = 1$ at $\mathfrak{X} = (z-2)/(z+2)$, which is the point P shown in Fig. 9.13. When K_e is calculated at P it proves to be less than K_c for all common lattices. Hence, as shown by the broken lines in Fig. 9.13, there are three distinct first-order transition curves, RB on the line $\mu = -\frac{1}{2}\varepsilon$ with QA and SC placed symmetrically on either side. Since $P = 0$ at Q, QA can be regarded as the vapour–liquid transition line. By symmetry, $\rho_1 + \rho_2 = 1$ on RB, where ρ_1 and ρ_2 are the densities of conjugate phases, and hence $\rho_d = \frac{1}{2}$ and there is no singularity in the diameter. This can be attributed to the fact that, for $w_1 = 0$, $\lambda\partial K_e/\partial\lambda = 0$ when $\mathfrak{Y} = 1$. However, $\lambda\partial K_e/\partial\lambda \neq 0$ on the lines QA and SC so that in both cases there is a singular diameter.

Though the symmetrical case is exceptional, it serves as a useful guide to the situation when $\varepsilon \neq -2zw/(z-2)$. It can be shown that the curve of $\exp(2L_e)$ against $\mathfrak{Y}$ is monotonically increasing for high temperatures in all cases but that as T is lowered, an interval of negative slope develops for $\mathfrak{Y} < 1$ when $-\varepsilon > 2zw/(z-2)$ and for $\mathfrak{Y} > 1$ when $-\varepsilon < 2zw/(z-2)$. Thus for $-\varepsilon > 2zw/(z-2)$ the $L_e = 0$ locus consists of a loop from Q to R and a single branch going from S to $T = \infty$. At $-\varepsilon = 2zw/(z-2)$ we have the intermediate stage shown in Fig. 9.13 whereas for $2zw/(z-2) > -\varepsilon > zw/(z-2)$ there is a closed loop from R to S and a single branch going from Q to $T = \infty$. As $-\varepsilon \to zw/(z-2) + 0$ the interval RS of intermediate state stability on the

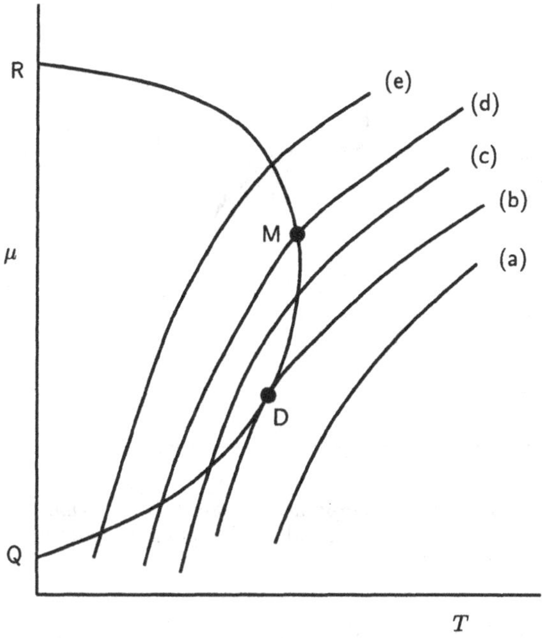

Fig. 9.14. QMR is the closed-loop branch of the $L_e = 0$ locus. The curves (a)–(e) show the relative positions of the lower part of the $K_e = K_c$ locus as $|\varepsilon|/w$ decreases.

μ-axis shrinks to zero and with it the RS loop of the $L_e = 0$ locus. For the simple cubic, body-centred cubic and face-centred cubic lattices, the loop QR of the $L_e = 0$ locus lies entirely in the $K_e > K_c$ region when $\varepsilon = -\infty$ (see Example 9.6).

The position of part of the lower half of the $K_e = K_c$ locus relative to the loop QR at successive stages as $-\varepsilon$ decreases from infinity towards $2zw/(z-2)$ is indicated by curves (a)–(e) in Fig. 9.14. (For simplicity, the loop QR is shown as fixed.) The point M on the loop where $dT/d\mu = 0$ is a *maxithermal point.* Curve (a) represents the case $\varepsilon = -\infty$. At the next stage (b), the two curves touch at D which is a *critical double-point.* The coexistence curves in the (ρ, T) plane in the neighbourhood of D and M at stage (b) are sketched in Fig. 9.15. At stage (c) the critical double-point is replaced by two critical points separated by a homogeneous region, and at (d) the $K_e = K_c$ curve intersects the loop QR at M, which becomes a *cuspoidal point.* The (ρ, T) plane configuration here is sketched in Fig. 9.15. By stage (e) the $K_e = K_c$ locus cuts the loop QR in two ordinary critical points. Except at stage (a) there is another first-order transition line, not shown in Fig. 9.14, which starts at S on the μ-axis and ends at an ordinary critical point on the $K_e = K_c$ locus. A number of accurate coexistence curves and other diagrams

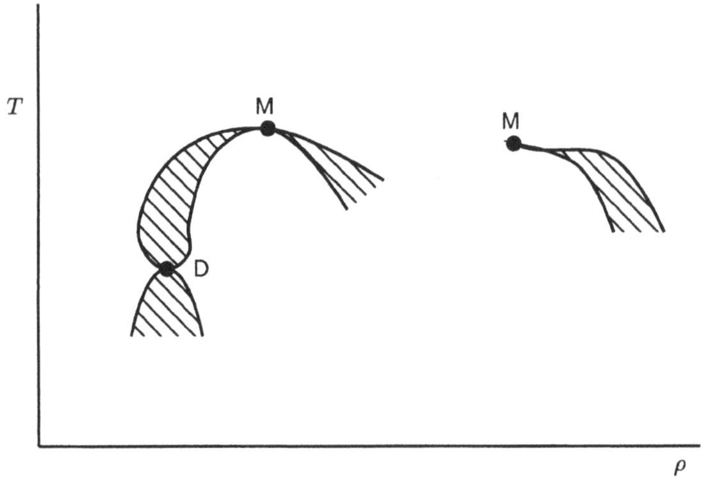

Fig. 9.15. Coexistence curves in the density–temperature plane. The left-hand diagram displays a maxithermal point M and a critical double-point D and the right-hand diagram shows a cuspoidal point M.

are given by Mulholland and Rehr (1974). These authors also discuss the phase diagram when $w < 0$ for the additive case $w_1 = 0$.

The water-like model treated in Sect. 9.12.2 was an example of the non-additive case, with $w_1 = w$. The assumption that the $L_e = 0$ locus for this model is a single connected curve is confirmed from the condition (9.159), which is not satisfied when $w_1 = w$ and $\varepsilon > 0$. The general non-additive model can be discussed only briefly. One important difference from the additive model is that the $K_e = K_c$ locus in the (T, μ) plane is no longer symmetrical about the line $\mu = -\frac{1}{2}\varepsilon$. However, for $w_1 < w$ the entire locus lies to the right of the line $T = \frac{1}{4}(w - w_1)/K_c$, which forms the $\mu \to \infty$ asymptote. When $w_1 = w$ the locus bends round to terminate on the μ-axis. For $w_1 > w$ there are regions where $K_e < 0$ and the properties of the decorated model depend on those of the Ising antiferromagnet.

The present model can be transcribed to a decorated mixture by using a restricted grand, rather than a grand, distribution and substituting $\mu_A - \mu_B$ for μ. The molecules of the present model are labelled A and the holes are replaced by molecules of another component B so that we can set $\rho = x$ where x denotes the mole fraction of A. In this transcription the maxithermal point in Fig. 9.15 becomes an *azeotropic point* whereas the cuspoidal point becomes a *critical azeotropic point*. Denote the values of x and T at such a point by x_c and T_c, and let x_1 and x_2 be the mole fractions in the coexisting phases when T is just less than T_c. Then $x_1 - x_c$ and $x_2 - x_c$ have the same sign, contrasting with the opposite signs found near an ordinary critical point.[12]

[12] A similar remark can be made about density differences near a cuspoidal point.

If the two-dimensional (x, T) or $(\mu_A - \mu_B, T)$ space is extended into three dimensions by adding an ε/w axis, the critical azeotropic point occurs where a line of azeotropic points ends on a line of critical points. In spaces of three intensive thermodynamic variables this phenomenon occurs experimentally at high pressures (Rowlinson and Swinton 1982, Sect. 6.2) and theoretically in the decorated lattice gas mixtures described in the next section.

9.13 Decorated Lattice Gas Mixtures

We now consider briefly two component decorated lattice gas models, taking the term 'lattice gas' to imply that both primary and secondary sites can be vacant. Thus for condition (i) of Sect. 9.1 to apply, molecules of only one of the components can occupy primary sites. This component is termed the *solvent* and the other component the *solute*. There are then two possibilities:

(1) Only solute molecules can occupy secondary sites, so that the pure solvent is a simple lattice gas as defined in Sect. 5.3.

(2) Both solute and solvent molecules can occupy secondary sites.

A solute component could be added to any of the lattice gas models of Sect. 9.12 to produce a system of type (2). From condition (ii) of Sect. 9.1 there is no direct solute-solute interaction so that there can be no phase separation in pure solute in either type (1) or type (2) models. Such models are, therefore, applicable only when the critical temperature of the solute is much lower than that of the solvent or when the solute mole fraction is fairly small. Work on models of type (1) was initiated by Widom (1967); (for further developments, see Wheeler (1977)). They can display a number of cooperative phenomena, including maximum critical solution temperatures and critical azeotropy. Bartis and Hall (1974) obtained critical double-points in the pressure–mole fraction plane from a model of type (2). They needed to use a solvent-solvent interaction which is attractive for primary–secondary pairs and repulsive for primary–primary pairs. This implies a potential which is negative at small separations and positive at larger ones. Sallouta and Bell (1976)[13] modelled aqueous solutions with a system of type (2), a solute being added to the decorated lattice gas water model discussed in Sect. 9.12.2. They found that with repulsive solute-water interaction the water structure breaks down at small solute concentrations. With attractive solute-water interaction the water structure is stable for quite high solute concentrations, and large attractive interactions give rise to azeotropy and critical azeotropy.

The decoration transformation has given exactly solvable models displaying a variety of critical phenomena. The two models solved by the star-triangle transformation (Sect. 9.8 and Example 9.4) yield results similar to those

[13] There is an error in the table on p. 341 of this paper. For conformity with the rest of the paper, $-u$ should be replaced by u and the statement $u > 0$ deleted.

of the corresponding decorated models (Sect. 9.5 and Example 9.1) respectively). However there are limits to the versatility of the decoration method. It has so far failed to produce any exact theories of phenomena dependent on three-phase equilibrium, such as triple points, tricritical points or critical end-points. This situation is discussed by Wheeler (1977) who attributes it to the fact that a decorated model must transcribe to a zero-field standard Ising model for exact results to be obtained. However this underlying Ising model is limited to two phase equilibrium. Although the ferrimagnetic models of Sect. 9.5 and 9.8 give, for appropriate parameter values, compensation points where four phases are in equilibrium, these phases become identical in pairs as $T \to T_{\text{comp}} \pm 0$. This is because the $\mathcal{H} = +0$ phase for $T = T_{\text{comp}} + 0$ becomes the $\mathcal{H} = -0$ phase for $T = T_{\text{comp}} - 0$ and vice versa. Each identical pair corresponds to a single phase in the underlying Ising model, so supporting Wheeler's argument.

Examples

9.1 The Syozi dilution model is a mixture of Ising spins A and magnetically inert atoms B on a decorated lattice. All primary sites are occupied by As but a secondary site can be occupied by either an A or a B. There are primary–secondary ferromagnetic interactions, with the constant denoted by J, between A atoms. Using a restricted grand partition function, show that, in zero field,

$$\exp(2K_{\text{e}}) = \frac{1 + 2\lambda \cosh(2G)}{1 + 2\lambda},$$

$$L_{\text{e}} = 0,$$

where $\lambda = \exp[(\mu_{\text{A}} - \mu_{\text{B}})/T]$, $G = J/T$.
Hence show that the fraction x of secondary sites occupied by A is given by

$$x = \frac{\lambda \cosh(2G)}{1 + \lambda \cosh(2G)}[1 + g_{\text{p}}(K_{\text{e}}, 0)] + \frac{\lambda}{1 + 2\lambda}[1 - g_{\text{p}}(K_{\text{e}}, 0)],$$

where $g_{\text{p}}(K_{\text{e}}, L_{\text{e}})$ is defined by (9.10). Show that the critical curve in the (T, x) plane is

$$x = \frac{[\exp(2K_{\text{c}}) - 1]}{2[\cosh(2G) - 1]} \times [\exp(-2K_{\text{c}})\cosh(2G)(1 + g_{\text{p}}(K_{\text{c}}, 0)) + 1 - g_{\text{p}}(K_{\text{c}}, 0)]$$

and deduce that the critical value of x is

$$x_{\text{c}} = \frac{1}{2}[1 - \exp(-2K_{\text{c}})][1 + g_{\text{p}}(K_{\text{c}}, 0)].$$

Show, using (8.13) and (8.17), that, if x_c^* denotes the critical value of x_c for the dual primary lattice, then $x_c + x_c^* = 1$ and deduce that $x_c = \frac{1}{2}$ for the square lattice.

9.2 For the decorated ferrimagnet of Sect. 9.5, show that

$$\left(r + \tfrac{1}{2}z\right) T \left(\frac{\partial m}{\partial \mathcal{H}}\right)_{\mathcal{H}=0} = \tfrac{1}{2}z\left\{1 - \tfrac{1}{2}[1 + g_p(K_c, 0)]\tanh^2(2G)\right\}$$
$$+ \left(\frac{\partial m}{\partial L_e}\right)^{(\text{I.M.})}_{L_e=0} \left[r - \tfrac{1}{2}z\tanh(2G)\right]^2 ,$$

where m and $g_p(K_c, L_e)$ are respectively defined by (9.77) and (9.10). Consider a $d = 1$ lattice with alternate sites labelled a and b. All sites are occupied by Ising spins and there is an interaction term $J_{ab}\sigma_i\sigma_j$, ($J_{ab} > 0$), for each nearest-neighbour ab pair. The spins on a and b sites have magnetic moments rm_0 and m_0 respectively. This model can be regarded as a decorated ferrimagnet with the b sites decorating the segments connecting the a sites and $J_{aa} = 0$. By using the formula deduced above and results from Chap. 2, show that

$$\left(\frac{\partial m}{\partial \mathcal{H}}\right)_{\mathcal{H}=0} = \frac{(r-1)^2\exp(2G) + (r+1)^2\exp(-2G)}{2(r+1)T},$$

where $G = J_{ab}/T$. Verify that this reduces to equation (4.5) when $r = 1$. Show that $(\partial\mathcal{H}/\partial m)_{\mathcal{H}=0}$ plotted against T has a horizontal tangent at $T = 0$. Show that the paramagnetic Curie temperature is $-4rJ_{ab}/(r^2+1)$.

9.3 For the $q = 1$ case of the competing interaction model of Sect. 9.10, show that

$$K_{\min} = -\tfrac{1}{4}[(1+\alpha)\ln(1+\alpha) + (1-\alpha)\ln(1-\alpha)].$$

Show that in the range $0 < \alpha < 1$, $K_{\min}$ decreases as α increases and that $K_{\min} \to -\frac{1}{2}\ln(2)$ as $\alpha \to 1 - 0$.

9.4 Using the notation of question 1, suppose that the black sublattice sites of the honeycomb lattice in Fig. 8.3 are all occupied by As but that each white site can be occupied by either an A or a B. There are ferromagnetic interactions with interaction constant J between a spin on a white site and the spins on the neighbouring black sites. Show that in zero field, the white sites can be removed by a star-triangle transformation with equation (8.25) replaced by

$$1 + 2\lambda\cosh[(\alpha+\beta+\gamma)G] = \mathrm{R}\exp[K(\alpha\beta + \beta\gamma + \gamma\alpha)].$$

Hence show that

$$\exp(4K) = (1 + 2\lambda c_3)/(1 + 2\lambda c_1)\,,$$

$$\mathsf{R}^4 = (1 + 2\lambda c_3)(1 + 2\lambda c_1)^3\,,$$

where $c_\ell = \cosh(\ell G)$. Show that the critical curve in the (x, T) plane, x being the fraction of white sites occupied by As, is

$$x = \frac{\exp(4K_c) - 1}{8(c_2 - 1)}\{3[1 - g_p(K_c, 0)] + (2c_2 - 1)[1 + 3g_p(K_c, 0)]\exp(-4K_c)\}\,,$$

K_c and $g_p(K_c, 0)$ having the values appropriate to the triangular lattice. Hence show that the critical mole fraction x_c is given by

$$x_c = \frac{1}{4}[1 - \exp(-4K_c)][1 + 3g_p(K_c, 0)] = \frac{1}{2}\,.$$

9.5 From equation (9.144) for the decorated lattice gas deduce that, when $w_1 = 0$ and $\varepsilon = -2zw/(z-2)$, equation (9.165) applies.

9.6 For the decorated lattice gas, show from (9.144) that, when $w_1 = 0$ and $\varepsilon = -\infty$,

$$\exp(4K_e) = \exp(w/T)(1 + \lambda)\,,$$

$$\exp(2L_e) = \lambda\exp[zw/(2T)](1 + \lambda)^{-z/2}.$$

By considering the two possibilities, $\lambda \to 0$ as $T \to 0$ and $\lambda \to \infty$ as $T \to 0$, show that for $w > 0$ the points $\mu = -\frac{1}{2}zw$ and $\mu = zw/(z-2)$ are the termini of the $L_e = 0$ locus on the μ-axis of the (T, μ) plane. Show that on the $L_e = 0$ locus there is a maximum value T_{max} of T given by

$$\exp[zw/(2T_{max})] = \tfrac{1}{2}z\left[\frac{\frac{1}{2}z}{\frac{1}{2}z - 1}\right]^{\frac{1}{2}z-1}$$

and that the $L_e = 0$ locus is a loop between the termini on the μ-axis. Show that on the $L_e = 0$ locus the minimum value of $\exp(4K_e)$ is $\{z/[(z-1)^{(z-1)/z}]\}^2$ and then, using Table 8.1, show that the entire $L_e = 0$ locus lies in the $K_e > K_c$ region of the (T, μ) plane for simple cubic, body-centred cubic and face-centred cubic primary lattices.

10. The Six-Vertex Model

10.1 Two-Dimensional Ice-Rule Models

In this chapter we consider exact solutions for a class of two-dimensional models which have completely different behaviour from the Ising model. These models are based on a square lattice version of a completely hydrogen-bonded network, like that of ice I (see Appendix A.3). With all bond configurations having equal energy, the free energy is related to the ground-state (zero-point) entropy of a $d = 2$ ice analogue (*square ice*), which was evaluated by Lieb (1967a). As well as water, certain other substances, such as KH_2PO_4 (potassium dihydrogen phosphate), crystallize as four-coordinated hydrogen-bonded networks. KH_2PO_4 is a *ferroelectric* with a spontaneous electrical polarization in zero field below a critical temperature. Square lattice analogues of ferroelectrics and antiferroelectrics can be constructed by appropriate assignment of energies to the various bond configurations. Lieb (1967b, 1967c) obtained expressions for the free energy of these models by an extension of the method used for square ice. In contrast to the Ising model, some results can be obtained for non-zero fields (electric fields in this case) (Sutherland 1967, Lieb 1969). The development in this chapter is based on the extensive review article of Lieb and Wu (1972).

We first introduce an *arrow* convention for the polarity of the bonds. A water molecule (see Appendix A.3) can be regarded as having two negative and two positive 'arms' (Fig. 10.1(a)), and two molecules are hydrogen bonded by the junction of a positive arm from one and a negative arm from the other. Now distinguish the positive arms, that is those corresponding to the positions of the two protons, by arrows directed away from the molecule's centre (Fig. 10.1(b)). There is then a single arrow on the segment between

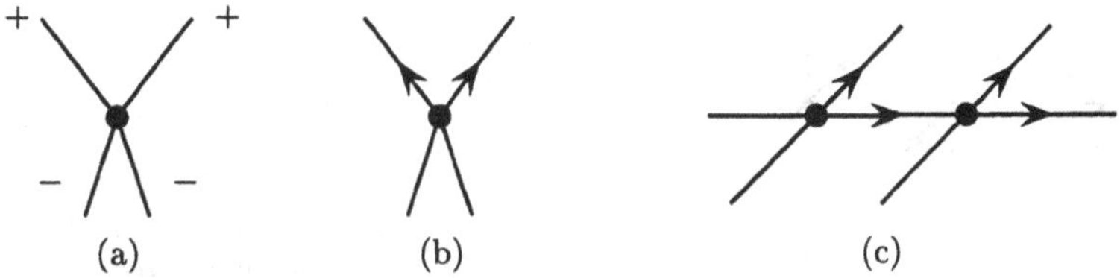

Fig. 10.1. Water molecule polarity and bonding: arrow notation.

the centres of any bonded pair of molecules (Fig. 10.1(c)). When a completely hydrogen-bonded network is represented by a graph with the centres of the molecules at the vertices and one arrow on each edge, there are inward arrows on two of the four edges meeting at each vertex and outward arrows on the other two. This 'two in, two out' law is known as the *ice rule.*

Since the coordination number $z = 4$ for the square lattice, the latter can be used as a two-dimensional analogue of a completely hydrogen-bonded network. With the application of the ice rule there are six permitted types of vertices and the model is referred to as the *six-vertex model.* The most general vertex model on the square lattice is the *sixteen-vertex model*, in which all 2^4 possible arrow configurations about a vertex are allowed. This model, which can be shown to be equivalent to an Ising model with two, three and four site interactions and an external field (Suzuki and Fisher 1971), is unsolved. The model in which the ice rule is modified to allow in addition the two vertices when all the arrows are pointing inwards or outwards is the *eight-vertex model.* The zero-field case of this model was solved by Baxter (1972). It is discussed in Volume 2, Chap. 5 where we show that both the zero-field Ising model and six-vertex models are special cases (see also Baxter 1982a, Lavis 1996).

Each site of the six-vertex model is identified with a vertex of one of the six types permitted by the ice rule. These are shown, with assigned index numbers, on the first line of Fig. 10.2. They fall into pairs (12), (34) and (56) such that reversal of all arrows interchanges the two members of each pair. In the second line of Fig. 10.2 the double-headed arrows indicate the resultant electric dipole moments. Vertices 1 and 2 have oppositely directed dipole moments as do 3 and 4 whereas 5 and 6 have no resultant dipole moments in the plane of the diagram. These *unpolarized* vertices reverse the 'flow' of arrows along any row of edges and so occur alternately (see Fig. 10.3). Thus, in the thermodynamic limit, the numbers of 5 and 6 in each row, and hence in the whole lattice, are equal. The configurational energy is assumed to be

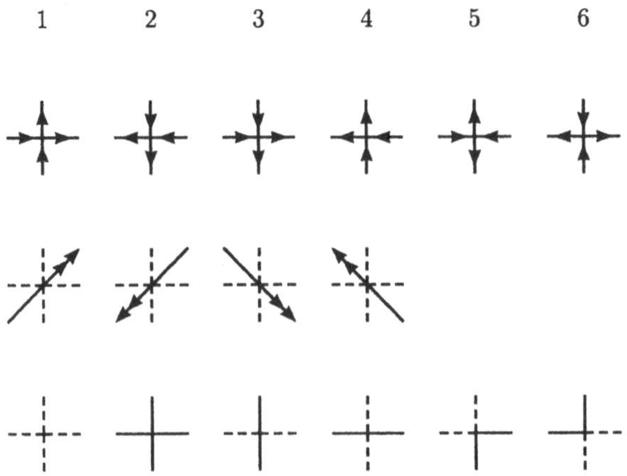

Fig. 10.2. Vertex types in the completely hydrogen-bonded square network.

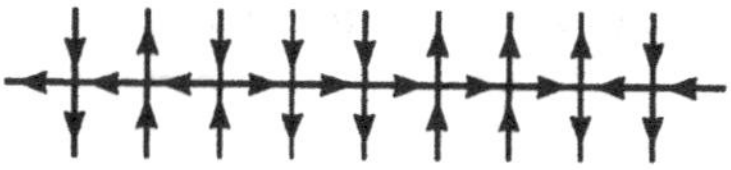

Fig. 10.3. Row of vertices showing reversal of horizontal arrow direction by types 5 and 6.

the sum of vertex energies $e_\mathfrak{p}$, $\mathfrak{p} = 1, 2, \ldots, 6$, one of which can always be taken as zero since this involves changing the total configurational energy by only a constant. We impose the conditions

$$e_1 = e_2 , \qquad e_3 = e_4 , \qquad e_5 = e_6 . \tag{10.1}$$

Since the numbers of 5 and 6 are equal, the last relation involves no loss of generality.[1] From the first two relations, the energy of a polarized vertex is unaltered by a 180° rotation of the dipole moment. These relations therefore imply the absence of an electric field.

For the ice problem, $e_\mathfrak{p} = 0$ for $\mathfrak{p} = 1, \ldots, 6$. The entropy of $d = 2$ ice is thus $k_\mathrm{B} \ln \Omega(N)$ where $\Omega(N)$ is the number of arrow configurations satisfying the ice rule at all the N vertices of the lattice. As will be seen below, the derivation of $\Omega(N)$ is not easy. However a good approximation was obtained by Pauling (1935). For large N the number of edges is $\frac{1}{2}\mathrm{z}N = 2N$ and so the number of unrestricted arrow configurations is 2^{2N}. Since the ice rule allows only six out of the sixteen possible configurations at each vertex, Pauling put

$$\Omega(N) = 2^{2N} \left(\frac{6}{16}\right)^N = \left(\frac{3}{2}\right)^N , \tag{10.2}$$

giving the value $k_\mathrm{B} \ln(3/2)$ for the entropy per molecule.

Other systems originally studied were the KDP (Lieb 1967c) and F (Lieb 1967b) models which respectively represent a ferroelectric and an antiferroelectric. As can be seen from Table 10.1, the vertex energies for the KDP model favour vertices 1 and 2 whereas those for the F model favour the unpolarized vertices 5 and 6.

10.2 Parameter Space

We consider the most general zero-field case with the $e_\mathfrak{p}$ satisfying (10.1). The vertex energies expressed in terms of ε_1, ε_2 ε_3 are given in the 'zero-field' line of Table 10.1. The model is referred to as a ferroelectric or antiferroelectric according to whether the lowest energy configuration (ground state) consists of polarized or unpolarized vertices.[2] We introduce the classification

[1] In fact it is often convenient, as in the models in Table 10.1, to impose *also without loss of generality* the condition $e_5 = e_6 = 0$.

[2] The terms *intrinsically ferroelectric* or *intrinsically antiferroelectric* are sometimes used to emphasize that it is the zero-field ground states which are the basis of the classification. However the word 'intrinsically seems no more necessary here than it is for the Ising model.

Table 10.1. Vertex energies for different realizations of the six-vertex model.

Model	Vertex Energies						
	e_1	e_2	e_3	e_4	e_5	e_6	
Zero Field	$\varepsilon_2 - \varepsilon_1$	$\varepsilon_2 - \varepsilon_1$	ε_2	ε_2	ε_3	ε_3	
Square Ice	0	0	0	0	0	0	$(\varepsilon_1 = \varepsilon_2 = \varepsilon_3 = 0)$
KDP	$-\varepsilon_1$	$-\varepsilon_1$	0	0	0	0	$(\varepsilon_1 > 0,\ \varepsilon_2 = \varepsilon_3 = 0)$
IKDP	$-\varepsilon_1$	$-\varepsilon_1$	0	0	0	0	$(\varepsilon_1 < 0,\ \varepsilon_2 = \varepsilon_3 = 0)$
F	ε_2	ε_2	ε_2	ε_2	0	0	$(\varepsilon_2 > 0,\ \varepsilon_1 = \varepsilon_3 = 0)$
IF	ε_2	ε_2	ε_2	ε_2	0	0	$(\varepsilon_2 < 0,\ \varepsilon_1 = \varepsilon_3 = 0)$

(i) $\varepsilon_1 > 0$, $\varepsilon_1 > \varepsilon_2 - \varepsilon_3$, ($e_1 < e_3$, $e_1 < e_5$: ferroelectric),

(ii) $\varepsilon_1 < 0$, $\varepsilon_2 < \varepsilon_3$, ($e_3 < e_1$, $e_3 < e_5$: ferroelectric),

(iii) $\varepsilon_2 > \varepsilon_3$, $\varepsilon_2 > \varepsilon_1 + \varepsilon_3$, ($e_5 < e_3$, $e_5 < e_1$: antiferroelectric).

For case (i), the ground-state vertices are of type 1 or 2, for case (ii) of type 3 or 4, and for case (iii) of type 5 or 6. From Table 10.1, the KDP and F models are examples of (i) and (iii) respectively. The IF (inverse F) model is transitional between (i) and (ii) whereas the IKDP (inverse KDP) model is transitional between (ii) and (iii). We now define the *vertex weights* $\mathfrak{Z}(\mathfrak{p}) = \exp(-e_{\mathfrak{p}}/T)$, $\mathfrak{p} = 1, \ldots, 6$ and using (10.1), the more compact notation,

$$\mathfrak{a} = \mathfrak{Z}(1) = \mathfrak{Z}(2) = \exp[(\varepsilon_1 - \varepsilon_2)/T]\,,$$
$$\mathfrak{b} = \mathfrak{Z}(3) = \mathfrak{Z}(4) = \exp(-\varepsilon_2/T)\,, \tag{10.3}$$
$$\mathfrak{c} = \mathfrak{Z}(5) = \mathfrak{Z}(6) = \exp(-\varepsilon_3/T)\,.$$

As indicated above we could, without loss of generality, set $\varepsilon_3 = 0$, $\mathfrak{c} = 1$. However, in view of the discussion of the staggered six-vertex model in Sect. 10.14 it is not convenient to do so. But, since $\mathfrak{a}$, $\mathfrak{b}$ and $\mathfrak{c}$ are all positive, the parameter space is the positive quadrant of the plane of the scaled variables $(\mathfrak{a}/\mathfrak{c}, \mathfrak{b}/\mathfrak{c})$ (see Fig. 10.4) where the point $(1,1)$ corresponds to $T = \infty$ for all ε_1 and ε_2. For any model (defined by prescribing values of ε_1 ε_2 and ε_3) a point $(\mathfrak{a}/\mathfrak{c}, \mathfrak{b}/\mathfrak{c})$ is specified by (10.3) at each value of T. As T varies from infinity to zero, the point traces out a trajectory in the appropriate part of parameter space. For case (iii), $\mathfrak{a} \leq \mathfrak{c}$, $\mathfrak{b} \leq \mathfrak{c}$, so that the unit square bounded by the lines $\mathfrak{a}/\mathfrak{c} = 1$ and $\mathfrak{b}/\mathfrak{c} = 1$ is the antiferroelectric domain. The origin $\mathfrak{a}/\mathfrak{c} = \mathfrak{b}/\mathfrak{c} = 0$ corresponds to $T = 0$ for any antiferroelectric model. The

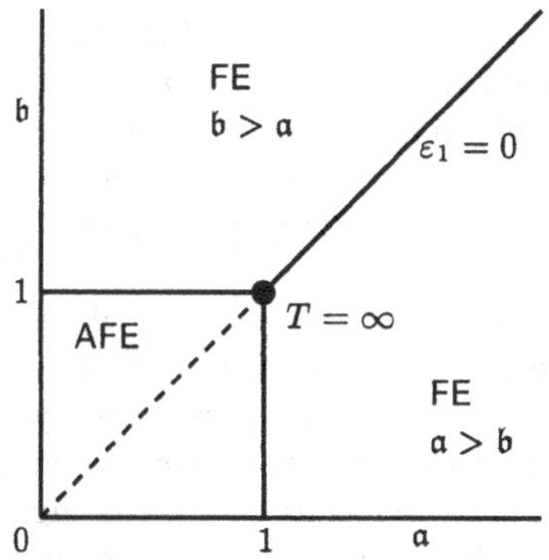

Fig. 10.4. Parameter space for the six-vertex model showing ferroelectric (FE) and antiferroelectric (AFE) regions.

region outside the unit square is the ferroelectric domain. It is divided into two by the line $\mathfrak{a} = \mathfrak{b}$, which corresponds to $\varepsilon_1 = 0$. Below this line, $\mathfrak{a} > \mathfrak{c}$ and $\mathfrak{a} > \mathfrak{b}$ so that the inequalities (i) apply, whereas above it, $\mathfrak{b} > \mathfrak{c}$ and $\mathfrak{b} > \mathfrak{a}$ so that the inequalities (ii) apply. For all ferroelectric models, $\sqrt{\mathfrak{a}^2 + \mathfrak{b}^2}/\mathfrak{c} \to \infty$ as $T \to 0$.

Consider a square lattice with an equal number of rows and columns and arrows on its edges in a particular ice-rule configuration. Suppose that the lattice is rotated in its own plane through an angle of 90° in an anticlockwise sense. This produces a different ice-rule configuration in which the original vertices of types 1, 2, 3, 4, 5 and 6 become vertices of types 4, 3, 1, 2, 6 and 5 respectively. Thus $\mathfrak{a}$ and $\mathfrak{b}$ are exchanged in the corresponding term in the partition function. Now when this rotation is applied to all ice-rule configurations successively, the result is to produce the same set of configurations in a different order. It follows that the partition function Z expressed in terms of $\mathfrak{a}$, $\mathfrak{b}$ and $\mathfrak{c}$ satisfies the symmetry condition

$$Z(\mathfrak{a}, \mathfrak{b}, \mathfrak{c}) = Z(\mathfrak{b}, \mathfrak{a}, \mathfrak{c}). \tag{10.4}$$

Thus no generality is lost by putting $\mathfrak{a} \geq \mathfrak{b}$ $(\varepsilon_1 \geq 0)$. Henceforth we shall take $\mathfrak{a} > \mathfrak{b}$ $(\varepsilon_1 > 0)$, unless stated otherwise, leaving the case $\mathfrak{a} = \mathfrak{b}$ $(\varepsilon_1 = 0)$ for separate consideration. So, for ferroelectric models, we consider only case (i), where the ground state consists of vertices of type 1 or 2 and the ground-state energy is $N(\varepsilon_2 - \varepsilon_1)$.

10.3 Graphical Representation and Ground States

In Chap. 8 we represented Ising model spin configurations by graphs on the lattice or its dual. For arrow models we obtain a configuration graph by putting a line on any horizontal edge with a left pointing arrow and on any vertical edge with a downward arrow. The corresponding line patterns at the six types of ice-rule vertex are shown on the bottom row of Fig. 10.2.

Before considering the form of the configuration graphs it is helpful to think of the square lattice as wrapped round a vertical cylinder in such a way that the last site of each row becomes a nearest neighbour of the first. The only boundary sites are now those on the top and bottom rows. Also we adopt the convention that the 'cross' pattern of vertex 2 is to be regarded as a pair of 'corners' like those at vertices 5 and 6.

A typical graph is shown on Fig. 10.5(a). It consists of three paths which can be regarded as followed in the sense of the arrows, which is the same as that of the original arrows on the lines making up the path. The apparently isolated piece of path in the bottom right-hand corner is a continuation on the cylinder of the left-hand path. From the permitted vertex patterns, no sequence of lines can terminate except at a boundary site. Only steps downward or to the left are allowed so that a path leaving the top row of sites must finish at the bottom. So any path passes through a given row of vertical edges exactly once. This has an important consequence. If n denotes the number of downward arrows on a row of vertical edges in the original arrow configuration, then in the configuration graph the number of paths passing through this row is n. Thus n is the same for all rows of vertical edges. In Fig. 10.5(a), for instance, $n = 3$ but the constant n rule also applies for alternative path patterns. Fig. 10.5(b) shows a configuration graph consisting of a single path spiralling round the cylinder. It can be seen that $n = 1$.

Any closed path must encircle the cylinder. Since no upward steps are allowed, no downward steps can occur either, and the path is confined to a single row of horizontal edges. Configuration graphs composed of these *ring paths* correspond to $n = 0$. The vertices on the ring paths are all of type 4 whereas the remaining rows consist of vertices of type 1. The *null graph* is a particular case where all vertices on the lattice are of type 1.

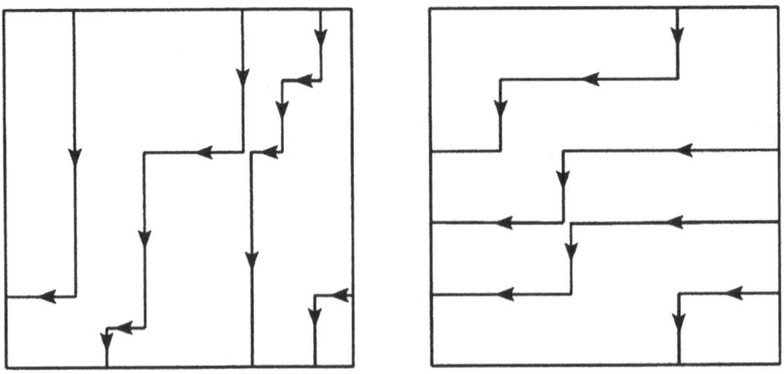

(a) (b)

Fig. 10.5. Configuration graphs on the cylinder wrapped square lattice.

In the ferroelectric ground state, all vertices are identical and are either of type 1 or of type 2. The null graph corresponding to the type 1 ground state can be changed only by the addition of a ring path or a path like one of those shown in Fig. 10.5. Thus no small perturbation of the ground state is possible. Since the energy is invariant under the reversal of all arrows, the same is true for a type 2 ground state. This contrasts with the Ising ferromagnetic ground state where the complete spin alignment can be perturbed by overturning one spin. From Sect. 8.2 this corresponds to a dual lattice configuration graph consisting of a single 1×1 square. No such closed polygons can occur in the configuration graphs of ice-rule models.

In the antiferroelectric ground state the configurational energy is zero and all vertices are of types 5 and 6. Since the latter must occur alternately along every row and column the only possible arrangement is one of sublattice order (see Appendix A.1) with one sublattice consisting of vertices of type 5 and the other of type 6. The configuration graph is shown by the full lines in Fig. 10.6. Small perturbations can occur, and one is shown in Fig. 10.6. The lines marked with a cross are replaced by the broken lines, and the numbers are the indices of the new vertices introduced by the perturbation. Although the new vertices are polarized ones, the resultant dipole moment is zero.

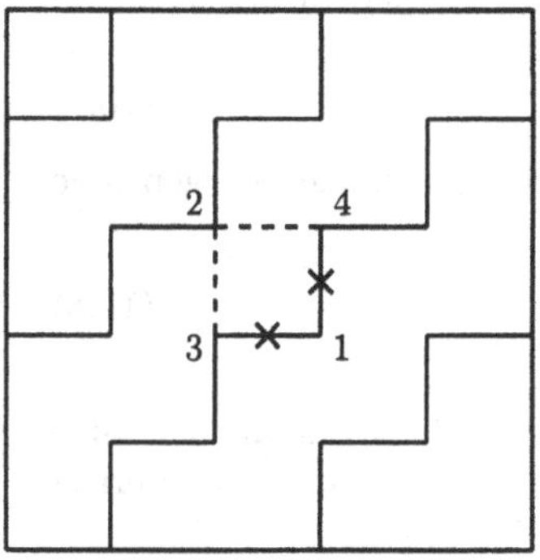

Fig. 10.6. Configuration graph for antiferroelectric (sublattice ordered) ground state with a perturbation.

10.4 Free Energy and Transfer Matrices

It has been shown that n, the number of downward pointing arrows, is the same for all rows of vertical edges in a given configuration. Denote the number of sites in each row by N_1 and the number in each column by N_2 so that the total number of sites $N = N_1 N_2$. Let ϕ_n be the sum of the terms in the zero electric field partition function which correspond to configurations with the given value of n. Using the procedure of Sect. 2.6 the whole partition function can be equated to the maximum ϕ_n. Since, for zero field, the configurational energy is unchanged by reversal of all arrows we have the symmetry relation

$$\phi_n = \phi_{N_1-n} \,. \tag{10.5}$$

Thus the maximum ϕ_n is unique only if it corresponds to $n = \frac{1}{2}N_1$. Numbering the vertical edges of row j from 1 to N_1, suppose that the n downward arrows are on edges $1 \le \eta_1 < \eta_2 < \cdots < \eta_n \le N_1$. The arrow distribution is then specified by the vector $\boldsymbol{\eta}_\mathrm{j} = (\eta_1, \eta_2, \ldots, \eta_n)$ and similarly that on the (j+1)-th row of vertical edges by a vector $\boldsymbol{\eta}_{\mathrm{j}+1}$. Taking the j-th row of vertices as lying between the j-th and (j + 1)-th rows of vertical edges, we define

$$V(\boldsymbol{\eta}_\mathrm{j}, \boldsymbol{\eta}_{\mathrm{j}+1}) = \sum \exp(-E_\mathrm{j}/T) \,, \tag{10.6}$$

where E_j is the sum of the energies of the j-th row of vertices, and the summation is over all vertex configurations on row j compatible with the given $\boldsymbol{\eta}_\mathrm{j}$ and $\boldsymbol{\eta}_{\mathrm{j}+1}$. The $V(\boldsymbol{\eta}_\mathrm{j}, \boldsymbol{\eta}_{\mathrm{j}+1})$ are elements of a transfer matrix $\boldsymbol{V}$ of dimension $N_1!/[n!(N_1 - n)!]$, and a method similar to that of Sect. 2.4.1 will be used though the single sites are now replaced by rows of vertices. In Sect. 2.4.2 the line of sites was converted into a ring by making the first and last sites into nearest neighbours. We now make the sites of row 1 into nearest neighbours of the corresponding sites of row N_2, thus bending the cylinder on which the lattice is wrapped into a torus. Then

$$\begin{aligned} \phi_n &= \sum_{\{\boldsymbol{\eta}_1, \boldsymbol{\eta}_2, \ldots, \boldsymbol{\eta}_{N_2}\}} V(\boldsymbol{\eta}_1, \boldsymbol{\eta}_2) V(\boldsymbol{\eta}_2, \boldsymbol{\eta}_3) \cdots V(\boldsymbol{\eta}_{N_2-1}, \boldsymbol{\eta}_{N_2}) V(\boldsymbol{\eta}_{N_2}, \boldsymbol{\eta}_1) \\ &= \mathrm{Trace}\{\boldsymbol{V}^{N_2}\} \,. \end{aligned} \tag{10.7}$$

The (right) eigenvector of $\boldsymbol{V}$ with elements $g(\boldsymbol{\eta})$ and corresponding eigenvalue λ satisfies the relation

$$\sum_{\{\boldsymbol{\eta}'\}} V(\boldsymbol{\eta}, \boldsymbol{\eta}') g(\boldsymbol{\eta}') = \lambda g(\boldsymbol{\eta}) \,, \tag{10.8}$$

where $\boldsymbol{\eta}$ and $\boldsymbol{\eta}'$ specify the distributions of downward arrows on successive rows of vertical edges. An important theorem due to Perron (Gantmacher 1979) has the following form:

> *A matrix of positive (non-zero) elements has an eigenvalue which is real and positive and of magnitude larger than that of any other eigenvalue. This eigenvalue has left and right eigenvectors with all elements real and positive (non-zero).*

Since $\boldsymbol{V}$ is a matrix of this type then it has a unique real maximum eigenvalue denoted by Λ_n. As we shall see below, only one right eigenvector can have all its elements real and positive, and we shall use this fact to identify Λ_n. Equation (10.7) yields an expression like (2.92) but with $N_1!/[n!(N_1 - n)!]$ terms instead of two in the sum. The leading term $(\Lambda_n)^{N_2}$ becomes dominant for large N_2 and

$$\lim_{N_2 \to \infty} \frac{\ln(\phi_n)}{N_2} = \ln(\Lambda_n). \tag{10.9}$$

Although the torus condition is convenient, it is not essential and, with a little more trouble, (10.9) can be proved for an open-ended cylinder.

Since the temperature and electric field have been taken as the independent intensive thermodynamic variables, with the field fixed at zero, $-T\ln(\Lambda_n)$ is a non-equilibrium free energy per row of sites, depending on the 'order variable' n. Minimizing $-T\ln(\Lambda_n)$ with respect to n is equivalent to finding the maximum term ϕ_n. In the thermodynamic limit, when $N_1 \to \infty$, n must be replaced by the relative vertical electrical polarization component p, defined by

$$p = \frac{N_1 - 2n}{N_1}, \tag{10.10}$$

which is also regarded as an order parameter. A zero-field non-equilibrium free energy $f(T,p)$ per site is defined by

$$f(T,p) = -T \lim_{N_1\to\infty} \left[\frac{\ln(\Lambda_n)}{N_1}\right]_{n=N_1(1-p)/2}, \tag{10.11}$$

the ratio n/N_1 being kept constant during the limiting process. From (10.9) and (10.11)

$$f(T,p) = -T \lim_{N_1\to\infty} \left[\frac{1}{N_1} \lim_{N_2\to\infty} \frac{\ln(\phi_n)}{N_2}\right]_{n=N_1(1-p)/2}. \tag{10.12}$$

It can be seen that the thermodynamic limit is approached in two stages.[3] From the symmetry relation (10.5) and from (10.10),

$$f(T,p) = f(T,-p). \tag{10.13}$$

To obtain the equilibrium zero-field free energy per site, $f(T,p)$ must be minimized with respect to p.[4]

10.5 Transfer Matrix Eigenvalues

Suppose that the lattice is wrapped on a vertical cylinder but that the step of bending the cylinder into a torus has not yet been taken. Let the cylinder be rotated by an angle $2\pi s/N_1$ about its axis, s being an integer. Then each $\eta_i \to \eta_i + s$, (mod N_1) and rearrangement of the resulting numbers in ascending order yields a new vector $\boldsymbol{\eta}^{(s)}$. Since the energies E_j are not changed by the rotation,

$$V(\boldsymbol{\eta}^{(s)}, \boldsymbol{\eta}'^{(s)}) = V(\boldsymbol{\eta}, \boldsymbol{\eta}'). \tag{10.14}$$

[3] See Lieb and Wu (1972), p. 360, for proof of validity of this procedure.

[4] It should be noted that if p, which is a density in the sense of Sect. 1.8, were taken as an independent variable instead of the field then $f(T,p)$ would become the equilibrium Helmholtz free energy density.

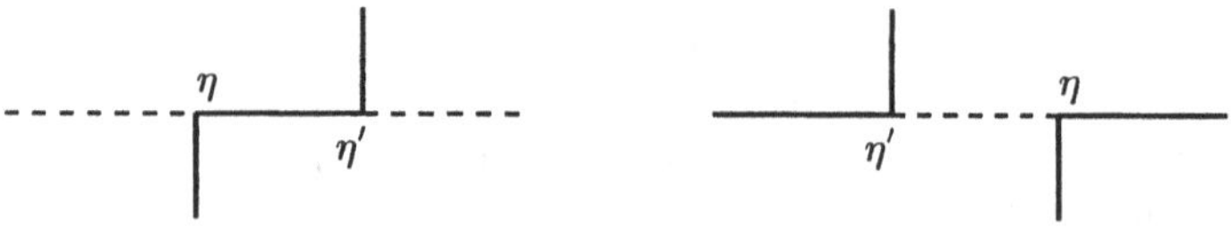

Fig. 10.7. Part of a configuration graph for one row of vertices: the $n = 1$ case.

It follows that, for all $\boldsymbol{\eta}$,

$$g(\boldsymbol{\eta}^{(s)}) = \zeta^{(s)} g(\boldsymbol{\eta}), \tag{10.15}$$

for some $\zeta^{(s)}$. Since $\boldsymbol{\eta}^{(N_1)} = \boldsymbol{\eta}$, $\zeta^{(N_1)} = 1$. Also $\zeta^{(s)}\zeta^{(r)} = \zeta^{(r+s)}$ so that there is a number ζ such that

$$\zeta^{(s)} = \zeta^{s}, \qquad \zeta^{N_1} = 1. \tag{10.16}$$

The determination of the eigenvalues of $\boldsymbol{V}$ for general n is complicated. In this section we give the derivation for the cases $n = 0$ and $n = 1$ but present only the results for general n. The detailed analysis is contained in Appendix A.4.

10.5.1 The Case $n = 0$

The matrix $\boldsymbol{V}$ is 1×1 and hence equal to Λ_0. On the intervening row of horizontal edges, either all arrows point to the right or all arrows point to the left, giving vertices of type 1 or 4 respectively. From (10.6) and (10.3)

$$\Lambda_0 = \mathfrak{a}^{N_1} + \mathfrak{b}^{N_1}. \tag{10.17}$$

The same result applies for $n = N_1$, in accordance with (10.5), but the vertices are of type 2 or 3. Since $\mathfrak{a} > \mathfrak{b}$, the first term on the right-hand side of (10.17) is dominant for large N_1. Hence, from (10.9), (10.11) and (10.13),

$$f(T, 1) = f(T, -1) = -T\ln(\mathfrak{a}) = \varepsilon_2 - \varepsilon_1, \tag{10.18}$$

corresponding to the ferroelectric ground state.

10.5.2 The Case $n = 1$

The vector $\boldsymbol{\eta}$ reduces to a scalar η, which denotes the position of the single downward arrow, and the matrix $\boldsymbol{V}$ is $N_1 \times N_1$. Equation (10.8) becomes

$$\sum_{\eta'=1}^{N_1} V(\eta, \eta') g(\eta') = \lambda g(\eta). \tag{10.19}$$

For $\eta \neq \eta'$ there is only one term in the sum on the right of (10.6) and its form depends on whether $\eta < \eta'$ or $\eta > \eta'$, the relevant portions of configuration graph for the two situations being shown in Fig. 10.7. (See also Fig. 10.5(b), which shows a particular $n = 1$ configuration.) It can be seen that the two situations are still distinct in the limiting cases where $\eta' = \eta$

and there are accordingly two terms in $V(\eta,\eta)$, corresponding respectively to rightward and leftward flows of arrows on the row of horizontal edges. For the eigenvector elements $g(\eta)$, we shall try the simplest form which satisfies (10.15) and (10.16):

$$g(\eta)=\zeta^{\eta}\,,\quad \zeta=\exp(\mathrm{i}k)\,,\quad k=2\pi\kappa/N_1\,,\quad \kappa=0,1,\ldots,N_1-1\,. \tag{10.20}$$

Each eigenvector is characterized by one of the N_1 possible values of the wave number k and we have to find the corresponding $\lambda(k)$. We now write

$$\sum_{\eta'=1}^{N_1} V(\eta,\eta')g(\eta')=\sum_{\eta'=1}^{N_1} V(\eta,\eta')\zeta^{\eta'}=\Omega^{(\mathrm{i})}(\eta)+\Omega^{(\mathrm{ii})}(\eta)\,, \tag{10.21}$$

where the index (i) corresponds to $1\le\eta'\le\eta$ and (ii) to $\eta\le\eta'\le N_1$. Using Figs. 10.2 and 10.7

$$\begin{aligned}\Omega^{(\mathrm{i})}(\eta)&=\mathfrak{a}\mathfrak{b}^{N_1-1}\zeta^{\eta}+\mathfrak{c}^2\sum_{\eta'=1}^{\eta-1}\mathfrak{a}^{\eta-\eta'-1}\mathfrak{b}^{N_1-\eta+\eta'-1}\zeta^{\eta'}\,,\\ \Omega^{(\mathrm{ii})}(\eta)&=\mathfrak{a}^{N_1-1}\mathfrak{b}\zeta^{\eta}+\mathfrak{c}^2\sum_{\eta'=\eta+1}^{N_1}\mathfrak{a}^{N_1-\eta'+\eta-1}\mathfrak{b}^{\eta'-\eta-1}\zeta^{\eta'}\,.\end{aligned} \tag{10.22}$$

After summing the geometrical progressions and some rearrangement,

$$\begin{aligned}\Omega^{(\mathrm{i})}(\eta)&=\overline{\mathfrak{M}}(\zeta)\mathfrak{b}^{N_1}\zeta^{\eta}+\mathfrak{c}^2\mathfrak{b}^{N_1}\left(\frac{\mathfrak{a}}{\mathfrak{b}}\right)^{\eta}\zeta[\mathfrak{a}(\mathfrak{a}-\mathfrak{b}\zeta)]^{-1}\,,\\ \Omega^{(\mathrm{ii})}(\eta)&=\overline{\mathfrak{L}}(\zeta)\mathfrak{a}^{N_1}\zeta^{\eta}-\mathfrak{c}^2\mathfrak{b}^{N_1}\left(\frac{\mathfrak{a}}{\mathfrak{b}}\right)^{\eta}\zeta^{N_1+1}[\mathfrak{a}(\mathfrak{a}-\mathfrak{b}\zeta)]^{-1}\,,\end{aligned} \tag{10.23}$$

where

$$\overline{\mathfrak{L}}(\zeta)=\frac{\mathfrak{b}^2-\mathfrak{c}^2-\mathfrak{a}\mathfrak{b}\zeta^{-1}}{\mathfrak{a}(\mathfrak{b}-\mathfrak{a}\zeta^{-1})}\,,\qquad \overline{\mathfrak{M}}(\zeta)=\frac{\mathfrak{a}^2-\mathfrak{c}^2-\mathfrak{a}\mathfrak{b}\zeta}{\mathfrak{b}(\mathfrak{a}-\mathfrak{b}\zeta)}\,. \tag{10.24}$$

Substituting (10.23) into (10.21) and using the second relation of (10.16),

$$\sum_{\eta'=1}^{N_1} V(\eta,\eta')\zeta^{\eta'}=\zeta^{\eta}[\overline{\mathfrak{L}}(\zeta)\mathfrak{a}^{N_1}+\overline{\mathfrak{M}}(\zeta)\mathfrak{b}^{N_1}]\,, \tag{10.25}$$

so that equation (10.19) is satisfied by the assumed form of $g(\eta)$.

The eigenvalue is the coefficient of ζ^{η} on the right-hand side of (10.25) and it is desirable to express it in terms of the wave number k. Accordingly we define

$$\mathsf{L}(k)=\overline{\mathfrak{L}}[\exp(\mathrm{i}k)]=\frac{2\Delta-\mathfrak{b}^{-1}\mathfrak{a}-\exp(-\mathrm{i}k)}{1-\mathfrak{b}^{-1}\mathfrak{a}\exp(-\mathrm{i}k)}\,, \tag{10.26}$$

$$\mathsf{M}(k)=\overline{\mathfrak{M}}[\exp(\mathrm{i}k)]=\frac{2\Delta-\mathfrak{a}^{-1}\mathfrak{b}-\exp(\mathrm{i}k)}{1-\mathfrak{a}^{-1}\mathfrak{b}\exp(\mathrm{i}k)}\,, \tag{10.27}$$

where

$$\Delta = \frac{\mathfrak{a}^2 + \mathfrak{b}^2 - \mathfrak{c}^2}{2\mathfrak{a}\mathfrak{b}} = \cosh(\varepsilon_1/T) - \frac{1}{2}\exp[(2\varepsilon_2 - 2\varepsilon_3 - \varepsilon_1)/T]. \tag{10.28}$$

The eigenvalues can now be written as

$$\lambda(k) = \mathsf{L}(k)\mathfrak{a}^{N_1} + \mathsf{M}(k)\mathfrak{b}^{N_1}, \tag{10.29}$$

where k can take any of the N_1 values specified in (10.20).

As indicated in Sect. 10.4, the largest eigenvalue can be deduced from Perron's theorem since the corresponding eigenvector can be taken to have real positive elements and from (10.20) this occurs only when $k = 0$ ($\kappa = 0$). Hence, by (10.29),

$$\Lambda_1 = \lambda(0) = \mathsf{L}(0)\mathfrak{a}^{N_1} + \mathsf{M}(0)\mathfrak{b}^{N_1}. \tag{10.30}$$

Since $\mathfrak{a} > \mathfrak{b}$, the first term in the last expression becomes dominant for large N_1, just as in Λ_0.

10.5.3 The General n Case

We assume the *Bethe ansatz* that the elements of a given eigenvector depend on n real wave numbers $k_1, k_2, \ldots, k_n$ according to the relationship

$$g(\eta) = \sum_{\{\Pi\}} A(\Pi) z_{\pi_1}^{\eta_1} z_{\pi_2}^{\eta_2} \cdots z_{\pi_n}^{\eta_n}, \tag{10.31}$$

where $z_m = \exp(\mathrm{i}k_m)$ and Π is a permutation $\{\pi_1, \pi_2, \ldots, \pi_n\}$ of the numbers $1, 2, \ldots, n$ and the sum is over all $n!$ permutations. In Appendix A.4 it is shown that

$$\text{(I)} \qquad z_1 z_2 \cdots z_n = \zeta, \tag{10.32}$$

where ζ is the quantity appearing in (10.16). Results for $n = 1$ are regained by putting $\eta_1 = \eta$, $z_1 = \zeta$ and $k_1 = k$.

In Appendix A.4 two important results are proved for the coefficients $A(\Pi)$. Defining operators $\mathfrak{R}$ and $\mathfrak{S}_{j,j+1}$ acting on the permutation Π by

$$\begin{aligned} \mathfrak{R}\Pi &= \{\pi_2, \pi_3, \ldots, \pi_n, \pi_1\}, \\ \mathfrak{S}_{j,j+1}\Pi &= \{\pi_1, \pi_2, \ldots, \pi_{j+1}, \pi_j, \ldots, \pi_n\}, \end{aligned} \tag{10.33}$$

these are

$$\text{(II)} \qquad A(\mathfrak{R}\Pi) = z_{\pi_1}^{-N_1} A(\Pi) = \exp(-\mathrm{i}k_{\pi_1} N_1) A(\Pi), \tag{10.34}$$

$$\text{(III)} \qquad A(\mathfrak{S}_{j,j+1}\Pi) = -A(\Pi)\exp[\mathrm{i}\theta(k_{\pi_j}, k_{\pi_{j+1}})], \tag{10.35}$$

where

$$\begin{aligned} \exp[\mathrm{i}\theta(q_1, q_2)] &= \frac{\mathsf{L}(q_2)\mathsf{M}(q_1) - 1}{\mathsf{L}(q_1)\mathsf{M}(q_2) - 1} \\ &= \frac{1 + \exp\{\mathrm{i}(q_2 + q_1)\} - 2\Delta\exp(\mathrm{i}q_2)}{1 + \exp\{\mathrm{i}(q_1 + q_2)\} - 2\Delta\exp(\mathrm{i}q_1)}. \end{aligned} \tag{10.36}$$

Since q_1 and q_2 are real, $|\exp\{\mathrm{i}\theta(q_1,q_2)\}| = 1$ so that $\theta(p,q)$ is real. Also, from (10.36),

$$\theta(q_1,q_2) = \begin{cases} 0, & q_1 = q_2, \\ -\theta(q_2,q_1), & q_1 \neq q_2. \end{cases} \tag{10.37}$$

The results (I) and (II) are direct consequences of symmetry with respect to rotation about the axis of the cylinder. Together with (III) they imply (see Appendix A.4) that the eigenvalue corresponding to the eigenvector (10.31) is

(IV) $$\lambda(k_1,k_2,\ldots,k_n) = \mathfrak{a}^{N_1}\mathsf{L}(k_1)\mathsf{L}(k_2)\cdots\mathsf{L}(k_n) + \mathfrak{b}^{N_1}\mathsf{M}(k_1)\mathsf{M}(k_2)\cdots\mathsf{M}(k_n). \tag{10.38}$$

Any permutation can be obtained from any other by a sequence of operations $\mathfrak{S}_{j,j+1}$. For instance, $\mathfrak{R} = \mathfrak{S}_{n-1,n}\mathfrak{S}_{n-2,n-1}\cdots\mathfrak{S}_{1,2}$ and comparing (10.34) and (10.35)

$$\exp(-\mathrm{i}k_{\pi_1}N_1) = (-1)^{n-1}\exp\Big[\mathrm{i}\sum_{\{\pi_m\neq\pi_1\}}\theta(k_{\pi_1},k_{\pi_m})\Big]. \tag{10.39}$$

However, k_{π_1} can be any of the wave numbers $k_1,\ldots,k_n$ and, from (10.37), $\theta(k_j,k_j) = 0$ so that (10.39) is equivalent to

$$\exp(\mathrm{i}k_jN_1) = (-1)^{n-1}\exp\Big[-\mathrm{i}\sum_{r=1}^{n}\theta(k_j,k_r)\Big]. \tag{10.40}$$

Equation (10.38) is a generalization of (10.29) and it might be expected that λ could be maximized by putting all $k_j = 0$. Unfortunately this is not so, except when $\Delta = 1$, and it will now be shown that, for $\Delta \neq 1$, all k_j must be distinct. For definiteness, consider k_1 and k_2. Let $\mathfrak{C}_{12}$ be an operator acting on a permutation Π which transposes the indices 1 and 2. Suppose that, in Π, 1 occurs before 2 and that they are separated by s other symbols. Then $\mathfrak{C}_{12}$ is the product of $2s+1$ adjacent transposition operators $\mathfrak{S}_{j,j+1}$, and, from (10.35), $A(\mathfrak{C}_{12}\Pi) = (-1)^{2s+1}A(\Pi)\exp(\mathrm{i}\Sigma)$ where Σ is the sum of $2s+1$ terms. One of these is $\theta(k_1,k_2)$ and the other terms can be arranged in pairs $\theta(k_1,k_i)+\theta(k_i,k_2)$. Then, if $k_1 = k_2$, $\theta(k_1,k_2) = 0$ and $\theta(k_1,k_i)+\theta(k_i,k_2) = 0$ from (10.37). Hence $\Sigma = 0$, $A(\mathfrak{C}_{12}\Pi) = -A(\Pi)$ and, since $z_1 = z_2$, the terms corresponding to Π and $\mathfrak{C}_{12}\Pi$ cancel in (10.31). As the whole set of $n!$ permutations can be divided into pairs differing by a $1 \leftrightarrow 2$ transposition it follows that $g(\eta) = 0$ when $k_1 = k_2$. The same reasoning is valid for any pair of wave numbers k_i, k_j. The exceptional case $\Delta = 1$ will be discussed in Sect. 10.6.

No method is known for finding and maximizing Λ_n for a finite lattice at an arbitrary value of T. However the problem can be solved in the thermodynamic limit and the equilibrium $f(T)$ obtained. In the thermodynamic limit it is necessary to consider a continuous spectrum of k values, and this theory is discussed in Sects. 10.7–10.11.

10.6 The Low-Temperature Frozen Ferroelectric State

In the present and subsequent sections it will be shown that the values $\Delta = \pm 1$ for the parameter Δ defined in (10.28) give the transition temperatures for ice-rule models. The line $\mathfrak{b} = \mathfrak{c} - \mathfrak{a}$ on which $\Delta = -1$ and the lines $\mathfrak{b} = \mathfrak{a} - \mathfrak{c}$ and $\mathfrak{b} = \mathfrak{a} + \mathfrak{c}$ on which $\Delta = 1$ are shown in Fig. 10.8. Comparison with Fig. 10.4 shows that each line is relevant to only one type of model, the $\Delta = -1$ line to the antiferroelectric case (iii) and the $\Delta = 1$ lines $\mathfrak{b} = \mathfrak{a} - \mathfrak{c}$ and $\mathfrak{b} = \mathfrak{a} + \mathfrak{c}$ to ferroelectric cases (i) and (ii) respectively.

We concentrate on the ferroelectric and, since we assume $\mathfrak{a} > \mathfrak{b}$, need consider only the line $\mathfrak{b} = \mathfrak{a} - \mathfrak{c}$. For given values of ε_1 and ε_2 the trajectory crosses this line at $T = T_c$ where, by (10.3),

$$\exp(-\varepsilon_2/T_c) = \exp[(\varepsilon_1 - \varepsilon_2)/T_c] - \exp(-\varepsilon_3/T_c)\,. \tag{10.41}$$

The trajectory for the KDP model, where $\mathfrak{b} = \mathfrak{c}$ for all T, is shown in Fig. 10.8. Now we look at the equations for general n given in Sect. 10.5. The value $\Delta = 1$ has a special significance for (10.36) since the right-hand side becomes indeterminate for $q_1 = q_2 = 0$. To find the limiting value we expand in a Maclaurin series. Then

$$\exp[\mathrm{i}\theta(q_1, q_2)] \simeq \frac{\mathrm{i}(q_1 - q_2) + \cdots}{\mathrm{i}(q_2 - q_1) + \cdots} \to -1 \qquad \text{as } q_1 \to 0,\ q_2 \to 0\ . \tag{10.42}$$

It follows that (10.40) is satisfied (both sides being equal to 1) when $k_j = 0$, $(j = 1, \ldots, n)$. Again, (10.35) reduces to $A(\mathfrak{S}_{j,j+1}\Pi) = A(\Pi)$ and the eigenvector $g(\eta)$ does not vanish, as it does for $\Delta \neq 1$ unless all the k_j are distinct. In fact, from (10.31), all eigenvector elements have an equal value which can be set equal to 1. Thus for $T = T_c$ ($\Delta = 1$) not only does the set of wave numbers $k_j = 0$ $(j = 1, \ldots, n)$ give an eigenvalue but, by Perron's theorem (see Sect. 10.4), this is the maximum eigenvalue. Hence, by (10.38),

$$\Lambda_n = \mathfrak{a}^{N_1}[\mathsf{L}(0)]^n + \mathfrak{b}^{N_1}[\mathsf{M}(0)]^n. \tag{10.43}$$

However, this is not all. From (10.26) and (10.27), $\mathsf{L}(0) = \mathsf{M}(0) = 1$ when $\Delta = 1$ and, by comparison with (10.17), we have

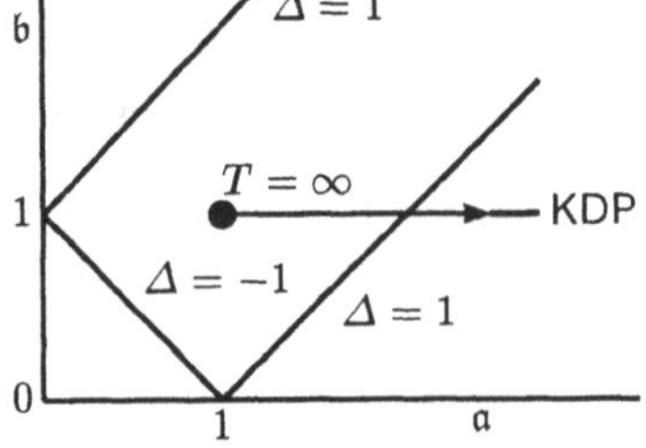

Fig. 10.8. The six-vertex parameter space with lines of critical points and KDP trajectory.

$$\Lambda_n = \Lambda_0\,, \qquad \text{for all } n\,. \tag{10.44}$$

From (10.11) it follows that, at $T = T_c$, $f(T,p)$ is independent of p.

Now suppose that $\Delta > 1$, implying a ferroelectric with $T < T_c$ (see Example 10.1). Take N_2 and N_1 so large that $\phi_n = \Lambda_n^{N_2}$ and $\Lambda_0 = \mathfrak{a}^{N_1}$ without significant error. Then

$$\phi_0 = \exp[-N(\varepsilon_2 - \varepsilon_1)/T]\,, \tag{10.45}$$

with $N(\varepsilon_2 - \varepsilon_1)$ being the ferroelectric ground state energy. Now for $n \neq 0$ and $n \neq N_1$ the terms in ϕ_n correspond to configurational energies $N(\varepsilon_2 - \varepsilon_1) + \Delta E$, where $\Delta E > 0$. It follows that the ratio ϕ_n/ϕ_0 is the sum of terms each depending on T through a factor $\exp(-\Delta E/T)$. Hence $\mathrm{d}(\phi_n/\phi_0)/\mathrm{d}T > 0$. However, at $T = T_c$, from (10.44),

$$\phi_n/\phi_0 = (\Lambda_n/\Lambda_0)^{N_2} = 1\,. \tag{10.46}$$

Thus, for $T < T_c$,

$$\phi_n < \phi_0 = \phi_{N_1} \qquad \text{when } n \neq 0 \text{ and } n \neq N_1\,. \tag{10.47}$$

Thus, from (10.12) and (10.13) we have, for the non-equilibrium free energy per site,

$$f(T,p) > f(T,1) = f(T,-1) \qquad \text{when } p \neq \pm 1\,. \tag{10.48}$$

The equilibrium free energy per site for the ferroelectric with $T < T_c$ is thus

$$f(T) = f(T,\pm 1) = -N^{-1}T\ln(\phi_0) = -T\ln(\mathfrak{a}) = \varepsilon_2 - \varepsilon_1\,. \tag{10.49}$$

The equilibrium state is thus one of perfect long-range order with either $p = 1$ (all vertices of type 1) or $p = -1$ (all vertices of type 2). This is a remarkable result. The kind of completely ordered configuration which exists only at $T = 0$ for the Ising ferromagnet is here 'frozen in' over the finite temperature range $(0, T_c)$. Although T_c is clearly a transition or critical temperature for the ferroelectric we know nothing about the type of singularity there until we have studied the equilibrium state for $T > T_c$.

10.7 Wave-Number Density

We can take $\Delta < 1$, which applies for a ferroelectric when $T < T_c$ and for an antiferroelectric at all temperatures (see Example 10.1). Now (10.40) implies

$$N_1 k_j = 2\pi\ell_j - \pi(n-1) - \sum_{r=1}^{n} \theta(k_j, k_r)\,, \qquad j = 1, \ldots, n\,, \tag{10.50}$$

where the ℓ_j are integers. Since, from (10.10), $n \to \infty$ in the thermodynamic limit for $p < 1$, we can take n as even. It has been shown that the k_j are distinct for $\Delta < 1$ and, as the required largest of the eigenvalues given by (10.38) is real, they must occur in equal and opposite pairs. The dominant term on the right-hand side of (10.38) then depends on the real products $\mathsf{L}(k)\mathsf{L}(-k)$. By (10.26),

$$|\mathsf{L}(k)|^2 = \mathsf{L}(k)\mathsf{L}(-k) = \frac{(2\Delta - \mathfrak{X})^2 + 1 - 2(2\Delta - \mathfrak{X})\cos(k)}{1 + \mathfrak{X}^2 - 2\mathfrak{X}\cos(k)}, \tag{10.51}$$

where

$$\mathfrak{X} = \mathfrak{a}/\mathfrak{b} = \exp(\varepsilon_1/T). \tag{10.52}$$

For $\Delta < 1$, $|\mathsf{L}(0)| > 1$ and $|\mathsf{L}(k)|$ decreases as k increases (see Example 10.2). So, for the largest eigenvalue, the n wave numbers k_j must be packed as closely as possible in a distribution symmetrical about $k = 0$. To this end we put $\ell_j = j - 1$ and the term $2\pi\ell_j - \pi(n-1)$ then varies from $-\pi(n-1)$ to $\pi(n-1)$. Since $\theta(-p,-q) = -\theta(p,q)$ it follows that $k_{n-j+1} = -k_j$.[5]

Now consider the thermodynamic limit in which N_1 and n tend to infinity while p, defined by (10.10), remains constant. From (10.50), with $\ell_j = j-1$, the number δj of values of j corresponding to the interval $(k, k+\delta k)$ is given, for large N_1, by

$$N_1\delta k = 2\pi\delta j - \sum_{r=1}^{n}[\theta(k+\delta k, k_r) - \theta(k, k_r)]. \tag{10.53}$$

In the thermodynamic limit there is a continuous wave-number distribution symmetrical about $k = 0$ and we introduce a density function $\rho(k)$, such that $\delta j = N_1\rho(k)\delta k$ and $\rho(-k) = \rho(k)$. Since there are $n = \frac{1}{2}N_1(1-p)$ wave numbers in total, the normalization relation

$$\int_{-q_0}^{q_0} \rho(q)\mathrm{d}q = \frac{1}{2}(1-p) \tag{10.54}$$

is satisfied, where $(-q_0, q_0)$ is the range of the wave numbers. In view of the condition $n < \frac{1}{2}N_1$ we must take $p \geq 0$ but, from the symmetry relation (10.13), no generality is lost. Dividing (10.53) by $N_1\delta k$ and letting $\delta k \to 0$ we obtain

$$2\pi\rho(k) = 1 + \int_{-q_0}^{q_0} \frac{\partial\theta(k,q)}{\partial k}\rho(q)\mathrm{d}q, \tag{10.55}$$

where summation with respect to r is replaced by integration with respect to q.

Since $\mathfrak{a} > \mathfrak{b}$ the right-hand side of (10.38) can be replaced by its first term for large N_1. As the distribution $\rho(k)$ corresponds to the largest eigenvalue, we replace λ by Λ_n. Then, from (10.11), the zero-field non-equilibrium free energy per site is given by

$$f(T,p) = -T\ln(\mathfrak{a}) - T\int_{-q_0}^{q_0} \ln|\mathsf{L}(q)|\rho(q)\mathrm{d}q. \tag{10.56}$$

[5] The problem of the k_j distribution is the same as for the one-dimensional Heisenberg Model; see the contribution by Barouch in Sect. IV D of Lieb and Wu (1972).

To minimize $f(T,p)$ with respect to the order variable p, q_0 is adjusted to maximize the integral in (10.56), subject to (10.55). Then the equilibrium value of p is deduced from (10.54). To solve (10.55) a transformation from k to another variable x is necessary. Define a transformed density $R(x)$ by

$$R(x) = 2\pi\rho(k)\frac{\mathrm{d}k}{\mathrm{d}x} \tag{10.57}$$

and the normalization equation (10.54) becomes

$$\frac{1}{2\pi}\int_{-x_0}^{x_0} R(x)\mathrm{d}x = \frac{1}{2}(1-p), \tag{10.58}$$

where x_0 corresponds to q_0. Suppose that $\theta(k,q)$ transforms to $\overline{\theta}(x,y)$ where q has the same relation to y as k has to x. The integral equation (10.55) then becomes

$$R(x) = \frac{\mathrm{d}k}{\mathrm{d}x} + \frac{1}{2\pi}\int_{-x_0}^{x_0} \frac{\partial\overline{\theta}(x,y)}{\partial x}R(y)\mathrm{d}y\,. \tag{10.59}$$

Fourier techniques can be used to solve (10.59) for suitable x_0 if $\overline{\theta}(x,y)$ depends on x and y through the difference $x-y$. The last term in (10.59) is then a convolution integral with a difference kernel, and the equation can be written as

$$\begin{aligned} R(x) &= \frac{\mathrm{d}k}{\mathrm{d}x} + \frac{1}{2\pi}\int_{-x_0}^{x_0} K(x-y)R(y)\mathrm{d}y\,, \\ K(x-y) &= \frac{\partial\overline{\theta}(x-y)}{\partial x}\,. \end{aligned} \tag{10.60}$$

If $\mathsf{L}(k)$ transforms to $\widetilde{\mathsf{L}}(x)$, then equation (10.56) becomes

$$f(T,p) = -T\ln(\mathfrak{a}) - \frac{T}{2\pi}\int_{-x_0}^{x_0} \ln|\widetilde{\mathsf{L}}(x)|R(x)\mathrm{d}x\,. \tag{10.61}$$

It turns out that the form of solution depends on whether $-1 < \Delta < 1$ or $\Delta < -1$. The value of Δ lies in the first range for both the ferroelectric and antiferroelectric models at high temperatures, but the second range is relevant only for the antiferroelectric (see Figs. 10.4 and 10.8).

10.8 The Solution of the Integral Equation for $-1 < \Delta < 1$

For this range, a real parameter μ can be defined by

$$\cos(\mu) = -\Delta, \qquad \text{with } 0 < \mu < \pi, \tag{10.62}$$

where $\mu \to \pi - 0$ as $\Delta \to 1 - 0$ and $\mu \to +0$ as $\Delta \to -1 + 0$. From (10.51), $|\mathsf{L}(k)| = 1$ when $\cos(k) = \Delta$. So $\ln|\mathsf{L}(k)| > 0$ for $|k| < \pi - \mu$ but changes sign when $|k| = \pi - \mu$ from which it follows that $q_0 = \pi - \mu$ is the largest value which q_0 can assume without the integrand in (10.56) becoming negative. Thus putting $q_0 = \pi - \mu$ minimizes $f(T,p)$, provided that this value of q_0 is compatible with sensible results for $\rho(k)$ and p from (10.54) and (10.55). We assume that $q_0 = \pi - \mu$ corresponds to the equilibrium state.

If $x_0 = \infty$ the convolution integral in (10.60) has, from (A.72), a Fourier transform which is the product of the transforms of $K(x)$ and $R(x)$. We try the transformation from k to x defined by

$$\exp(x) = \frac{\exp(\mathrm{i}k) + \exp(\mathrm{i}\mu)}{1 + \exp[\mathrm{i}(k+\mu)]}, \tag{10.63}$$

which yields a real x for real μ and k. Also $x = \pm\infty$ for $k = \pm(\pi - \mu)$ so that $q_0 = \pi - \mu$ corresponds to $x_0 = \infty$. From (10.63),

$$\cos(k) = \frac{1 - \cos(\mu)\cosh(x)}{\cosh(x) - \cos(\mu)}, \qquad \sin(k) = \frac{\sinh(x)\sin(\mu)}{\cosh(x) - \cos(\mu)} \tag{10.64}$$

and hence,

$$\frac{\mathrm{d}k}{\mathrm{d}x} = \frac{\sin(\mu)}{\cosh(x) - \cos(\mu)}. \tag{10.65}$$

Using (10.63) and a similar relation with y and q replacing x and k respectively, (10.36) becomes

$$\exp[-\mathrm{i}\theta(k,q)] = \exp[-\mathrm{i}\bar{\theta}(x,y)] = \frac{1 - \exp(y-x)\exp(2\mathrm{i}\mu)}{\exp(y-x) - \exp(2\mathrm{i}\mu)}. \tag{10.66}$$

Since the last expression depends on x and y through the difference $x - y$, the aim of a difference kernel has been achieved. After logarithmic differentiation of this expression with respect to x and substitution from (10.65), equation (10.60) becomes

$$R(x) = S(\mu, x) - \frac{1}{2\pi}\int_{-\infty}^{\infty} S(2\mu, x-y)R(y)\mathrm{d}y, \tag{10.67}$$

where

$$S(\mu, x) = \frac{\sin(\mu)}{\cosh(x) - \cos(\mu)}. \tag{10.68}$$

The Fourier transform $S^*(\mu, \omega)$ of this function is, from (A.70), given by

$$S^*(\mu, \omega) = \frac{2\pi\sinh[\omega(\pi - \mu)]}{\sinh(\pi\omega)}. \tag{10.69}$$

and using (A.72) with(10.67)

$$R^*(\omega) = \frac{2\pi S^*(\mu,\omega)}{2\pi + S^*(2\mu,\omega)}. \tag{10.70}$$

This gives

$$R^*(\omega) = \pi \mathrm{sech}(\mu\omega) \tag{10.71}$$

and inverting the transformation

$$R(x) = \frac{\pi}{2\mu}\mathrm{sech}\left(\frac{\pi x}{2\mu}\right). \tag{10.72}$$

With $x_0 = \infty$, the left-hand side of (10.58) is equal to $R^*(0)/(2\pi)$. However, from (10.71), $R^*(0) = \pi$. So it follows from by (10.58) that the equilibrium value of $p = 0$. By (10.10), this corresponds to $n = \frac{1}{2}N_1$ or an equal number of up and down arrows on each row of vertical edges. The domain of the $(\mathfrak{a}/\mathfrak{c}, \mathfrak{b}/\mathfrak{c})$ plane for which $-1 < \Delta < 1$ (see Fig. 10.8) thus represents the disordered state, which is reasonable since it corresponds to the high-temperature range for both ferroelectric and antiferroelectric, with infinite temperature corresponding to $\mathfrak{a} = \mathfrak{b} = \mathfrak{c}$ where $\Delta = \frac{1}{2}$.

10.9 The Free Energy of the Disordered State

We now derive the equilibrium zero-field free energy for the disordered state with $p = 0$, $(-1 < \Delta < 1)$. We first define a real temperature dependent parameter φ by

$$\exp(\mathrm{i}\varphi) = \frac{1 + \mathfrak{X}\exp(\mathrm{i}\mu)}{\mathfrak{X} + \exp(\mathrm{i}\mu)}, \tag{10.73}$$

where $\mathfrak{X}$ is given by (10.52) and μ by (10.62). This is equivalent to

$$\cos(\varphi) = \frac{2\mathfrak{X} - \Delta(1 + \mathfrak{X}^2)}{1 + -2\Delta\mathfrak{X} + \mathfrak{X}^2}. \tag{10.74}$$

From (10.74), $\cos(\varphi) \geq \cos(\mu)$ allowing the condition $0 \leq \varphi \leq \mu$. At $T = \infty$, where $\mathfrak{X} = 1$ and $\Delta = \frac{1}{2}$, $\varphi = 0$ and $\mu = \frac{2}{3}\pi$. For the ferroelectric at $T = T_c$, where $\Delta = 1$, $\varphi = \mu = \pi$. To obtain $|\tilde{\mathrm{L}}(x)|$, substitute from (10.64) into (10.51). The resulting expression can be simplified by deriving the real part of $\exp(\mathrm{i}\varphi - 2\mathrm{i}\mu)$ from (10.73) and also using (10.74). This gives

$$|\tilde{\mathrm{L}}(x)|^2 = \frac{\cosh(x) - \cos(2\mu - \varphi)}{\cosh(x) - \cos(\varphi)}. \tag{10.75}$$

Substituting (10.75) and (10.72) into (10.61) and putting $x_0 = \infty$, which corresponds to $p = 0$, the equilibrium free energy per site takes the form

$$f(T) = f(T, 0) = -T\ln(\mathfrak{a}) + T\phi(\mu, \varphi), \tag{10.76}$$

where

$$\phi(\mu, \varphi) = -\frac{1}{8}\int_{-\infty}^{\infty} \frac{1}{\mu}\mathrm{sech}\left(\frac{x\pi}{2\mu}\right)\ln\left[\frac{\cosh(x) - \cos(2\mu - \varphi)}{\cosh(x) - \cos(\varphi)}\right] \mathrm{d}x. \tag{10.77}$$

This applies to both the ferroelectric and antiferroelectric in temperature ranges for which $-1 < \Delta < 1$. Replacing x by $2x\mu$ gives the alternative form

$$\phi(\mu,\varphi) = -\frac{1}{4}\int_{-\infty}^{\infty} \operatorname{sech}(\pi x)\ln\left[\frac{\cosh(2\mu x)-\cos(2\mu-\varphi)}{\cosh(2\mu x)-\cos(\varphi)}\right] dx\,. \tag{10.78}$$

Another form for $\phi(\mu,\varphi)$ can be derived directly from (10.61) with $x_0 = \infty$. The Fourier transform of $R(x)$ is given by (10.71) and

$$\mathfrak{F}_\omega\{\ln|\widetilde{\mathrm{L}}(x)|\} = \frac{2\pi\sinh[\omega(\pi-\mu)]\sin[\omega(\mu-\varphi)]}{\omega\sinh(\pi\omega)} \tag{10.79}$$

(Baxter 1982a, p. 148). It follows from Parseval's formula (A.73) that

$$\phi(\mu,\varphi) = -\frac{1}{2}\int_{-\infty}^{\infty}\frac{\sinh[y(\pi-\mu)]\sinh[y(\mu-\varphi)]}{y\cosh(\mu y)\sinh(\pi y)}dy. \tag{10.80}$$

We now consider two particular cases:

10.9.1 The KDP Model

Here $\varepsilon_2 = 0$ and, by (10.28) and (10.52), $\mathfrak{X} = 2\Delta$ so that

$$\cos(\varphi) = 3\Delta - 4\Delta^3 = \cos(3\mu)\,. \tag{10.81}$$

Hence we put

$$\varphi = 3\mu - 2\pi \tag{10.82}$$

which yields $\varphi = 0$ for $\mu = \frac{2}{3}\pi$ and $\varphi = \pi$ for $\mu = \pi$. Glasser (1969) showed that for the KDP model the integral in (10.80) can be expressed in terms of elementary functions to give, from (10.76),

$$f(T) = -\varepsilon_1 - T\ln\left|\frac{2\mu}{\pi\sin(\mu)}\cot\left(\frac{\pi^2}{2\mu}\right)\right|. \tag{10.83}$$

10.9.2 Square Ice

Here $\varepsilon_1 = \varepsilon_2 = 0$ and $\mathfrak{a} = \mathfrak{b} = \mathfrak{c}$, so that square ice represents the infinite temperature limit of any ice-rule model. Accordingly, putting $\mu = \frac{2}{3}\pi$ in (10.83), the entropy of the square ice model on N sites is given by Lieb's (1967a) formula

$$k_{\mathrm{B}}\ln[\Omega(N)] = Nk_{\mathrm{B}}\ln\left|\frac{4\cot(3\pi/4)}{3\sin(2\pi/3)}\right| = \frac{3}{2}Nk_{\mathrm{B}}\ln\left(\frac{4}{3}\right), \tag{10.84}$$

which yields

$$\Omega(N) = \left(\frac{4}{3}\right)^{3N/2} = (1.5396)^N, \tag{10.85}$$

a result which is remarkably close to the Pauling estimate (10.2).

10.9.3 Non-Zero Polarization

An explicit form for $f(T,p)$ is available only for $p = 0$. However, an intricate perturbation analysis (Lieb and Wu 1972) yields, for $-1 < \Delta < 1$,

$$f(T,p) - f(T,0) = \frac{1}{4}T(\pi - \mu)\cos\left(\frac{\pi\varphi}{2\mu}\right)p^2 + O(p^4). \qquad (10.86)$$

Since $\mu < \pi$ and $\varphi < \mu$ the coefficient of p^2 is positive, so that $f(T,p)$ has a minimum at $p = 0$, as would be expected. The zero-field susceptibility can be obtained from (10.86). With a non-zero vertical electric field[6] $\mathcal{E}$, the constant-field non-equilibrium free energy per site becomes

$$f(T,\mathcal{E},p) = f(T,p) - \mathcal{E}p. \qquad (10.87)$$

Minimizing $f(T,\mathcal{E},p)$ with respect to the order variable p, we obtain the equilibrium condition

$$\mathcal{E} = \left(\frac{\partial f(T,p)}{\partial p}\right)_T. \qquad (10.88)$$

Substitution from (10.86) yields

$$\left(\frac{\partial p}{\partial \mathcal{E}}\right)_{\mathcal{E}=0} = \lim_{\mathcal{E}\to 0}\frac{p}{\mathcal{E}} = \frac{2}{T(\pi - \mu)\cos(\pi\varphi/2\mu)}. \qquad (10.89)$$

The equilibrium free energy $f(T,\mathcal{E})$ can be obtained by substituting the value of p given by (10.88) in (10.87). The alternative symmetry condition

$$f(T,\mathcal{E}) = f(T,-\mathcal{E}) \qquad (10.90)$$

follows from (10.87), (10.88) and (10.13).

10.10 The Ferroelectric Transition in Zero Field

In Sect. 10.6 it was shown that for $T < T_c$, ($\Delta > 1$) a ferroelectric is frozen into a completely ordered configuration with $p = \pm 1$. For $T > T_c$, the equilibrium polarization is $p = 0$ so that p switches from 0 to ± 1 at $T = T_c$, implying a first-order transition. We now look at other aspects of this transition. Consider a temperature just above T_c putting

$$T = T_c + \delta T, \qquad T_c \gg \delta T > 0. \qquad (10.91)$$

Since $\mu \to \pi - 0$ and $\varphi \to \pi - 0$ as $\delta T \to 0$,

$$\gamma = \pi - \mu \ll 1, \qquad \gamma' = \pi - \varphi \ll 1 \qquad (10.92)$$

in this limit and from (10.62),

[6] Here $\mathcal{E}$ is a 'field' in the sense of Sect. 1.7, and is the vertical electric field multiplied by the vertical electric dipole moment component of a vertex of type 1 or 4.

$$-\Delta(T) = \cos(\mu) = -\cos(\gamma) = -1 + \frac{1}{2}\gamma^2 + \mathrm{O}(\gamma^4), \tag{10.93}$$

$$\cos(\varphi) = -\cos(\gamma') = -1 + \frac{1}{2}(\gamma')^2 + \mathrm{O}([\gamma']^4). \tag{10.94}$$

It is necessary to relate γ' to γ. From (10.74) and (10.93)

$$\begin{aligned}\cos(\varphi) &= \frac{-(\mathfrak{X}-1)^2 + \frac{1}{2}(1+\mathfrak{X}^2)\gamma^2 + \mathrm{O}(\gamma^4)}{(\mathfrak{X}-1)^2 + \mathfrak{X}\gamma^2 + \mathrm{O}(\gamma^4)} \\ &= -1 + \frac{1}{2}\gamma^2\left(\frac{\mathfrak{X}+1}{\mathfrak{X}-1}\right)^2 + \mathrm{O}(\gamma^4)\end{aligned} \tag{10.95}$$

and comparison with (10.94) then yields

$$\gamma' = \gamma\left(\frac{\mathfrak{X}+1}{\mathfrak{X}-1}\right) + \mathrm{O}(\gamma^3) \tag{10.96}$$

and

$$\cos(2\mu - \varphi) = -\cos(2\gamma - \gamma') = -1 + \frac{1}{2}(2\gamma - \gamma')^2 + \mathrm{O}(\gamma^4). \tag{10.97}$$

We now need a relation between γ and δT. Using a Taylor expansion,

$$1 - \Delta(T) = \Delta(T_c) - \Delta(T) = \frac{1}{2}A\delta T + \mathrm{O}(\delta T^2), \tag{10.98}$$

where, by (10.28) and (10.41),

$$A = -2\left(\frac{\mathrm{d}\Delta}{\mathrm{d}T}\right)_{T=T_c} = 2T_c^{-2}(\mathfrak{X}_c - 1)[\varepsilon_1 - \varepsilon_2(1 - \mathfrak{X}_c^{-1})] > 0, \tag{10.99}$$

with $\mathfrak{X}_c = \exp(\varepsilon_1/T_c)$. Comparison with (10.93) then yields

$$\gamma = A^{1/2}(\delta T)^{1/2}[1 + \mathrm{O}(\delta T)]. \tag{10.100}$$

We now investigate the behaviour of the free energy $f(T)$ as $T \to T_c + 0$. Using (10.94)–(10.97), the expansion of the logarithmic term in the integrand of (10.77) yields

$$\ln\left[\frac{\cosh(x) - \cos(2\mu - \varphi)}{\cosh(x) - \cos(\varphi)}\right] = \frac{4\gamma^2}{(\mathfrak{X}-1)[\cosh(x)+1]} + \mathrm{O}(\gamma^4). \tag{10.101}$$

To derive the essential results about the transition we need the γ^3 term in the expansion of $\phi(\mu, \varphi)$. Accordingly, using a binomial expansion of $(1-\gamma/\pi)^{-1}$ and a Taylor expansion of the sech function, we expand the other factor in the integrand to obtain

$$\begin{aligned}\frac{1}{\mu}\operatorname{sech}\left(\frac{x\pi}{2\mu}\right) &= \frac{1}{\pi}\left(1 - \frac{\gamma}{\pi}\right)^{-1}\operatorname{sech}\left[\frac{1}{2}x\left(1 - \frac{\gamma}{\pi}\right)^{-1}\right] \\ &= \frac{1}{\pi}\operatorname{sech}(x/2)\left\{1 + \frac{\gamma}{2\pi}[2 - x\operatorname{sech}(x/2)\sinh(x/2)]\right\} \\ &\quad + \mathrm{O}(\gamma^2).\end{aligned} \tag{10.102}$$

Substituting from (10.101) and (10.102) and changing the variable of integration from x to $y = \frac{1}{2}x$,

$$\begin{aligned}\phi(\mu,\varphi) &= -\frac{\gamma^2}{2(\mathfrak{X}-1)\pi}\int_{-\infty}^{\infty}\frac{1}{\cosh^3(y)}\left[1+\frac{\gamma}{\pi}-\frac{y\gamma\sinh(y)}{\pi\cosh(y)}\right]\mathrm{d}y + \mathrm{O}(\gamma^4)\\ &= \frac{\gamma^2}{4(\mathfrak{X}-1)}\left(1+\frac{2\gamma}{3\pi}\right)+\mathrm{O}(\gamma^4).\end{aligned} \tag{10.103}$$

From (10.100), an expression accurate to terms in $(\delta T)^{3/2}$ can be obtained by replacing γ in (10.103) by $A^{1/2}(\delta T)^{1/2}$ and $\mathfrak{X}$ by $\mathfrak{X}_c$. From (10.77) and (10.103),

$$\begin{aligned}\frac{f(T)}{T} &= -\ln(\mathfrak{a}) + \phi(\mu,\varphi)\\ &= \frac{\varepsilon_2-\varepsilon_1}{T} - \frac{A}{4(\mathfrak{X}_c-1)}\left[\delta T + \frac{2A^{1/2}}{3\pi}(\delta T)^{3/2}\right] + \mathrm{O}(\delta T^2).\end{aligned} \tag{10.104}$$

By comparison with (10.49) the free energy is, as would be expected, continuous at $T = T_c$. At zero field the configurational energy and enthalpy are equal. Dividing both sides of (1.66) by N and substituting from (10.104), the configurational energy per site u is given by

$$\begin{aligned}u &= -T^2\frac{\partial}{\partial T}\left(\frac{f(T)}{T}\right)\\ &= \varepsilon_2-\varepsilon_1+\frac{AT_c^2}{4(\mathfrak{X}_c-1)}\left[1+\frac{A^{1/2}}{\pi}(\delta T)^{1/2}\right]+\mathrm{O}(\delta T).\end{aligned} \tag{10.105}$$

Since $u = \varepsilon_2 - \varepsilon_1$ for $T < T_c$, there is a finite discontinuity in u, giving a latent heat per site

$$\begin{aligned}\Delta u = u(T_c+0) - u(T_c-0) &= \frac{AT_c^2}{4(\mathfrak{X}_c-1)}\\ &= \frac{1}{2}[\varepsilon_1 - \varepsilon_2(1-\mathfrak{X}_c^{-1})].\end{aligned} \tag{10.106}$$

For the KDP model, where $\varepsilon_2 = 0$, $\Delta u = \frac{1}{2}\varepsilon_1$. For $T_c \gg \delta T$ the leading term in the zero-field configurational heat capacity c_0 per site is given by

$$c_0 = k_B\frac{\mathrm{d}u}{\mathrm{d}T} = \frac{k_B A^{3/2}T_c^2}{8\pi(\mathfrak{X}_c-1)}(\delta T)^{-1/2}. \tag{10.107}$$

Thus, c_0 diverges as $(\delta T)^{-1/2}$ when $T \to T_c + 0$ and, since $c_0 = 0$ for $T < T_c$, it has an infinite discontinuity at $T = T_c$. From (10.89) and (10.92),

$$\left(\frac{\partial p}{\partial \mathcal{E}}\right)_{\mathcal{E}=0} = 2\left\{T\gamma\cos\left[\frac{\pi(\pi-\gamma')}{2(\pi-\gamma)}\right]\right\}^{-1} = \frac{2}{T\gamma\sin[\frac{1}{2}(\gamma'-\gamma)]}. \tag{10.108}$$

Using (10.96), the leading term is given by

$$\left(\frac{\partial p}{\partial \mathcal{E}}\right)_{\mathcal{E}=0} \simeq \frac{2(\mathfrak{X}_c-1)}{T_c\gamma^2} \simeq \frac{2(\mathfrak{X}_c-1)}{T_c A}(\delta T)^{-1}. \tag{10.109}$$

The zero-field (electric) susceptibility thus diverges as $(\delta T)^{-1}$ when $T \to T_c + 0$. One very unusual feature of the foregoing analysis is that exact results have been derived for the behaviour of thermodynamic functions near a first-order transition.

10.11 The Antiferroelectric State

The range $\Delta < -1$ is relevant only for an antiferroelectric (see Sect. 10.6 and Example 10.1) and then corresponds to $T < T_c'$ where T_c' is given, from (10.3), by

$$\begin{aligned} &\mathfrak{a}(T_c') + \mathfrak{b}(T_c') = \mathfrak{c}(T_c'), \\ &\exp[(\varepsilon_1 - \varepsilon_2)/T_c'] + \exp(-\varepsilon_2/T_c') = \exp(-\varepsilon_3/T_c'). \end{aligned} \tag{10.110}$$

For $\Delta < -1$, $|\mathsf{L}(k)| > 1$ for all real k (see Example 10.2) and hence $f(T,p)$, as given by (10.56), is minimized by putting $q_0 = \pi$. Because of this change in the equilibrium value of q_0, the analysis is rather different from that applicable when $-1 < \Delta < 1$. For $\Delta < -1$ the values of μ and φ, given by (10.62) and (10.73) respectively, become imaginary and we define new real parameters

$$\lambda = -\mathrm{i}\mu, \qquad \vartheta = -\mathrm{i}\varphi, \tag{10.111}$$

where $\lambda > 0$. From (10.62) and (10.73),

$$\begin{aligned} &\cosh(\lambda) = -\Delta, \\ &\exp(\vartheta) = \frac{1 + \mathfrak{X}\exp(\lambda)}{\mathfrak{X} + \exp(\lambda)}. \end{aligned} \tag{10.112}$$

To keep the variables real, we replace the variable x by $\mathrm{i}x$ and the functions R and K by $\mathrm{i}^{-1}R$ and $\mathrm{i}^{-1}K$ respectively. The equations (10.57)–(10.61) are unchanged in form and (10.63) becomes

$$\exp(\mathrm{i}x) = \frac{\exp(\lambda + \mathrm{i}k) + 1}{\exp(\lambda) + \exp(\mathrm{i}k)}. \tag{10.113}$$

It can be seen that $x = \pm\pi$ for $k = \pm\pi$. Hence $q_0 = \pi$ transforms to $x_0 = \pi$. The integral equation (10.67) is thus replaced by

$$R(x) = \Sigma(\lambda, x) - \frac{1}{2\pi}\int_{-\pi}^{\pi} \Sigma(2\lambda, x - y)R(y)\mathrm{d}y, \tag{10.114}$$

where

$$\Sigma(\lambda, x) = \frac{\sinh(\lambda)}{\cosh(\lambda) - \cos(x)} = \mathrm{i}^{-1}S(\mathrm{i}\lambda, \mathrm{i}x). \tag{10.115}$$

With x, y, μ and φ replaced by $\mathrm{i}x$, $\mathrm{i}y$, $\mathrm{i}\lambda$ and $\mathrm{i}\vartheta$ respectively, (10.75) becomes

$$|\widetilde{\mathrm{L}}(x)|^2 = \frac{\cosh(2\lambda - \vartheta) - \cos(x)}{\cosh(\vartheta) - \cos(x)} . \tag{10.116}$$

Although the integral in (10.114) still has a difference kernel, the change in the range of integration to $(-\pi, \pi)$ means that it is necessary to use Fourier series rather than Fourier integrals. Applying the series transformation (A.74)–(A.76) to (10.114) gives

$$R_n^* = \frac{2\pi \Sigma_n^*(2\lambda)}{2\pi + \Sigma_n^*(\lambda)}, \tag{10.117}$$

where the transform of $\Sigma(\lambda, x)$ is

$$\Sigma_n^*(\lambda) = 2\pi \exp(\lambda|n|) . \tag{10.118}$$

Thus

$$R_n^* = \pi \mathrm{sech}(\lambda n), \qquad \text{for all } n. \tag{10.119}$$

When $x_0 = \pi$ is substituted into (10.58) the left-hand side becomes equal to $R_0^*/(2\pi)$. Since $R_0 = \pi$ by (10.119) this implies $p = 0$ which is consistent with the antiferroelectric ground state, where the two sublattices are composed of unpolarized vertices of types 5 and 6 respectively (see end of Sect. 10.3).

Substitution of $x_0 = \pi$ and the expression for $|\widetilde{\mathrm{L}}(x)|$ given by (10.116) into (10.61) yields the equilibrium free energy in the form

$$f(T) = f(T, 0) = -T\ln(a) + T\phi(\lambda, \vartheta) , \tag{10.120}$$

where

$$\phi(\lambda, \vartheta) = -\frac{1}{4\pi}\int_{-\pi}^{\pi} R(x) \ln\left[\frac{\cosh(2\lambda - \vartheta) - \cos(x)}{\cosh(\vartheta) - \cos(x)}\right] \mathrm{d}x . \tag{10.121}$$

It can be shown (Example 10.4) that $f(0) = 0$, as would be expected for the antiferroelectric. The Fourier coefficients of the logarithmic part of the integrand in (10.121) are given by

$$\mathfrak{F}_n\left\{\ln\left[\frac{\cosh(2\lambda - \vartheta) - \cos(x)}{\cosh(\vartheta) - \cos(x)}\right]\right\} = \begin{cases} 4\pi(\lambda - \vartheta) , & \text{for } n = 0, \\ \dfrac{4\pi \sinh[n(\lambda - \vartheta)]}{n \exp(\lambda|n|)} , & \text{for } n \neq 0, \end{cases} \tag{10.122}$$

and thus, from Parseval's formula (A.77),

$$\phi(\lambda, \vartheta) = -\tfrac{1}{2}(\lambda - \vartheta) - \sum_{n=1}^{\infty} \frac{\exp(-\lambda n)\sinh[n(\lambda - \vartheta)]}{n\cosh(\lambda n)} \tag{10.123}$$

for the antiferroelectric when $T < T_\mathrm{c}'$, $(\Delta < -1)$. From (10.120) and (10.123) the free energy depends on T both directly and through the parameters λ and ϑ so that the configurational entropy is non-zero. So the antiferroelectric zero-field ground state with perfect 5–6 sublattice order is only attained at

$T = 0$. This contrasts with an ferroelectric in zero field when $0 < T < T_c$ where the ground state is frozen in with constant configurational free energy and hence zero configurational entropy. This difference is related to the fact that, as we saw in Sect. 10.3, the antiferroelectric ground state admits small perturbations whereas the ferroelectric ground state does not. The zero-field correspondence in the Ising model between orientationally ordered and sublattice ordered states below the critical temperature clearly does not exist for ice-rule models.

Now consider a perturbation of the polarization from zero to a value p, where $0 < p \ll 1$. Suppose that x_0 changes from π to $\pi - \delta$ and that $R(x)$ is modified to $R(x) + \delta R(x)$. Then (10.58) becomes

$$\frac{1}{2\pi}\int_{-\pi+\delta}^{\pi-\delta}[R(x)+\delta R(x)]\mathrm{d}x = \frac{1}{2}(1-p)\,. \tag{10.124}$$

Subtracting this from the corresponding relation for $p = 0$, we have to the first order, denoting the $n = 0$ Fourier coefficient of $\delta R(x)$ by δR_0^*,

$$-\delta R_0^* + R(\pi)(\delta/\pi) = \frac{1}{2}p\,. \tag{10.125}$$

Performing a similar variation on (10.60) and on (10.61) and using (A.74)–(A.77) we obtain, for the non-equilibrium free energy $f(T,p)$, the relation

$$f(T,p) - f(T,0) = |p|\Xi(T)\,, \qquad |p| \ll 1\,, \tag{10.126}$$

where

$$\Xi(T) = -\frac{1}{2}T(\lambda - \vartheta) + T\ln\left[\frac{\cosh\left(\lambda - \frac{1}{2}\vartheta\right)}{\cosh\left(\frac{1}{2}\vartheta\right)}\right] - T\sum_{n=1}^{\infty}\frac{(-1)^n\exp(-2\lambda n)\sinh[n(\lambda-\vartheta)]}{n\cosh(n\lambda)}\,. \tag{10.127}$$

The replacement of positive p by $|p|$ is justified by the symmetry relation (10.13). Further analysis using elliptic functions shows (Lieb and Wu 1972) that $\Xi(T) > 0$ for $0 \le T < T_c'$ and that, as $\Delta \to -1 - 0$ ($\lambda \to +0$),

$$\Xi(T) \simeq 4T\sin\left(\frac{\pi}{\mathfrak{X}+1}\right)\exp\left(-\frac{\pi^2}{2\lambda}\right), \tag{10.128}$$

so that $\Xi(T) \to 0$ as $T \to T_c' - 0$ and $\lambda \to +0$.

Since $\Xi(T) > 0$ for $\Delta < -1$, (10.126) shows that $f(T,p)$ has a downward pointing cusp at $p = 0$. This is quite unlike the smooth minimum which $f(T,p)$ has at $p = 0$ when $-1 < \Delta < 1$ (see 10.86)). It can be shown (Lieb and Wu 1972) that the next term in the Maclaurin expansion of $f(T,p)$ is $\frac{1}{3}\Xi_3(T)|p|^3$, where $\Xi_3(T) > 0$. Hence, for $\Delta < -1$, $f(T,p)$ has the form sketched in Fig. 10.9. Its least value occurs at the tip of the cusp, which corresponds to the equilibrium state with $p = 0$. For comparison, the limiting form at $T = 0$ (see Example 10.4) and a sketch of $f(T,p)$ for the disordered region where $-1 < \Delta < 1$ are also given in Fig. 10.9.

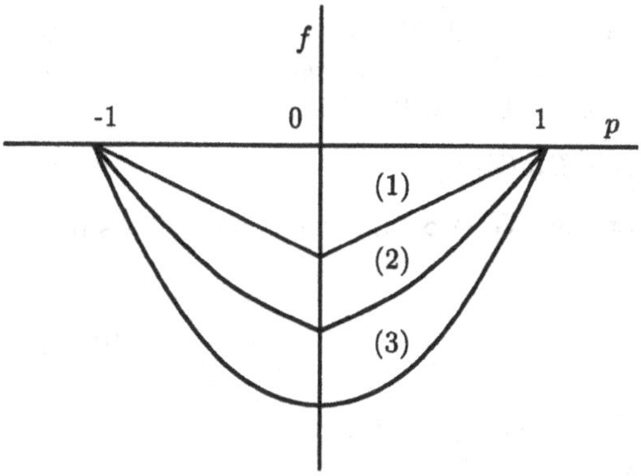

Fig. 10.9. Dependence of free energy density on reduced polarization: (1) $\Delta = -\infty$, (2) $\Delta < -1$, (3) $-1 < \Delta < 1$.

The investigation of the antiferroelectric transition at $T = T_c'$, requires analysis based on regarding $\phi(\lambda, \vartheta)$ as a function of the complex variable λ. (Lieb and Wu 1972, Glasser et al. 1972). It transpires that there is only a very weak singularity at $T = T_c'$ and that all derivatives of $f(T)$ with respect to T are continuous there. The transition is said to be of *infinite order*. If the zero-field heat capacity is plotted against T, the curve resembles that found for the $d = 1$ Ising model, with a maximum at a temperature well below T_c'. The reader may feel that the antiferroelectric transition is a non-event, in contrast to the ferroelectric transition, which could hardly be more dramatic. However, significant changes in the properties of the antiferroelectric do occur at $T = T_c'$. In zero field for $T < T_c'$, there is sublattice ordering of the vertices of types 5 and 6. Also the cusped form of $f(T, p)$ has important implications for the response to an applied electric field, as will be seen in Sect. 10.13.

10.12 Field-Induced Transitions to the Completely Polarized State

From Sect. 10.6 the completely polarized state $p = \pm 1$ is attained in zero field at a non-zero temperature for some ice-rule models. Such a state can also be attained at a finite value of $|\mathcal{E}|$ if the right-hand side of (10.88) remains finite at $p = \pm 1$. Taking $\mathcal{E} > 0$ for definiteness, there is then a curve in the $(T, \mathcal{E})$ plane,

$$\mathcal{E} = \mathcal{E}_1(T) = \left(\frac{\partial f(T,p)}{\partial p} \right)_{p=1}, \tag{10.129}$$

where p reaches the value 1. Since p varies with $\mathcal{E}$ for $\mathcal{E} < \mathcal{E}_1(T)$ but is fixed at 1 for $\mathcal{E} > \mathcal{E}_1(T)$, (10.129) defines a transition curve. From the symmetry condition (10.90) there is a mirror image curve $\mathcal{E} = -\mathcal{E}_1(T)$ in the lower half of the $(T, \mathcal{E})$ plane.

We now evaluate the right-hand side of (10.129) and check that it is finite. Suppose that the number N_1 of sites in each row is so large that $-TN_1^{-1}\ln \Lambda_n$ can be equated to $f(T,p)$, p being given by (10.10). Then

$$f(T,1) - f(T,1-2N_1^{-1}) = N_1^{-1}T\ln(\Lambda_1/\Lambda_0)\,. \tag{10.130}$$

Now divide both sides by $2N_1^{-1}$ and let $N_1 \to \infty$. Since $\mathfrak{a} > \mathfrak{b}$ it follows from (10.17) and (10.30) that

$$\left(\frac{\partial f(T,p)}{\partial p}\right)_{p=1} = \frac{1}{2}T\ln[\mathrm{L}(0)]\,. \tag{10.131}$$

Substituting (10.131) into (10.129) and using (10.26) and (10.52),

$$\mathcal{E}_1(T) = \frac{1}{2}T\ln\left[\frac{1+\mathfrak{X}-2\Delta}{\mathfrak{X}-1}\right]\,. \tag{10.132}$$

We assume $\varepsilon_1 > 0$ so that $\mathfrak{X} > 1$ and hence $\mathcal{E}_1(T) > 0$ for $-\infty < \Delta < 1$. It may be shown (Example 10.5) that $\mathrm{d}\mathcal{E}_1/\mathrm{d}T > 0$ for $-\infty < \Delta < 1$. Also, since $\Delta \to \frac{1}{2}$ as $T \to \infty$, it follows from (10.132) that $\mathcal{E}_1(T) \simeq \frac{1}{2}T\ln(T/\varepsilon_1) \to \infty$.

To investigate the transition on the line $\mathcal{E} = \mathcal{E}_1(T)$, information is needed about further derivatives of f with respect to p. From (10.58), $p = 1$ corresponds to $x_0 = 0$. By differentiating equations (10.58), (10.60) and (10.61) with respect to p and then letting $x_0 \to 0$ it is possible to reproduce (10.131) and to show that, where $-\infty < \Delta < 1$,

$$\left(\frac{\partial^2 f(T,p)}{\partial p^2}\right)_{p=1} = 0\,, \qquad 2B(T) = \left(\frac{\partial^3 f(T,p)}{\partial p^3}\right)_{p=1} < 0\,, \tag{10.133}$$

where the last relation defines $B(T)$, (Lieb and Wu 1972). Putting $p = 1-\delta p$, where $\delta p \ll 1$, and using (10.129) a Taylor expansion then gives

$$f(T,1-\delta p) = \varepsilon_2 - \varepsilon_1 - \mathcal{E}_1(T)\delta p - \frac{1}{3}B(T)(\delta p)^3 + \mathrm{O}(\delta p^4)\,, \tag{10.134}$$

since $f(T,1) = \varepsilon_2 - \varepsilon_1$. From (10.88), the equilibrium relation for δp is

$$\mathcal{E} = \mathcal{E}_1(T) + B(T)(\delta p)^2 + \mathrm{O}(\delta p^3)\,. \tag{10.135}$$

Thus as $\mathcal{E} \to \mathcal{E}_1(T)$,

$$1 - p = \delta p \simeq [\mathcal{E}_1(T) - \mathcal{E}]^{1/2}|B(T)|^{-1/2} \tag{10.136}$$

and it follows that

$$\left(\frac{\partial p}{\partial \mathcal{E}}\right)_T \simeq \frac{1}{2}|B(T)|^{-1/2}[\mathcal{E}_1(T) - \mathcal{E}]^{-1/2}\,. \tag{10.137}$$

Let $\mathcal{E}^* = \mathcal{E}_1(T^*)$ so that $(T^*,\mathcal{E}^*)$ is a point on the transition line $\mathcal{E} = \mathcal{E}_1(T)$ in the $(T,\mathcal{E})$ plane. Suppose that this point is approached from below on the line $T = T^*$ (see Fig. 10.10 for a sketch of the ferroelectric case). Putting $T = T^*$ and $\mathcal{E} = \mathcal{E}^* - \delta\mathcal{E}$, $\delta\mathcal{E} > 0$, in (10.137) it follows that $\partial p/\partial\mathcal{E} \sim (\delta\mathcal{E})^{-1/2}$ as $\mathcal{E} \to \mathcal{E}^* - 0$. For $T = T^*$, $\mathcal{E} > \mathcal{E}^*$ the system is completely polarized with

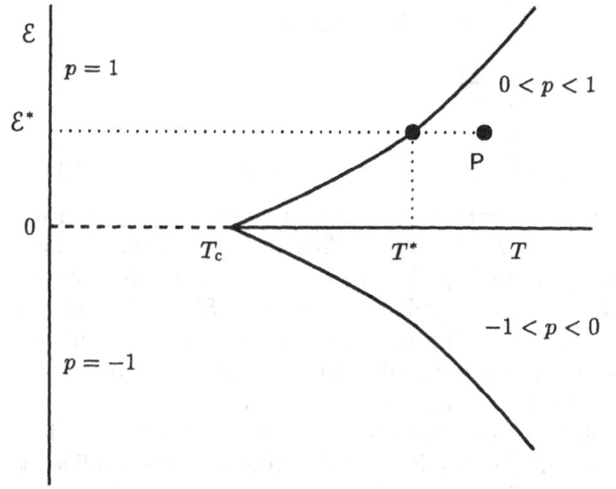

Fig. 10.10. Transition curves in the temperature–electric field plane for a ferroelectric.

$p = 1$ so that $\partial p/\partial \mathcal{E} = 0$. There is therefore an infinite discontinuity in $\partial p/\partial \mathcal{E}$ at $\mathcal{E} = \mathcal{E}^*$ although p itself is continuous. Thus a second-order transition on the curve $\mathcal{E} = \mathcal{E}_1(T)$ is indicated. This will be confirmed by considering an approach along the line $\mathcal{E} = \mathcal{E}^*$.

Take a point P in Fig. 10.10 with coordinates $(T^* + \delta T, \mathcal{E}^*)$ where $0 < \delta T \ll T^*$. By a Taylor expansion

$$\mathcal{E}_1(T^* + \delta T) - \mathcal{E}^* = A^* \delta T + \mathrm{O}(\delta T^2)\,,$$
$$A^* = \left(\frac{\mathrm{d}\mathcal{E}_1}{\mathrm{d}T}\right)_{T=T^*} > 0\,. \tag{10.138}$$

With $T = T^* + \delta T$, $\mathcal{E} = \mathcal{E}^*$ in (10.136) and (10.137)

$$\delta p \simeq \left(\frac{A^*}{|B^*|}\right)^{1/2} (\delta T)^{1/2} \qquad \text{as } T \to T^* + 0\;, \tag{10.139}$$

$$\left(\frac{\partial p}{\partial \mathcal{E}}\right)_T \simeq \frac{1}{2}(A^*|B^*|)^{-1/2}(\delta T)^{-1/2} \qquad \text{as } T \to T^* + 0\;, \tag{10.140}$$

where $B^* = B(T^*)$. From (10.87), the non-equilibrium free energy is

$$f(T^* + \delta T, \mathcal{E}^*, 1 - \delta p) = f(T^* + \delta T, 1 - \delta p) - (1 - \delta p)\mathcal{E}^*\,. \tag{10.141}$$

Substituting the equilibrium value of δp from (10.139) and using (10.134) and (10.138), the equilibrium free energy at $T = T^* + \delta T$ and $\mathcal{E} = \mathcal{E}^*$ is

$$f(T^* + \delta T, \mathcal{E}^*) = \varepsilon_2 - \varepsilon_1 - \mathcal{E}^*$$
$$- \frac{2}{3}(A^*)^{3/2}|B^*|^{-1/2}(\delta T)^{3/2} + \mathrm{O}(\delta T^2)\,. \tag{10.142}$$

The configurational enthalpy per site is then given by

$$\begin{aligned} h(T^* + \delta T, \mathcal{E}^*) &= -T^2 \frac{\partial[f(T)/T]}{\partial T}, \\ &= \varepsilon_2 - \varepsilon_1 - \mathcal{E}^* \\ &\quad + T^*(A^*)^{3/2}|B^*|^{-1/2}(\delta T)^{1/2} + \mathrm{O}(\delta T). \end{aligned} \tag{10.143}$$

It follows that the constant-field configurational heat capacity is of the order of $(\delta T)^{-1/2}$ in the limit $T \to T^* + 0$ on the line $\mathcal{E} = \mathcal{E}^*$. From (10.140) this behaviour is similar to that of $\partial p/\partial \mathcal{E}$. For $\mathcal{E} = \mathcal{E}^*$, $T < T^*$, the system is completely polarized with $p = 1$ and $h = \varepsilon_2 - \varepsilon_1 - \mathcal{E}^* =$ constant so that the heat capacity is zero. Thus h is continuous at the transition but the constant-field heat capacity has an infinite discontinuity there. The transition is second-order and the lines $\mathcal{E} = \pm\mathcal{E}_1(T)$ are critical curves.

This treatment of behaviour near the critical curves applies to both ferroelectric and antiferroelectric, but the form of the critical curves is different for the two types of model. We now specialize to the ferroelectric, where $\Delta \to 1 - 0$ as $T \to T_c + 0$. Hence $\mathcal{E}_1 = 0$ for $T = T_c$ so that the mirror image critical curves meet at the ferroelectric critical point on the T-axis (see Fig. 10.10). There is, therefore, a second-order transition on any line along which $\mathcal{E}$ has a non-zero constant value, and the transition temperature increases with $|\mathcal{E}|$. The first-order transition, with latent heat, on the $\mathcal{E} = 0$ axis (see Sect. 10.10) is thus an exceptional case.

From the form of the curve of f against p when $-1 < \Delta < 1$ (see Fig. 10.9), the relative polarization p increases steadily with $\mathcal{E}$ from $p = 0$ at $\mathcal{E} = 0$ to $p = 1$ at $\mathcal{E} = \mathcal{E}_1(T)$ for any temperature $T > T_c$. For $T < T_c$, $p = 1$ when $\mathcal{E} > 0$ and $p = -1$ when $\mathcal{E} < 0$ so that the segment $0 < T < T_c$ of the $\mathcal{E} = 0$ axis in the $(T, \mathcal{E})$ plane is a first-order transition line. This line bifurcates into the critical (second-order transition) curves $\mathcal{E} = \pm\mathcal{E}_1(T)$ at the point $(T_c, 0)$. The latter thus resembles a *bicritical point* (Aharony 1983), a type of multicritical point we have not previously encountered. However, the situation is abnormal in that inside the ordered regions of the phase plane there is a frozen state of perfect order with $p = \pm 1$.

10.13 An Antiferroelectric in an Electric Field

We consider the case $\mathcal{E} > 0$, $p > 0$. From (10.90) the part of the phase diagram in the lower half of the $(T, \mathcal{E})$ plane is the mirror image of that in the upper half. We first look at the situation when $T = 0$. In the perfectly ordered antiferroelectric state (see the end of Sect. 10.3), the configurational energy and polarization are zero and hence the configurational enthalpy per site, $h = u - p\mathcal{E} = 0$ for any $\mathcal{E}$. For a completely polarized state with all vertices of type 1, $p = 1$ and

$$h = h_1 = u - p\mathcal{E} = \varepsilon_2 - \varepsilon_1 - \mathcal{E}. \tag{10.144}$$

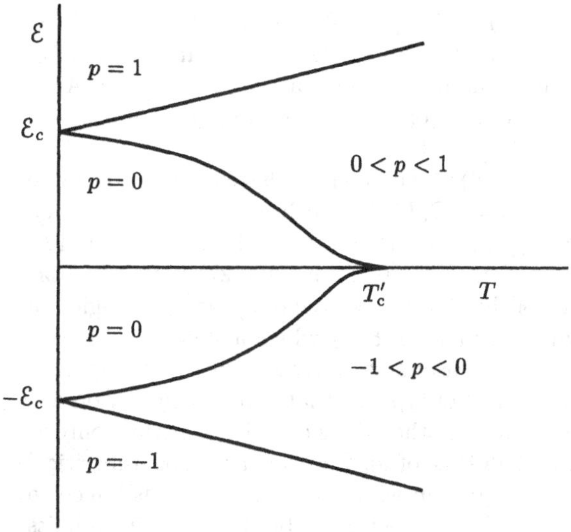

Fig. 10.11. Transition curves in the temperature–electric field plane for an antiferroelectric.

Recalling that $\varepsilon_2 - \varepsilon_1 > 0$ for an antiferroelectric (case (iii) of Sect. 10.2) we have $h_1 > 0$ for $\mathcal{E} < \mathcal{E}_c$ and $h_1 < 0$ for $\mathcal{E} > \mathcal{E}_c$ where

$$\mathcal{E}_c = \varepsilon_2 - \varepsilon_1 \,. \tag{10.145}$$

Since the equilibrium state at $T = 0$ is that of minimum h, the state of perfect sublattice order is stable for $\mathcal{E} < \mathcal{E}_c$, and the completely polarized state for $\mathcal{E}_c < \mathcal{E}$. This behaviour is similar to that of the Ising antiferromagnet at $T = 0$, but there are important differences for $T > 0$, as we shall see.

For an antiferroelectric, the parameter Δ varies from $\frac{1}{2}$ at $T = \infty$ to $-\infty$ at $T = 0$. Hence, by (10.132), the critical curve for transitions to the $p = 1$ state never reaches the $\mathcal{E} = 0$ axis. It can easily be deduced from the inequality $\varepsilon_2 > \varepsilon_1 > 0$ that

$$\mathcal{E}_1(T) - (\varepsilon_2 - \varepsilon_1) \simeq \frac{1}{2} T \exp(-\varepsilon_1/T) \qquad \text{as } T \to 0 \,. \tag{10.146}$$

So $\mathcal{E}_1 = \varepsilon_2 - \varepsilon_1$ and $\mathrm{d}\mathcal{E}_1/\mathrm{d}T = 0$ at $T = 0$. Thus, by (10.145), the critical curve terminates at the critical field point $\mathcal{E} = \mathcal{E}_c$ on the axis $T = 0$ and the tangent there is horizontal (see Fig. 10.11). When $T > T_c'$, where T_c' is the antiferroelectric critical temperature given by (10.110), $\Delta > -1$ and the form of the curve of $f(T, p)$ against p is the same as for a ferroelectric when $T > T_c$ (see Fig. 10.9). So p increases continuously from 0 at $\mathcal{E} = 0$ to 1 at $\mathcal{E} = \mathcal{E}_1(T)$. However, when $T < T_c'$, $\Delta < -1$ and the $f(T, p)$ curve has the cusped form shown in Fig. 10.9. At $p = +0$, $(\partial f/\partial p)_T = \Xi(T)$ where $\Xi(T)$, given by (10.127), is positive and is the least value of the slope of the $f(T, p)$ curve in the range $0 \le p \le 1$. Thus for $\mathcal{E} < \Xi(T)$ there is no solution to the equilibrium relation (10.88). It is easy to see why by considering the form of the non-equilibrium constant-field free energy $f(T, \mathcal{E}, p)$, defined by (10.87).

When $0 < \mathcal{E} < \Xi(T)$, the term $-\mathcal{E}p$ makes $f(T,p) - \mathcal{E}p$ unsymmetrical about $p = 0$ but its least value is still at the tip of the cusp, that is at $p = 0$. It follows that the equilibrium value of p is zero, as it is for $\mathcal{E} = 0$. As $\mathcal{E}$ increases through $\Xi(T)$, the least value of $f(T,p) - \mathcal{E}p$ moves from $p = 0$ to a minimum in the range $0 < p < 1$.

There is thus a region of the $(T, \mathcal{E})$ plane, bounded by the transition curve $\mathcal{E} = \Xi(T)$ and its mirror image $\mathcal{E} = -\Xi(T)$, in which $p = 0$ (see Fig. 10.11). Now, for an antiferroelectric, $2\varepsilon_2 - \varepsilon_1 > \varepsilon_1$. Hence, from (10.28) and (10.112), $\lambda \simeq (2\varepsilon_2 - \varepsilon_1)/T$ and $\vartheta \simeq \varepsilon_1/T$ as $T \to 0$. From (10.127) it follows that $\Xi(0) = \varepsilon_2 - \varepsilon_1 = \mathcal{E}_{\mathrm{c}}$. The part of the $T = 0$ axis bounding the $p = 0$ region of the $(T, \mathcal{E})$ plane coincides with the range of $\mathcal{E}$ for which a state of sublattice order is stable at $T = 0$. It is thus reasonable to regard the $p = 0$ region as that of sublattice order, with vertices of type 5 situated preferentially on one sublattice and vertices of type 6 on the other. This ordering is perfect only at $T = 0$; there is no frozen state like that of an ferroelectric below its critical temperature. The existence of a region in which $p = 0$ for $\mathcal{E} \neq 0$ is linked to the fact that, as we saw in Sect. 10.3, the perfect sublattice ordering admits small perturbations which carry zero polarization.

It can be shown from (10.112) and (10.128) that

$$\Xi(T) \sim \exp[-C(T_{\mathrm{c}}' - T)^{-1/2}] \qquad \text{as } T \to T_{\mathrm{c}}' - 0 \,, \tag{10.147}$$

C being a constant. It follows that all derivatives of $\Xi(T)$, as well as $\Xi(T)$ itself, approach zero as $T \to T_{\mathrm{c}}' - 0$. The boundary $\mathcal{E} = \pm\Xi(T)$ of the $p = 0$ region in the $(T, \mathcal{E})$ plane thus has an infinitely sharp cusp at $T = T_{\mathrm{c}}'$ on the T-axis. The weakness of the singularity at the zero-field critical point is presumably connected with the very rapid decrease in the width of the ordered region as $T \to T_{\mathrm{c}}' - 0$. This contrasts with the Ising antiferromagnet where the boundary of the sublattice ordered region in the $(T, \mathcal{H})$ plane has a vertical tangent at $T = T_{\mathrm{c}}$, $\mathcal{H} = 0$ (see Sects. 4.3 and 9.4). Also, for the Ising antiferromagnet, $m \neq 0$ for all $T > 0$, $\mathcal{H} \neq 0$ so that there is no analogue of the zero-polarization region.

On the $p = 0$ side of the curve $\mathcal{E} = \Xi(T)$ in the $(T, \mathcal{E})$ plane the equilibrium free energy is $f(T, 0)$ while all thermodynamic functions depend on T only and have no singularities for $T < T_{\mathrm{c}}'$. On the other side where $0 < p \ll 1$, a Maclaurin expansion yields, using results from Sect. 10.11,

$$f(T,p) = f(T,0) + \Xi(T)p + \frac{1}{3}\Xi_3(T)p^3 + \mathrm{O}(p^4) \,, \quad \Xi_3(T) > 0 \,. \tag{10.148}$$

Thus, from (10.88), the equilibrium value of p satisfies the relation

$$\mathcal{E} = \Xi(T) + \Xi_3(T)p^2 + \mathrm{O}(p^3). \tag{10.149}$$

These relations are of the same form as (10.134) and (10.135), and by arguments like those in Sect. 10.12 it can be shown that the thermodynamic functions behave similarly on the disordered sides of the respective curves $\mathcal{E} = \mathcal{E}_1(T)$ and $\mathcal{E} = \Xi(T)$. Hence $\mathcal{E} = \Xi(T)$ is a line of second-order transitions or critical curve. In particular the configurational enthalpy is continuous at any point $(T^*, \mathcal{E}^*)$ on the curve $\mathcal{E} = \Xi(T)$, but the heat capacity diverges as $(T - T^*)^{-1/2}$ when the point is approached on a line $\mathcal{E} = \mathcal{E}^* > 0$. The continuity of the heat capacity along the line $\mathcal{E} = 0$ is thus exceptional.

Hitherto we have assumed $\varepsilon_1 > 0$, but now we consider the F model, which is an antiferroelectric with $\varepsilon_1 = 0$. When $\varepsilon_1 = 0$, $\mathfrak{X} - 1 = 0$ for all $T > 0$ and (10.132) gives $\mathcal{E}_1(T) = \infty$ so that the transition lines $\mathcal{E} = \pm\mathcal{E}_1(T)$ are absent in the F model. From Table 10.1 the KDP and F models look like the simplest forms of ferroelectric and antiferroelectric respectively. However, whereas the KDP model is a typical ferroelectric, the F model is an untypical antiferroelectric. The reason is that the condition $\varepsilon_1 = 0$ puts it on the symmetry line for two-dimensional ice-rule models (see Fig. 10.4). For any value of $\mathcal{E}$, vertices 1 and 4 are equally probable as are vertices 2 and 3. So, as we have just seen, there is no transition to a completely polarized state at any finite $\mathcal{E}$ because the system is unable to 'choose' between the alternative configurations available. A rather similar situation exists for the zero-field IF model, (see Example 10.3), where no transition of any kind occurs.

By symmetry the reaction of an ice-rule model to a horizontal field is the same as to a vertical field. The phenomena associated with a field with both a horizontal and a vertical component are interesting but too complicated to discuss here (Lieb and Wu 1972).

10.14 The Potts Model

The Potts model was defined in Example 2.2. In this section we consider the anisotropic Potts model on a square lattice $\mathcal{N}$ of N sites. With the parameter R of Example 2.2 taking different values $R_{\rm h}$ and $R_{\rm v}$ for nearest-neighbour horizontal and vertical pairs respectively, the Hamiltonian can be expressed in the form

$$\widehat{H}(R_{\rm h}, R_{\rm v}; \sigma_i) = -R_{\rm h} \sum_{\{i,j\}}^{(\rm h.n.n.)} \delta^{\rm Kr}(\sigma_i - \sigma_j) - R_{\rm v} \sum_{\{i,j\}}^{(\rm v.n.n.)} \delta^{\rm Kr}(\sigma_i - \sigma_j)\,, \tag{10.150}$$

where the first and second summations are over all nearest-neighbour horizontal and vertical pairs respectively and $\delta^{\rm Kr}$ is the Kronecker delta function. The variable σ_i at any lattice site takes one of ν distinct integer values. In this section we demonstrate, using the method of Baxter (1973b, 1982b), the equivalence between this anisotropic square lattice ν-state Potts model and a *staggered six-vertex model*, with different vertex weights on the two equivalent interpenetrating sublattices of the square lattice. This equivalence was first demonstrated by Temperley and Lieb (1971).

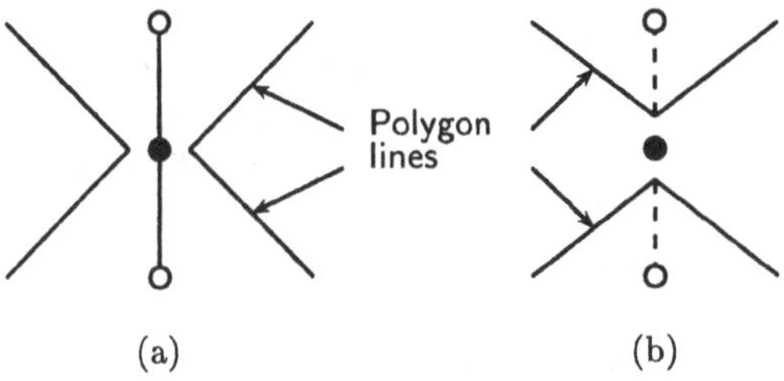

Fig. 10.12. Polygon construction on $\mathcal{N}$ (*white sites*) from the medial graph $\mathcal{N}'$ (*black sites*). **(a)** corresponds to the cases where a (*heavy black*) line of $\mathfrak{g}$ passes through the site of $\mathcal{N}'$ and **(b)** where it does not.

10.14.1 The Staggered Six-Vertex Model

It is not difficult to show that the partition function with Hamiltonian (10.150) is

$$Z(\tau_{\rm h},\tau_{\rm v}) = \sum_{\Lambda} \left\{ \prod_{\{i,j\}}^{(\rm h.n.n.)} [1+\sqrt{\nu}\,\tau_{\rm h}\delta^{\rm Kr}(\sigma_i-\sigma_j)] \right\} \times \left\{ \prod_{\{i,j\}}^{(\rm v.n.n.)} [1+\sqrt{\nu}\,\tau_{\rm v}\delta^{\rm Kr}(\sigma_i-\sigma_j)] \right\}, \qquad (10.151)$$

where the summation is over all ν^N points in the space Λ of values of the variables σ_i and

$$\tau_{\rm h} = [\exp(R_{\rm h}/T)-1]/\sqrt{\nu}\,, \qquad \tau_{\rm v} = [\exp(R_{\rm v}/T)-1]/\sqrt{\nu}\,. \qquad (10.152)$$

This form for the partition function is similar to the high-temperature series form for the Ising model in Sect. 8.3 and, as in that case, a binomial expansion of the products in (10.151) gives a sum of terms. Each term corresponds to a subgraph $\mathfrak{g}$ of the full lattice (graph) $\mathcal{N}$. The subgraph is obtained by connecting, by a line, each nearest-neighbour pair with microsystems in the same state. Suppose that $\mathfrak{g}$ has $\mathsf{q}_{\rm h}$ horizontal lines, $\mathsf{q}_{\rm v}$ vertical lines and c components. Then the degeneracy associated with $\mathfrak{g}$ will be $\nu^{\rm c}$ and the partition function can be expressed in the form

$$Z(\tau_{\rm h},\tau_{\rm v}) = \sum_{\{\mathfrak{g}\}} \nu^{{\rm c}+(\mathsf{q}_{\rm h}+\mathsf{q}_{\rm v})/2}\tau_{\rm h}^{\mathsf{q}_{\rm h}}\tau_{\rm v}^{\mathsf{q}_{\rm v}}. \qquad (10.153)$$

The *medial lattice* $\mathcal{N}'$ of $\mathcal{N}$ is defined to be the set of sites given by placing a site at the centre of each nearest-neighbour pair of $\mathcal{N}$. This new square lattice, which has $2N$ sites, divides in a natural way into two sublattices, of N sites each, denoted by $\mathfrak{h}$ and $\mathfrak{v}$, with the sites classified according to whether they respectively lie on a horizontal or vertical line of $\mathcal{N}$. From the nearest-neighbour lattice lines of $\mathcal{N}'$ we now construct a set of polygons in the following way. If a line of $\mathfrak{g}$ passes through a site of $\mathcal{N}'$ the lattice lines are moved as shown in Fig. 10.12(a) and otherwise they are moved as shown in Fig. 10.12(b). The final effect of this procedure is to give a set of closed polygons. For a finite lattice without toroidal boundary conditions the number $\mathfrak{p}$ of polygons constructed in this way is given by

$$\mathfrak{p} = 2\mathfrak{c} + \mathfrak{q}_{\mathrm{h}} + \mathfrak{q}_{\mathrm{v}} - N. \qquad (10.154)$$

To see the truth of equation (10.154) begin with the extreme case where no members of a nearest-neighbour pair on $\mathcal{N}$ are in the same state. Then $\mathfrak{c} = N$, $\mathfrak{q} = \mathfrak{q}_{\mathrm{h}} + \mathfrak{q}_{\mathrm{v}} = 0$ and $\mathfrak{p} = N$. Any subgraph $\mathfrak{g}$ can be constructed from this extreme situation by a systematic alteration in the states of the microsystems on $\mathcal{N}$. Suppose a particular site is in a state different from all its neighbours. If these neighbours are themselves all in different states then changing the centre site to one of these will decrease $\mathfrak{c}$ and $\mathfrak{p}$ by one and increase $\mathfrak{q}$ by one which leaves (10.154) satisfied. If two neighbours are in the same state and the centre site is changed to this state $\mathfrak{q}$ is increased by two. If the two neighbours belonged to the same component one polygon would be destroyed and one created leaving $\mathfrak{p}$ unchanged with $\mathfrak{c}$ decreased by one. If the two neighbours belonged to different components $\mathfrak{c}$ and $\mathfrak{p}$ would both be decreased by two. In each case (10.154) continues to hold. The reasoning for three and four neighbours in the same state is left as an exercise for the reader. It is clear that any subgraph $\mathfrak{g}$ corresponds to a polygon decomposition of the type described above and, using (10.154), the partition function (10.153) can be re-expressed in the form

$$Z(\tau_{\mathrm{h}}, \tau_{\mathrm{v}}) = \nu^{\frac{1}{2}N} \sum^{(\mathrm{p.d.})} \nu^{\frac{1}{2}\mathfrak{p}} \tau_{\mathrm{h}}^{\mathfrak{q}_{\mathrm{h}}} \tau_{\mathrm{v}}^{\mathfrak{q}_{\mathrm{v}}}, \qquad (10.155)$$

where the summation is over all polynomial decompositions. The factor $\tau_{\mathrm{h}}^{\mathfrak{q}_{\mathrm{h}}} \tau_{\mathrm{v}}^{\mathfrak{q}_{\mathrm{v}}}$ in (10.155) is 'local' in the sense that it is made up of a contribution of either 1 or τ_{h} from every site of $\mathfrak{h}$ and 1 or τ_{v} from every site of $\mathfrak{v}$. We now need to reduce the factor $\nu^{\frac{1}{2}\mathfrak{p}}$ to a product of local contributions. Arrows are placed on the edges of all the polygons so that around any polygon all the arrows are pointing in the same direction. The arrow configurations at each site of $\mathcal{N}'$ obey the 'two in, two out' ice rule of the six-vertex model (Sect. 10.1). Now consider the partition function

$$Z(\tau_{\mathrm{h}}, \tau_{\mathrm{v}}, \omega) = \nu^{\frac{1}{2}N} \sum^{(\mathrm{p.d.})} \sum^{(\mathrm{a.c.})} \Big(\prod^{(\mathrm{p.c.})} \omega' \Big) \tau_{\mathrm{h}}^{\mathfrak{q}_{\mathrm{h}}} \tau_{\mathrm{v}}^{\mathfrak{q}_{\mathrm{v}}}, \qquad (10.156)$$

where the inner summation is over all allowed arrow configurations and the product is over all polygon corners, with $\omega' = \omega$ if, following in the direction of the arrows, the corner represents a 90° turn to the left and ω^{-1} if it represents a 90° turn to the right. It is clear that the effect of the new summation and product in (10.156) is to give a factor $\omega^4 + \omega^{-4}$ for every polygon and the partition functions (10.155) and (10.156) are equal if

$$\nu^{\frac{1}{2}} = \omega^4 + \omega^{-4}. \qquad (10.157)$$

Consider a site of sublattice $\mathfrak{h}$. There are eight possibilities for the configuration of arrows and the presence or not of a line of $\mathfrak{g}$. If the polygons are reconnected at the site, there are six vertex configurations with the weights

$$\begin{aligned} \mathfrak{Z}_{\mathrm{h}}(1) &= \mathfrak{Z}_{\mathrm{h}}(2) = \tau_{\mathrm{h}}\,, & \mathfrak{Z}_{\mathrm{h}}(3) &= \mathfrak{Z}_{\mathrm{h}}(4) = 1\,, \\ \mathfrak{Z}_{\mathrm{h}}(5) &= \omega^2\tau_{\mathrm{h}} + \omega^{-2}\,, & \mathfrak{Z}_{\mathrm{h}}(6) &= \omega^{-2}\tau_{\mathrm{h}} + \omega^{2}\,. \end{aligned} \tag{10.158}$$

The ordering of the vertex configurations is that of Fig. 10.2. A similar analysis of a site of $\mathfrak{v}$ yields

$$\begin{aligned} \mathfrak{Z}_{\mathrm{v}}(1) &= \mathfrak{Z}_{\mathrm{v}}(2) = 1\,, & \mathfrak{Z}_{\mathrm{v}}(3) &= \mathfrak{Z}_{\mathrm{v}}(4) = \tau_{\mathrm{v}}\,, \\ \mathfrak{Z}_{\mathrm{v}}(5) &= \omega^{-2}\tau_{\mathrm{v}} + \omega^{2}\,, & \mathfrak{Z}_{\mathrm{v}}(6) &= \omega^{2}\tau_{\mathrm{v}} + \omega^{-2}\,. \end{aligned} \tag{10.159}$$

The two summations in (10.156) are together equivalent to one summation over arrow configurations at the sites of $\mathcal{N}'$ which satisfy the ice rule and

$$Z(\tau_{\mathrm{h}},\tau_{\mathrm{v}},\omega) = (\omega^4+\omega^{-4})^N Z_{\mathrm{s6v}}(\tau_{\mathrm{h}},\tau_{\mathrm{v}},\omega)\,, \tag{10.160}$$

where $Z_{\mathrm{s6v}}(\tau_{\mathrm{h}},\tau_{\mathrm{v}},\omega)$ is the partition function on the medial lattice of $2N$ sites of a model which differs in two important respects from the standard six-vertex model:

(i) The model is a staggered six-vertex model with different vertex weights on the two sublattices $\mathfrak{h}$ and $\mathfrak{v}$.

(ii) The vertex weights for vertices 5 and 6 are different, unless $\omega = \pm 1$ ($\nu = 4$).

For the lattice as a whole, with toroidal boundary conditions, we know that the numbers $\mathfrak{n}_5$ and $\mathfrak{n}_6$ of vertices 5 and 6 are equal. It is, however, not in general the case that $\mathfrak{n}_{5\mathrm{h}} = \mathfrak{n}_{6\mathrm{h}}$, $\mathfrak{n}_{5\mathrm{v}} = \mathfrak{n}_{6\mathrm{v}}$ and the inequality of the vertex weights 5 and 6 for each individual sublattice is a genuine generalization of the model. Replacing τ_{h} and τ_{v} by τ_{h}^{-1} and τ_{v}^{-1} in (10.158) and (10.159) is equivalent to dividing the weights in (10.158) by τ_{h} and in (10.159) τ_{v} and interchanging the sublattices. It follows that the partition function satisfies the duality relationship

$$Z(\tau_{\mathrm{h}},\tau_{\mathrm{v}},\omega) = (\tau_{\mathrm{h}}\tau_{\mathrm{v}})^N Z(\tau_{\mathrm{h}}^{-1},\tau_{\mathrm{v}}^{-1},\omega)\,, \tag{10.161}$$

which maps the model between high and low temperatures. Except in two special cases the staggered six-vertex model with weights (10.158) and (10.159) has not been solved. Let

$$\xi_{\mathrm{h}} = \frac{\tau_{\mathrm{h}}\omega^4+1}{\tau_{\mathrm{h}}+\omega^4}\,, \qquad \xi_{\mathrm{v}} = \frac{\tau_{\mathrm{v}}\omega^4+1}{\tau_{\mathrm{v}}+\omega^4}\,. \tag{10.162}$$

In terms of these variables, it was shown by Baxter (1982b) that the Bethe ansatz can be extended to solve the cases with constraints

$$\xi_{\mathrm{h}}\xi_{\mathrm{v}} = \pm 1\,, \tag{10.163}$$

which we now treat separately

10.14.2 The Solvable Case $\xi_h \xi_v = 1$

The constraint here reduces to

$$\tau_h \tau_v = 1 , \tag{10.164}$$

which was the case considered by Baxter (1973b). The vertex weights (10.158) and (10.159) are proportional to each other and with τ_h and τ_v positive the model is equivalent to the regular six-vertex model. The corresponding Potts model is ferromagnetic with its temperature fixed by the condition

$$\nu = [\exp(R_h/T) - 1][\exp(R_v/T) - 1] . \tag{10.165}$$

When $R_h = R_v$ (10.165) gives the critical temperature of the ferromagnetic Potts model on a square lattice. That this is the case can be seen by imposing the isotropy condition $\tau = \tau_h = \tau_v$ on (10.161) giving

$$Z(\tau, \omega) = \tau^{2N} Z(\tau^{-1}, \omega) . \tag{10.166}$$

The mapping $\tau \to \tau^{-1}$ is the duality transformation first obtained by Potts (1952) for his model. Given that the ferromagnetic Potts model has a phase transition between a low-temperature ordered state and the high-temperature disordered state and assuming that there is a single phase transition on the T-axis, it will correspond to a singularity in the partition function, which from (10.166) must occur at $\tau = 1$ which gives the critical temperature

$$T_c = R/\ln(\sqrt{\nu} + 1) , \qquad R > 0 . \tag{10.167}$$

For $\nu = 2$ and $R = 2J$ this reduces to the formula (8.20) for the critical temperature of the square lattice Ising model. At the critical temperature the staggered nature of the equivalent six-vertex model disappears. From (10.158) and (10.159),

$$\mathfrak{a} = \mathfrak{b} = 1 , \qquad \mathfrak{c} = \omega^2 + \omega^{-2} = (2 + \sqrt{\nu})^{1/2} \tag{10.168}$$

and from (10.28)

$$\Delta = -\sqrt{\nu}/2 . \tag{10.169}$$

From Table 10.1, (10.3) and (10.168) $\varepsilon_1 = \varepsilon_2 = 0$, $\varepsilon_3 < 0$. By subtracting ε_3 from each vertex energy we see that this is the case of an F model. At its critical temperature the ν-state ferromagnetic Potts model can now be interpreted with variations of the parameter ν playing the role of temperature variation. In doing so we must remember that the six-vertex model is on the medial lattice of $2N$ sites.

The Range $0 < \nu < 4$. From (10.169) this gives $-1 < \Delta < 1$, which is the disordered region considered in Sects. 10.8 and 10.9. From (10.52) and (10.73) $\varphi = 0$ and, from (10.62), μ is given by

$$\sqrt{\nu} = 2\cos(\mu) , \qquad 0 < \mu < \frac{1}{2}\pi . \tag{10.170}$$

In deriving the free energy per site of the Potts model at the critical temperature from (10.76) and (10.80) we must remember that the corresponding six-vertex model has $2N$ lattice sites and we must also include the factor $-\frac{1}{2}\ln(\nu)$ which arises from the prefactor of $(\omega^4+\omega^{-4})^N = \nu^{\frac{1}{2}N}$ in (10.160). This finally gives

$$f(T_c) = -\frac{1}{2}T_c\ln(\nu) - 2T_c\int_0^\infty \frac{\sinh[y(\pi-\mu)]\tanh(\mu y)}{y\sinh(y\pi)}\mathrm{d}y\,. \tag{10.171}$$

The Range $\nu > 4$. From (10.169) this gives $\Delta < -1$, which is the disordered region considered in Sect. 10.11. From (10.52) and (10.112) $\vartheta_0 = 0$ and λ is given by

$$\sqrt{\nu} = 2\cosh(\lambda)\,, \qquad \lambda > 0\,. \tag{10.172}$$

From (10.120) and (10.123),

$$f(T_c) = -\frac{1}{2}T_c\ln(\nu) - \lambda T_c - 2T_c\sum_{n=1}^{\infty}\frac{\exp(-n\lambda)\tanh(n\lambda)}{n}\,. \tag{10.173}$$

The Boundary $\nu = 4$. In this case the free energy can be obtained by taking the limit $\mu \to 0$ in equation (10.171) or $\lambda \to 0$ in (10.173). Using (10.167), the result is

$$\begin{aligned} f(T_c) &= -T_c\ln(2) - 2T_c\int_0^\infty \exp(-y)\tanh(y)\mathrm{d}y\,,\\ &= -\frac{8R}{\ln(3)}\ln\left[\frac{\Gamma\left(\frac{1}{4}\right)}{2^{\frac{5}{8}}\pi^{\frac{1}{2}}}\right], \end{aligned} \tag{10.174}$$

(Gradshteyn and Ryzhik 1980, p.327).

10.14.3 The Solvable Case $\xi_h\xi_v = -1$

Using (10.157), the constraint here reduces to

$$(\tau_h\sqrt{\nu}+2)(\tau_v\sqrt{\nu}+2) = 4-\nu\,, \tag{10.175}$$

which, from (10.152), fixes the temperature of the anisotropy Potts model with the condition

$$4-\nu = [\exp(R_h/T)+1][\exp(R_v/T)+1]\,. \tag{10.176}$$

Although it is of interest to consider non-physical situations (Baxter 1982b), (10.176) has a physical solution only when $\nu \le 3$ and then only when at least one of R_h and R_v is negative. When $R_h = R_v = R < 0$ (10.176) gives the critical temperature

$$T_c = R/\ln(\sqrt{4-\nu}-1)\,, \qquad R < 0\,. \tag{10.177}$$

of the antiferromagnetic square lattice Potts model. For $\nu = 2$ and $R = 2J$ this formula gives the result (8.20) for the antiferromagnetic Ising model.

For $\nu = 3$ the critical temperature of the antiferromagnetic 3-state Potts model is $T_c = 0$. Unlike the ferromagnetic case the staggered nature of the underlying six-vertex model does not disappear at the critical temperature and evaluation of the free energy in this case is rather more complicated (Baxter 1982b).

10.14.4 The Polarization and Internal Energy

It is clear from the above discussion that the ν-state Potts model differs in its behaviour according to whether ν is greater of less that four. One way to understand more clearly the effect on the nature of the phase transition is to calculate the internal energy as the critical point is approached from above and below. For the ferromagnetic case this has been achieved by Baxter (1973b). In our present notation

$$u = -R(1 + \nu^{-\frac{1}{2}})[1 \pm p_0 \tanh(\lambda/2)], \tag{10.178}$$

where p_0 is the spontaneous polarization and the $+$ and $-$ signs in (10.178) apply according to whether the critical point is approached from below or above. The spontaneous polarization was calculated by Baxter (1973a, 1973c) and is zero when $0 < \nu \leq 4$ with

$$p_0 = \prod_{n=1}^{\infty} \tanh^2(n\lambda), \qquad \nu > 4. \tag{10.179}$$

Twice the second term in (10.178), therefore, represents the latent heat at the transition for $\nu > 4$, the transition being first-order. In all cases the first term in (10.178) gives the internal energy per site averaged over the values below and above the transition. This was first calculated by Potts (1952).

10.14.5 Critical Exponents

For $0 < \nu \leq 4$ it has been conjectured by den Nijs (1979) that

$$\alpha = \frac{2\pi - 8\mu}{3\pi - 6\mu} \tag{10.180}$$

and independently by Nienhuis et al. (1980) and Pearson (1980) that

$$\beta = \frac{\pi + 2\mu}{12\pi}, \tag{10.181}$$

where μ is given by (10.170). Using the scaling laws (3.77) and (3.87) this gives

$$\gamma = \frac{7\pi^2 - 8\mu\pi + 4\mu^2}{6\pi(\pi - 2\mu)}, \qquad \delta = \frac{15\pi^2 - 16\mu\pi + 4\mu^2}{\pi^2 - 4\mu^2}. \tag{10.182}$$

The formula (10.180) for α has now been establish by Black and Emery (1981) by further exploiting the relationship to the staggered F model. From

(10.170), with $\nu = 2$, $\mu = \pi/4$ and the exponent values for the Ising model, derived in Sect. 8.9, are recovered. When $\nu = 3$, $\mu = \pi/6$

$$\alpha = \tfrac{1}{3}, \quad \beta = \tfrac{1}{9}, \quad \gamma = \tfrac{13}{9}, \quad \delta = 14. \tag{10.183}$$

These values are strongly supported by the finite-size scaling analysis of Blöte and Nightingale (1982) and by conformal invariance arguments (Friedan et al. 1984, Cardy 1996, see also Volume 2, Sect. 2.17). They are the same as those of the hard hexagon model (Baxter 1982a) and support the argument of Alexander (1975) that the 3-state Potts model and hard hexagon model lie in the same universality class (see Volume 2, Sect. 5.12).

Examples

10.1 With Δ defined by (10.28), show that:

(a) For case (iii) of Sect. 10.2, Δ increases steadily with T from $-\infty$ at $T = 0$ to $\frac{1}{2}$ at $T = \infty$.

(b) For case (i) of Sect. 10.2, with $\varepsilon_2 \leq \frac{1}{2}\varepsilon_1$, Δ decreases steadily, from $+\infty$ at $T = 0$ to $\frac{1}{2}$ at $T = \infty$.

(c) For case (i) of Sect. 10.2, with $\varepsilon_2 > \frac{1}{2}\varepsilon_1$, Δ decreases steadily from $+\infty$ at $T = 0$ to a value $\Delta_{\min}$, where $0 < \Delta_{\min} < \frac{1}{2}$, and then increases to $\frac{1}{2}$ at $T = \infty$.

10.2 From (10.28) and (10.52), show that, given $\varepsilon_1 > 0$,

$$\mathfrak{X} - \Delta > 0, \qquad 1 + \mathfrak{X} - 2\Delta > 0, \qquad \mathsf{L}(0) > 0,$$

for all T. From (10.51) show that

$$|\mathsf{L}(k)|^2 = 1 + \frac{4(\mathfrak{X} - \Delta)[\cos(k) - \Delta]}{1 + \mathfrak{X}^2 - 2\mathfrak{X}\cos(k)}.$$

Hence show that $|\mathsf{L}(k)|$ decreases as k increases. For $-1 < \Delta < 1$, show that $|\mathsf{L}(k)| > 1$ for $0 < k < \pi - \mu$ and $|\mathsf{L}(k)| < 1$ for $\pi - \mu < k < \pi$, where μ is defined by (10.62). For $\Delta < -1$, show that $|\mathsf{L}(k)| > 1$ for all k.

10.3 Using a diagram similar to Fig. 10.8, draw trajectories for the F, IKDP and IF models, showing that no transition occurs in the two latter models with $T > 0$. Sketch trajectories for models where (A) $\varepsilon_2 = \varepsilon_1$, $\varepsilon_1 > 0$; (B) $\varepsilon_2 = 2\varepsilon_1$, $\varepsilon_1 < 0$; (C) $\varepsilon_2 = 2\varepsilon_1$, $\varepsilon_1 > 0$. Verify that (A), (B) and (C) belong respectively to the categories (i), (ii) and (iii) of Sect. 10.2. Show that, for (A), (B) and (C), the critical temperature is $|\varepsilon_1| / \ln[(\sqrt{5} + 1)/2]$.

10.4 For an antiferroelectric with $\varepsilon_1 > 0$, show that

$$\lim_{T\to 0}\left(\lambda - \frac{2\varepsilon_2 - \varepsilon_1}{T}\right) = 0\,, \qquad \lim_{T\to 0}\left(\frac{\vartheta - \varepsilon_1}{T}\right) = 0\,,$$

where λ and ϑ are defined by (10.112). Hence show, using (10.58) with $x_0 = \pi$ and $p = 0$ and (10.121), that $f(T) \to 0$ as $T \to 0$. Also show, using (10.58) and (10.61) with $|\widetilde{\mathrm{L}}(x)|$ given by (10.116), that

$$f(T,p) = (\varepsilon_2 - \varepsilon_1)|p| \qquad \text{at } T = 0\ .$$

Hence show that

$$f(O,\mathcal{E},p) = (\varepsilon_2 - \varepsilon_1)|p| - \mathcal{E}p = (\varepsilon_2 - \varepsilon_1 - |\mathcal{E}|)|p|$$

is minimized by $p = 0$, for $-\mathcal{E}_c < \mathcal{E} < \mathcal{E}_c$, by $p = 1$, for $\mathcal{E}_c < \mathcal{E}$, and by $p = -1$, for $\mathcal{E} < -\mathcal{E}_c$, where $\mathcal{E}_c = \varepsilon_2 - \varepsilon_1$.

10.5 Show that equation (10.132) can be expressed as

$$\mathcal{E}_1(T) = \tfrac{1}{2}T\ln\left[\frac{1 - \exp(-\varepsilon_1/T) + \exp\{(2\varepsilon_2 - \varepsilon_1)/T\}}{\exp(\varepsilon_1/T) - 1}\right].$$

Derive the relation

$$\frac{\mathrm{d}\mathcal{E}_1}{\mathrm{d}T} = \frac{\mathcal{E}_1}{T} + \frac{\varepsilon_1\exp(\varepsilon_1/T)}{2T[\exp(\varepsilon_1/T) - 1]} - \frac{(2\varepsilon_2 - \varepsilon_1)\exp[(2\varepsilon_2 - \varepsilon_1)/T] + \varepsilon_1\exp(-\varepsilon_1/T)}{2T\{1 + \exp[(2\varepsilon_2 - \varepsilon_1)/T] - \exp(-\varepsilon_1/T)\}}$$

and hence show that for an ferroelectric ($\varepsilon_1 > 0$, $\varepsilon_1 > \varepsilon_2$), $\mathrm{d}\mathcal{E}_1/\mathrm{d}T > 0$ for $T \ge T_c$. Show that, for a ferroelectric at $T = T_c$,

$$\frac{\mathrm{d}\mathcal{E}_1}{\mathrm{d}T} = T_c^{-1}\{\varepsilon_1 - \varepsilon_2[1 - \exp(-\varepsilon_1/T_c)]\}\,.$$

Prove that an alternative expression for $\mathcal{E}_1$ is

$$\mathcal{E}_1(T) = \varepsilon_2 - \varepsilon_1 + \tfrac{1}{2}T\ln\left\{\frac{1 + \exp[(\varepsilon_1 - 2\varepsilon_2)/T] - \exp(-2\varepsilon_2/T)}{1 - \exp(-\varepsilon_1/T)}\right\}$$

and hence that $\mathcal{E}_1(T) > \varepsilon_2 - \varepsilon_1$ for all $T > 0$. Derive equation (10.146) for the antiferroelectric ($\varepsilon_2 > \varepsilon_1 > 0$) and show that $\mathrm{d}\mathcal{E}_1/\mathrm{d}T > 0$ for all $T > 0$ by differentiating the alternative expression with respect to T.

10.6 Consider the parameter value $\Delta = 0$, ($\mathfrak{a}^2 + \mathfrak{b}^2 = \mathfrak{c}^2$) which corresponds to a disordered state of the antiferroelectric. Show that $\theta(q_1, q_2)$, given by (10.36), is zero for all q_1 and q_2, $\rho(k) = 1/(2\pi)$ and the limits of integration $\pm q_0$ in equation (10.54) are given by $q_0 = \frac{1}{2}\pi(1 - p)$. Hence show that

$$f(T,p) = \varepsilon_2 - \varepsilon_1 - \frac{T}{2\pi}\int_0^{2\pi(1-p)/2} \ln\left[\frac{\mathfrak{X}^2 + 1 + 2\mathfrak{X}\cos(k)}{\mathfrak{X}^2 + 1 - 2\mathfrak{X}\cos(k)}\right] \mathrm{d}k\,.$$

Prove that the relative vertical polarization p and the vertical field $\mathcal{E}$ are connected by

$$\mathcal{E} = \frac{1}{4}T\ln\left[\frac{\mathfrak{X}^2 + 1 + 2\mathfrak{X}\sin(\frac{1}{2}\pi p)}{\mathfrak{X}^2 + 1 - 2\mathfrak{X}\sin(\frac{1}{2}\pi p)}\right].$$

Verify (10.132) and (10.89) for the case $\Delta = 0$. Putting $p = 1 - \delta p$, show that, as $\delta p \to 0$,

$$\mathcal{E}_1 - \mathcal{E} \simeq \frac{1}{3}T\pi^2\frac{\mathfrak{X}(\mathfrak{X}^2 + 1)}{(\mathfrak{X}^2 - 1)^2}(\delta p)^2\,.$$

A. Appendices

A.1 Regular Lattices

This appendix contains diagrams of the one-, two- and three-dimensional regular lattices. Sites are shown by black or white circles and nearest-neighbour pairs are connected by line segments. In graph theory language the sites are the *vertices* and the segments are the *edges* of what is called the *full lattice graph.*[1] Throughout the book d denotes the dimension of the lattice and z the *coordination number*, which is the number of nearest neighbours of a lattice site.

The body-centred cubic lattice of Fig. A.7 can be obtained from the simple cubic of Fig. A.6 by placing a new site at the centre of each cube of eight sites. The face-centred cubic lattice of Fig. A.8 can be obtained from the simple cubic by placing a new site at the centre of each square of four sites. The diamond lattice of Fig. A.9 can be obtained from the body-centred cubic by eliminating alternate sites. A *loose-packed lattice* is one which can be divided

Fig. A.1. The linear lattice; $d = 1$, $z = 2$.

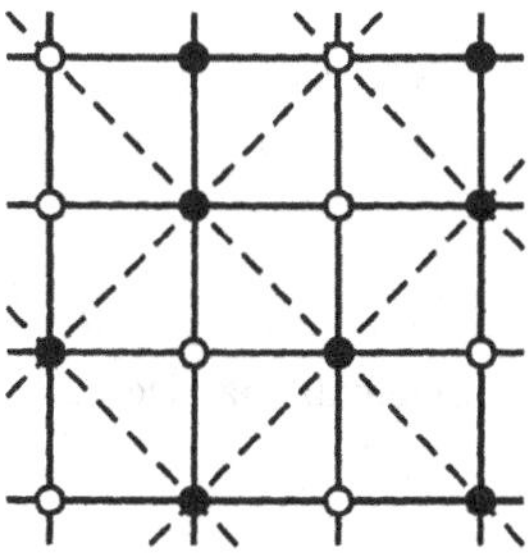

Fig. A.2. The square lattice; $d = 2$, $z = 4$.

into two equal sublattices $\mathfrak{a}$ and $\mathfrak{b}$ in such a way that all the nearest neighbours of a site of one sublattice belong to the other sublattice; a *close-packed*

[1] For detailed accounts of graph theory terminology see Temperley (1981) and Volume 2, Appendix A.7.

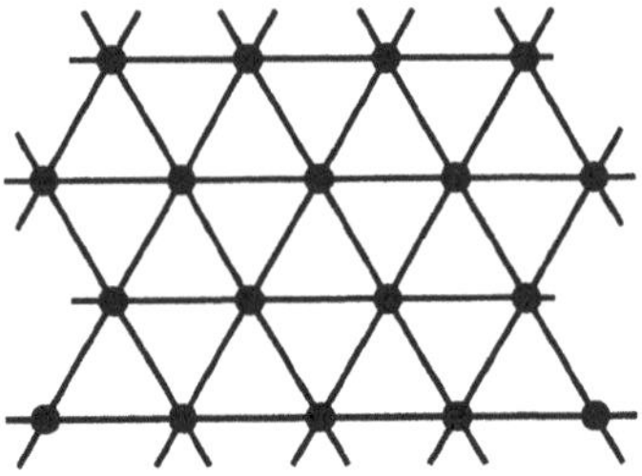

Fig. A.3. The triangular lattice; $d = 2$, $z = 6$.

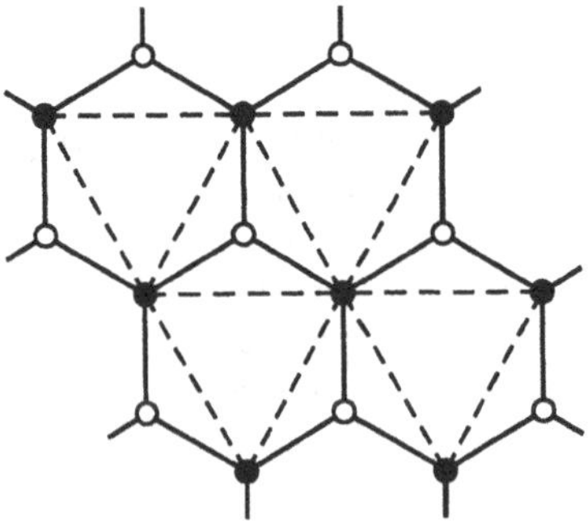

Fig. A.4. The honeycomb; $d = 2$, $z = 3$.

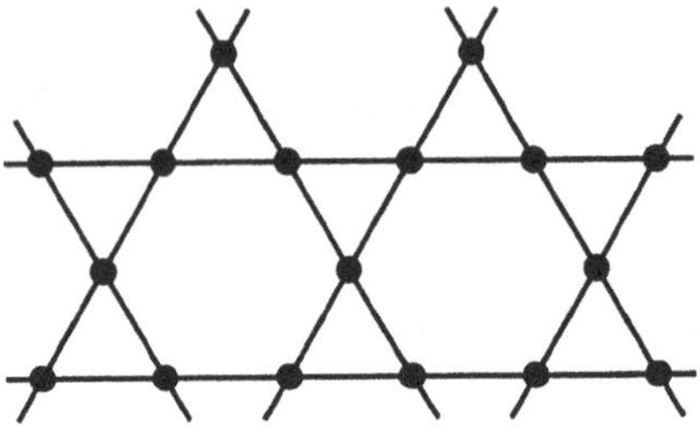

Fig. A.5. The kagomé lattice; $d = 2$, $z = 4$.

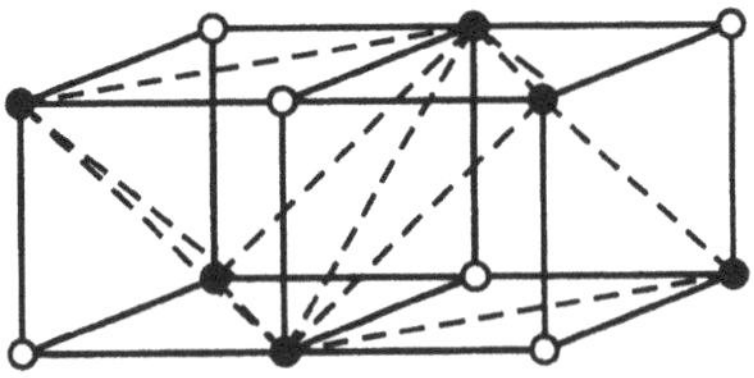

Fig. A.6. The simple cubic lattice; $d = 3$, $z = 6$.

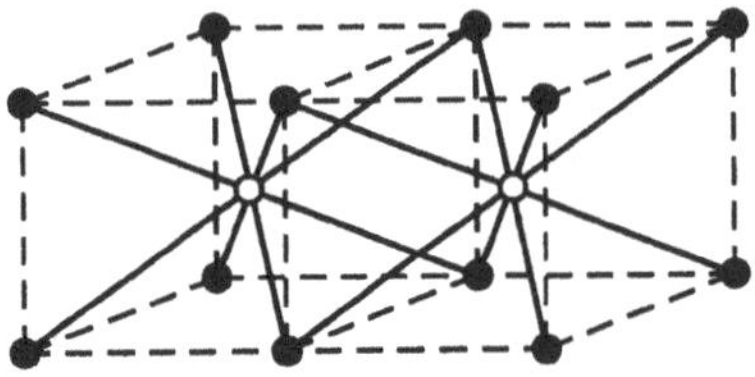

Fig. A.7. The body-centre cubic lattice; $d = 3$, $z = 8$.

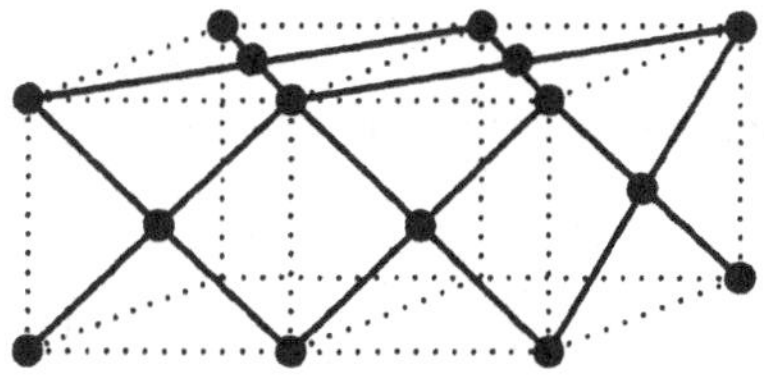

Fig. A.8. The face-centred cubic lattice; $d = 3$, $z = 12$.

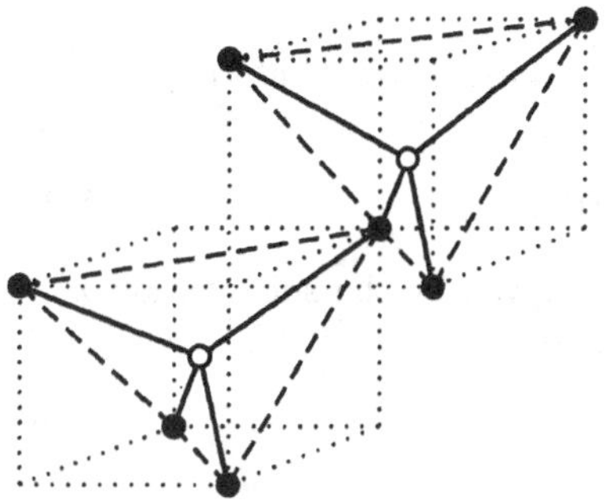

Fig. A.9. The diamond lattice; $d = 3$, $z = 4$.

lattice is one for which this is not possible. The linear, square, honeycomb, simple cubic, body-centred cubic and diamond lattices are loose-packed and the triangular, Kagomé and face-centred cubic lattices are close-packed. In the diagrams of the loose-packed lattices the sites of sublattice $\mathfrak{a}$ are denoted by • and those of sublattice $\mathfrak{b}$ by ∘. Nearest-neighbour pairs of sites are linked by full line segments and second-neighbour $\mathfrak{a}\mathfrak{a}$ pairs by broken line segments. The corresponding $\mathfrak{b}\mathfrak{b}$ links are not shown.

A.2 Elliptic Integrals and Functions

Certain results in the theory of elliptic integrals and functions are needed for the treatment the Ising model in Chap. 8 and these together with additional material on *nome series* are needed for the discussion of the eight-vertex model in Volume 2, Chap. 5. For convenience a brief discussion of this material is presented in this appendix and further results are given in Volume 2, Appendix A.6. For a extensive treatments the reader is referred to Bowman (1953), Lawden (1989) or Baxter (1982a) Chap. 15. Comprehensive lists of formulae are given by Gradshteyn and Ryzhik (1980) and Abramowitz and Segun (1965).

A.2.1 Elliptic Integrals

Elliptic integrals of the first and second kind, respectively, are defined by

$$\begin{aligned} F(\theta|k) &= \int_0^\theta \frac{\mathrm{d}\omega}{\sqrt{1-k^2\sin^2(\omega)}}, \\ E(\theta|k) &= \int_0^\theta \sqrt{1-k^2\sin^2(\omega)}\,\mathrm{d}\omega, \end{aligned} \tag{A.1}$$

where $0 \le k < 1$ is the *modulus*. *Complete elliptic integrals of the first and second kind*, respectively, are defined by

$$\mathcal{K}(k) = F\left(\tfrac{1}{2}\pi|k\right), \qquad \mathcal{E}(k) = E\left(\tfrac{1}{2}\pi|k\right). \tag{A.2}$$

Alternative forms for $F(\theta|k)$ and $E(\theta|k)$ are given by the substitution $x = \sin(\omega)$, $x_0 = \sin(\theta)$. Then

$$F(\theta|k) = \int_0^{x_0} \frac{\mathrm{d}x}{\sqrt{(1-x^2)(1-k^2x^2)}}. \tag{A.3}$$

From (A.1) and (A.2) $\mathcal{K}(0) = \mathcal{E}(0) = \frac{1}{2}\pi$. By binomial expansion of the integrands

$$\begin{aligned} \mathcal{K}(k) &= \frac{\pi}{2}\left[1 + \left(\frac{1}{2}\right)^2 k^2 + \left(\frac{1.3}{2.4}\right)^2 k^4 + \cdots\right], \\ \mathcal{E}(k) &= \frac{\pi}{2}\left[1 - \left(\frac{1}{2}\right)^2 k^2 - \frac{1}{3}\left(\frac{1.3}{2.4}\right)^2 k^4 + \cdots\right]. \end{aligned} \tag{A.4}$$

As $k \to 1$, $\mathcal{K}(k) \to \infty$. It can be shown (Bowman 1953, pp. 21 and 22) that

$$\mathcal{K}(k) = \ln\left(\frac{4}{k'}\right) + \delta(k), \tag{A.5}$$

where $\delta(k) \to 0$ as $k \to 1$ and k' is the *complementary modulus* defined by

$$k' = \sqrt{(1-k^2)}. \tag{A.6}$$

It also follows from (A.1) and (A.2) that

$$\mathcal{E}(1) = 1. \tag{A.7}$$

From (A.1) and (A.2), by differentiating under the integral sign,

$$\frac{\mathrm{d}\mathcal{K}(k)}{\mathrm{d}k} = \frac{\mathcal{E}(k) - k'^2\mathcal{K}(k)}{kk'^2}, \qquad \frac{\mathrm{d}\mathcal{E}(k)}{\mathrm{d}k} = \frac{\mathcal{E}(k) - \mathcal{K}(k)}{k}. \tag{A.8}$$

Interchanging k and k' and using (A.6) gives

$$\frac{\mathrm{d}\mathcal{K}(k')}{\mathrm{d}k} = \frac{k^2\mathcal{K}(k') - \mathcal{E}(k')}{kk'^2}, \qquad \frac{\mathrm{d}\mathcal{E}(k')}{\mathrm{d}k} = \frac{k[\mathcal{K}(M') - \mathcal{E}(k')]}{k'^2}. \tag{A.9}$$

From (A.8) and (A.9) it can be verified that $\mathcal{K}(k)$ and $\mathcal{K}(k')$ are solutions of the differential equation

$$\frac{\mathrm{d}}{\mathrm{d}k}\left(kk'^2\frac{\mathrm{d}f}{\mathrm{d}k}\right) = kf. \tag{A.10}$$

The general solution is therefore $\zeta\mathcal{K}(k)+\eta\mathcal{K}(k')$ for arbitrary constants ζ and η. An important formula relating complete elliptic integrals is the *Legendre relation*

$$\mathcal{E}(k)\mathcal{K}(k') + \mathcal{E}(k')\mathcal{K}(k) - \mathcal{K}(k)\mathcal{K}(k') = \frac{1}{2}\pi\,. \tag{A.11}$$

That the left-hand side is indeed a constant can be established from (A.8) and (A.9). That the value of the constant is $\pi/2$ can be established by taking the limit $k \to 0$ and using (A.4), (A.5) and (A.7).

There are a number of formulae connecting elliptic integrals of different moduli. In Chaps. 8 and 10 we need the *Landen transformation* given by

$$k_1 = \frac{2\sqrt{k}}{1+k}, \tag{A.12}$$

for which

$$\mathcal{K}(k_1) = (1+k)\mathcal{K}(k)\,. \tag{A.13}$$

From (A.6) and (A.12),

$$k_1' = \frac{1-k}{1+k} \tag{A.14}$$

and

$$\mathcal{K}(k_1') = \frac{1}{2}(1+k)\mathcal{K}(k')\,. \tag{A.15}$$

A.2.2 Elliptic Functions

In (A.3), F is defined as a function of x_0. Writing

$$\mathrm{u} = F(\arcsin(x_0)|k) \tag{A.16}$$

the inverse relation defines the *Jacobian elliptic function* sn. That is to say

$$x_0 = \mathrm{sn}(\mathrm{u}|k)\,. \tag{A.17}$$

The associated Jacobian elliptic cn and dn functions are defined by

$$\mathrm{cn}(\mathrm{u}|k) = \sqrt{1-\mathrm{sn}^2(\mathrm{u}|k)}\,, \qquad \mathrm{dn}(\mathrm{u}|k) = \sqrt{1-k^2\mathrm{sn}^2(\mathrm{u}|k)}\,. \tag{A.18}$$

From (A.3), (A.16) and (A.17)

$$\mathrm{sn}(\mathrm{u}|0) = \sin(\mathrm{u})\,, \qquad \mathrm{sn}(\mathrm{u}|1) = \tanh(\mathrm{u}) \tag{A.19}$$

and thus, from (A.18),

$$\mathrm{cn}(\mathrm{u}|0) = \cos(\mathrm{u})\,, \qquad \mathrm{cn}(\mathrm{u}|1) = \mathrm{sech}(\mathrm{u})\,, \tag{A.20}$$

$$\mathrm{dn}(\mathrm{u}|0) = 1, \qquad \mathrm{dn}(\mathrm{u}|1) = \mathrm{sech}(\mathrm{u}). \tag{A.21}$$

In the limit $k \to 0$ the elliptic functions reduce to trigonometric functions and in the limit $k \to 1$ to hyperbolic functions. A generalization of the duplication formulas for sin and tanh is

$$\mathrm{sn}(2u|k) = \frac{2\mathrm{sn}(u|k)\mathrm{cn}(u|k)\mathrm{dn}(u|k)}{1 - k^2\mathrm{sn}^2(u|k)} . \tag{A.22}$$

The elliptic integral of the second kind, defined in (A.1), can be regarded as a function $E^*(u|k)$ of u through the relation $\theta = \arcsin(x_0) = \arcsin[\mathrm{sn}(u|k)]$. It then satisfies the duplication formula

$$E^*(2u|k) = 2E^*(u|k) - k^2\mathrm{sn}^2(u|k)\mathrm{sn}(2u|k) . \tag{A.23}$$

A.2.3 Results Required for Chapter 8

Many integrals can be reduced to standard forms involving elliptic integrals. These include those with integrand $1/\sqrt{Y(y)}$ where $Y(y)$ is a quartic in y, positive over the range of integration with respect to y. The substitution $y = \tan(\alpha)$ in equations (8.90) gives, when $k < 1$,

$$A(K|k) = F(\theta|k') ,$$
$$\tag{A.24}$$
$$B(K|k) = k'^{-2}[F(\theta|k') - E(\theta|k')] ,$$

where $\tan(\theta) = \sinh(2K)$. When $k > 1$, an alternative treatment is needed. Putting $\ell = 1/k$ and $y = \ell\tan(\beta)$

$$A(K|k) = \ell F(\chi|\ell') ,$$
$$\tag{A.25}$$
$$B(K|k) = -C(K|k) + [E(\chi|\ell') - \ell^2 F(\chi|\ell')]/\ell'^2 ,$$

when

$$\tan(\chi) = \ell^{-1}\sinh(2K) = k\sinh(2K) ,$$

$$\ell'^2 = 1 - \ell^2 ,$$
$$\tag{A.26}$$

$$C(K|k) = \frac{\tanh(2K)}{\sqrt{1 + k^2\sinh^2(2K)}} .$$

If k is given by (8.74) then

$$C(K|k) = \tanh(2K)\tanh(2G) . \tag{A.27}$$

Two other integrals needed in Chap. 8 are

$$\int_{-1}^{1} \frac{dx}{\sqrt{(1 - x^2)(\alpha - x)(\beta - x)}} = \frac{2\mathcal{K}(k)}{\sqrt{(\alpha - 1)(\beta + 1)}} , \tag{A.28}$$

where

$$k^2 = \frac{2(\alpha - \beta)}{(\alpha - 1)(\beta + 1)}, \tag{A.29}$$

α and β are real and the intervals $[\beta, \alpha]$ and $[-1, 1]$ are disjoint, and

$$\int_{-1}^{1} \frac{\mathrm{d}x}{\sqrt{(1 - x^2)(\alpha + x)(\beta + x)}} = \frac{2\mathcal{K}(k)}{\sqrt{\frac{1}{2}\lambda(\alpha + \beta)}}, \tag{A.30}$$

where

$$\begin{aligned} k^2 &= \frac{(\alpha + \beta)(1 + \frac{1}{2}\lambda) + 2 - \lambda^2}{\lambda(\alpha + \beta)}, \\ \lambda &= \frac{1}{2}\sqrt{(\alpha + \beta)(\alpha + \beta + 8)} \end{aligned} \tag{A.31}$$

and α and β are conjugate complex numbers with positive real part.

For the thermodynamic functions of the triangular and honeycomb lattices we need relations between $A(K|k)$ and $A(G|k)$, and between $B(K|k)$ and $B(G|k)$, where K and G are connected by the star-triangle formula (8.27) and k is given by (8.73). Writing $A(K|k)$ simply as A, it follows from (A.24), (A.16) and (A.17) that

$$\mathrm{sn}(A|k') = \sin(\theta) = \tanh(2K) \tag{A.32}$$

and thus, from (A.18) and (8.73),

$$\mathrm{cn}(A|k') = \mathrm{sech}(2K), \tag{A.33}$$

$$\mathrm{dn}(A|k') = \sqrt{1 - k'^2 \tanh^2(2K)} = \mathrm{sech}(2K)\coth(2G). \tag{A.34}$$

Substituting from (A.32)–(A.34) into (A.22) and using (8.73)

$$\mathrm{sn}(2A|k') = \frac{2\cosh(2G)\sinh(2G)\tanh(2K)}{\sinh^2(2G) + \tanh^2(2K)\cosh^2(2G)}. \tag{A.35}$$

Now the star-triangle relation (8.27) can be expressed in the form

$$\tanh(2K) = \frac{\cosh(2G) - 1}{\cosh(2G)} \tag{A.36}$$

and substituting in (A.35) we have

$$\mathrm{sn}(2A(K|k)|k') = \tanh(2G), \tag{A.37}$$

which is equivalent to

$$2A(K|k) = F(\theta_1|k'), \tag{A.38}$$

where $\tan(\theta_1) = \sinh(2G)$ and $\sin(\theta_1) = \tanh(2G)$. From (A.24)

$$2A(K|k) = A(G|k). \tag{A.39}$$

From (A.24), (A.32), (A.37) and (A.38)

$$\begin{aligned} B(G|k) &= k'^{-2}[2A - E^*(2A|k')]\,, \\ B(K|k) &= k'^{-2}[A - E^*(A|k')]\,. \end{aligned} \tag{A.40}$$

Then, from (A.23),

$$B(G|k) = 2B(K|k) + \tanh^2(2K)\tanh(2G)\,. \tag{A.41}$$

The derivations of (A.39) and (A.41) are valid only when $k < 1$. It may, however, be shown using a similar method and (A.25) that the results are also valid for $k > 1$.

A.3 The Water Molecule and Hydrogen Bonding

The hydrogen (H) nucleus is a single proton and the ground state of the single electron is designated **1s**, that is principal quantum number unity and angular quantum number zero. The oxygen (O) nucleus has an eight proton charge, and of the eight electrons two are in the **1s** state with, by the Pauli exclusion principle, oppositely directed spins and form an 'inner layer'. There is an 'outer layer' of six electrons which have principal quantum number two, two electrons being in the **2s** state and four in the **2p** state, with angular momentum quantum number unity. When chemical combination takes place between oxygen and hydrogen atoms, the wave function solutions of Schrödinger's equation for the original atoms are, of course, no longer valid. The resulting situation can be represented approximately by regarding the **1s** orbital of each hydrogen atom as mixing or hybridizing with one of the oxygen **2p** orbitals to form a molecular orbital occupied by two electrons with opposite spins. A *covalent bond* is then said to exist.

In the water molecule there are two covalent O–H bonds with the two vectors from the oxygen nucleus to the hydrogen nuclei each of length 0.957×10^{-10} m and at an angle of 104.5° to each other. Owing to the shifting of the negative charge probability density, which surrounded the hydrogen nuclei in the atomic state, there is now a resultant positive charge in the neighbourhood of each hydrogen nucleus. The two covalent bonds take up four electrons, but the wave functions of the four remaining electrons from the oxygen outer layer must also be modified. They now occupy two so-called *lone-pair orbitals*, the electrons of each pair having opposite spins. The charge probability distribution due to each lone pair has its densest part in the form of a 'lobe' symmetrically distributed with respect to the plane bisecting the H–O–H angle and on the opposite side of the oxygen nucleus from the hydrogen nuclei. The two negatively charged lobes are mirror images of each other in the plane of the oxygen and hydrogen nuclei, and the angle between them is about 120°.

The two negative and two positive outer charges in the water molecule can thus be picturesquely regarded as situated on four 'arms' radiating out from the oxygen nucleus in roughly the directions of the vertices of a tetrahedron with the oxygen nucleus at its centre.[2] From this description of the water molecule it is obvious that it is highly polar, with significant electric dipole and quadrupole moments. What is not so obvious is that specific and directed links, called *hydrogen bonds*, can form between water molecules. Such a bond is created when the oxygen nuclei of two molecules are about 2.75×10^{-10} m apart and positive arm of one molecule and a negative arm of the other are approximately directed along the line segment connecting the oxygen nuclei. Each bond therefore involves one hydrogen nucleus which is situated roughly in the line of the two oxygen nuclei, but is much nearer the oxygen nucleus with which it has a covalent link. Since the positive hydrogen nucleus of one molecule is close to the negative lone-pair charge region of the other molecule, the hydrogen bond energy contains a large negative electrostatic term. There is also a covalent or *electron delocalization* term, due to distortion of the molecular electronic wave functions, and a dispersion interaction term, both of which are negative. They are partially offset by a positive term due to overlap or repulsive forces (Rao 1972). Hydrogen bonds owe their stability to the very small size of a hydrogen atom when partially stripped of its single electron. They can form between other molecules containing hydrogen atoms and electronegative regions, for instance KH_2PO_4.

Any water molecule can form four hydrogen bonds with other water molecules, two of which involve its own hydrogen atoms. It is therefore possible for a completely hydrogen-bonded water crystal to have a tetrahedrally coordinated array of oxygen nuclei. This means that any oxygen nucleus has four nearest-neighbour oxygen nuclei situated at the vertices of a tetrahedron with itself at the centre. From Appendix A.1, the oxygen nuclei could then be on the sites of a diamond-type lattice, and this is the case in ice Ic, which is a metastable form occurring between $-120\,°C$ and $-140\,°C$. The structure is an open one with a good deal of empty space. However, the form called ice I, which is stable under normal conditions, has a wurtzite-type lattice, which is also tetrahedrally coordinated. Diamond has cubic symmetry, and the sites can be regarded as arranged in layers perpendicular to an axis which makes equal angles with the three cubic directions indicated by the broken lines in the diamond lattice diagram of Appendix A.1. These layers are composed of puckered rings which are like the hexagons of sites in a honeycomb lattice except that, owing to the tetrahedral angles of the edges, the six sites of a ring are not coplanar. A shift in the relative position of successive layers changes the lattice to the hexagonally symmetric wurzite type. The density of ice I is nearly equal to that of ice Ic so it is an equally open structure, but the empty spaces now become shafts perpendicular and parallel to the layers. Again, the nearest-neighbour environments of a water molecule are the same

[2] The vectors from the centre of the tetrahedron to the vertices make angles of 109.47° with each other. Note that in Fig. 6.2, either the four $\mathfrak{b}_1$ sites or the $\mathfrak{b}_2$ sites are the vertices of a tetrahedron with the $\mathfrak{a}_1$ site as centre.

in the two forms, the differences starting with second neighbours (Eisenberg and Kauzmann 1969, Fletcher 1970). A water molecule bonded tetrahedrally to four other water molecules can take up six orientations, depending on the pair of bonds where its own hydrogen atoms are situated. However, in a completely hydrogen-bonded network, any reorientation must be cooperative since there must always be exactly one hydrogen on each bond. Even with this restriction there are about $(1.507)^M$ ways of placing the protons on the bonds of a network of M molecules, so that ice I has a residual configurational reduced entropy per molecule, $s = 0.410$ (Nagle 1966).

There are a number of high pressure polymorphs of ice, in all of which each water molecule is still hydrogen bonded to four others. Ices II-V achieve a greater density than that of ice I by distortion of the hydrogen bonds. In ice II the positions of the protons, and hence the orientations of the molecules, are ordered and the residual entropy is absent. The highest density forms are ices VI, VII and VIII, which are composed of two intertwining fully-bonded networks. In ice VII the oxygen nuclei occupy the sites of a body-centred cubic lattice and are linked by hydrogen bonds into two diamond networks.[3] To force the molecules of one bonded network into the interstices of another involves a good deal of work against repulsive forces and hence ice VII is stable only at pressures above about 22,000 atm. Ice VIII is similar to ice VII except that the proton positions are ordered.

In a condensed water phase the rupture of some of the hydrogen bonds decreases the amount of interstitial space and increases the density. So ice I at its melting point is less dense than the liquid with which it is in equilibrium. Ice I, therefore, floats on its melt, and its melting point temperature decreases as the pressure increases. However, just above the melting point temperature, large regions of ice-like open structure persist in the liquid. As the temperature increases, more hydrogen bonds are broken, and the open structure regions diminish in size. The consequent closer packing offsets the usual decrease in density and results in the density increasing with temperature until a maximum is reached. At 1 atm pressure this occurs at 4 °C. Again, instead of the steady increase in compressibility with temperature observed in other liquids, the compressibility of water decreases until a minimum is attained, this being at 46 °C for 1 atm pressure. These anomalous properties of fluid water persist up to pressures well above the critical value of 218 atm but eventually disappear.

A.4 Results for the Six-Vertex Model

The proofs are presented for the results I–IV given in equations (10.32), (10.34), (10.35) and (10.38) respectively.

[3] In Fig. 6.2 it is possible to construct one diamond sublattice from the sites of the $\mathfrak{a}_1$ and $\mathfrak{b}_1$ face-centred cubic sublattices, while the $\mathfrak{a}_2$ and $\mathfrak{b}_2$ sites form the other diamond sublattice.

A.4.1 The Proof of I

Equation (10.31) is valid for all vectors $\boldsymbol{\eta}$ of the type defined in Sect. 10.4. In particular, if $\eta_n < N_1$,

$$g(\boldsymbol{\eta}^{(1)}) = \sum_{\{\Pi\}} A(\Pi) z_{\pi 1}^{\eta_1+1} z_{\pi 2}^{\eta_2+1} \cdots z_{\pi n}^{\eta_n+1} = z_1 z_2 \cdots z_n \, g(\boldsymbol{\eta}), \tag{A.42}$$

and so, from (10.15) and (10.16),

$$z_1 \cdots z_n = \zeta. \tag{A.43}$$

A.4.2 The Proof of II

Now suppose we choose $\eta_1 < s < \eta_2$, then

$$\begin{aligned} g(\boldsymbol{\eta}^{(-s)}) &= \sum_{\{\Pi\}} A(\Pi) z_{\pi 1}^{\eta_2 - s} \cdots z_{\pi n-1}^{\eta_n - s} z_{\pi n}^{N_1+\eta_1-s} \\ &= \sum_{\{\Pi\}} A(\mathfrak{R}\Pi) z_{\pi 1}^{N_1+\eta_1-s} z_{\pi 2}^{\eta_2-s} \cdots z_{\pi n}^{\eta_n-s}, \end{aligned} \tag{A.44}$$

which, from (A.42) and (A.43), gives

$$\begin{aligned} g(\boldsymbol{\eta}^{(-s)}) &= \zeta^{-s} g(\boldsymbol{\eta}) \\ &= z_{\pi 1}^{N_1} \zeta^{-s} \sum_{\{\Pi\}} A(\mathfrak{R}\Pi) z_{\pi 1}^{\eta_1} z_{\pi 2}^{\eta_2} \cdots z_{\pi n}^{\eta_n}. \end{aligned} \tag{A.45}$$

Comparing this with (10.31) we have

$$A(\mathfrak{R}\Pi) = z_{\pi 1}^{-N_1} A(\Pi). \tag{A.46}$$

A.4.3 The Proof of III

There are two possibilities for the components of the vector $\boldsymbol{\eta}'$ in (10.8):

(i) $1 \le \eta_1' \le \eta_1, \quad \eta_{j-1} \le \eta_j' \le \eta_j, \qquad j = 2, \ldots, n.$

(ii) $\eta_j \le \eta_j' \le \eta_{j+1}, \quad j = 1, \ldots, n-1, \qquad \eta_n \le \eta_n' \le N_1,$

and, as for the case $n = 1$ (see (10.21)) we divide the left-hand side of (10.8) into two parts corresponding to these two cases.

$$\sum_{\{\boldsymbol{\eta}'\}} V(\boldsymbol{\eta}, \boldsymbol{\eta}') g(\boldsymbol{\eta}') = \sum_{\{\Pi\}} A(\Pi)[\Omega^{(\mathrm{i})}(\Pi, \boldsymbol{\eta}) + \Omega^{(\mathrm{ii})}(\Pi, \boldsymbol{\eta})], \tag{A.47}$$

where

$$\Omega^{(m)}(\Pi, \boldsymbol{\eta}) = \sum_{\{\boldsymbol{\eta}'\}}^{(m)} V(\boldsymbol{\eta}, \boldsymbol{\eta}') z_{\pi 1}^{\eta_1'} \cdots z_{\pi n}^{\eta_n'}, \qquad m = \mathrm{i}, \mathrm{ii} \tag{A.48}$$

are, the respective sums restricted as indicated above. They are generalizations of $\Omega^{(\mathrm{i})}(\eta)$ and $\Omega^{(\mathrm{ii})}(\eta)$ defined in (10.22) for the case $n=1$. In order to avoid separate evaluations of $\Omega^{(\mathrm{i})}$ and $\Omega^{(\mathrm{ii})}$ we define the transformation

$$\mathfrak{L}=\begin{cases} \mathfrak{a}\to\mathfrak{b}\,, \\ \mathfrak{b}\to\mathfrak{a}\,, \\ z_{\pi_j}\to 1/z_{\pi_{n+1-j}}\eta_j\to N_1+1-\eta_{n+1-j} \end{cases} \tag{A.49}$$

and we have

$$\Omega^{(\mathrm{ii})}(\Pi,\boldsymbol{\eta})=\zeta\,\mathfrak{L}\Omega^{(\mathrm{i})}(\Pi,\boldsymbol{\eta})\,, \tag{A.50}$$

$$\mathfrak{L}\overline{\mathsf{M}}(z_j)=\overline{\mathsf{L}}(z_{n+1-j})\,, \tag{A.51}$$

the functions $\overline{\mathsf{L}}(z)$ and $\overline{\mathsf{M}}(z)$ being defined in (10.24). For simplicity we consider $\Omega^{(i)}(\mathrm{I};\boldsymbol{\eta})$, where I is the identity permutation. The summation in (A.48) is a geometrical progression as in (10.22), and after summation we have an expression of the form

$$\Omega^{(i)}(\mathrm{I};\boldsymbol{\eta})=\mathfrak{b}^{N-\eta_n}\sum\mathsf{H}_1(z_1)\cdots\mathsf{H}_n(z_n)\,, \tag{A.52}$$

where the sum is over the different forms for the functions $\mathsf{H}_j(z)$ given by

$$\mathsf{H}_1(z)=\begin{cases} (\mathfrak{a}/\mathfrak{b})\mathfrak{b}^{\eta_1}z^{\eta_1}\equiv\alpha_1\,, \\ \mathfrak{b}^{\eta_1}z^{\eta_1}\left[\overline{\mathsf{M}}(z)-\mathfrak{a}/\mathfrak{b}\right]+\mathfrak{a}^{\eta_1}\left[\overline{\mathsf{L}}(z)-\mathfrak{b}/\mathfrak{a}\right]\equiv\beta_1\,, \end{cases} \tag{A.53}$$

$$\mathsf{H}_j(z)=\begin{cases} (\mathfrak{a}/\mathfrak{b})\mathfrak{b}^{\eta_j-\eta_{j-1}}z^{\eta_j}\equiv\alpha_j, \\ \mathfrak{b}^{\eta_j-\eta_{j-1}}z^{\eta_j}\left[\overline{\mathsf{M}}(z)-\mathfrak{a}/\mathfrak{b}\right] \\ \quad+\mathfrak{a}^{\eta_j-\eta_{j-1}}z^{\eta_{j-1}}\left[\overline{\mathsf{L}}(z)-\mathfrak{b}/\mathfrak{a}\right]\equiv\beta_j, \\ (\mathfrak{b}/\mathfrak{a})\mathfrak{a}^{\eta_j-\eta_{j-1}}z^{\eta_{j-1}}\equiv\gamma_j, \end{cases} \qquad N_1\geq j>1. \tag{A.54}$$

In (i) and (ii) the condition that neighbouring members of the set $\eta_1',\dots,\eta_n'$ cannot be equal must hold and this is ensured in (A.52) by imposing the rule that an α function cannot be followed on the right by a γ function. Every term in (A.52) contains a particular set of factors $\mathfrak{a}^{\eta_j-\eta_{j-1}}$, $\mathfrak{b}^{\eta_j-\eta_{j-1}}$ and we gather together those terms with the same set of factors to give

$$\Omega^{(\mathrm{i})}(\mathrm{I};\boldsymbol{\eta})=\sum_{\{Q_r\}}\mathsf{f}_r(\mathrm{I};Q_r)\mathfrak{p}^{\eta_{q_1}}\cdots\mathfrak{a}^{\eta_{q_r}-\eta_{q_{r-1}}}\mathfrak{b}^{N_1-\eta_{q_r}}\,, \tag{A.55}$$

where $\mathfrak{p}=\mathfrak{a}$ or $\mathfrak{b}$ according as r is even or odd. The summation in (A.55) is over all subsets $Q_r=\{q_1,q_2,\dots,q_r\}$ of the integers $1,\dots,n$ including the empty subset denoted by Q_0. Let

$$\mathsf{W}(\mathrm{I};j) = (z_{j+1}z_j)^{\eta_j}\left[\overline{\mathsf{L}}(z_{j+1})\overline{\mathsf{M}}(z_j) - 1\right], \tag{A.56}$$

$$\mathsf{X}(\mathrm{I};j_1,j_2) = \begin{cases} 1 & j_2 < j_1, \\ \prod_{j=j_1}^{j_2}\left[z_j^{\eta_j}\overline{\mathsf{M}}(z_j)\right] & 1 \le j_1 \le j_2 \le n, \end{cases} \tag{A.57}$$

$$\mathsf{Y}(\mathrm{I1};j_1,j_2) = \begin{cases} 1 & j_2 < j_1, \\ \overline{\mathsf{L}}(z_1) - \mathfrak{b}/\mathfrak{a}, & j_1 = j_2 = 1, \\ \left[\overline{\mathsf{L}}(z_1) - \mathfrak{b}/\mathfrak{a}\right]\prod_{j=2}^{j_2}\left[z_j^{\eta_j - 1}\overline{\mathsf{L}}(z_j)\right], & 1 = j_1 < j_2 \le n, \\ \prod_{j=j_1}^{j_2}\left[z_j^{\eta_j-1}\overline{\mathsf{L}}(z_j)\right], & 1 \le j_1 \le j_2 \le n, \end{cases} \tag{A.58}$$

with corresponding expressions for the permutation Π. Then, from (A.52)–(A.58),

$$\mathsf{f}_0(\mathrm{I};Q_0) = \mathsf{X}(\mathrm{I};1,n), \tag{A.59}$$

$$\mathsf{f}_1(\mathrm{I};Q_1) = \mathsf{Y}(\mathrm{I};1,q_1)\mathsf{X}(\mathrm{I};q_1+1,n), \tag{A.60}$$

$$\begin{aligned}\mathsf{f}_r(\mathrm{I};Q_r) &= \mathsf{X}(\mathrm{I};1,q_1-1)\mathsf{W}(\mathrm{I};q_1)\mathsf{Y}(\mathrm{I};q_1+2,q_2) \\ &\times\left[\prod_{j=2}^{r/2}\mathsf{X}(\mathrm{I};q_{2j-2}+1,q_{2j-1}-1)\mathsf{W}(\mathrm{I};q_{2j-1})\mathsf{Y}(\mathrm{I};q_{2j-1}+2,q_{2j})\right] \\ &\times\mathsf{X}(1;q_r+1,n), \quad r \ge 2 \text{ and even,}\end{aligned} \tag{A.61}$$

$$\begin{aligned}\mathsf{f}_r(\mathrm{I};Q_r) &= \mathsf{Y}(\mathrm{I};1,q_1) \\ &\times\left[\prod_{j=1}^{(r-1)/2}\mathsf{X}(\mathrm{I};q_{2j-1}+1,q_{2j}-1)\mathsf{W}(\mathrm{I};q_{2j})\mathsf{Y}(\mathrm{I};q_{2j}+2,q_{2j+1})\right] \\ &\times\mathsf{X}(\mathrm{I};q_r+1,n) \quad r \ge 3 \text{ and odd.}\end{aligned} \tag{A.62}$$

The primary aim of this appendix is to establish the conditions for the right-hand side of (A.48) to be of the form of the right-hand side of (10.8). From (A.55)–(A.62) it can be seen that a sufficient condition for this to be the case would be

$$\sum_{\{\Pi\}} A(\Pi) f_r(P, Q_r) = 0\,, \tag{A.63}$$

for all $r \geq 1$. In fact, as we shall see below, this is stronger than is needed. It is sufficient that (A.63) is true for all $r > 1$. We see from (A.61) and (A.62) that all the functions $f_r(\mathrm{I}; Q_r)$ for $r > 1$ (and similarly when the permutation Π replaces I) contain W functions. For (A.63) to hold for $r > 1$, it is therefore sufficient that the coefficients $A(\Pi)$ satisfy the condition

$$A(\mathfrak{C}_{j,j+1}\Pi) \left[\overline{\mathrm{L}}(z_{\pi_j})\overline{\mathrm{M}}(z_{\pi_{j+1}}) - 1\right] = A(\Pi) \left[\overline{\mathrm{M}}(z_{\pi_j})\overline{\mathrm{L}}(z_{\pi_{j+1}}) - 1\right]. \tag{A.64}$$

This is equivalent to (10.35)

A.4.4 The Proof of IV

From (A.55), (A.59)–(A.63)

$$\begin{aligned}\sum_{\{\Pi\}} A(\Pi)\Omega^{(\mathrm{i})}(\Pi;\eta) = \mathfrak{b}^{N_1} &\sum_{\{\Pi\}} A(\Pi) \prod_{j=1}^{n} \left[z_{\pi_j}^{\eta_j}\overline{\mathrm{M}}(z_{\pi_j})\right] \\ &+ \sum_{\{\Pi\}} A(\Pi) \sum_{r=1}^{n} \mathfrak{b}^{N_1-\eta_r}\mathfrak{a}^{\eta_r}\mathrm{Y}(\Pi;1,r)\mathrm{X}(\Pi;r,n)\end{aligned} \tag{A.65}$$

and from (A.49)–(A.50)

$$\begin{aligned}\sum_{\{\Pi\}} A(\Pi)\Omega^{(\mathrm{ii})}(\Pi;\eta) = \mathfrak{a}^{N_1} &\sum_{\{\Pi\}} A(\Pi) \prod_{j=1}^{n} \left[z_{\pi_j}^{\eta_j}\overline{\mathrm{L}}(z_{\pi_j})\right] \\ &+ \zeta \sum_{\{\Pi\}} A(\Pi) \sum_{r=1}^{n} \mathfrak{L}\left\{\mathfrak{b}^{N_1-\eta_r}\mathfrak{a}^{\eta_r}\mathrm{Y}(\Pi;1,r)\mathrm{X}(\Pi;r,n)\right\}.\end{aligned} \tag{A.66}$$

Now, from (10.24),

$$\overline{\mathrm{L}}(z) - \mathfrak{b}/\mathfrak{a} = z(\mathfrak{b}/\mathfrak{a})\left[\mathfrak{a}/\mathfrak{b} - \overline{\mathrm{M}}(z)\right] \tag{A.67}$$

and using this result together with (10.33), (A.46) and (A.49) it can be shown, after some manipulation, that

$$\begin{aligned}\zeta \sum_{\{\Pi\}} A(\Pi) \sum_{r=1}^{n} &\mathfrak{L}\left\{\mathfrak{b}^{N_1-\eta_r}\mathfrak{a}^{\eta_r}\mathrm{Y}(\Pi;1,r)\mathrm{X}(\Pi;r,n)\right\} = \\ &- \sum_{\{\Pi\}} A(\Pi) \sum_{r=1}^{n} \mathfrak{b}^{N_1-\eta_r}\mathfrak{a}^{\eta_r}\mathrm{Y}(\Pi;1,r)\mathrm{X}(\Pi;r,n).\end{aligned} \tag{A.68}$$

From (A.47), (A.65), (A.66) and (A.68) we then have

$$\sum_{\{\eta'\}} V(\eta,\eta')g(\eta') = \left[\mathfrak{a}^{N_1}\prod_{j=1}^{n}\mathfrak{L}(z_j) + \mathfrak{b}^{N_1}\prod_{j=1}^{n}\overline{\mathfrak{M}}(z_j)\right] \times \sum_{\{\Pi\}} A(\Pi)\prod_{j=1}^{n} z_j^{\eta_j}, \tag{A.69}$$

which is equivalent to (10.8) with λ given by (10.38).

A.5 Fourier Transforms and Series

We summarize the Fourier transform and series formulae needed for Chap. 10. A more detailed discussion of these transforms in d dimensions is given in Volume 2, Appendix A.1.

A.5.1 Fourier Transforms

The Fourier transform $\mathfrak{F}_\omega\{G(x)\}$ of a function $G(x)$ is defined by

$$G^*(\omega) = \mathfrak{F}_\omega\{G(x)\} = \int_{-\infty}^{\infty} G(x)\exp(-\mathrm{i}x\omega)\mathrm{d}x\,, \tag{A.70}$$

where

$$G(x) = \frac{1}{2\pi}\int_{-\infty}^{\infty} G^*(\omega)\exp(\mathrm{i}\omega x)\mathrm{d}\omega\,. \tag{A.71}$$

The transform of a convolution integral of two such functions $F(x)$ and $G(x)$ is given by

$$\mathfrak{F}_\omega\left\{\int_{-\infty}^{\infty} F(x-y)G(y)\mathrm{d}y\right\} = F^*(\omega)G^*(\omega)\,, \tag{A.72}$$

which gives *Parseval's formula*

$$\int_{-\infty}^{\infty} F(x-y)G(y)\mathrm{d}y = \frac{1}{2\pi}\int_{-\infty}^{\infty} F^*(\omega)G^*(\omega)\exp(\mathrm{i}\omega x)\mathrm{d}\omega\,. \tag{A.73}$$

A.5.2 Fourier Series

The coefficients $\mathfrak{F}_n\{g(x)\}$ for the complex Fourier series of a piecewise continuous function $g(x)$ defined in a range $(-\pi,\pi)$ are defined by

$$g_n^* = \mathfrak{F}_n\{g(x)\} = \int_{-\pi}^{\pi} g(x)\exp(-\mathrm{i}nx)\mathrm{d}x\,, \tag{A.74}$$

where

$$g(x) = \frac{1}{2\pi}\sum_{n=-\infty}^{\infty} g_n^*\exp(\mathrm{i}nx)\,. \tag{A.75}$$

The coefficients of a convolution integral of two such functions $f(x)$ and $g(x)$ are given by

$$\mathfrak{F}_n\left\{\int_{-\pi}^{\pi} f(x-y)g(y)\mathrm{d}y\right\} = f_n^* g_n^* , \tag{A.76}$$

which gives *Parseval's formula*

$$\int_{-\pi}^{\pi} f(x-y)g(y)\mathrm{d}y = \frac{1}{2\pi}\sum_{n=-\infty}^{\infty} f_n^* g_n^* \exp(\mathrm{i}nx) . \tag{A.77}$$

References and Author Index

The page numbers where the references are cited in the text are given in square brackets.

Abraham D.B. (1973): Susceptibility and fluctuations in the Ising ferromagnet. Phys. Lett. A **43**, 163–166, [229].

Abramowitz M. and Segun I.A. (1965): Handbook of Mathematical Functions. Dover, New York, U.S.A. [337].

Adkins C.J. (1983): Equilibrium Thermodynamics. C.U.P., Cambridge, U.K. [1].

Aharony A. (1983): Multicritical points. In Critical Phenomena (Ed. F.J.W. Hahne), 209–258, Springer, Berlin, Heidelberg, Germany, [162,322].

Albrecht O., Gruber H. and Sackmann E. (1978): Polymorphism of phospholipid monolayers. J. de Phys. **39**, 301–313, [187].

Alexander S. (1975): Lattice gas transitions of He on grafoil. A continuous transition with cubic terms. Phys. Lett. A **54**, 353–354, [332].

Anderson G.R. and Wheeler J.C. (1978a): Directionality dependence of lattice models for solutions with closed-loop coexistence curves. J. Chem. Phys. **69**, 2082–2088, [275].

Anderson G.R. and Wheeler J.C. (1978b): Theory of lower critical solution points in aqueous mixtures. J. Chem. Phys. **69**, 3403–3413, [276]

Baker G.A. and Essam J.W. (1971): Statistical mechanics of a compressible Ising model with application to β brass. J. Chem. Phys. **55**, 861–879, [251].

Barker J.A. and Fock W. (1953): Theory of upper and lower critical solution temperatures. Disc. Faraday Soc. **15**, 188–195, [270,276].

Bartis J.T. and Hall C.K. (1974): A lattice model of gas-gas equilibria in binary mixtures. Physica **78**, 1–21, [278,289].

Banville M., Albinet G. and Firpo J.L. (1986): Fatty alcohol monolayers at the air–water interface – a spin-one model. J. Chim. Phys. et Phys. Chim. Biol. **83**, 511–517, [187].

Baret J.F. and Firpo J.L. (1983): A spin-1 Ising model to describe amphile monolayer phase transitions. J. Coll. Inter. Sci. **94**, 487–496, [187].

Baxter R.J. (1972): Partition function of the eight-vertex lattice model. Ann. Phys. **70**, 193–228, [294].

Baxter R.J. (1973a): Spontaneous staggered polarization of the F model. J. Phys. C: Solid State Phys. **6**, L94–L96, [331].

Baxter R.J. (1973b): Potts model at the critical temperature. J. Phys. C: Solid State Phys. **6**, L445–L448, [165,325,329,331].

Baxter R.J. (1973c): Spontaneous staggered polarization of the F-model. J. Stat. Phys. **9**, 145–182, [331].

Baxter R.J. (1975): Triplet order parameter of the triangular Ising model. J. Phys. A: Math. Gen. **8**, 1797–1805, [266,268].

Baxter R.J. (1982a): Exactly Solved Models in Statistical Mechanics. Academic Press, London, U.K. [174,198,205,229,231,294,312,332,337].

Baxter R.J. (1982b): Critical antiferromagnetic square lattice Potts model. Proc. Roy. Soc. A **383**, 43–54, [325,328,330,331].

Baxter R.J. and Enting I.G. (1978): 339th solution of the Ising model. J. Phys. A: Math. Gen. **11**, 2463–2473, [205,214,218–220].

Baxter R.J., Temperley H.N.V. and Ashley S.E. (1978): Triangular Potts model at its transition temperature, and related problems. Proc. Roy. Soc. A **358**, 535–559, [165].

Bell G.M. (1953): Statistical thermodynamics of regular ternary assemblies. Trans. Faraday Soc. **49**, 122–132, [151,157].

Bell G.M. (1958): Dilution effects in regular assemblies. Proc. Phys. Soc. **72**, 649–660, [166,167].

Bell G.M. (1969): One-dimensional bonded fluids. J. Math. Phys. **10**, 1753–1760, [128,130].

Bell G.M. (1972): Statistical mechanics of water: lattice model with directed bonding. J. Phys. C: Solid State Phys. **5**, 889–905, [187,188,190–192].

Bell G.M. (1974a): Ising ferrimagnetic models I. J. Phys. C: Solid State Phys. **7**, 1174–1188, [110,266].

Bell G.M. (1974b): Ising ferrimagnetic models II. J. Phys. C: Solid State Phys. **7**, 1189–1205, [256,258,259,261,266].

Bell G.M. (1975): The diluted Ising model on linear and Bethe lattices. J. Phys. C: Solid State Phys. **8**, 669–682, [168].

Bell G.M. (1980): A continuous one-dimensional fluid as the limit of a lattice fluid. J. Phys. A: Math. Gen. **13**, 659–673, [124,128,131].

Bell G.M., Combs L.L. and Dunne L.J. (1981): Theory of cooperative phenomena in lipid systems. Chem. Rev. **81**, 15–48, [187].

Bell G.M. and Fairbairn W.M (1961a): The critical temperatures of binary alloys with one magnetic component. Phil. Mag. **6**, 907–928, [168].

Bell G.M. and Fairbairn W.M (1961b): Regular models for solid hydrogen: I. Mol. Phys. **4**, 481–489, [270].

Bell G.M. and Fairbairn W.M (1967): Orientable ordering in solid hydrogen. Phys. Rev. **158**, 530–536, [270,276].

Bell G.M. and Lavis D.A. (1965): Sublattice order in binary alloys with one magnetic component. Phil. Mag. **11**, 937–953, [168].

Bell G.M. and Lavis D.A. (1968): Statistical effects of super-exchange in binary mixtures with one magnetic component. Phys. Rev. **168**, 543–549, [168].

Bell G.M. and Lavis D.A. (1970): Two-dimensional bonded lattice fluids: II. J. Phys. A: Math. Gen. **3**, 568–581, [187,192,202].

Bell G.M. and Lavis D.A. (1989): Statistical Mechanics of Lattice Models. Vol.1: Closed-Form and Exact Theories of Cooperative Phenomena. Ellis Horwood, Chichester, U.K. [VI].

Bell G.M. and Sallouta H. (1975): An interstitial model for fluid water: accurate and approximate results. Mol. Phys. **29**, 1621–1637, [281–283].

Bell G.M. and Salt D.W. (1976): Three-dimensional lattice model for the water/ice system. J. Chem. Soc. Faraday Trans. II **72**, 76–86, [191].

Bell G.M., Mingins J. and Taylor R.A.G. (1978): Second-order phase changes in phospholipid monolayer at the oil–water interface and related phenomena. J. Chem. Soc. Faraday Trans. II **74**, 223–234, [182,187].

Bethe H.A. (1935): Statistical theory of superlattices. Proc. Roy. Soc. A **150**, 552–575, [173].

Bidaux R., Carrara P. and Vivet B. (1967): Antiferromagnetisme dans un champ magnetique I. Traitement de champ moleculaire. J. Phys. Chem. Solids **28**, 2453–2469, [108].

Binder K. (1987): Theory of first-order transitions. Rep. Prog. Phys. **50**, 783–859, [28,196].

Black J.L. and Emery V.J. (1981): Critical properties of two-dimensional models. Phys. Rev. B **23**, 429–432, [331].

Blöte H.W.J. and Nightingale M.P. (1982): Critical behaviour of the two-dimensional Potts model with a continuous number of states; a finite size scaling analysis. Physica A **112**, 405–465, [332].

Blume M. (1966): Theory of the first-order magnetic phase change in UO_2. Phys. Rev. **141**, 517–524, [151].

Blume M., Emery V.J. and Griffiths R.B. (1971): Ising model for the λ transition and phase separation on ^3He-^4He mixtures. Phys. Rev. A **4**, 1071–1077, [151,157,160].

Bowman F. (1953): Introduction to Elliptic Functions with Applications. E.U.P., London, U.K. [337,338].

Bragg W.L. and Williams E. (1934): The effect of thermal agitation on atomic arrangement in alloys. Proc. Roy. Soc. A **145**, 699–730, [139].

Brankov J.G. and Zagrebnov V.A. (1983): On the description of the phase transition in the Husimi–Temperley model. J. Phys. A: Math. Gen. **16**, 2217–2224, [77].

Buchdahl H.A. (1966): The Concepts of Thermodynamics. C.U.P., Cambridge, U.K. [1].

Burley D.M. (1972): Closed-form approximations for lattice systems. In Phase Transitions and Critical Phenomena. (Eds. C. Domb and M. S. Green), Vol. 2, 329–374. Academic Press, London, U.K., [173].

Capel H.W. (1966): On the possibility of first-order phase transitions in Ising systems of triplet ions with zero-field splitting. Physica **32**, 966–988, [151].

Cardy J.L. (1996): Scaling and Renormalization in Statistical Physics. C.U.P., Cambridge, U.K. [332].

Chandler D. (1987): Introduction to Modern Statistical Mechanics. O.U.P., Oxford, U.K. [31].

Danielian A. (1961): Ground state of an Ising face-centred cubic lattice. Phys. Rev. Lett. **6**, 670–671, [196].

De Boer J. (1974): Van der Waals in his time and the present revival. Physica **73**, 1–27, [9].

De Neef T. and Enting I.G. (1977): Series expansions from the finite-lattice method. J. Phys. A: Math. Gen. **10**, 801–805, [173].

Den Nijs M.P.M. (1979): A relationship between the temperature exponents of the eight-vertex and q-state Potts model. J. Phys. A: Math. Gen. **12**, 1857–1868, [331].

De Jongh L.J. and Miedema (1974): Experiments on simple magnetic model systems. Adv. in Phys. **23**, 1–260, [44].

Domb C. (1960): On the theory of cooperative phenomena in crystals. Phil. Mag. Supp. **9**, 149–361, [173,197,227,229,231,237].

Domb C. (1974): Ising model. In Phase Transitions and Critical Phenomena. (Eds. C. Domb and M. S. Green), Vol. 3, 357–484. Academic Press, London, U.K., [75,103,104,211,227].

Domb C., (1985): Critical phenomena: a brief historical survey. Contemporary Physics **26**, 49–72, [V].

Domb C. and Sykes M.F. (1961): Use of series expansions for the Ising model susceptibility and excluded volume problem. J. Math. Phys. **2**, 63–67, [88].

Eggarter T.P. (1974): Cayley trees, the Ising problem and the thermodynamic limit. Phys. Rev. B **9**, 2989–2992, [198].

Ehrenfest P. (1933): Phase changes in the ordinary and extended sense classified according to the corresponding singularity of the thermodynamic potential. Proc. Acad. Sci. Amsterdam **36**, 153–157, [72].

Eisenberg D. and Kauzmann W. (1969): The Structure and Properties of Water, Clarendon Press, Oxford, U.K. [188,344].

Elcock E.W. (1958): Order-Disorder Phenomena. Methuen, London, [141].

Enting I.G. (1977): Triplet order parameters in triangular and honeycomb Ising models. J. Phys. A: Math. Gen. **10**, 1737–1743, [266].

Essam J.W. (1972): Percolation and cluster size. In Phase Transitions and Critical Phenomena. (Eds. C. Domb and M. S. Green), Vol. 2, 197–270. Academic Press, London, U.K., [167,250].

Essam J.W. (1980): Percolation theory. Rep. Prog. Phys. **43**, 833–912, [167].

Essam J.W. and Fisher M.E. (1963): Padé approximant studies of the lattice gas and Ising ferromagnet below the critical point. J. Chem. Phys. **38**, 802–812, [86,88].

Essam J.W. and Garelick H. (1967): Critical behaviour of a soluble model of dilute ferromagnetism. Proc. Phys. Soc. **92**, 136–149, [250].

Firpo J.L., Dupin J.J., Albinet G., Baret J.F. and Caille A. (1981): Model for the tricritical point of the chain-melting transition in a monomolecular layer of amphiphilic molecules. J. Chem. Phys. **74**, 2569–2575, [187].

Fisher M.E. (1963): Perpendicular susceptibility of the Ising model. J. Math. Phys. **4**, 124–135, [105].

Fisher M.E. (1959a): Transformations of Ising models. Phys. Rev. **113**, 969–981, [218,241,243].

Fisher M.E. (1959b): The susceptibility of the plane Ising model. Physica **25**, 521–524, [229].

Fisher M.E. (1960): Lattice statistics in a magnetic field I. A two-dimensional super-exchange antiferromagnet. Proc. Roy. Soc. A **254**, 66–85, [252].

Fisher M.E. (1967): Theory of equilibrium critical phenomena. Rep. Prog. Phys. **30**, 615–730, [73].

Fisher M.E. (1968): Renormalization of critical exponents by hidden variables. Phys. Rev. **176**, 257–272, [250,256].

Fisher M.E. and Essam J.W. (1961): Some cluster size and percolation problems. J. Math. Phys. **2**, 609–619, [167].

Fisher M.E. and Scesney P.E. (1970): Visibility of critical-exponent renormalization. Phys. Rev. A **2**, 825–835, [250].

Fisher M.E. and Sykes M.F. (1962): Antiferromagnetic susceptibilities on the simple cubic and body-centred cubic Ising lattices. Physica **28**, 939–956, [104].

Fleming M.E. and Gibbs J.N. (1974): An adaption of the lattice gas to the water problem. J. Stat. Phys. **101**, 157–173, 351–378, [192].

Fletcher N.H. (1970): The Chemical Physics of Ice. C.U.P., Cambridge, U.K. [188,344].

Fowler R. and Guggenheim E.A. (1949): Statistical Thermodynamics. C.U.P., Cambridge, U.K. [76,145].

Friedan D., Qiu Z. and Shenker S. (1984): Conformal invariance, unitarity and critical exponents in two dimensions. Phys. Rev. Lett. **52**, 1575–1578, [332].

Gantmacher F.R. (1979): Theory of Matrices. Chelsea, New York, U.S.A. [300].

Gotoh O. (1983): Prediction of melting profiles and local helix stability for sequenced DNA. Adv. in Biophys. **16**, 1–52, [58,62].

Georges A., Hansel D., Le Doussal P. and Maillard J.M. (1986): The vicinity of disorder varieties: a systematic expansion. J. Phys. A: Math. Gen. **19**, 1001–1005, [273].

Glasser M.L. (1969): Evaluation of the partition functions for some two-dimensional ferroelectric models. Phys. Rev. **184**, 539–542, [312].

Glasser M.L. Abraham D.B. and Lieb E.H. (1972): Analytic properties of the free energy for the "ice" model. J. Math. Phys. **13**, 887–900, [319].

Gradshteyn I.S. and Ryzhik I.M. (1980): Table of Integrals, Series, and Products. (4th Ed.) Academic Press, London, U.K. [330,337].

Green H.S. and Hurst C.A. (1964): Order-Disorder Phenomena. Interscience, London, [205,229,231].

Griffiths R.B. (1967a): Thermodynamic functions for fluids and ferromagnets near the critical point. Phys. Rev. **158**, 176–187, [81].

Griffiths R.B. (1967b): First-order phase transitions in spin-one Ising systems. Physica **33**, 689–690, [270].

Griffiths R.B. (1970): Thermodynamics near the two-fluid critical mixing point in He^3-He^4. Phys. Rev. Lett. **24**, 715–717, [151,161].

Griffiths R.B. (1972): Rigorous results and theorems. In Phase Transitions and Critical Phenomena. (Eds. C. Domb and M. S. Green), Vol. 1, 7–109. Academic Press, London, U.K., [33].

Griffiths R.B. and Wheeler J.C. (1970): Critical points in multicomponent systems. Phys. Rev. A **2**, 1047–1064, [16].

Gunton J.D., San Miguel M. and Sahai P.S. (1983): Dynamics of first-order transitions. In Phase Transitions and Critical Phenomena. (Eds. C. Domb and J. L. Lebowitz), Vol. 8, 269–466. Academic Press, London, U.K., [28].

Guggenheim E.A. (1935): The statistical mechanics of regular solutions. Proc. Roy. Soc. A **148**, 304–312, [173].

Guggenheim E.A. (1952): Mixtures. O.U.P., Oxford, U.K. [176].

Guggenheim E.A. and McGlashan M.L. (1951): Statistical mechanics of regular mixtures. Proc. Roy. Soc. A **206**, 335–353, [173].

Hemmer P.C. and Stell G. (1970): Fluids with several phase transitions. Phys. Rev. Lett. **24**, 1284–1287, [82].

Hemmer P.C. and Lebowitz J.L. (1976): Systems with weak long-range potentials. In Phase Transitions and Critical Phenomena. (Eds. C. Domb and M. S. Green), Vol. 5b, 107–203. Academic Press, London, U.K., [77].

Hijmans J. and de Boer J. (1955): An approximation method for order-disorder problems I–III. Physica **21**, 471–484,485–498,499–498, [151,161].

Hijmans J. and de Boer J. (1956): An approximation method for order-disorder problems IV–V. Physica **22**, 408–428,429–442, [173].

Houska C.R. (1963): A theoretical treatment of atomic configurations in some iron-aluminum solid solutions. J. Phys. Chem. Solids **24**, 95–107, [168].

Huang K. (1963): Statistical Mechanics. Academic Press, London, U.K. [31].

Husimi K. (1953): Statistical mechanics of condensation. In Proc. Int. Conf. on Theoretical Physics, 531–533, Kyoto and Tokyo, [77].

Irani R.S. (1972): The thermodynamics and mechanisms of ordering systems. Contemporary Physics **13**, 559–583, [147,197].

Izyumov Yu. A. and Skryabin Yu. N. (1988): Statistical Mechanics of Magnetically Ordered Systems, Plenum Press, New York, U.S.A. [42].

Jancel R. (1969): Foundations of Classical and Quantum Statistical Mechanics. Pergamon, Oxford, U.K. [32].

Kac M. (1968): Mathematical mechanisms of phase transitions. In Statistical Physics, Phase Transitions and Superfluidity. Vol. 1 (Eds. M. Cretien et al.). 241–305, Gordon and Breach, New York [77].

Kac M. and Ward J.C. (1952): A combinatorial solution of the two-dimensional Ising model. Phys. Rev. **88**, 1332–1337, [205].

Kasteleyn P.W. and van Kranendonk J (1956a): Constant-coupling approximation for Heisenberg ferromagnetism. Physica **22**, 317–337, [173].

Kasteleyn P.W. and van Kranendonk J (1956b): Constant-coupling approximation for antiferromagnetism. Physica **22**, 367–385, [173].

Kasteleyn P.W. and van Kranendonk J (1956c): Constant-coupling approximation for Ising spin systems. Physica **22**, 387–396, [173].

Kaufman B. (1949): Crystal statistics. II Partition function evaluated by spinor analysis. Phys. Rev. **76**, 1232–1243, [205].

Kaufman B. and Onsager L. (1949): Crystal statistics. III Short-range order in a binary Ising lattice. Phys. Rev. **76**, 1244–1252, [205].

Kikuchi R. (1951): A theory of cooperative phenomena. Phys. Rev. **81**, 988–1003, [173].

Kikuchi R. (1974): Superposition approximation and natural iteration calculation in cluster-variation method. J. Chem. Phys. **60**, 1071–1080, [196].

Kincaid J.M. and Cohen E.G.D. (1975): Phase diagrams of liquid helium mixtures and metamagnets: experiment and mean field theory. Phys. Rep. **22**, 57–143, [162].

Kirkwood J.G. (1943): Phase transitions in monolayers due to hindered molecular rotation. Pub. Am. Assoc. Advant. Sci. **21**, 157–160, [182].

Klein M.J. (1974): The historical origins of the van der Waals equation. Physica **73**, 28–47, [9].

Knobler C.M. and Scott R.L. (1984): Multicritical points in fluid systems. In Phase Transitions and Critical Phenomena. (Eds. C. Domb and J. L. Lebowitz), Vol. 9, 163–231. Academic Press, London, U.K., [164].

Landsberg P.T. (1978): Thermodynamics and Statistical Mechanics. O.U.P., Oxford, U.K. [1].

Lavis D.A. (1996): The derivation of the free energy of the Ising model from that of the eight-vertex model. Rep. Math. Phys. **37**, 147–155, [294].

Lavis D.A. and Bell G.M. (1967): The effect of sublattice order in binary alloys with one magnetic component: III. Phil. Mag. **15**, 587–601, [168].

Lavis D.A. and Bell G.M. (1999): Statistical Mechanics of Lattice Systems. Vol.2: Exact, Series and Renormalization Group Methods. Springer, Berlin, Heidelberg, Germany, [V].

Lavis D.A. and Christou N.I. (1977): The dielectric constant of a bonded lattice model for water. J. Phys. A: Math. Gen. **10**, 2153–2169, [191].

Lavis D.A. and Christou N.I. (1979): A two-dimensional bonded lattice model for water. J. Phys. A: Math. Gen. **12**, 1869–1890, [192].

Lavis D.A. and Fairbairn W.M. (1966): The effect of sublattice order in binary alloys with one magnetic component. Phil. Mag. **13**, 477–492, [168].

Lavis D.A. and Quinn A.G. (1983): An Ising ferrimagnetic model on a triangular lattice. J. Phys. C: Solid State Phys. **16**, 3547–3562, [266,268].

Lavis D.A. and Quinn A.G. (1987): Transfer matrix and phenomenological renormalization methods applied to a triangular Ising ferrimagnetic model. J. Phys. C: Solid State Phys. **20**, 2129–2138, [266].

Lavis D.A. and Southern B.W. (1984): Renormalization group study of a three-dimensional lattice model with directional bonding. J. Stat. Phys. **35**, 489–506, [192].

Lawden D.F. (1989): Elliptic Functions and Applications. Springer, Berlin, Heidelberg, Germany, [337].

Lawrie I.D. and Sarbach S. (1984): Theory of tricritical points. In Phase Transitions and Critical Phenomena. (Eds. C. Domb and J. L. Lebowitz), Vol. 9, 2–161. Academic Press, London, U.K., [162].

Lebowitz J.L. and Penrose O. (1966): Rigorous treatment of the van der Waals-Maxwell theory of the liquid-vapour transition. J. Math. Phys. **7**, 98–113, [77].

Le Guillou J.C. and Zinn-Justin J. (1980): Critical exponents from field theory. Phys. Rev. B **21**, 3976–3998, [88].

Li Y.Y. (1949): Quasi-chemical theory of order for the copper gold alloy system. J. Chem. Phys. **17**, 447–452, [173,196].

Lieb E.H. (1966): Quantum-mechanical extension of the Lebowitz-Penrose theorem on the van der Waals theory. J. Math. Phys. **7**, 1016–1024 [77]

Lieb E.H. (1967a): Residual entropy of square ice. Phys. Rev. **162**, 162–171, [293,312].

Lieb E.H. (1967b): Exact solution of the F model of an antiferroelectric. Phys. Rev. Lett. **18**, 1046–1048, [293,295].

Lieb E.H. (1967c): Exact solution of the two-dimensional Slater KDP model of a ferroelectric. Phys. Rev. Lett. **19**, 108–110, [293,295].

Lieb E.H. (1969): Two dimensional ice and ferroelectric models. In Lectures in Theoretical Physics, Vol. 11 (Eds. K.T. Mantappa and W.E. Brittin), 329–354, Gordon and Breach, New York, U.S.A. [293].

Lieb E.H. and Mattis D.C. (Eds.) (1966): Mathematical Physics in One Dimension. Academic Press, New York, U.S.A. [124].

Lieb E.H. and Wu F.W. (1972): Two-dimensional ferroelectric models. In Phase Transitions and Critical Phenomena. (Eds. C. Domb and M. S. Green), Vol. 1, 332–490. Academic Press, London, U.K., [293,301,308,313,318–320,325].

Ma S.K. (1985): Statistical Mechanics, World Scientific, Philadelphia, U.S.A. [31].

Maillard J.M. (1985): Star-triangle and inversion relations in statistical mechanics. In Critical Phenomena (Eds. V. Ceausescu et al.), Progress in Physics, Vol. 11, 375–401, Birkhauser, Boston, U.S.A. [205].

Martin D.H. (1967): Magnetism in Solids. Iliffe, London, U.K. [42,44,96,105,110].

Mattis D.C. (1985): Theory of Magnetism II. Springer, Berlin, Heidelberg, Germany [42,44,105].

Mattis D.C. (Ed.) (1993): The Many-Body Problem: An Encyclopedia of Exactly Solved Models in One Dimension. World Scientific, Singapore, [124]

McCoy B.M. (1972): Exact calculation on a random Ising model. In Phase Transitions and Critical Phenomena. (Eds. C. Domb and M. S. Green), Vol. 2, 161–195. Academic Press, London, U.K., [167].

McCoy B.M. and Wu T.T. (1973): The Two-Dimensional Ising Model. Harvard U.P., Cambridge, Mass. U.S.A. [205,228,229].

Meijering J.L. (1950): Segregation in regular ternary mixtures, I. Philips Res. Rep. **5**, 333–356, [151].

Meijering J.L. (1951): Segregation in regular ternary mixtures, II. Philips Res. Rep. **6**, 183–210, [151].

Mermin N.D. (1971a): Solvable model of a vapour-liquid transition with a singular coexistence-curve diameter. Phys. Rev. Lett. **26**, 169–172, [82,281].

Mermin N.D. (1971b): Lattice gas with short-range pair interactions and a singular coexistence-curve diameter. Phys. Rev. Lett. **26**, 957–959, [82,281].

Mermin N.D. and Rehr J.J. (1971): Generality of the singular diameter of the liquid-vapour coexistence curve. Phys. Rev. Lett. **26**, 1155–1156, [82].

Morita T. (1984): Convergence of the cluster variation method for random Ising models in the thermodynamic limit. Prog. Theor. Phys. Supp. **80**, 103–107, [173].

Mouritsen O.G. (1984): Computer Studies of Phase Transitions and Critical Phenomena. Springer, Berlin, Heidelberg, Germany, [238].

Mukamel D. and Blume M. (1974): Ising model for tricritical points in ternary mixtures. Phys. Rev. A **10**, 610–617, [162].

Mulholland G.W. and Rehr J.J. (1974): Coexistence curve properties of Mermin's decorated lattice gas. J. Chem. Phys. **60**, 1297–1306, [285,288].

Nagle J.F. (1966): Lattice statistics of hydrogen bonded crystals. I The residual entropy of ice. J. Math. Phys. **7**, 1484–1491, [344].

Nagle J.F. (1975): Chain model theory of lipid monolayer transitions. J. Chem. Phys. **63**, 1255–1261, [182].

Nagle J.F. (1986): Theory of lipid monolayer and bilayer chain-melting phase transitions. Faraday Disc. Chem. Soc. **81**, 151–162, [182,187].

Nakono H. (1968): Ordering in certain statistical systems of Ising spins. Prog. Theor. Phys. **39**, 1121–1132, [273].

Naya S. (1954): On the spontaneous magnetization of honeycomb, Kagomé and Ising lattices. Prog. Theor. Phys. **11**, 53–62, [264].

Nienhuis B., Riedel E.K. and Schick M. (1980): Magnetic exponents of the two-dimensional q-state Potts model. J. Phys. A: Math. Gen. **13**, L189–L192, [331].

Onsager L. (1944): Crystal statistics. I Two-dimensional model with order-disorder transition. Phys. Rev. **65**, 117–149, [205].

O'Reilly D.E. (1973): Scaled particle-quasilattice model of liquid water. Phys. Rev. A **7**, 1659–1661, [192].

Pallas N.R. and Pethica B.A. (1985): Liquid-expanded to liquid-condensed transitions in lipid monolayers at the air water interface. Langmuir **1**, 509–513, [182].

Palmer R.G. (1982): Broken ergodicity. Adv. in Phys. **31**, 669–735, [57,75].

Parsonage N.G. and Staveley L.A.K. (1978): Disorder in Crystals. O.U.P., Oxford, U.K. [197].

Pauling L. (1935): The structure and entropy of ice and other crystals with some randomness of atomic arrangement. J. Am. Chem. Soc. **57**, 2680–2684, [295].

Pearson R. (1980): Conjecture for the extended Potts model magnetic eigenvalue. Phys. Rev. B **22**, 2579–2580, [331].

Perram J.W. (1971): Interstitial models of water in the random mixing approximation. Mol. Phys. **20**, 1077–1085, [281].

Phani M.K., Lebowitz J.L. and Kalos M.H. (1980): Monte Carlo studies of a face-centred cubic Ising antiferromagnet with nearest- and next-nearest-neighbour interactions. Phys. Rev. B **21**, 4027–4037, [196].

Pippard A.B. (1957): Elements of Classical Thermodynamics. C.U.P., Cambridge, U.K. [1,26,72].

Potts R.B. (1952): Some generalized order-disorder transformations. Proc. Camb. Phil. Soc. **48**, 106–109, [164,329,331].

Pynn R. and Skjeltorp A. (Eds.) (1984): Multicritical Phenomena. Plenum Press, New York, U.S.A. [162].

Rao C.N.R. (1972): Theory of hydrogen bonding in water. In Water; A Comprehensive Treatise., Vol. 1 (Ed. F. Franks). 93–114, Plenum Press, New York, U.S.A. [343].

Rapaport D.C. (1972): The Ising ferromagnet with impurities: a series expansion approach, I. J. Phys. C: Solid State Phys. **5**, 1830–1858, [245].

Rigby M. (1970): The van der Waals fluid: a renaissance. Quart. Rev. Chem. Soc. **24**, 416–432, [9].

Rowlinson J.S. (1973): Legacy of van der Waals. Nature **244**, 414–417, [9].

Rowlinson J.S. (1988): English translation (with introductory essay) of J.D. van der Waals: On the continuity of the gaseous and liquid states of matter, Studies in Statistical Mechanics, Vol. 14. North-Holland, Amsterdam, New York, [362].

Rowlinson J. and Swinton F.L. (1982): Liquids and Liquid Mixtures. Butterworths, London, U.K. [22,119,164,281,289].

Ruelle D. (1969): Statistical Mechanics. Benjamin, New York, U.S.A. [33].

Runnels L.K. (1967): Phase transition of a Bethe lattice gas of hard molecules. J. Math. Phys. **8**, 2081–2087, [198].

Runnels L.K. (1972): Lattice gas theories of melting. In Phase Transitions and Critical Phenomena. (Eds. C. Domb and M. S. Green), Vol. 2, 305–328. Academic Press, London, U.K., [120].

Rushbrooke G.S. and Scoins H.I. (1955): On the Ising problem and Mayer's cluster sums. Proc. Roy. Soc. A **230**, 74–90, [197].

Sallouta H. and Bell G.M. (1976): An interstitial model for aqueous solutions of non-electrolytes. Mol. Phys. **32**, 839–855, [289].

Sato H., Arrott A. and Kikuchi R. (1959): Remarks on magnetically dilute systems. Phys. Chem. Sol. **10**, 19–34, [167].

Schlijper A.G. (1983): Convergence of the cluster variation method in the thermodynamic limit. Phys. Rev. B **27**, 6841–6848, [173].

Shinmi M. and Huckaby D. (1986): A lattice gas model for carbon tetrachloride. J. Chem. Phys. **84**, 951–955, [192].

Shockley W. (1938): Theory of order for the copper gold alloy system. J. Chem. Phys. **6**, 130–144, [147].

Sinha K. and Kumar N. (1980): Interactions in Magnetically Ordered Solids, O.U.P., Oxford, U.K. [42].

Southern B.W. and Lavis D.A. (1980): Renormalization group study of a two-dimensional lattice model with directional bonding. J. Phys. A: Math. Gen. **13**, 251–262, [192].

Stanley H.E. (1987): Introduction to Phase Transitions and Critical Phenomena. O.U.P., Oxford, U.K. [47,83].

Stephenson J. (1970): Ising model with antiferromagnetic next-nearest-neighbour coupling: spin correlations and disorder points. Phys. Rev. B **1**, 4405–4409, [273].

Stillinger E.H. (1975): Theory and molecular models for water. Adv. Chem. Phys. **31**, 1–101, [192].

Stinchcombe R.B. (1983): Dilute magnetism. In Phase Transitions and Critical Phenomena. (Eds. C. Domb and J. L. Lebowitz), Vol. 7, 152–280. Academic Press, London, U.K., [167].

Straley J.P. and Fisher M.E. (1973): Three-state Potts model and anomalous tricritical points. J. Phys. A: Math. Gen. **6**, 1310–1326, [166].

Sutherland B. (1967): Exact solution of a two-dimensional model for hydrogen-bonded crystals. Phys. Rev. Lett. **19**, 103–104, [293].

Suzuki M. and Fisher M.E. (1971): Zeros of the partition function for the Heisenberg, ferroelectric and general Ising models. J. Math. Phys. **12**, 235–246, [294]

Syozi I. (1972): Transformation of Ising models. In Phase Transitions and Critical Phenomena. (Eds. C. Domb and M. S. Green), Vol. 1, 270–329. Academic Press, London, U.K., [205,241,250,256,264,273].

Syozi I. and Nakono H. (1955): Statistical models of ferrimagnetism. Prog. Theor. Phys. **13**, 69–78, [256].

Temperley H.N.V. (1954): The Mayer theory of condensation tested against a simple model of the imperfect gas. Proc. Phys. Soc. **67**, 233–238, [77].

Temperley H.N.V. (1965): An exactly soluble lattice model of the fluid-solid transition. Proc. Phys. Soc. **86**, 185–192, [198].

Temperley H.N.V. (1972): Two-dimensional Ising models. In Phase Transitions and Critical Phenomena. (Eds. C. Domb and M. S. Green), Vol. 1, 227–267. Academic Press, London, U.K., [205].

Temperley H.N.V. (1981): Graph Theory and Applications. Ellis Horwood, Chichester, U.K. [197,206,207,335].

Temperley H.N.V. and Lieb E.H. (1971): Relations between the percolation and colouring problems on the plane square lattice. Some exact results for the percolation problem. Proc. Roy. Soc. A **322**, 251–280, [325].

Thompson C.S. (1972a): One-dimensional models – short range forces. In Phase Transitions and Critical Phenomena. (Eds. C. Domb and M. S. Green), Vol. 1, 177–226. Academic Press, London, U.K., [45].

Thompson C.S. (1972b): Mathematical Statistical Mechanics. Macmillan, New York, U.S.A. [77].

Thompson C.S. (1985): Random spin systems. In Critical Phenomena (Eds. V. Ceausescu et al.) Progress in Physics Vol. 11. 136–186, Birkhauser, Boston, U.S.A. [167].

Thompson C.S. (1988): Classical Equilibrium Statistical Mechanics. O.U.P., Oxford, U.K. [31].

Van der Waals J.D. (1873) Over de Continuiteit van der Gas- en Vloeistoftoestand, Doctoral Thesis, Leiden; (for an English translation see Rowlinson (1988)), [9].

Vdovichenko N.V. (1965a): A calculation of the partition function for a plane dipole lattice. Soviet Phys. JETP. **20**, 477–479, [205].

Vdovichenko N.V. (1965b): Spontaneous magnetization of a plane diploe lattice. Soviet Phys. JETP. **21**, 350–352, [205].

Wada A., Yubaki S. and Husimi Y. (1980): Fine structure in the thermal denaturation of DNA: high temperature-resolution spectrophotometric studies. CRC Crit. Rev. Biochem. **9**, 87–144, [60].

Wannier G.H. (1950): Antiferromagnetism: the triangular net. Phys. Rev. **79**, 357–364, [237].

Wartell R.M. and Benight A.S. (1985): Thermal denaturation of DNA molecules: a comparison of theory and experiment. Phys. Rep. **126**, 67–107, [58,60].

Wartell R.M. and Montroll E.W. (1972): Equilibrium denaturation of natural and of periodic synthetic DNA molecules. Adv. Chem. Phys. **22**, 129–203, [58,60].

Weres O. and Rice S.A. (1972): A new model for liquid water. J. Am. Chem. Soc. **94**, 8983–9002, [191].

Wheeler J.C. (1975): "Exactly solvable" two component lattice solution with upper and lower critical solution temperatures. J. Chem. Phys. **62**, 433–439, [275].

Wheeler J.C. (1977): Decorated lattice-gas models of critical phenomena in fluids and fluid mixtures. Ann. Rev. Phys. Chem. **28**, 411–443, [82,273,275,281,289,290].

Wheeler J.C. and Anderson G.R. (1980): Some interesting new phase diagrams in hydrogen-bonded liquid mixtures. J. Chem. Phys. **73**, 5778–5785, [276].

Wheeler J.C. and Widom B. (1970): Phase equilibrium and critical behaviour in a two-component Bethe lattice gas or a three-component Bethe lattice solution. J. Chem. Phys. **52**, 5334–5343, [151].

Whitehouse J.S., Christou N.I., Nicholson D. and Parsonage N.G. (1984): A grand ensemble Monte Carlo investigation of the Bell lattice model for water. J. Phys. A: Math. Gen. **17**, 1671–1682, [192].

Widom B. (1964): Degree of the critical isotherm. J. Chem. Phys. **41**, 1633–1634, [85].

Widom B. (1967): Plait points in two- and three-component liquid mixtures. J. Chem. Phys. **49**, 3324–3333, [250,289].

Widom B. and Rowlinson J. S. (1970): New model for the study of liquid-vapour phase transitions. J. Chem. Phys. **52**, 1670–1684, [82].

Wiegel F.W. (1983): Models of conformal phase transitions. In Phase Transitions and Critical Phenomena. (Eds. C. Domb and J. L. Lebowitz), Vol. 7, 101–149. Academic Press, London, U.K., [62].

Wilson G.L. and Bell G.M. (1978): Lattice model for aqueous solutions of non-electrolytes. J. Chem. Soc. Faraday Trans. II **74**, 1702–1722, [191].

Wood D.W. (1975): Thermodynamic behaviour in the critical region. In Statistical Mechanics, Vol. 2 (Ed. K. Singer), 55–187, Chem. Soc. Specialist Periodical Report, London, U.K. [44,76].

Wu F.Y. (1982): The Potts model. Rev. Mod. Phys. **54**, 235–268, [164–166].

Yang C.N. (1945): A generalization of the quasi-chemical method in the statistical theory of superlattices. J. Chem. Phys. **13**, 66–76, [173].

Young A.P. (1981): The Kikuchi approximation: a reformulation and application to the percolation problem. J. Phys. A: Math. Gen. **14**, 873–881, [173].

Ziman J. (1979): Models of Disorder. C.U.P., Cambridge, U.K. [63,67,173].

Subject Index

Springer and the environment

At Springer we firmly believe that an international science publisher has a special obligation to the environment, and our corporate policies consistently reflect this conviction.
We also expect our business partners – paper mills, printers, packaging manufacturers, etc. – to commit themselves to using materials and production processes that do not harm the environment. The paper in this book is made from low- or no-chlorine pulp and is acid free, in conformance with international standards for paper permanency.

Springer